Convenient Approximate Equivalences

Quantity	USCS	SI Exact	
		Units	
	USCS	SI Exact	
Length	1 in.	25.4 mm	
	1 ft	0.304 8 m	
Mass	1 lb (mass)	0.453 6 kg	
Area	1 ft^2	0.092 90 m^2	0.1 m^2
Volume	1 yd^3	0.765 m^3	0.75 m^3
Force	1 lb	4.448 N	4.5 N
	1 kip	4.448 kN	4.5 kN
	1 lb/ft	14.59 N/m	15 N/m
	1 kip/ft	14.59 kN/m	15 kN/m
Moment	1 ft-lb	1.356 N · m	1.4 N · m
	1 ft-kip	1.356 kN · m	1.4 kN · m
Pressure	1 psi	6.895 kPa	7 kPa
	1 psf	47.88 Pa	48 Pa
Some typical values:			
Water density	62.4 lb/ft^3	1 000 kg/m^3	1 000 kg/m^3
Concrete density	150 lb/ft^3	2 402.81 kg/m^3	2 400 kg/m^3
Water (weight/volume)	62.4 lb/ft^3	9.81 kN/m^3	10 kN/m^3
Concrete (weight/volume)	150 lb/ft^3	23 576 N/m^3	24 kN/m^3
Concrete stress	3,000 psi	20.67 MPa	21 MPa
Steel stress	25,000 psi	172.38 MPa	170 MPa
E steel	29×10^6 psi	199.9 GPa	200 GPa
E concrete	3.5×10^6 psi	24.13 GPa	24 GPa
Coeff. of thermal expansion for steel	6.5×10^{-6}/°F	1.2×10^{-5}/°C	1.2×10^{-5}/°C

SI Prefixes

Factor	Prefix Name	Symbol
10^{18}	exa	E
10^{15}	peta	P
10^{12}	tera	T
10^{9}	giga	G
10^{6}	mega	M
10^{3}	kilo	k
10^{2}	hecto	h
10^{1}	deca	da
10^{-1}	deci	d
10^{-2}	centi	c
10^{-3}	milli	m
10^{-6}	micro	μ
10^{-9}	nano	n
10^{-12}	pico	p
10^{-15}	femto	f
10^{-18}	atto	a

Fundamentals
of Structural Analysis

George Washington Bridge, spanning the Hudson River between New Jersey and New York (courtesy The Port Authority of New York and New Jersey).

Fundamentals of Structural Analysis

Harry H. West
Professor Emeritus of Civil Engineering
The Pennsylvania State University

and

Louis F. Geshwindner
Professor of Architectural Engineering
The Pennsylvania State University

John Wiley & Sons, Inc.

Cover: East Huntington Bridge, Huntington, West Virginia

Acquisitions Editor	Wayne Anderson
Production Management	Jeanine Furino
Production Editor	Sandra Russell
Marketing Manager	Katherine Hepburn
Designer	Madelyn Lesure
Production Management Services	Ingrao Associates
Cover Photo by David P. Billington	

This book was typeset in 10/12 Janson by Publication Services and printed and bound by Hamilton Printing Company. The cover was printed at The Lehigh Press, Inc.

The paper in this book was manufactured by a mill whose forest management programs include sustained yield harvesting of its timberlands. Sustained yield harvesting principles ensure that the number of trees cut each year does not exceed the amount of new growth.

This book is printed on acid-free paper. ⊗

Library of Congress Cataloging-in-Publication Data

West, Harry H., 1936–
Fundamentals of structural analysis / Harry H. West, Louis F. Geschwindner.--2nd ed.
p. cm.
Includes bibliographical references and index.

ISBN 0-471-35556-9 (cloth : alk. paper)
1. Structural analysis (Engineering)   I. Geschwindner, Louis F.   II. Title.

TA645.W44 2002
624.1'71--dc21   2001026852

Printed in the United States of America.

10 9 8 7 6 5 4 3 2 1

Dedication

In his book, *The Tower and the Bridge*, David Billington makes the following observation:

> *While automation prospers, our roads, bridges, and urban works rot. Children control computers while adults weave between potholes. The higher that high technology sails the worse seem our earthbound services for water, transportation, and shelter. Yet civilization is civil works and insofar as these deteriorate so does society, our high technology notwithstanding. We forget that technology is as much structures as it is machines, and that these structures symbolize our common life as much as machines stand for our private freedoms. Technology is frequently equated only with machines, those objects that save labor, multiply power, and increase mobility. In reality, machines are only one half of technology, the dynamic half, and structures are the other, static half—objects that create a water supply, permit transportation and provide shelter.*

This textbook is about the "static half" of technology—structures—and it is dedicated to practicing structural engineers. History has shown that structural engineers have designed and built structures whose shape and utility have changed the face of the world. Upon the completion of the Brooklyn Bridge in 1883, Montgomery Schuyler wrote in *Harper's Weekly*:

> *It so happens that the work which is most likely to be our most durable monument, and to convey some knowledge of us to the most remote posterity, is a work of bare utility; not a shrine, not a fortress, not a palace, but a bridge.*

This observation has certainly proved to be prophetic as it relates to the Brooklyn Bridge because it stands as one of the most durable monuments of our built environment. And the structural engineer has continued to reach horizontally with spans that are now more than four times that of the Brooklyn Bridge.

However, the towers of the Brooklyn Bridge did not dominate the skyline of New York City for long. By the turn of the twentieth century, structural engineers had begun to reach vertically, first in Chicago and then in New York, and in 1906 H. G. Wells, on viewing for the first time the prow of the Flatiron Building in Manhattan, wrote, "I found myself agape, admiring a skyscraper." Viewers continue to find themselves agape as they admire skyscrapers more than five times the height of the Flatiron Building.

Whether the task is to span horizontally or to stretch vertically, it is structural engineers who do it. They design and build the "static half" of our technology—the structures that "create a water supply, permit transportation, and provide shelter." It is the authors' hope that this textbook will be part of what will beckon students to practice the noble profession of structural engineering.

Preface

The title of this textbook underscores our belief that certain *fundamental* concepts must be mastered by any serious student of structural analysis. We further believe that many of these concepts are best learned and understood by applying them to the solution of problems through use of the so-called classical methods of analysis, which focus on specific modes of structural response and behavior, before proceeding with the more general matrix methods. No attempt is made to develop detailed and unique schemes for the application of the classical methods to large and complicated structures—such an approach would be totally out of step with current practice. Instead, the emphasis is on discrete problems of limited scope to illustrate foundational principles that will subsequently enhance the understanding of more comprehensive and powerful methods. Of course, matrix methods are central to modern computer applications, and they are addressed in the book because they, too, are a vital part of any fundamental treatment of structural analysis. However, the generality of these methods tends to obscure some of the basic concepts and, therefore, they should not be central to the teaching of fundamentals.

The book is divided into five parts. Part One, which consists of Chapters 1 and 2, provides an orientation to the study of structural analysis. Chapter 1 gives a general introduction to the subject; it includes topics such as a description of the design process, a discussion of the various structural forms that are employed in practice, and an enumeration of the array of loading conditions to which a structure may be subjected. Chapter 2 presents some of the basic concepts that are essential to the study of structural analysis. This material is set apart because of its global nature, and it is referred to throughout the book in support of the detailed development of various formulations. Students should be encouraged to read Chapter 2 early on to gain an appropriate overview, but instructors will find it advantageous to reassign portions of it for further study in the context of later topics of study.

Part Two, which includes Chapters 3 through 6, treats statically determinate structures. The topic of reactions is covered in Chapter 3, which focuses on the requirements of external equilibrium. Chapters 4 and 5, which look at the internal requirements of equilibrium, cover the determination of member forces in truss-type frameworks and beam and frame structures, respectively. Throughout Chapters 3 through 5, the emphasis is on planar structures; however, some simple nonplanar cases are discussed. The final chapter of Part Two, Chapter 6, presents influence lines for the purpose of analyzing structures subjected to variable loading situations. In this chapter, consideration is limited to planar structures.

Part Three includes Chapters 7 and 8, and these chapters examine, in turn, the determination of elastic deflections for truss frameworks and beam and frame structures. The emphasis in these chapters is, again, on planar structures, but some elementary nonplanar configurations are considered.

Part Four, which includes Chapters 9 through 12, examines the analysis of statically indeterminate structures. Chapter 9 is foundational in nature and, hence, discusses the general characteristics of this class of structures. Chapter 10 presents the method of consistent deformations and other compatibility methods, whereas Chapter 11 examines the slope deflection method and other equilibrium techniques. The final chapter in Part Four, Chapter 12, presents the moment distribution method, and the mathematical basis for this iterative method is described in detail.

Part Five contains the fully generalized matrix methods. Force-displacement relations are introduced in Chapter 13, and the equilibrium and compatibility methods are then developed in their elemental forms as the stiffness and flexibility methods, respectively, in Chapters 14 and 15.

One problem that is encountered in a textbook covering both classical and matrix methods is the dramatically different jargon for the two approaches. Some terms that have relevance in the classical approaches are less important in the matrix environment, and there are new ideas introduced in the matrix formulations that are foreign to the classical domain. For instance, the distinction between statically determinate and indeterminate structures, which is central in the classical methods, is of far less significance in the matrix approaches. However, there are behavioral differences between the two types of structures that are explained qualitatively by considering statical classification and, therefore, the retention of the distinction serves sound pedagogical objectives.

There is no formal instruction in matrix algebra in the book. Most engineering curricula require a course in linear algebra during the first two years of a student's education, and structural analysis is typically taught in junior- and senior-level courses. It is suggested that instructors review some of the basic matrix operations as they are needed; it has been our experience, however, that students are fairly well prepared to handle the elementary ideas that are needed for the material covered in this book.

The first edition of this book contained some computer software in support of selected topics. However, it has been found that opinions vary greatly regarding the use of computers in introductory courses, with respect to both extent and content. Furthermore, most instructors who include computer applications have their own software packages. Therefore, software is not included in this edition.

The status of SI units continues to be transitional at this time. In some disciplines, the conversion from conventional units to SI has moved ahead rapidly, but structural engineering is not on the leading edge of this change. Both systems are used in this book. The conventional system is used to keep students in touch with what continues to be a part of current practice, and SI units are employed to prepare them for the inevitable change. The emphasis is on the development of familiarity with each set of units and not on training for a mere conversion capability. It is strongly urged that the student develop a dual capacity in this area, because structural engineers are going to be forced to operate in both systems for many years. A brief explanation of the SI system is presented in Chapter 1 for students who are not familiar with its use, and some tables are given inside the front cover of the book for convenient reference.

The illustrations in the book cover the spectrum from realistic to idealistic representations. Problem statements generally include a fairly realistic representation of the structure to assist the student in visualizing what the real structure looks like—member type, connection details, support hardware, and so forth, However, the problem solutions then show the transformation to the idealistic representation that is customarily used in engineering practice, which incorporates the normal assumptions embraced in the analysis process.

Most universities currently offer a single course in structural analysis, and the content of this course will vary greatly from institution to institution. This textbook is not written to suit the specific needs of any of these possibilities. Instead, it offers a comprehensive coverage of the fundamentals of structural analysis with the belief that an instructor can select those topics that best meet the objectives of a given course.

In writing this textbook, we have endeavored to develop a strong teaching tool with an emphasis on the fundamental concepts that undergird the subject of structural analysis. Our goal has been to provide an instrument for teaching fundamentals rather than to prepare a reference work for the practicing engineer. Our hope is that it will be an asset for those who engage in teaching as well as for those who seek to learn.

State College, Pennsylvania

Harry H. West
Louis F. Geschwindner

Acknowledgments

This second edition of *Fundamentals of Structural Analysis* constitutes a reshaping of the first edition, which was published in 1993. That edition was, in turn, a revamping of a previous work entitled *Analysis of Structures: An Integration of Classical and Modern Methods,* which was first published in 1980 and revised in 1989. All three of the earlier editions were under the sole authorship of the senior author of this work. Although this edition, under joint authorship, includes some changes in content, organization, and design when compared with the earlier versions, considerable material has been brought forward. Therefore, it is appropriate to acknowledge again the contributions of those who were cited earlier, with each category expanded appropriately to reflect the joint authorship of this edition—former professors who patiently taught us and provided the foundation of knowledge on which we could build, colleagues who contributed to a stimulating environment for the refinement of ideas, students who presented the challenge to communicate effectively, and family members who consistently provided encouragement. Each individual within these groups has made unique contributions to the collective resources upon which we have drawn as authors. In addition, the users of the earlier versions have provided many suggestions for improvements, as have the reviewers of the various manuscripts. We have incorporated many of their ideas into this edition, and all of these contributions are acknowledged with thanks. And finally, we wish to thank all of those at John Wiley who worked tirelessly to produce this work—the acquisitions editor, the copy editors, designers and illustrators, production and manufacturing personnel, and marketing people.

Harry H. West
Louis F. Geschwindner

Contents

Part One

Orientation

Chapter 1

Introduction

The Empire State Building, New York City (courtesy NYC & Company—
the Convention and Visitors Bureau).

1.1 STRUCTURE

The word *structure* describes much of what is seen in nature. Living plants, from the frailest of ferns to the most rugged of trees, possess a structural form consistent with their needs. In each case, as can be seen in Fig. 1.1, the plant is the recipient of a structure—a gift of Providence—that supports its life. Insects and animals play a more active role in building the structures that they need. The delicate web of the spider, the vastly complicated societal complex of the termite, the carefully articulated dam or lodge of the beaver—each structure is fabricated to support the creature's activities. However, despite the intricacies of these structures, as can be seen in Fig. 1.2, the evidence clearly indicates that these creatures build from instinct and not from design.

Humans, too, are builders of structures; but more than that, they are conceivers and designers. If we suppose that the first structure twas a tree that conveniently fell

Figure 1.1 Plant structures. *(a)* Fern (photo by Lew Sheckler). *(b)* Shrub with ribbed branches (photo by Michael Houtz). *(c)* Oak tree (photo by Harry G. West). *(d)* American elm trees (photo by Harry G. West).

(a)

Figure 1.2 Structures of animals and insects. *(a)* The structure of a spider (courtesy of Richard B. Mansfield). *(b)* The structure of the compass termite (Australian Information Service Photograph). *(c)* Beaver dam (foreground) and lodge (background) (Neg. No. 238208 (photo by B. M. DeCon) courtesy the Library, American Museum of Natural History).

(b)

(c)

across a chasm and was subsequently used as a bridge, then since that meager and accidental beginning, humans have indeed advanced in their ability to design and build structures. When the structures of humans began to reflect their ability to conceive and design them as well as to construct them, structural engineering was born, and it has grown in sophistication as it has endeavored to meet the demands of humanity.

These demands may be highly functional and related to the basic needs of society, or they may be cosmetic and, therefore, related to the aesthetic or emotional sensitivities of humans. These two extremes are illustrated in Figs. 1.3 and 1.4, in which the basic functionality of the material-handling structure is contrasted against the symbolic arch monument. Of course, most structures fall between these two extremes in that they serve a specific function and yet are designed in harmony with the aesthetic sensitivities of humanity. Most bridges and buildings are in this category, and Figs. 1.5 and 1.6 show two classic cases in which the structure is highly functional and yet aesthetically pleasing. Each of the structures cited above, designed to meet its own set of demands, is the product of structural engineering.

1.2   STRUCTURAL ENGINEERING

The area of structural design might casually be associated with either science or engineering; however, there are important differences between the roles played by these two disciplines. These differences are perhaps best summarized by noting that science involves the investigation of what exists, whereas engineering engages in synthesis, to form that which does not exist. Although engineering requires the intelligent application of scientific principles, the creative nature of the discipline makes it an art. The list of modifiers that prefix the term *engineering* has become long over the years, as the number of disciplines to which the engineering approach is applied has grown.

Structural engineering centers about the conception, design, and construction of the structural systems that are needed in support of human activities. Although structural engineering is most directly associated with civil and architectural engineering, it interfaces with any engineering discipline that requires a structural system or component in meeting its objectives. Specific projects that involve structural engineering include bridges, buildings, dams, transportation facilities, liquid or gas storage and transmission facilities, power generation and transmission units, water and sewage treatment plants, industrial factories and plants, vehicular frames, and machine components. Each of these projects requires structural systems or components that must be conceived to meet the needs for which they are being built, designed to safely and serviceably carry the loads that will impinge upon them, and constructed to provide a final product consistent with the conception and design.

1.3   HISTORY OF STRUCTURAL ENGINEERING

The evolution of structural engineering to its current form involved the development of several individual areas of endeavor: the development of the theories of mechanics of materials and structural analysis; the formulation of the computational techniques necessary to solve the governing equations of these theories; the introduction of new building materials; the application of the theories and materials to the creation of new structural forms; and the inventive development of construction techniques. Some of these areas required the analytic talents of the mathematician, scientist, or engineer, whereas others required the daring and artistic skills of the entrepreneur or builder. Although each of these areas has its own historical chronology, their juxtaposition shows how the individual areas are nested—how a development in one area sparked a need

Figure 1.3 Stacker/reclaimer handling coal (courtesy Dravo Corporation).

Figure 1.4 The Jefferson National Gateway Arch, St. Louis, Missouri (photo by Louis F. Geschwindner).

Figure 1.5 Sunshine Skyway Bridge, Tampa Bay, Florida (courtesy Figg Engineering Group).

Figure 1.6 Dulles International Airport Terminal, Washington, D.C. (photo by Louis F. Geschwindner).

and, therefore, a subsequent development in another area. However, the presentation of a complete history of structural engineering is not the purpose of this section. Such treatments are available elsewhere. The intent, instead, is to provide a glimpse of the skeleton of such a history. A few of the people and events that served as benchmarks will be noted—especially some of those that are related directly to the topics in this book.

A beginning point for structural engineering as we currently view it would be at about 500 B.C. From that point until the time of Christ, the Greeks primarily used stone to build post and lintel structures, that is, structures whose columns supported short beams. An example of this form is shown in Fig. 1.7a. Even though experience

Figure 1.7 Notable historical structures. *(a)* Greek Parthenon (438 B.C.)—an example of post and lintel construction (courtesy Greek National Tourist Organization). *(b)* Pont du Gard (13 B.C.)—a Roman aqueduct in southern France (courtesy French Government Tourist Office). *(c)* The Cathedral of Notre Dame, Paris (13th century)—an example of Gothic buttress system (courtesy French Government Tourist Office). *(d)* Coalbrookdale Iron Bridge over Severn at Shropshire, England (1776–1779)—first use of iron on a major scale (photo by Harry G. West).

and empirical rules formed the basis of this structural activity, Aristotle (384–322 B.C.) and Archimedes (287–212 B.C.) were establishing the beginnings of the principles of statics. Although some metals and wood were introduced, stone and masonry continued as the primary building material of the Romans until about A.D. 500. They introduced new structural forms such as the arch, vault, dome, and even the wooden truss. Some of these structures, such as the aqueduct shown in Fig. 1.7*b*, remain with us today as monuments of that era. However, the Romans were not analytic in their approach, but rather were builders who concentrated on certain structural forms.

During the Middle Ages (500–1500), much of what the Greeks and Romans had developed was lost. The only major structural accomplishment during this time was achieved by the Gothic builders; witness their splendid cathedrals, characterized by pointed arches stabilized by "flying buttresses." Figure 1.7c provides an example of such a cathedral.

Following the inactivity of the Middle Ages, the Renaissance saw new impetus in many areas, including structural engineering. In the early part of this period, Leonardo da Vinci (1452–1519) formulated the beginning of structural theory. However, Galileo (1564–1642), who published *Two New Sciences,* is generally credited with originating the mechanics of materials. He studied the failure of a cantilever beam, and even though his writings were not wholly correct, they did establish an important beginning. Europe was rife with the activity of rebirth, which spawned more than a few significant analytic accomplishments. The most important of these for our purposes were those of A. Pallidio (1518–1580), who introduced the modern truss; R. Hooke (1635–1703), who established the law governing the linear behavior of materials; Johann Bernoulli (1667–1748), who stated the principle of virtual displacements; Daniel Bernoulli (1700–1782), who contributed to the understanding of elastic curves and the strain energy of flexure; Leonard Euler (1701–1783), who examined column buckling and energy methods; and Louis Navier (1785–1836), who followed up the earlier work of C. A. de Coulomb (1736–1806) and published a book on strength of materials that dealt with the elastic analysis of beam flexure.

These accomplishments were paralleled by new developments in building and construction materials. Timber was used by German and Swiss engineers to construct bridges up to 300 feet long. Iron arrived with revolutionary impact. As a material, it exhibited elastic properties much better than those of wood or stone, and thus the new theories could be applied to enable more daring structural forms to be used with confidence. A whole host of "firsts" in both form and dimension followed—cast iron arch bridges (Fig. 1.7d), iron trusses, suspension bridges, etc.

However, the golden age of structural engineering is considered to be 1800–1900. During this period, most of the present-day theories of mechanics of materials and structural analysis were developed. A few relevant developments are as follows: S. Whipple (1804–1888), K. Culmann (1821–1881), and J. W. Schwedler (1823–1894) formulated the principles of statically determinate trusses; B. P. E. Clapeyron (1799–1864) established the three-moment equation; J. C. Maxwell (1831–1879) developed the method of consistent displacements and the reciprocal theorem of deflections; O. Mohr (1835–1918) presented the method of elastic weights and worked on influence lines; A. Castigliano (1847–1884) stated the theorems that would carry his name; C. E. Greene (1842–1903) formulated the moment–area method; H. Müller-Breslau (1851–1925) published his principle for influence line construction; and A. Föppl (1854–1924) worked in the area of space frame analysis.

This golden age also saw new materials on the scene. Portland cement appeared early in the 1800s, and the first reinforced concrete bridge was constructed before the end of that century. Iron rolling mills made iron more usable, and quantity steel production was introduced by H. Bessemer. These developments led to new structural forms. In fact, the theoretical developments of the mid-1800s paved the way for the analysis of continuous beams and frames, and these forms grew in popularity by the early 1900s.

The twentieth century brought in some modest advancements in structural theory and some significant developments in solution techniques. A few are as follows: G. Maney (1888–1947) introduced the slope deflection method, which was the forerunner of modern displacement methods; H. Cross (1885–1959) contributed the moment distribution method, and R. Southwell (1888–1970) presented the more general relaxation method (these two developments allowing systematic solution of statically indeterminate structures and serving as the cornerstone of frame analysis for a quarter of a century); and several analysts contributed to the merging of matrix algebra and frame and continuum analysis to form the modern matrix and finite element methods of analysis. At the same time, the areas of inelastic analysis and strength methods were introduced.

The 1900s also saw a host of new materials, techniques, and structural forms introduced. Material developments brought forth aluminum, high-strength steels and concretes, special cements, plastics, laminated timber, and composites. Developments in technique include the introduction of experimental research, the use of electric welding and pretressed concrete, the development of improved construction methods, and the introduction of electronic computation in the 1950s. New advances in structural form included the perfection of long-span bridges of many configurations, record-breaking heights in buildings, and newer forms such as shells, panels, and stress-skin structures.

1.4 THE ENGINEERING DESIGN PROCESS

The *engineering design process* encompasses much more than structural design. Although the primary role of the structural engineer is in structural design, he or she is necessarily enmeshed in the entire design process. This is illustrated by the following breakdown of the engineering process as it is related to a typical civil or architectural engineering project in which a structural engineer is involved.

Conceptual Stage. Any engineering project must be directed toward the satisfaction of a unique set of objectives. During the conceptual or planning stage, the specific needs are identified and the objectives are carefully articulated to meet these needs. These objectives must be consistent with the desires of the client and the interests of other involved parties.

This stage requires input from the client, architects, planners, the public as represented by elected officials, governmental regulatory agencies or civic organizations, and the engineer. During this stage, the engineer frequently serves as a resource person regarding the engineering feasibility and the economic soundness of the various alternatives under consideration.

The conceptual stage should bring forth a plan that maximizes the satisfaction of the stated objectives while minimizing any objectionable features of the project.

Preliminary Design Stage. The plan that emerges from the conceptual stage frequently includes several alternatives that are to be investigated through the preparation of individual preliminary designs.

The preliminary designs are of vital importance, and the structural engineer plays a central role during this stage. It is during this stage that the creative talent of the engineer is vitally important as he or she considers the options brought forward from the conceptual stage. Yet, the engineer must keep in mind that the structure he or she designs has to be built, and thus the construction and fabrication aspects must be carefully considered. In many cases, the most severe loading conditions occur during fabrication. Figures l.8*a* and 1.8*b* show the erection of a large arch structure and a cantilever truss, respectively. The loading on these partially erected structures is vastly different from the loading that the final structures will support.

Figure 1.8 Structures during fabrication. *(a)* New River Gorge Arch Bridge under construction, Ansted, West Virginia (courtesy U.S. Steel Group, a unit of USX Corporation). *(b)* Greater New Orleans Mississippi River Bridge No. 2 during fabrication (courtesy Modjeski and Masters, Inc., Consulting Engineers).

Each preliminary design involves a thorough consideration of the loads and actions that the structure will have to support, including the conditions that will occur during fabrication. For each case, a structural analysis is necessary—that is, the forces and deformations throughout the structure must be determined. It is this area, the structural analysis of the system, that is examined in detail in this textbook.

Frequently, the preliminary designs are based on approximate theories of structural analysis in order to minimize the time and effort invested in the preliminary phase. This phase must produce sufficient detail so that intelligent decisions can be made in the final selection of one of the alternatives that was proposed in the conceptual stage.

Selection Stage. Once the preliminary designs are completed, a selection must be made. At this point, the parties involved in the conceptual stage are reconvened so that they may participate in the selection process.

The structural engineer is concerned at this stage with the relative economies of the alternatives, the impact that any unique features of each alternative might have on the structural behavior or practicality of construction, and many other areas related to the decision.

The result of this stage is usually a decision to proceed with one of the alternatives for which a preliminary design has been prepared. A growing trend in bridge engineering is to carry two competing designs forward to the final design stage. Here, a final decision is delayed until after bid prices are established for the two alternatives. However, our discussion will proceed on the basis of the selection of a single design.

In some cases, all preliminary designs are rejected, and a return to the preliminary design stage is in order. However, before proceeding to the next stage, a selection among the alternatives is necessary.

Final Design Stage. The results of the preliminary design stage constitute a starting point for the final design stage; however, the structural engineer must proceed with greater care from this point. Here the loads are determined with greater accuracy than was necessary during preliminary design, and all plausible loading conditions and combinations must be considered. The structural analysis required for this stage, which is the focus of this textbook, must be carried out with great precision, and many of the approximations of the preliminary design stage must be eliminated. Each member is proportioned and the connections are detailed to ensure that the structure will behave in accordance with the assumptions made in the structural analysis.

The results of the final design stage are capsulated in a set of complete design drawings, which give a graphic portrayal of the details of the entire system. These are generally accompanied by written specifications that stipulate the materials to be used, the quality of workmanship, the pertinent codes to be employed, and many other items.

Construction Stage. The goal of this stage is to bring into existence that which was described in the final design stage. The completed documents of the final design stage serve as the basis for bidding by the prospective building contractors. The successful bidder frequently prepares additional drawings related to the fabrication of the structure. The contractor must have on staff competent structural engineers who can work closely with those who designed the structure. The role of the structural engineer is vital here, as is evidenced by the fact that is not uncommon to have structural failures occur during construction. In fact, some of the most challenging problems in structural analysis are related to fabrication and erection.

This discussion is related specifically to the role of the structural engineer in a civil or architectural engineering project. This procedure is not followed in all engineering disciplines. Even in civil and architectural engineering projects, where the design and construction are done by a single organization, the sequential operations may be different. However, the key ingredients of conception, preliminary design, final design, and construction are present in any engineering design process.

1.5 STRUCTURAL ANALYSIS

Structural analysis is the process by which the structural engineer determines how a structure responds to specified loads or actions. This response is usually measured by establishing the forces and deformations throughout the structure. A given method of structural analysis is commonly expressed as a mathematical algorithm; however, it is based on information gained through the application of engineering mechanics theory laboratory research, model and field experimentation, experience, and engineering judgment.

The earliest demands for sophisticated analysis, coupled with some serious limitations on computational capability, led to a host of special techniques for solving a corresponding set of special problems. These so-called *classical methods* incorporated some ingenious innovations and served the needs of the structural engineer very well for many years. However, the advent and subsequent development of the digital computer increased computational capabilities by several orders of magnitude and thus obviated the need for special techniques for many problems, especially for large and complicated ones. The ingenious specializations of the classical methods were replaced by the sweeping generalities of the *modern matrix methods.*

The transition from the classical methods to the modern matrix methods has triggered some revolutionary changes in structural engineering and in the education of structural engineers. Although matrix methods have become the foundation of modern structural analysis as it is employed in the practice of structural engineering, classical methods continue to play a vital role in the educational process because they introduce the fundamentals of structural analysis. Therefore, the strategy in this textbook is to introduce fundamental concepts through the use of the classical methods and to employ these methods in the solution of small problems. Later in the book, the modern matrix methods are introduced and their application to larger problems is illustrated. This approach is intended to equip the student with the knowledge and understanding to proceed with hand calculations by classical methods for the solution of small problems (or for checking portions of the solution of large problems), and to employ the modern computer methods for the solution of large problems.

By either classical or matrix methods, the analysis process can be a part of preliminary design, final design, or construction, as was described in the preceding section. However, it is important to note that structural analysis plays a limited role in the structural design process and an even smaller role in the overall design process. Furthermore, the role that it plays is entirely supportive of the design process. That is, structural analysis is not an end in itself. It is particularly important to understand the supportive nature of structural analysis as one studies the subject. It is easy for an engineering student to become enamored of this intriguing subject to the point of aspiring to become a structural analyst. However, such a goal is unrealistic. Good structural engineers are necessarily good structural analysts, and they will use their analytic ability intelligently as they fulfill their primary responsibility as structural engineers.

1.6 STRUCTURAL FORM

The form that a structure is to take depends on many considerations. Frequently, the functional requirements of a structure will narrow the possible forms that can be considered. Other factors such as the aesthetic requirements, foundation conditions, availability of materials, and economic limitations may play important roles in establishing the structural form. The structural forms that are available, along with their respective features, will be discussed only briefly here. More complete treatments are available elsewhere. The student is encouraged to become sensitive to the various types of structures that appear in nature or have been built by human beings; this discussion is intended to promote the student's general awareness of structural form.

1.6.1 Tension and Compression Structures

Tension and compression structures are composed of members that are subjected to pure tension or compression. Such an arrangement provides for highly efficient material usage because there is a constant stress level over the entire cross-sectional area of each member. This is particularly true for tension elements, where the stress level is limited only by the material strength coupled with an appropriate factor of safety. Compression elements, however, are susceptible to buckling, which can limit the allowable stress to a level lower than that dictated by material considerations.

One of the simplest structural forms in which the elements are in pure tension is the *cable-supported structure*. Structures of this type range from simple guyed or stayed structures to large cable-supported bridge and roof systems. Two simple cable systems are shown in Fig. 1.9 along with their force-carrying mechanisms, and several examples of actual structures of this form are given in Fig. 1.10. Each of these structures contains cable elements that serve as tension members; some are primary members of the structure, whereas others are secondary hangers or bracing-type members.

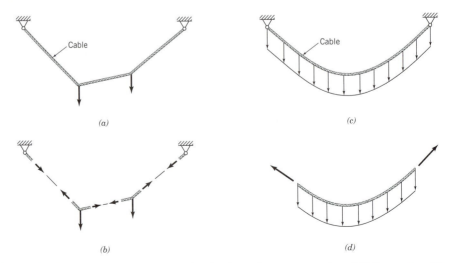

Figure 1.9 Force-carrying mechanisms for simple tension structures. *(a and b)* Concentrated loads on cable. *(c and d)* Uniform load on cable.

Figure 1.10 Typical tension structures. *(a)* Pipeline suspension bridge, Portsmouth, Ohio (courtesy Bethlehem Steel Corporation). *(b)* Radar-radio telescope, Arecibo, Puerto Rico (courtesy Bethlehem Steel Corporation). *(c)* Roof suspension system for Madison Square Garden, New York (courtesy Bethlehem Steel Corporation). *(d)* Golden Gate suspension bridge, San Francisco, Calif. (courtesy Bethlehem Steel Corporation).

The most common structure that carries pure compression in its primary element is the *arch*. Structures of this type, which have the configuration of an inverted cable, have a pure compressive thrust along the rib of the arch under the specific loading condition for which it was designed. Variations in this loading will introduce some bending in the arch, but compression will remain the dominant mode of action. Examples of arch action are shown in Fig. 1.11. Secondary members, which attach other portions of the structure to the arch, are compression members when they attach from above the arch or tension members when they attach from below the arch. Figure 1.12 gives several examples of structures in which the arch plays a

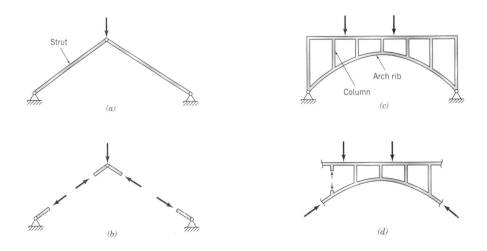

Figure 1.11 Force-carrying mechanisms for simple arch-type structures. *(a and b)* Concentrated loads on simple strut structure. *(c and d)* Arch loaded by columns from above.

Figure 1.12 Typical compression structures. *(a)* Tunkhannock Creek Viaduct, Nicholson, Pennsylvania (photo by Harry H. West). *(b)* New River Gorge steel arch bridge, Ansted, West Virginia (courtesy U.S. Steel Group, a unit of USX Corporation). *(c)* Fort Pitt Bridge, Pittsburgh, Pennsylvania (courtesy Richardson, Gordon and Associates, Consulting Engineers). *(d)* Laminated wood arches (courtesy American Institute of Timber Construction).

major role. Note that the arch can be formed by a single rib or it can be built up as a trussed rib.

Of course, many structures have individual elements that carry compression. The most common of these elements is the *column,* which is a primary compressive member. These elements appear in nearly all structural forms. For instance, the towers in Fig. 1.10*a* and the vertical members between the roadway and arch in Fig. 1.12*a* are column elements.

The *pin-connected truss* is a common structural form in which tension and compression elements are combined. Here, each element carries pure tension or compression and acts in concert with other elements to form a stable structural system, as shown in Fig. 1.13. As will be seen later, true pin-connected frameworks are rare; however, under certain restrictions on member sizes and connections, and when the structure receives loads only at its joints, the tension–compression behavior is an appropriate assumption.

Truss-type structures are frequently composed of repetitive planar units that are analyzed as *planar trusses.* These are usually connected by beam systems, which transfer the loads from the slab or deck to the truss units. In other cases, the spatial relationships must be accounted for in the analysis; structures of this type are *space trusses.* Truss structures can take a variety of arrangements, as illustrated by the examples in Fig. 1.14

1.6.2 Flexural Beam and Frame Structures

A flexural element is one that is subjected to a bending action, as opposed to the pure tension or compression that was previously discussed. The reader will recall from mechanics that bending induces compression on one side of the element and tension on the other side and that a transverse shear may also be present in the member. The simplest structural member that displays this mode of response is the *beam* element, which is shown in Fig. 1.15. Flexural *frame structures* are formed from a combination of beam and column elements, some of which are subjected to pure flexure while others carry a combination of flexure and tension or compression. A typical frame structure is shown in Fig. 1.16.

Structures that are composed of flexural elements can receive along their member lengths loads that are directed normal to the orientation of the member. They also are distinguished from pin-connected trusses by the possibility of some flexural-resistant connections that ensure continuity at the member ends. The concept of continuity is important when dealing with flexural beam and frame structures. This topic is discussed fully in Chapter 9.

Structures of this type can take a variety of configurations, a few of which are shown in Fig. 1.17 (see page 20). As was true with truss structures, typical building or bridge frames may be composed of repetitive planar units that are tied

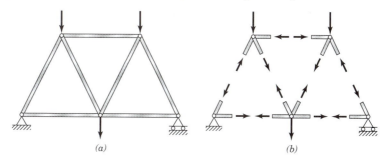

(a)   (b)

Figure 1.13 *(a)* Simple truss. *(b)* Load-carrying mechanism for a truss.

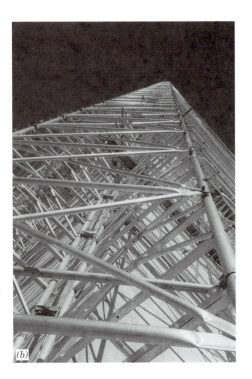

Figure 1.14 Typical truss-type structures. *(a)* Truss bridge across Ohio River, St. Marys, West Virginia (courtesy U.S. Steel Group, a unit of USX Corporation). *(b)* Space frame scaffolding for Washington Monument renovation (courtesy James Madison Cutts Structural Engineers, Inc.). *(c)* Example of open-web steel joist trusses (courtesy Bethlehem Steel Corporation). *(d)* Electrical transmission tower under erection—both tower and crane boom are truss-type structures (courtesy Bethlehem Steel Corporation).

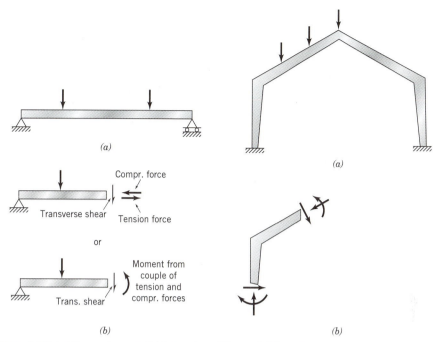

Figure 1.15 *(a)* Beam element. *(b)* Internal forces in a beam.

Figure 1.16 *(a)* Frame structure. *(b)* Internal forces in a frame.

together by secondary systems. Figure 1.18 illustrates this arrangement for a highway bridge. The view here is from the underside of the structure. There are four main *girders* that span longitudinally between the *bents* that serve as the bridge piers. The transverse members that connect a pair of girders are called *floorbeams,* and the small longitudinal members that are placed atop the floorbeams are *stringers.* The floorbeam–stringer combination serves to transfer the applied loads from the deck to the girders, which are the primary planar units of the structure. The cross *bracing* between floorbeams is composed of tension and compression members to resist the lateral loadings caused by the wind. There are cases where repetitive planar units are not employed. In such cases, the full spatial assemblages must be taken into account.

1.6.3 Surface Structures

The structures that have been considered thus far are composed of individual elements, each of which carries pure tension or compression, pure flexure, or a combination of flexure coupled with tension or compression. These elements are connected to form a skeletal structure in space that serves to support itself and everything that is attached to it.

Surface structures derive their spatial configuration through continuous three-dimensional surfaces, and the loads are resisted by the surfaces themselves. These structures carry tension, compression, and in-plane shear within the surface as membrane forces. Bending and transverse shear are carried either normal to or within the surfaces, depending on the loading and the surface orientation.

Structures that fall into the category of surface structures are *slabs, folded plates, shells, domes, skin-type structures,* and *inflatable membranes.* This class of structures provides some of the most efficient structural systems in terms of material usage, and

Figure 1.17 Typical beam and frame structures. *(a)* Slant-leg frame bridge, Charlottesville, Virginia. (courtesy Bethlehem Steel Corporation). *(b)* Continuous plate-girder bridge over Quinnipiac River, New Haven, Connecticut (courtesy Parsons Transportation Group). *(c)* Typical rigid frame construction (courtesy Lincoln Electric Company, Cleveland, Ohio). *(d)* Moment-resistant frame structure, Dresser Tower, Houston, Texas (courtesy Bethlehem Steel Corporation).

this leads to inherent economies. Another major advantage of surface structures is that they give the designer tremendous flexibility to create an aesthetically pleasing structure. The manner in which a folded plate, a floor slab, and a thin shell support gravity loading is shown in Fig. 1.19, and Fig. 1.20 illustrates some typical examples of surface structures.

1.6.4   Closing Notes on Structural Form

Most structures are composed of a combination of several structural forms in which each substructure serves in a unique way and the combination meets the functional objectives required of the structure. For instance, consider the Golden Gate Bridge,

Figure 1.18 Girder–floorbeam–stringer system for typical highway bridge (courtesy U.S. Steel Group, a unit of USX Corporation).

shown in Fig. 1.10d. The main cable and the hangers form a structural system of pure tension elements; the deck is a truss-type structure that stiffens the roadway and is referred to as the stiffening truss; the near approach span is supported by a trussed frame that contains a trussed arch in its low portion; the towers are frame structures that carry massive compressive forces coupled with some bending and torsion.

The structure of Fig. 1.14c is a second example of how different structural forms are used in a single structure. Here the roof surface is to be supported by the steel roof trusses; the roof trusses are supported by a system of beams and girders, which transfer their loads through flexural action to the columns; the columns act primarily as compression elements by carrying their loads to the foundation.

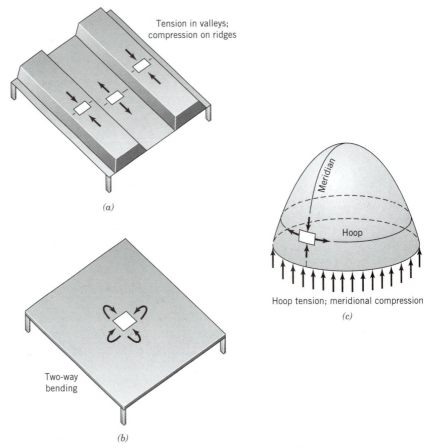

Figure 1.19 Surface structures under gravity loading. *(a)* Folded plate. *(b)* Floor slab. *(c)* Thin shell.

This textbook considers the analysis of the first two structural forms: tension and compression structures and flexural beam and frame structures. In these cases, the structure is composed of discrete structural elements. The third structural form is not treated in this textbook. It is much more difficult to analyze because the surface geometry and the three-dimensional material properties must be taken into account. Modern computer methods discretize these structures into a system of individual elements, which can then be treated by an extension of the concepts that are used in frame analysis. This approach is referred to as *finite element analysis*.

1.7 SIMPLIFICATIONS FOR PURPOSES OF ANALYSIS

A careful study of the photographs in the previous section reveals that the makeup of a structural system is rife with details concerning the individual members and their interconnection. In most cases, the overall structural analysis, which establishes the member forces and deformations, will ignore much of this detail. This is done to simplify the analysis; however, the process of simplification must be tempered with judgment to ensure that the results do not depart drastically from those in the real structure. There is no substitute for experience in establishing the limits in the simplification process.

Figure 1.20 Typical surface structures. *(a)* Folded
plate roof structure (courtesy Portland Cement
Association). *(b)* Hyperbolic paraboloid roof structure
(courtesy Portland Cement Association).
(c) Concrete dome, folded plate roof structure;
University of Illinois Assembly Hall, Champaign,
Illinois (courtesy Portland Cement Association).
(d) Membrane roof supported by air pressure, Pontiac
Stadium, Pontiac, Michigan (courtesy Bethlehem Steel
Corporation).

 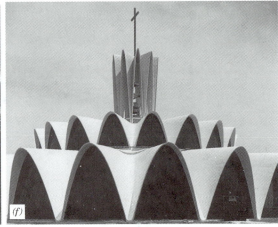

Figure 1.20 (cont.) *(e)* Floor slab system (courtesy Portland Cement Association). *(f)* St. Louis Priory Chapel, system of parabolic shells (courtesy Portland Cement Association).

For common truss- and frame-type structures, the member lengths are appreciably greater than their transverse dimensions. Also, the dimensions of the connections are usually small when compared with the member lengths. In such cases, the model that is adopted for purposes of analysis will be composed of line elements between the joints. This model is referred to as a *line diagram*. The lengths of the line elements are determined by the distances between the joint centers, which, in turn, are established by the intersections of the centroidal axes of the members that frame into a joint.

The member connections for truss-type structures are taken as frictionless pins with no moment-resisting capacity. For frame structures, rigid connections may be assumed for some joints, in which all member ends that frame into that joint have a common rotation. The support conditions are also idealized as frictionless pins or rollers, or points of complete fixity. More is said about the simplifications of member connections and supports in later chapters.

For cases where the members are nonprismatic—that is, their cross-sectional characteristics change along their lengths—these variations eventually must be taken into account in establishing the member characteristics. However, these refinements are frequently ignored in the early stages of analysis.

A casual review of real structures shows that they are three-dimensional assemblages of members. However, in many cases there are planar components of the structure that essentially function as independent units and can thus be isolated and analyzed as planar structures. There are, of course, cases where this is not possible or where the final analysis will require a full three-dimensional analysis, but there are numerous cases where the planar analysis of substructures is in order.

As an example of a simplification for analysis, consider the truss structure of Fig. 1.14a. The initial step in the simplification is to consider the structure to be composed of two longitudinal trusses that are connected by transverse-trussed frames. One such truss is shown in Fig. 1.21a. In representing these longitudinal trusses, the camber (slight curvature of the roadway) is ignored, the members are represented by straight-line elements that are located along their respective longitudinal axes, and pins are assumed to connect the members at their ends. The supports are taken as either pins or rollers. The effect of these assumptions leads to the representation shown in Fig. 1.21b. This figure shows only part of the truss, but the effects of the approximation are clear.

As a second example of how a structure is simplified for analysis, consider the slant-leg frame structure of Fig. 1.17a. This bridge is composed of five parallel frames, and a separate planar analysis of each is appropriate. The details of the connection between the inclined column and the girders are shown in Fig. 1.22a. This structure can be modeled as a simplified frame, with the dimensions shown in Fig. 1.22b. The members are assumed to be rigidly connected, and the supports are simplified in the manner indicated.

The internal member forces that correspond to the points where the members interface with the haunched connections can be taken as the forces on the connection detail for an analysis of the connection itself. Also, it is possible to account for the member haunching (increase in depth) and the column taper if such refinements are desired.

1.8   LOADING CONDITIONS

Establishment of the loads that act on a structure is one of the most difficult and yet important steps in the overall process of design. The computer has made it possible to analyze structures that could not be analyzed a mere decade ago, but the accuracy of the results of the analysis is directly dependent on the accuracy of the loads used.

The loads that enter a system are of three different types. *Concentrated loads* are those that are applied over a relatively small area. A single vehicular wheel load and a load that one member transfers to another member are examples of concentrated loads. *Line loads* are distributed along a narrow strip of the structure. The weight of a member itself and the weight of a wall or partition are examples of this type of load. *Surface loads* are loads that are distributed over an area. The loads on a warehouse floor and the snow load on a roof are examples of surface loads.

The loads that act on a structure can be grouped according to three categories: *dead loads, live loads,* and *environmental loads.* These categories can be further divided according to the specific nature of the loading. Because the method of

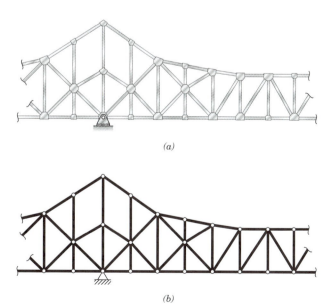

(a)

(b)

Figure 1.21 Simplified representation of the bridge truss of Fig. 1.14a. *(a)* Realistic structure. *(b)* Simplified model for analysis.

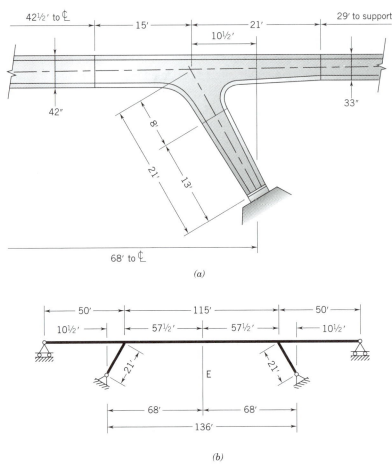

Figure 1.22 Simplified representation of the slant-leg frame bridge of Fig. 1.17a. *(a)* Connection detail. *(b)* Simplified model for analysis.

analysis is the same for each category of loading, all loads could be combined before the analysis is performed. However, separate analyses for the individual loading cases are usually carried out to facilitate the consideration of various combined loadings.

1.8.1 Dead Loads

Dead loads are those that act on the structure as a result of the weight of the structure itself and of the components of the system that are permanent fixtures. As a result, dead loads are characterized as having fixed magnitudes and positions. Examples of dead loads are the weights of the structural members themselves, such as beams and columns, the weights of roof surfaces, floor slabs, ceilings, or permanent partitions, and the weights of fixed service equipment.

The dead loads associated with the structure can be determined if the materials and sizes of the various components are known. Standard material unit weights, such as those given in Table 1.1, are used for calculating these dead loads. A complete list of minimum design loads for various building components, such as typical flooring, ceiling, and roofing materials, is given in applicable

Table 1.1 Unit Weights of Typical Building Materials

Material	(lb/ft^3)	(kN/m^3)
Aluminum	165	25.9
Brick	120	18.9
Concrete		
Reinforced with stone aggregate	150	23.6
Block, 60 percent void	87	13.7
Steel, rolled	490	77.0
Wood		
Fir	32–44	5.0–6.9
Plywood	36	5.7

codes and standards. The loads associated with service equipment are normally available from the manufacturers.

Whether these loads are treated as concentrated loads, line loads, or surface loads depends on the application of the loads and the arrangement of the structural members. Sometimes secondary structural analyses are necessary to determine the loads that are transferred from a substructure to a primary structure.

Accounting for the dead loads in design requires an iterative procedure because the load magnitudes depend on the member sizes, which are not known prior to the design. Approximate methods are available to estimate the dead weight of a structure, and these methods, or experience gained from previous designs, can be used as a guide for preliminary design. Each subsequent cycle of the iteration uses the latest information on member sizes to establish the dead loads, and, eventually, the design will reflect member sizes that are consistent with the assumed dead loads.

1.8.2 Live Loads

In a general sense, live loads are considered to include all loads on the structure that are not classified as dead loads. However, it has become common to narrow the definition of live loads to include only loads that are produced through the construction, use, or occupancy of the structure and not to include environmental or dead loads.

These loads are dynamic in character in that they are fixed in neither magnitude nor position. Live loads where the dynamic nature has significance because of the rapidity with which change in position occurs are called *moving loads,* whereas live loads in which change occurs over an extended period of time, or where there is the potential for change whether exercised or not, are referred to as *movable loads.* Moving loads include vehicular loads on bridges or crane loads in industrial buildings. Examples of movable loads are stored material in a warehouse and movable partitions in an office building. Another type of live load is a variable load or a *time-dependent load*—that is, one whose magnitude changes with time, such as a load induced through the operation of machinery.

Recommended live load magnitudes are available through a carefully assembled record of experience in the form of building codes, design specifications, and research reports. For unusual cases, where suggested values are not available, the designer must embark on an independent study to establish the loading to be used.

1.8.2.1   Occupancy Loads for Buildings

Occupancy live loads for buildings are usually specified in terms of the minimum values that must be used for design purposes. Some representative values are given in Table 1.2. Although this listing is incomplete, it is included to give the reader some appreciation of the range of loadings that is used. These values are extracted from *Minimum Design Loads for Buildings and Other Structures* (ASCE-7), published by the American Society of Civil Engineers (ASCE), which is cited in Section 1.11.

1.8.2.2   Traffic Loads for Bridges

Bridges must be designed to support the vehicular loads associated with their functional use, and minimum loads are mandated for designed purposes. In the case of highway bridges, these loads are specified by the American Association of State Highway and Transportation Officials (AASHTO) in their publication *LRFD Bridge Design Specifications,* which is cited in Section 1.11. The approach is to specify the weights and spacings of axles and wheels for a design truck, the characteristics of which are illustrated in Fig. 1.23. Also specified is a design tandem, composed of a pair of 25.0-kip axles spaced 4.0 ft apart with a transverse wheel spacing of 6.0 ft. In addition, a design lane load is prescribed. It consists of a load of 0.64 kip/ft, uniformly distributed in the longitudinal direction. Transversely, the design lane load is uniformly distributed over a 10.0 ft width.

These loadings provide for a set of concentrated loads (which represent a truck-type loading) and a uniform load (which simulates a line of vehicles). The appropriate combinations of these loads is applied to the structure to produce the maximum value of the particular response quantity. Each response quantity of interest is treated

Figure 1.23   Characteristics of AASHTO Design Truck (From *AASHTO LRFD Bridge Design Specifications,* Copyright 1998, by the American Association of State Highway and Transportation Officials, Washington, D.C. Used by permission.)

Table 1.2   Minimum Uniformly Distributed Live Loads for Building Design[a]

Occupancy or Use	(lb/ft^2)	(kPa $=$ kN/m^2)
Residential dwellings, apartments, hotel rooms, school classrooms	40	1.91
Offices	50	2.39
Auditoriums (fixed seats)	60	2.87
Retail stores	73–100	3.59–4.79
Bleachers	100	4.79
Library stacks	150	7.18
Heavy manufacturing and warehouses	250	11.97

[a]Data are taken from ASCE-7.

separately, and the critical loading arrangement must be established for each one. Problems of this type are discussed in Chapter 6.

A similar approach is used in the design of railroad bridges. Here the standard loading consists of a series of concentrated loads, separated by prescribed distances, followed by a uniform loading. This loading is described in *Specifications for Steel Railway Bridges,* which is published by the American Railway Engineering Association (AREA).

Bridges might also be subjected to longitudinal forces associated with the stopping action of a truck or train; or transverse forces can be imparted by the wind acting on vehicles or by the centrifugal action of vehicles on a curved bridge.

1.8.2.3   Impact Loads

Loads that are applied over a very short period of time have a greater effect on the structure than would occur if the same loads were applied statically. The manner in which a load varies with time and the time over which the full load is placed on the structure will determine the factor by which the static response should be increased to obtain the dynamic response.

For building occupancy loads, the minimum design loads normally include adequate allowance for ordinary impact conditions. However, provisions must be made in the structural design for uses and loads that involve unusual vibrations and impact forces.

One situation in which an impact effect is routinely applied is for moving vehicular loads on a highway bridge. Although no effort is made to apply a rigorous analytic procedure, AASHTO requires that the static effects be increased by a specified percentage. Therefore, the factor to be applied to the static load is given by

$$\text{Factor Applied to Static Load} = (1 + IM/100) \tag{1.1}$$

where IM is defined as the dynamic load allowance and is given for various conditions in Table 1.3. The dynamic load allowance is not to be applied to centrifugal and braking forces, nor to the design lane load.

1.8.3   Environmental Loads

Structures experience numerous loading conditions as a result of the environment in which they exist. Several of these conditions are described in the following sections.

Table 1.3   AASHTO Dynamic Load Allowance, IM

Component	IM
Deck joints—all limit states	75%
All other components	
• Fatigue and fracture limit state	15%
• All other limit states	33%

Source: (From *AASHTO LRFD Bridge Design Specifications,*
Copyright 1998, by the American Association of State Highway and
Transportation Officials, Washington, D.C. Used by permission.)

1.8.3.1   Snow and Ice Loads

The procedure for establishing the static snow loads on a building is normally based on ground snow loads and an appropriate ground-to-roof conversion. A typical approach is that outlined by ASCE-7. Here, ground snow loads corresponding to a 50-year mean recurrence interval are specified by isolines on a map of the contiguous United States. These mapped data must be used with care—mountainous regions, high country, or other local conditions require adjustments. In some areas, the local variations are so extreme that values are given only for elevations below those specified on the map. Table 1.4 gives some representative ground snow loads from the ASCE-7 map to illustrate the range of loading.

The distribution of snow on a roof is complex, and many different approaches are used. According to the ASCE-7 method, the snow load on an unobstructed flat roof, p_f, is given by

$$p_f = 0.7 C_e C_t I p_g \tag{1.2}$$

where p_g is the site-specific ground snow load, C_e is a dimensionless exposure factor that ranges from 0.7 for a wind-exposed setting to 1.3 for a wind-sheltered area, C_t is a dimensionless thermal factor that ranges from 1.0 for a heated structure to 1.2 for an unheated structure, and I is a dimensionless importance factor that ranges from 0.8 for structures that represent low hazard to human life to 1.2 for essential structures. A value of 1.0 is used for most cases.

The sloped roof load acting on a horizontal projection of the roof surface, p_s, is given by

$$p_s = C_s p_f \tag{1.3}$$

Table 1.4   Typical Ground Snow Loads, p_g^a

Location	(lb/ft^2)	(kPa $=$ kN/m^2)
Portland, Maine	60	2.87
Minneapolis, Minnesota	50	2.39
Hartford, Connecticut	30	1.44
Chicago, Illinois	25	1.20
St. Louis, Missouri	20	0.96
Raleigh, North Carolina	15	0.72
Atlanta, Georgia	5	0.24

[a] Data are taken from ASCE-7.

where p_f is given by Eq. 1.2 and C_s is a dimensionless slope factor that ranges from 0.0 to 1.0 and depends on the roof slope and whether the roof is warm or cold (C_t factor). For normal roof surfaces, $C_s = 1.0$ for warm ($C_t = 1.0$) slopes up to 30° or for cold ($C_t > 1.0$) slopes up to 45°.

Numerous other factors must be considered in calculating snow loads. These include, but are not limited to, the effects of unloaded portions of roof, unbalanced or nonuniform loads on various roof configurations, drifting, sliding snow, and extra loads induced by rain on snow.

Snow loads are not normally considered in bridge design because they are usually small when compared with other loadings on the structure. However, ice loads can be appreciable on bridge structures. The icing not only creates loads on the structure but also increases the member sizes, which, in turn, increases the magnitude of the wind-induced loads.

1.8.3.2 Rain Loads

Roof loads that result from the accumulation of rainwater on flat roofs can be a serious problem. This condition is produced by the ponding that occurs when the water accumulates faster than it runs off, either because of the intensity of the rainfall or because of the inadequacy or blockage of the drainage system. The real danger is that as ponding occurs the roof deflects into a dished configuration, which can accommodate more water, and thus greater loads result.

The best way to prevent the problem is to provide a modest slope to the roof (0.25 in. per ft or more) and to design an adequate drainage system. In addition to the primary drainage, there should be a secondary system to preclude the accumulation of standing water above a certain level. ASCE-7 suggests that roofs be designed to sustain rainwater loads corresponding to the elevation of the secondary system plus 5.2 psf (0.24 kN/m^2) for each inch of water that might accumulate above the level of the secondary drainage system.

1.8.3.3 Wind Loads

The wind loads that act on a structure result from movement of the air against the obstructing surfaces. As was the case with snow loads, there is no single code that is uniformly applied throughout the United States. However, the procedure suggested by ASCE-7 will again be used to illustrate a typical approach. Wind effects induce forces, vibrations, and in some cases instabilities in the overall structure as well as its non-structural components. These wind effects depend on the wind speed, mass density of the air, location and geometry of the structure, and vibrational characteristics of the system.

Values for 3-second wind speed gusts have been accumulated over the contiguous United States, corrected to a standard height of 33 ft (10 m), and the highest values for a mean recurrence interval of 50 years have been mapped by ASCE using interpolation. Table 1.5 gives a few representative wind velocities from the ASCE map. Of course, as was true for snow loading, extreme local variations can occur, and the engineer must be cognizant of these possibilities.

The *design wind pressure* that is used to establish the wind load on a structure is directly related to *velocity pressure* (q), which is given by

$$q = \frac{\rho V^2}{2} \tag{1.4}$$

Table 1.5 Representative Wind Velocities and Resulting Dynamic Pressures[a]

Location	Wind Velocity (mph)	Velocity Pressure (lb/ft^2)	Velocity Pressure (kN/m^2)
Miami	145	53.8	2.58
Houston	120	36.9	1.77
New York City	105	28.2	1.35
Chicago	90	20.7	0.99
San Francisco	85	18.5	0.89

[a]Data are taken from ASCE-7.

where ρ is the mass density of air and V is the wind velocity. Taking the unit weight of air to be 0.07651 lb/ft^3, Eq. 1.4 can be expressed as

$$q = 0.00256V^2 \tag{1.5}$$

where q is in pounds per square foot and the velocity is in miles per hour. ASCE-7 modifies this equation to take the form

$$q_z = 0.00256K_zK_{zt}K_dV^2I \tag{1.6}$$

where the z subscript identifies the height above the ground, K_z is a dimensionless exposure coefficient that ranges from 0.32 for a low building in a sheltered location to 1.89 for a tall building in an unobstructed location, K_{zt} is a dimensionless topographic factor taken as 1.0 when the effects of special topographic features are minimal, K_d is a directionality factor taken as 0.85 for building structures, and I is a dimensionless importance factor that ranges from 0.77 to 1.15. Table 1.5 gives the velocity pressures for the corresponding wind velocities assuming values of unity for K_z, K_{zt}, K_d, and I.

The external design wind pressure for structural design of the main wind-force resisting system is given by

$$p = qGC_p - q_h(GC_{pi}) \tag{1.7}$$

where q is evaluated from Eq. 1.6 at a designated height, z, above the ground for each surface, G is a dimensionless gust factor that is taken as 0.85 for a rigid structure, C_p is a dimensionless pressure coefficient, q_h is the velocity pressure at the mean roof height, h, and (GC_{pi}) is the internal pressure coefficient. The pressure coefficients may be positive for direct pressure or negative for suction. A typical set of pressure coefficients for a structure with a gabled frame is shown in Fig. 1.24. The resulting force on any surface is determined by multiplying p by the appropriate area.

1.8.3.4 Earthquake Loads

A common dynamic loading that structures must resist is that associated with earthquake motions. Here, loads are not applied to the structure in the normal fashion. Instead, the base of the structure is subjected to a sudden movement. Since the upper portion of the structure resists motion because of its inertia, a deformation is induced in the structure. This deformation, in turn, induces a horizontal vibration that causes horizontal shear forces throughout the structure.

The resulting earthquake loads are dependent on the nature of the ground movement and the inertia response characteristics of the structure. In the United States, the ground motion expectation is provided through contour maps.

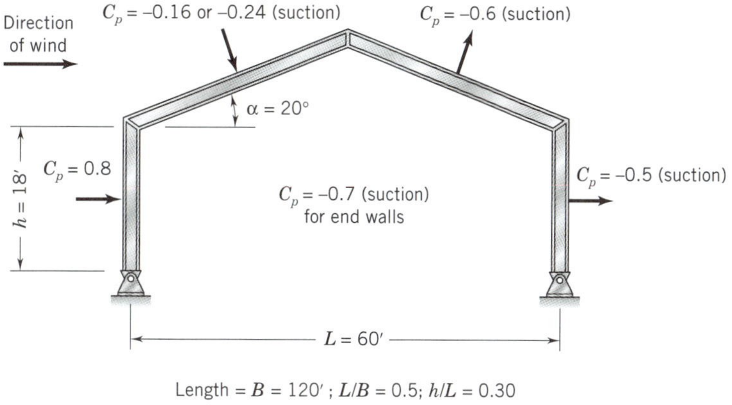

Length = $B = 120'$; $L/B = 0.5$; $h/L = 0.30$

Figure 1.24 Pressure coefficients for typical gabled frame with wind normal to ridge.

Current trends are toward using theories of structural dynamics to analyze a structure subjected to time-dependent earthquake motions. However, common practice continues to use a static approach. Here, the earthquake action is represented by a set of equivalent static forces, and a static analysis is used to establish the resulting member forces.

For example, ASCE–7 requires that the building be designed for a minimum total lateral seismic base shear force, V, given by the expression

$$V = C_s W \qquad (1.8)$$

where W is the total dead load of the structure in the same units as V, and C_s is the seismic response coefficient, which need not be greater than

$$C_s = \frac{S_{D1}}{T(R/I)} \qquad (1.9)$$

where R is the response modification factor, which depends on the ductility of the system and seismic force resisting system selected and which varies from 1.25 for the least ductile system to 8.0 for the most ductile system, T is the fundamental period of vibration of the structure, I is the occupancy importance factor, which varies from 1.0 to 1.5, and S_{D1} is the design spectral response acceleration.

EXAMPLE 1.1

Consider the gabled frame shown in Fig. 1.24. Assume that it is located in a suburb of Chicago, Illinois.

(a) Determine the roof snow load, p_s, acting on a horizontal projection if the structure is heated, is considered to be a regular structure, and is situated in a wind-sheltered, rural environment.

Determination of the load on a flat roof, p_f

For a flat roof, Eq. 1.2 gives

$$p_f = 0.7 C_e C_t I p_g$$

where

$C_e = 1.3$ for wind-sheltered, rural environment

$C_t = 1.0$ for heated structure

$I = 1.0$ for a regular structure

$p_g = 25$ lb/ft^2 for Chicago, Illinois (Table 1.4)

Therefore,

$$p_f = 0.7 \times 1.3 \times 1.0 \times 1.0 \times 25 = 22.8 \text{ lb/ft}^2$$

Conversion of the load from flat roof to sloped roof, p_s

For a sloped roof, such as the one in Fig. 1.24, Eq. 1.3 states

$$p_s = C_s p_f$$

where

$C_s = 1.0$ for a warm, normal roof surface with a slope $< 30°$

Thus,

$$p_s = 1.0 \times 22.8 = 22.8 \text{ lb/ft}^2$$

(b) Determine the design wind pressure for the main wind force resisting system if the wind acts normal to the ridge line (as shown in Fig. 1.24). Again, assume a rural, sheltered environment and a regular structure.

Determination of the velocity pressure, q_z

The velocity pressure is determined from Eq. 1.6, where z is taken as the vertical distance to the midheight of the roof. Thus, we have

$$q_z = 0.00256 K_z K_{zt} K_d V^2 I$$

where

$z = 18 + 15 \times \tan 20° = 18 + 5.5 = 23.5$ feet

$K_z = 0.65$ for given exposure at $z = 23.5$ feet (ASCE–7)

$K_{zt} = 1.0$ for no topographic impact

$K_d = 0.85$ for building

$I = 1.0$ for a regular structure

$V = 90$ mph for Chicago, Illinois (Table 1.5)

Therefore,

$$q_z = 0.00256 \times 0.65 \times 1.0 \times 0.85 \times 90^2 \times 1 = 11.5 \text{ lb/ft}^2$$

Determine design wind pressure, p, on the windward wall

The design pressure is obtained from Eq. 1.7 (modified for low-rise building), which in this case gives

$$p = qGC_p$$

where

$q = 11.5$ lb/ft for $z = 23.5$ feet; $GC_p = 0.71$ (ASCE–7)

Therefore,

$$p = 11.5 \times 0.71 = 8.2 \text{ lb/ft}^2 \text{ pressure on windward side.}$$

1.8.4   Load Combinations

The final design of a structure must be consistent with the most critical combination of loads that the structure is to support. However, some judgment is necessary in selecting loading conditions that can reasonably be combined. Obviously, the maximum effects of all loading conditions should not be combined because it is unlikely that they will all occur simultaneously. In fact, certain combinations are highly unlikely: full snow load is not likely to occur with full wind load because the wind would undoubtedly blow much of the snow off the structure; likewise, an earthquake is unlikely to occur at the instant that the structure is subjected to full wind load. The load combinations that must be considered are normally specified by the governing codes; in some cases, designers must exercise their judgment.

For example, ASCE–7 permits two different design approaches, and each carries its own patterns for local combinations. The first is *allowable-stress design* or *service-load design,* where the computed elastic stress associated with any load combination must not exceed a designated value. Some typical load combinations to be considered are

1. Dead
2. Dead + live + snow
3. Dead + live + wind or 0.7 × earthquake

The second design approach is *strength design* or *load-factor design.* Here, the strength of the structure must be adequate to support the most critical combination of factored loads, where a factored load results from multiplying a given load by a prescribed load factor. Some typical load combinations with the corresponding load factors are

1. 1.4 (dead)
2. 1.2 (dead) + 1.6 (live) + 0.5 (snow)
3. 1.2 (dead) + 0.5 (live) + [1.6 (wind) or 1.0 (earthquake)]

Case 2 reflects the assumption that the full snow load is unlikely to occur at the same time that the maximum live load occurs. Case 3 reflects a similar line of reasoning regarding the simultaneous occurrence of wind or earthquake with the full live load.

These load combinations relate to the unlikelihood of different loads reaching their maximum level at the same time. Another unlikely situation would be for every element of a structure to be fully loaded with the maximum live load at the same time. In order to account for this situation, ASCE–7 permits the designer to reduce the live load for members supporting an influence area of 400 ft^2 or more. Thus, the live load applied to a structural member is given by

$$L = L_o\left(0.25 + \frac{15}{\sqrt{A_i}}\right) \tag{1.10}$$

where

L = the reduced design live load

L_o = code specified design live load

A_i = influence area supported by the member under consideration. For beams the influence area is twice the tributary area. For columns the influence area is four times the tributary area.

Limitations on the use of this live load reduction are spelled out in ASCE–7 while the definition of tributary area will be discussed in Section 1.8.5.

1.8.5   Effects of Loads on Members

The loads described in the previous sections are imposed on the structure; they generally enter the structure by being applied on some surface, such as the roof, siding, or slab, and are then distributed to supporting members as forces. These forces are, in turn, transmitted to other members, and this succession continues until the effects of the imposed loads are delivered to the foundations. The response of the structure and the individual members is studied in detail in the remainder of the book. Our immediate concern is the manner in which the imposed loads filter through the structure.

The floor system shown in Fig. 1.25 is assumed to behave in the following manner: The *uniform surface load, q* (force per unit area), is applied to the slab, and the supporting beams then, in turn, receive loads from the slab. A useful concept for this discussion is that of *tributary area,* which is the area of the slab whose load contributes to the load transferred to a given beam. This area has a length equal to that of the beam and a width given by the sum of the half distances to each of its adjacent beams. Thus, for the floor system of Fig. 1.25a, beam *cd* has a tributary area with a length l and a width of $2(h/2) = h$, as shown in Fig. 1.25b, and the total load received by beam *cd* is, therefore, *qhl.* It is sometimes useful to treat this total load as if it were a *uniformly distributed line load* (force per unit length) along the beam length. In the case of member *cd,* the total load is thus divided by the length l to obtain the uniformly distributed load of *qh* as shown in Fig. 1.25c.

Expanding the concept of tributary area, and applying it to a uniformly loaded beam, results in the recognition that half of the load on the beam is carried by the support at each end of the beam. Thus, since beam *cd* is supported at *c* and *d* by girders *ae* and *bf,* respectively, a *concentrated force* of $(1/2)qhl$ is applied at *c* and *d* as shown in Fig 1.25c.

Since the load on this floor system is applied only to panel *a-b-f-e-a,* beams *ab* and *ef* receive load from only one side and, therefore, they each carry a total load of $q(h/2)l$ and a uniformly distributed load of $qh/2$. Each of these beams then applies half of its total load, $(1/4)qhl,$ to each of its supports as shown in Fig. 1.25c.

The concentrated loads from beams *ab, cd,* and *ef* are then applied to girders *ace* and *bdf,* as shown in Fig. 1.25c, and since these girders are loaded symmetrically, each girder will transfer half of its total load to its supports. Thus, concentrated loads of $qhl/2$ are applied at the columns at points *a, e, b,* and *f,* as shown in Fig. 1.25c.

In this discussion, only the superimposed load, *q,* has been considered. One should also consider the effects of the dead load from each component: slab load would be treated similarly to *q,* since it is a distributed surface load; beam dead load would add an additional component of uniform load along the beam; and girder dead weight would, likewise, enter as a uniform load on the girder.

Some floor systems are more complicated than the one described here. For the bridge system shown in Fig. 1.18, the slab rests on stringers, the stringers are supported by transverse floor beams, and the floor beams frame into girders, which are mounted atop the bridge piers.

The building system of Fig. 1.26 provides yet another example. Snow or wind loads that act on the roof surface are transferred to the supporting frames by *purlins.* For wind loads on the sides of the building, the surface loads are picked up by the *girts* and carried to the frames. Figure 1.17c provides an illustration of this type of system.

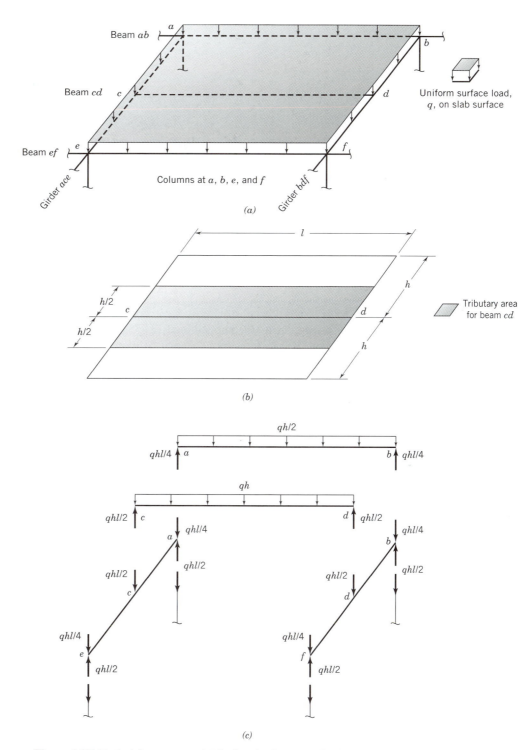

Figure 1.25 Typical floor system. *(a)* Surface load on slab. *(b)* Tributary area. *(c)* Beam, girder, and column loads.

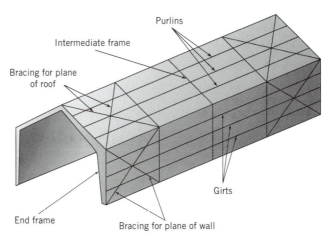

Figure 1.26 Typical skeletal system for frame building.

More complicated structures have correspondingly more complicated mechanisms for supporting the applied loads. However, the basic concepts hold—there is a progression of transfers from member to member until the full effects of the loads have been carried to the foundations.

EXAMPLE 1.2

For the floor system shown in Fig. 1.25, assume that the surface load, q, is applied on panel *abef*. If $q = 100$ lb/ft^2, determine the resulting effects on beams *ab, cd, ef,* girder *ae,* and the columns at *a* and *e.* Assume $l = 20$ ft and $h = 10$ ft.

Beam *ab* and Beam *ef*

$$\text{Tributary area} = (h/2)l = (10/2)20 = 100 \text{ ft}^2$$
$$\text{Total load} = q(\text{Tributary area}) = 100(100) = 10,000 \text{ lb}$$
$$\text{Uniform load} = w = (\text{Total load})/l = 10,000/20 = 500 \text{ lb/ft}$$
$$\text{End support load} = \frac{\text{Total load}}{2} = \frac{10,000}{2} = 5,000 \text{ lb}$$

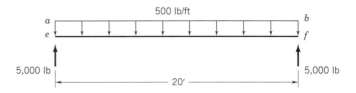

Beam *cd*

$$\text{Tributary area} = 2(h/2)l = 2(10/2)20 = 200 \text{ ft}^2$$
$$\text{Total load} = q(\text{Tributary area}) = 100(200) = 20,000 \text{ lb}$$
$$\text{Uniform load} = w = (\text{Total load})/l = 20,000/20 = 1,000 \text{ lb/ft}$$
$$\text{End support load} = \frac{\text{Total load}}{2} = \frac{20,000}{2} = 10,000 \text{ lb}$$

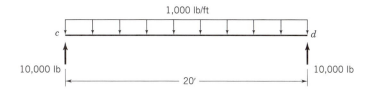

Girder *ae*

The end support loads from the beams at points *a, c,* and *e* must be applied to the girder.

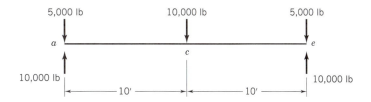

$$\text{End support load} = \frac{\text{Total load}}{2} = \frac{20{,}000}{2} = 10{,}000 \text{ lb}$$

Column Loads at *a* and *e*

The girder end support loads for girder *ae* are applied to the column tops. Column load at *a* and *e* = 10,000 lb.

Notes:

- The end support loads for both beams and girders are equal to half the total load because the members are *symmetrically* loaded. Procedures for *unsymmetrical* loading are covered in Chapter 3.

- If loading extended beyond beams *ab* and *ef,* the uniform loads on these beams would be increased.

- If loading extended beyond girders *ae* and *bf,* additional beam loads would enter the girders.

1.9   BUILDING MATERIALS

For a structure to function satisfactorily, it must have sufficient strength to support the applied forces without experiencing excessive deformations. The ability of the structure to satisfy these requirements depends largely on the material from which it is constructed. Although most structures experience complicated stress states, the important material properties can be determined through simple specimens loaded in tension or compression.

1.9.1   Steel

Steel is one of the most commonly used structural materials. It possesses essentially the same properties in both tension and compression, and a simple tension test on a small specimen is normally used to establish these properties. A simplified stress–strain curve for a mild carbon steel (A36 structural steel) is shown in Fig. 1.27, and this curve shows a yield stress of 36 ksi (248 MPa) and an ultimate stress of 58 ksi (400 MPa). Other

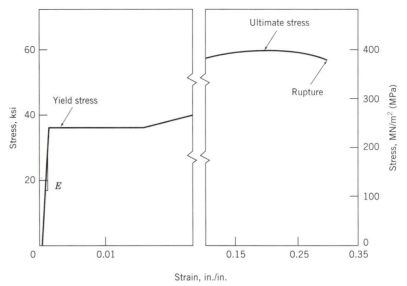

Figure 1.27 Simplified stress–strain curve for steel.

steels exhibit diff.erent yield characteristics and ultimate stress values; however, they all exhibit the same initial slope of the stress–strain curve. This property is called the *modulus of elasticity, E,* which has the value 29×10^3 ksi (200 GPa).

One of the major advantages of steel is its *ductility,* which is reflected by the large deformation (approximately 25 percent) that occurs after yielding, before rupture. However, steel can exhibit a brittle mode of failure under certain conditions, such as repetitive loading or cold temperatures. Also, steel is susceptible to corrosion.

1.9.2 Concrete

Structural concrete possesses considerable strength in compression, but it is very weak in tension. A typical stress–strain curve for a compression test of a standard cylindrical specimen is shown in Fig. 1.28. The compressive strength, f_c', varies from 3 to 7 ksi (20 to 48 MPa), and the initial portion of the stress–strain curve reflects the near-elastic behavior. For this reason, concrete structures are normally analyzed by elastic methods. For concrete of normal weight, the modulus of elasticity is approximated by

$$E = 57,000\sqrt{f_c'} \tag{1.11}$$

where f_c' and E are in units of pounds per square inch. For $f_c' = 3,000\,\text{psi}$, $E = 3,120,000$ psi (21.5 GPa), which is about 10 percent of the modulus for steel.

Figure 1.28 shows that there is no distinct yield point, nor is there any ductile behavior. However, the strains corresponding to the maximum strength, $\epsilon = 0.002$ in./in., and the point of rupture, $\epsilon = 0.004$ in./in., remain essentially the same for varying concrete strengths.

The tension strength of concrete is quite low—in the range of 0.5 ksi (3.4 MPa). For this reason, steel reinforcing bars are embedded in concrete members where tension exists, to form reinforced concrete.

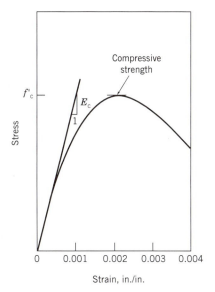

Figure 1.28 Stress–strain curve for concrete.

1.9.3 Wood

Wood is a material whose properties depend on the grain orientation and the cell structure. Tests show that wood exhibits linear behavior over the useful stress–strain range. The modulus of elasticity varies depending on the species and grade, but a typical value for Douglas Fir is 1,500 ksi (10.3 GPa).

Wood does not behave ductilely, but it has a high strength-to-weight ratio. Modern glue-lamination procedures enable the fabrication of members of widely varying shapes and sizes and have greatly broadened the range of usefulness of wood products.

1.10 SYSTEMS OF UNITS

The U.S. system of weights and measures, which is referred to as the U.S. Customary System (USCS), was officially established by action of the U.S. Treasury Department in the 1830s. Since this system is at variance with a host of other systems throughout the world, there has been much confusion in international exchanges. To resolve the problems that have resulted from this confusion, the International System of Units (Système International d'Unités), commonly referred to as SI, is being adopted throughout the world as a uniform measurement system. Although this is a metric system, it is not the metric system that has been used in various forms throughout Europe and other parts of the world, but is instead a completely new system.

The U.S. Congress passed the Metric Conversion Act in 1975. Because the changeover is voluntary, it seems clear that there will be a time when structural engineers will have to be conversant with both systems of units and be able to convert from one to the other. For these reasons, this textbook presents, in both systems of units, complete problem solutions that are designed to develop students' confidence in their ability to use either system.

The fundamentals concerning SI are briefly outlined in this section.

1.10.1 Base Units, Supplementary Units, and Derived Units

SI is founded upon seven *base units* on which the entire system is built. These base units represent quantities of length, mass, time, thermodynamic temperature, amount of substance, electric current, and luminous intensity. Each of these units, their definitions, and their symbols have complete international agreement. Only the first four of these base units are of specific interest to the structural engineer. These base units are listed in the table inside the front cover in both USCS and SI along with the conversion factor relating one to the other.

Two *supplementary units* represent the quantities of plane angle and solid angle. Plane angle is the only one of interest to the structural engineer, and it is entered in the table along with the units of both USCS and SI. The base units and supplementary units are combined to produce the required *derived units*. There are 17 of these derived units that have been given special names in SI. The derived units that are of most interest to the structural engineer are given in the table for both systems. Of the eight derived units that are listed, four represent quantities that have special names for their SI units.

1.10.2 Special Features of SI

Several features of SI deserve mention. First, SI is a *coherent* system in which all units are related to each other by unity. For instance, a force of one newton (N) results from a mass of one kilogram (kg) being given an acceleration of one meter per second squared (m/s^2). Or, a force of one newton (N) exerted over a distance of one meter (m) produces energy of one joule (J). Second, SI is an *absolute* system in which there is no dual meaning associated with any unit. The USCS uses the unit of pound (lb) to represent both mass and force. However, SI uses the kilogram (kg) as the base unit of mass (not to be confused with the kilogram force of the old metric system) and the newton (N) as the derived unit of force. Weight is not directly employed in SI, but this is understood to be the force caused as gravity acts on the mass. Third, SI is a *unique* system in that each quantity is represented by a single unit. This is not evident for the units given for structural engineers in the table inside the front cover, but it is an important feature in other disciplines. For instance, power in SI is given in the unit of watt (J/s) regardless of whether it is related to mechanical, thermal, or electrical systems.

There is a set of 16 prefixes for use with SI units in forming multiples and submultiples of these units. These prefixes are given inside the front cover, and they are used to advantage in eliminating insignificant digits and in providing a convenient substitute for writing powers of 10 in computations.

As an example of conversion from USCS to SI and of the use of prefixes, consider the modulus of elasticity of steel. Here,

$$E = 29 \times 10^6 \text{ psi}$$
$$= 29 \times 10^6 \times 6895 \simeq 200\,000 \times 10^6 \text{ Pa (N/m}^2)$$
$$= 200\,000 \text{ MPa (MN/m}^2) = 200 \text{ GPa (GN/m}^2)$$

Prefixes should not be used in the denominator of compound units. Thus, although $E = 200\,000$ MN/m^2 is equivalent to $E = 200\,000$ N/mm^2, the former is preferred to the latter.

The kilogram is a special case. Here, since the prefix kilo is part of the base unit itself, it may appear in the denominator. Also, since double prefixes are not to be used, the prefixes are to be used with gram (symbol g) and not with kilogram. Thus, 8×10^3 kg $= 8 \times 10^6$ g $= 8$ Mg.

It should also be noted that the prefixes centi, deci, deca, and hecto are to be avoided where possible in favor of the prefixes that represent multiples of 1000. This is done to encourage consistency in engineering practice and to avoid errors in calculations. Thus, the conventional units for length are meter (m) and millimeter (mm), and although centimeter (cm) was customarily used in the old metric system, it is discouraged in SI.

1.10.3 Additional Notes on the Use of SI

Care must be taken to avoid confusion between the use of a prefix and a derived compound unit. Thus, a moment in newton meters is represented by $N \cdot m$, with the dot representing multiplication, whereas a millinewton is mN without a dot. Similarly, a stress in newtons per square meter is given by N/m^2, in which the slash represents division.

Decimal points are used in the same fashion for USCS and SI, but commas are not used in SI to separate multiples of 1000. Instead, a space is used on both sides of the decimal point. Thus, 83,471.321,05 in USCS becomes 83 471.321 05 in SI.

1.10.4 Some Useful Equivalences

As was mentioned earlier, the emphasis in this book is not on conversion from one unit system to another. However, in translating the experience that one has in USCS to SI, it is helpful to note the approximate equivalences given in the table inside the front cover.

1.11   REFERENCES

Minimum Design Loads for Buildings and Other Structures, ASCE–7, American Society of Civil Engineers, Reston, VA, 1998.

LRFD Bridge Design Specifications (2nd ed.). American Association of State Highway and Transportation Officials, Washington, D.C., 1998.

1.12 SUGGESTED PROBLEMS

1.1 In order to carry out the analysis of a structure and its components, the analyst must be able to visualize the structure in a simplified form. Examples were shown in Figs. 1.21 and 1.22. For the structures in the figures listed below, construct a simplified sketch showing the main structural elements and the actions transmitted by these elements. Use the following key on our sketch to represent the *action a member induces on an adjacent joint.*

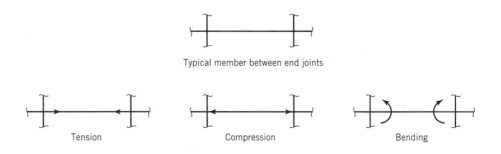

Typical member between end joints

Tension   Compression   Bending

(a) Pipeline bridge of Fig. 1.10a. Consider each of the following loading conditions separately.
- Uniform dead load of pipeline
- Uniform wind load on pipeline

(b) Suspension bridge of Fig. 1.10d. Consider the self weight of the road surface.

(c) Reinforced concrete arch bridge of Fig. 1.12a. Assume a uniform live load along the roadway.

(d) Steel arch bridge of Fig. 1.12c. Assume a uniform live load along the roadway.

(e) Wood arch church roof of Fig. 1.12d. Assume that the roof live load is carried to the arches by the purlins.

(f) Typical gable frame of Fig. 1.17c. Assume that the live load is transmitted by a roof deck, not shown, to the purlins and from these purlins to the inclined girders of the frame.

(g) Two stories of columns and beams for the right-most frame of the building shown in Fig. 1.17d. Consider each of the following loading conditions separately.
- Uniform dead load of the floor system on the girders
- A concentrated wind force at the left most column-beam intersections

1.2 Consider the corner of a reinforced concrete building as shown in plan view. Reinforced concrete girders (12 in. wide and 20 in. deep) span along line *A*, *B*, and *C*, and reinforced concrete beams (8 in. wide and 14 in. deep) run along lines 1 through 7. A 5-in. thick reinforced concrete floor slab is supported by the beams.

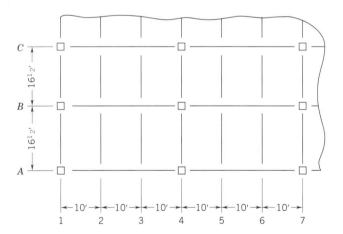

(a) Determine the dead load carried by the following members:
- Beam along line 1 between *A* and *B*.
- Beam along line 3 between *A* and *B*.
- Girder along line *A* between 1 and 4.
- Girder along *B* between 1 and 4.
- Columns at intersections *A*-1, *B*-1, and *B*-4.

(b) Determine the live load for these same members if the building is designed to support an office live load.

1.3 Consider a small, flat-roof steel building with a roof framing plan shown. The girders along lines *A*, *B*, and *C* and along lines 1, 2, and 3 each weigh 45 lb/ft. All intermediate beams weigh 26 lb/ft. A 4-in. deep concrete slab is supported by these intermediate members.

(a) Determine the total dead load carried by each intermediate beam. Note that these intermediate beams have different lengths and tributary widths.

(b) Determine the dead load carried by the girders along line *B*.

(c) If a roof snow load for Chicago is applied in an application where C_e, C_t, and *I* are all 1.0, determine the snow load carried by the columns at the intersection of *A*-1 and *B*-2.

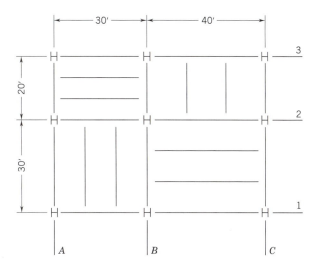

1.4 Consider the structure shown, which has a flat roof with an octagonal plan. If this building is to be constructed in Portland, Maine, determine the distribution and magnitude of the load caused by snow on the typical radial girder, *ab*. Assume that the roof surface is composed of a series of deck boards resting on the radial girders.

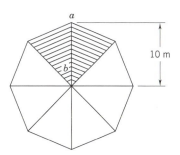

1.5 Consider the following gabled frame structure. The frames are spaced at 30-ft intervals along the building, and purlins and girts span between the frames as shown. It is assumed that each frame supports a section of roof that extends midway to the next frame in either direction. Likewise, purlins and girts receive loads from tributary areas that extend midway to the next purlin or girt, respectively.

(a) Determine the snow load, p_s, acting on a horizontal projection of the roof surface based on the following information:

- Structure is located in a suburb of Minneapolis, Minnesota.
- Building has occupancy of more than 300 people ($I = 1.1$).

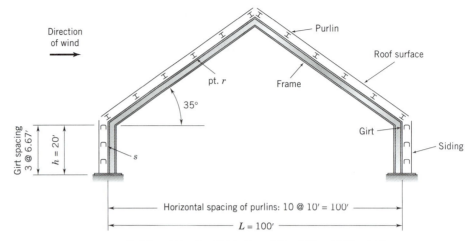

Direction of wind

Purlin

Roof surface

pt. r

Frame

35°

Girt

Siding

Girt spacing 3 @ 6.67'

$h = 20'$

s

Horizontal spacing of purlins: 10 @ 10' = 100'

$L = 100'$

Building width = B = 300 ft. Thus, $L/B = 0.33$ and $h/L = 0.20$

- Building is heated $(C_t = 1.0)$ and the roof is warm $(C_s = 0.9)$.
- Structure is in a windy, exposed site $(C_e = 0.8)$.

(b) For the loading of part (a), determine the concentrated load transmitted to the frame at point r.

(c) The wind pressure, p, on the windward wall is 18 lb/ft^2. Determine the concentrated load transmitted to the frame at point s.

(d) The wind pressure p, on the roof is 22 lb/ft^2. Determine the concentrated load transferred to the frame at point r.

Chapter 2

Basic Concepts of Structural Analysis

The Ganter Bridge, Simplon Road above Brig, Switzerland (photo by David P. Billington).

2.1   FORCES

Whatever the structure, it is subjected to a loading condition that results from its own dead weight, from the weight and functional input of the elements that it supports, and from the impact of the environment in which it exists. This loading may be very complicated and difficult to determine, as was explained in Chapter 1. With the assistance of building code information, research, and testing, however, a loading condition is finally assumed. Further assumptions might be necessary regarding how this loading condition is applied to the structure. But eventually, the result of these considerations is a set of *applied forces* of determined magnitudes, acting in specified directions and at designated points on the structure. In this context, these applied forces are of a general type in that they may include moments; however, whatever the nature of these forces, they are countered by *reaction forces* at the support points, as illustrated for the two structures in Fig. 2.1.

Note that the forces that act on a structure may be either static or dynamic. Static forces are of fixed magnitudes and are independent of time, whereas dynamic forces have time-dependent magnitudes. This book treats only static loading conditions or dynamic loading conditions that can be approximated by static loading.

2.2   SPECIFICATION OF A FORCE

The structures of Fig. 2.1 are each acted upon by two different types of forces—some are action forces that are applied to the structure, whereas others are reaction forces that act to resist the applied forces and thereby support the structure. Each of these forces is completely specified by its magnitude, direction, and point of application.

Figure 2.2a isolates the force P_i of the planar, two-dimensional structure shown in Fig. 2.1a. For convenience, the force P_i is replaced by its components P_{ix} and P_{iy} along the x and y axes, respectively, that are given by

$$P_{ix} = P_i \cos \alpha$$
$$P_{iy} = P_i \sin \alpha$$

$$(2.1)$$

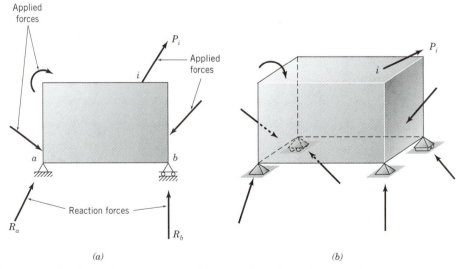

Figure 2.1 Applied and reaction forces. *(a)* Planar, two-dimensional structure. *(b)* Three-dimensional structure.

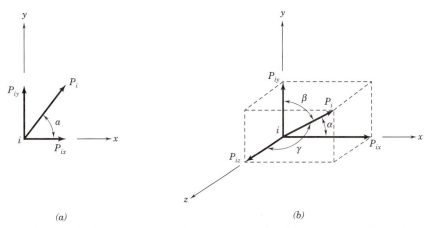

Figure 2.2 Specification of a force. *(a)* Planar, two-dimensional force. *(b)* Three-dimensional force.

These force components define the magnitude and direction of the force through the relationships

$$P_i = \sqrt{P_{ix}^2 + P_{iy}^2}$$

$$\alpha = \tan^{-1}(P_{iy}/P_{ix})$$

(2.2)

However, the point of application is still required for a complete specification of the force. This is accomplished by specifying a set of coordinates through which the force acts. Thus, for the planar, two-dimensional case, four quantities are needed to completely specify a force—the magnitude, P_i, the angle of inclination, α, and the two coordinates of the point of application, point i.

   In a similar fashion, Fig. 2.2b shows the force P_i of the three-dimensional structure of Fig. 2.1b. Again, the force is completely specified by its magnitude, direction, and point of application. Here, the direction is given by the angles α, β, and γ, which position the force with respect to the three coordinate axes. The components of P_i are

$$P_{ix} = P_i \cos \alpha$$

$$P_{iy} = P_i \cos \beta$$

$$P_{iz} = P_i \cos \gamma$$

(2.3)

where $\cos \alpha$, $\cos \beta$, and $\cos \gamma$ are referred to as the *direction cosines*.

   In this case, the magnitude of the force is given by

$$P_i = \sqrt{P_{ix}^2 + P_{iy}^2 + P_{iz}^2}$$

(2.4)

The angles α, β, and γ can be determined from Eqs. 2.3 and 2.4; however, it is clear from these equations that these angles are not mutually independent. That is, if any two are specified, the third can be determined.

   The point of application must also be specified, and this is done by designating a set of coordinates through which the force acts. Therefore, for the three-dimensional case, six quantities are required to specify a force—the magnitude, P_i, two direction cosines to fix the inclination of the force with respect to the coordinate axes, and the three coordinates of the point of application, point i.

   Each of the applied forces is part of a prescribed loading condition and, therefore, all of the quantities are known that are needed to specify its magnitude, direction, and point of application. For a reaction force, however, while certain quantities are specified because of the nature of the support mechanism, others are unknown and must be determined.

For instance, consider the reaction force at point a for the planar structure of Fig. 2.1a. Clearly, this force must act through the support point a; however, the magnitude and direction are unknown. These two unknowns can be expressed as R_{ax} and R_{ay}, and then Eqs. 2.2 can be used to determine the magnitude and angle of inclination. For the reaction at point b of the same structure, the point of application is again known. Also, in this case, the direction is known to be vertical because the roller mechanism precludes the presence of a horizontal force component. Thus, here there is a single unknown quantity—the magnitude of the reaction force R_b, which in this case is equal to its vertical compotent R_{by}. Similar arguments can be applied to the three-dimensional case shown in Fig. 2.1b to determine how many unknowns are associated with each reaction force.

Reaction forces are studied in Chapter 3, and the known and unknown quantities associated with a variety of support mechanisms are discussed in detail.

2.3   FREE-BODY DIAGRAMS

In structural analysis, one of the most useful tools at the analyst's disposal is the free-body diagram. A *free-body diagram* is simply a sketch of a structural component with all of the appropriate forces, both known and unknown, acting on it. It may be of an entire structure or of a subsection of a larger structure.

For instance, consider the planar structure shown in Fig. 2.1a. Figure 2.3a shows a free-body diagram of the entire structure together with the applied forces and the reaction forces. If the structure is cut along the line *c–d*, a free-body diagram of either section must show the internal forces that act on the cut face. These internal forces act in an equal and opposite fashion on the two free-body diagrams of Fig. 2.3b. If the two separate free-body diagrams are brought together, these internal forces cancel, and the total free-body diagram of Fig. 2.3a is restored.

Consideration of a free-body diagram of an entire structure leads to relationships between the external forces on the structure, whereas consideration of free-body diagrams

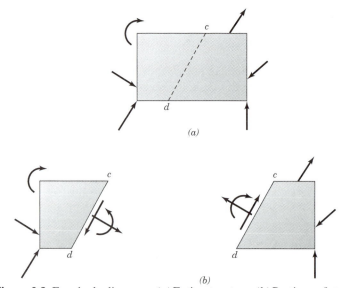

(a)

(b)

Figure 2.3 Free-body diagrams. (*a*) Entire structure. (*b*) Portions of structure.

of subsections of the structure lead to relationships between the external and internal forces that act on the free body. In any case, each free-body diagram represents a structural unit that is subject to all of the considerations of structural analysis. The judicious selection of free-body diagrams and the subsequent analysis of each structural component are fundamental to the field of structural analysis. These topics are examined in detail in this book.

The example discussed is for a planar structure, but the same concepts carry over to three-dimensional structures.

2.4 EQUATIONS OF EQUILIBRIUM

A structure is said to be in *equilibrium* when it is initially at rest and remains at rest as it is acted upon by a set of forces. When a structure satisfies this test for equilibrium, any constituent part of the structure will also be found to be in equilibrium.

For the condition of rest to be satisfied, there must be a balance in the force tendencies that would act to disturb the structure in any way. If we consider the planar structure shown in Fig. 2.4, with the cartesian coordinate axes shown, this balance can be expressed by

$$\sum P_x = 0; \quad \sum P_y = 0; \quad \sum M_z = 0 \tag{2.5}$$

where the first two equations refer to the summation of force components in the x and y directions, respectively, and the third equation gives the summation of moments about the z axis. When these *equations of static equilibrium* are satisfied, the body is in balance, at rest, and thus in equilibrium.

Equations 2.5 represent the most common form of the equilibrium equations for a planar structure; however, alternate forms are possible. For instance, we could use

$$\sum P_y = 0; \quad \sum M_{zo} = 0; \quad \sum M_{zp} = 0 \tag{2.6}$$

where o and p are two points in the xy plane as shown in Fig. 2.4, and $\sum M_{zo}$ and $\sum M_{zp}$ represent the summation of moments about a z axis through points o and p, respectively. This procedure is valid as long as the line connecting points o and p is not perpendicular to the axis along which the forces are summed, in this case the y axis.

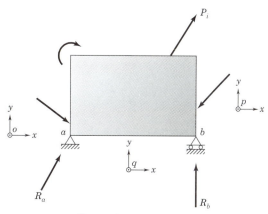

Note: z axis is normal to page.

Figure 2.4 Equilibrium of a planar structure.

Or in yet another form, we could have

$$\sum M_{zo} = 0; \quad \sum M_{zp} = 0; \quad \sum M_{zq} = 0 \qquad (2.7)$$

where point q also lies in the xy plane as depicted in Fig. 2.4, and $\sum M_{zq}$ is the summation of moments about a z axis through point q. These equations sufficiently express the equilibrium condition as long as points o, p, and q do not lie along a straight line.

Two special cases for planar structures deserve mention. The first deals with a *concurrent force system* in which all forces pass through a single point as illustrated in Fig. 2.5a. Here, the condition that $\sum M_z = 0$ is automatically satisfied, as is evident by taking moments about a z axis through the point of intersection for the forces, and Eqs. 2.5 reduce to

$$\sum P_x = 0; \quad \sum P_y = 0 \qquad (2.8)$$

A second special case deals with a *parallel force system*, such as the one shown in Fig. 2.5b. In this case, the summation of forces normal to the direction of the forces is automatically zero, and Eqs. 2.5 reduce to

$$\sum P = 0; \quad \sum M_z = 0 \qquad (2.9)$$

The term $\sum P$ represents the summation of forces in the direction corresponding to the direction of the parallel forces. In both of these special cases, alternate forms of the equations are possible, just as Eqs. 2.6 and 2.7 were alternate forms of Eq. 2.5.

For three-dimensional structures, the equations of equilibrium are expanded to

$$\sum P_x = 0; \quad \sum P_y = 0; \quad \sum P_z = 0$$
$$\sum M_x = 0; \quad \sum M_y = 0; \quad \sum M_z = 0 \qquad (2.10)$$

The first set of three equations gives the summation of all force components in the directions of the x, y, and z axes, and the second set represents the summation of moments about these three axes, respectively. Once again, it is possible to employ alternate forms of these equations as was demonstrated for planar structures. Also, there are special cases, such as a concurrent or parallel force system, where the total number of equations needed to ensure equilibrium is reduced. Some of these cases will be addressed in subsequent sections of the book, but they will not be developed in detail at this point.

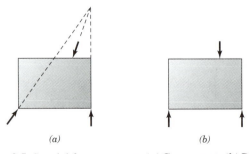

(a)   (b)

Figure 2.5 Special force systems. (*a*) Concurrent. (*b*) Parallel.

All of the formulations considered here may be applied to the structure as a whole or to a portion of the structure. In each case, the portion under consideration is isolated as a free-body diagram, and the application of the appropriate equilibrium equations will yield the conditions that the forces must satisfy for equilibrium to exist.

2.5   CONDITION EQUATIONS

As demonstrated in the previous section, there is a limited number of independent equations of equilibrium. Since these equations are used to solve for the unknown reaction force components, it follows that only a limited number of reaction components can be determined from the considerations of equilibrium. Where special internal conditions of construction exist, however, it is possible to write additional statical equations for use in determining the unknown reaction forces. These equations are referred to as *condition equations*, and each condition increases by one the number of reaction components that can be determined through the considerations of static equilibrium.

As an example, consider again the planar structure of Fig. 2.1*a*. It was noted in Section 2.2 that there are three unknown reaction components (R_{ax}, R_{ay}, and R_{by}), and these components can be determined through the application of Eqs. 2.5. If this structure is altered by appending the element *ef* with a roller support at point *f*, as shown in Fig. 2.6*a*, an additional reaction component, R_{fy}, is added to the structure.

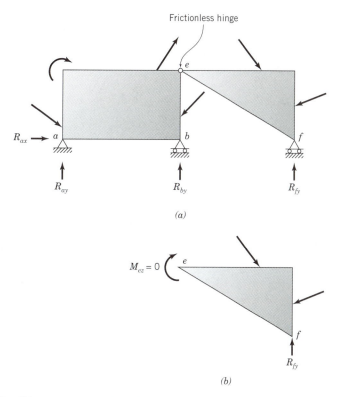

Figure 2.6 Condition equation. (*a*) Total structure with condition of construction at *e*. (*b*) Free-body diagram of *ef* making use of condition equation.

Since there are now four unknown reaction components, Eqs. 2.5 are no longer sufficient for their determination. In this case, however, the frictionless hinge at point *e* provides a unique condition of construction because it is unable to transmit a moment. Therefore, by taking a separate free-body diagram of element *ef* as shown in Fig. 2.6*b*, and noting that the moment at point *e* about a *z* axis must be zero ($M_{ez} = 0$), we introduce a condition equation that augments Eqs. 2.5 and allows for the solution of the four reaction components. Other conditions of construction are possible for planar structures, and clearly, the concept can be expanded to three-dimensional structures.

The use of condition equations, and the appropriate selection of free-body diagrams, are vital parts in the analysis of structures, and the application of these concepts will be examined in detail later in the book.

2.6   DISPLACEMENTS

In response to the forces that act on it, a structure undergoes *deformations*. At any point throughout the structure these deformations are manifested by a set of *displacements*. Consider again the planar structure of Fig. 2.1*a*, which is shown in Fig. 2.7 with the selected displacements identified. For instance, point *c* experiences a translational displacement, Δ_c, which is composed of the components Δ_{cx} and Δ_{cy}, along the *x* and *y* axes, respectively, and a rotation θ_c, about the *z* axis. At the support points, certain displacements are restricted because of the nature of the support mechanisms. Deformations and the accompanying displacements are important in evaluating the response of a structure because, just as the structure must have adequate strength to withstand the forces that act on it, it must also have sufficient rigidity so as not to become unserviceable because of unacceptable deformations.

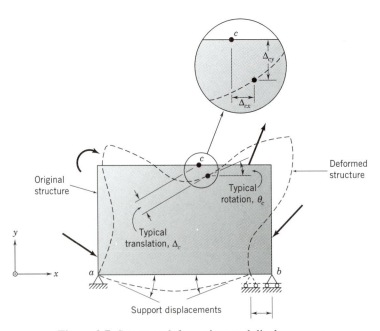

Figure 2.7 Structure deformations and displacements.

For the planar case shown, there are three independent components of displacement at a point—translational components in the x and y directions, and rotation about the z axis. In the case of a three-dimensional structure, there are six independent displacement components—one translational component along each of the three coordinate axes, and a rotational component about each axis.

The dual considerations of forces and displacements are of major importance in structural analysis. In the previous sections, we briefly discussed how the equilibrium and condition equations are used to calculate the reaction forces that result from the designated applied forces. In subsequent chapters, these problems will be studied in detail, and methods will also be introduced for the determination of the internal forces carried by the individual elements of the structure. Additionally, techniques will be developed for determining element deformations and the resulting structure displacements that are associated with the element forces and the prescribed set of applied forces.

At this point, it is sufficient to realize that displacements do occur as part of the response of a structure to a given loading condition. These displacements are of importance and are determinable.

2.7   COMPATIBILITY

As explained in Section 2.4, equilibrium requires that the forces on a structure satisfy certain relationships for the structure to be at rest. On the other hand, *compatibility* places certain demands on the displacements of the system to ensure that the constituent parts of the structure fit together without voids or overlap.

To illustrate the concept of compatibility, consider again the planar structure of Fig. 2.l*a* as it was separated into two sections in Fig. 2.3*b*. Figure 2.7 shows how the entire structure deforms under the action of the applied forces, and Figure 2.8 shows the deformed shapes of the two separate sections. Compatibility requires that the displacement components be equal for common points on the separate sections. That is, at point c, Δ_c, and θ_c must be the same for each section, and similarly for point d, Δ_d and θ_d must be equal on each side of the severed structure. Figure 2.8 focuses on points c and d for illustrative purposes, but for compatibility, there must be equality in the displacement components for all points along line c–d.

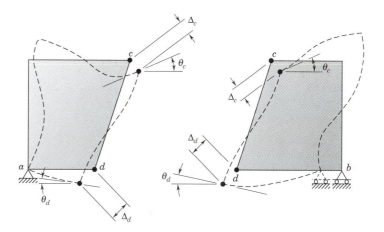

Figure 2.8 Structure compatibility.

When compatibility is satisfied, there are no abrupt changes in the displacement components. If a given displacement component were plotted as a function of position within the structure, the resulting curve would be continuous. For this reason, compatibility is sometimes referred to as *continuity*, which underscores the continuous nature of the displacements. Of course, there are cases where discontinuities occur. For instance, the frictionless hinge at point *e* for the structure of Fig. 2.6*a* permits a discontinuity in rotation. That is, there is no rotational compatibility at this point between the two sections of the structure.

The concept of compatibility is easily extended to three-dimensional structures by requiring continuity in all of the appropriate displacement components. In both planar and three-dimensional structures, it is clear that the ensurance of compatibility requires the computation of displacement quantities. The techniques for computing the required displacements for various structure types are treated in detail in later chapters.

2.8   BOUNDARY CONDITIONS

Structures are sustained in equilibrium and restrained against movement through their interface with support points. The details concerning how a structure connects to its supports are enumerated by *boundary conditions*. Some boundary conditions are related to constraints that the supports provide on movement and are referred to as *displacement boundary conditions*. Other boundary conditions express conditions relative to the forces that can be developed at a support point. These are referred to as *force boundary conditons*.

For instance, consider the frictionless point support at *a* or the planar structure of Fig. 2.1*a*. which is shown in Fig. 2.9*a*. This support imposes a displacement boundary condition that constrains point *a* against translation. Thus, a reaction force must develop that will preclude translation at this point. However, the existence of a frictionless point support implies that there is no resistance against rotation and, therefore, no support moment. This force boundary condition requires that sufficient rotation occur to render the support free of any moment.

Similar reasoning can be applied to the roller support at point *b* of the same structure, which is isolated in Fig. 2.9*b*. This support imposes the displacement boundary

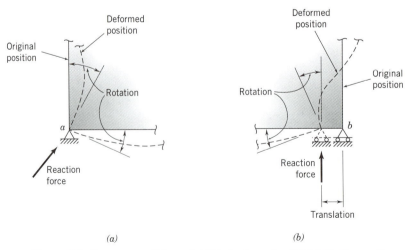

(a)   *(b)*

Figure 2.9 Support conditions. *(a)* Frictionless pin. *(b)* Frictionless roller.

condition that point b cannot translate normal to the surface on which the roller is mounted, and the vertical reaction force must sustain this restraint. However, there is no resistance to rotation or translation parallel to the support surface. These two force boundary conditions will require sufficient rotation and translation, respectively, to free the support of moment or force parallel to the surface.

From these two cases, it is clear that a pattern exists that can be summarized as follows:

1. A displacement boundary condition places a constraint on a displacement quantity; however, the corresponding force quantity must be allowed to take on whatever value is required to ensure the specified displacement.

2. A force boundary condition places a constraint on a force quantity: however, the corresponding displacement quantity must be allowed to take on whatever value is required to ensure the specified force.

Care must be exercised not to overspecify boundary conditions. If a displacement condition is specified, the corresponding force condition cannot be specified, whereas if a force condition is specified, the corresponding displacement condition cannot be specified.

2.9 PRINCIPLE OF SUPERPOSITION

Much in structural analysis is based on the *principle of superposition*, which states that the effects of individual actions can be combined to determine the total effects.

Consider the planar structure shown in Fig. 2.10a, which is subjected to the load P_i. The response of the structure includes the reaction forces, as represented by R_a, and the displacements, of which Δ_c is typical. In Fig. 2.10b, the structure is subjected to loadings of two different levels for P_i, namely, $P_{i1} = P$ and $P_{i2} = 2P$, and the individual reactions, R_{a1} and R_{a2}, and the displacements, Δ_{c1} and Δ_{c2}, result. According to the principle of superposition, the individual responses can be superimposed to obtain the total response. Therefore, for $P_i = P_{i1} + P_{i2} = 3P$, we have

$$R_a = R_{a1} + R_{a2} \tag{2.11}$$

$$\Delta_c = \Delta_{c1} + \Delta_{c2} \tag{2.12}$$

Implicit in these relationships is the assumption that R_a and Δ_c increase linearly with increasing P_i; however, there are limitations regarding this assumption.

Consider first the case of displacements as represented by Eq. 2.12. If Δ_c is linearly related to the magnitude of P_i, then the load–displacement diagrams given in Fig. 2.10c are valid. Here it is clear that the individual displacement for each of the cases in Fig 2.1b can be superimposed to obtain the displacement under the combined loading. But if the load–displacement relationship is not linear, as shown in Fig. 2.10d, then the individual displacements would clearly not sum to the total displacement under the combined loading. This latter case occurs in two situations. The first is associated with instances where the geometry of the structure is appreciably altered under the load. In this case, when load is applied to the structure, it changes its configuration such that subsequent loads are applied to a modified structure with a different load–displacement characteristic, or different structural stiffness. Structures of this type are classified as *geometrically nonlinear* or *geometrically unstable*. The second case where superposition is not valid is when the structure is constructed of material for which the stress is not linearly related to the strain. Here, the material stiffness changes as the load is increased.

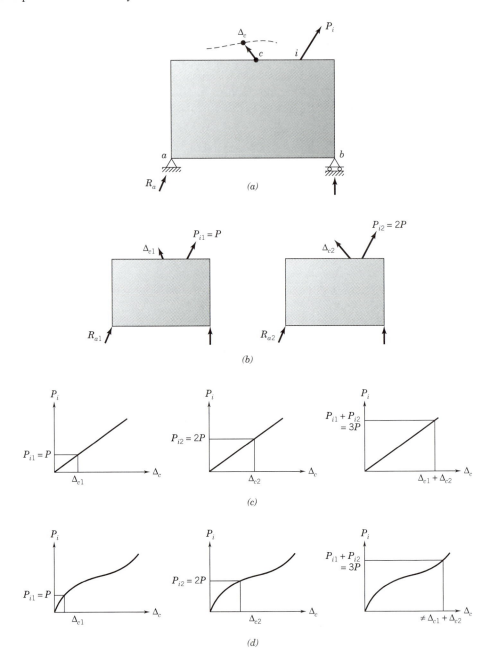

Figure 2.10 Superposition. *(a* and *b)* Loading cases. *(c)* Linear load–displacement response. *(d)* Nonlinear load–displacement response.

Consider next the superposition of force quantities as depicted by Eq. 2.11. If the reactions are determinable solely on the basis of the equations of equilibrium, then the superposition of force quantities is valid without exception. Structures with this property are classified as *statically determinate*. However, if in the determination of the reactions, the equations of equilibrium must be augmented by equations of compatibility (which involve displacement quantities), then the same restrictions on superposition apply as prevailed in the discussion of the superposition of displacement quantities. Such structures are classified as *statically indeterminate*. Here, force quantities cannot be superimposed if there are geometric nonlinearities or the material possesses a nonlinear stress–strain relationship.

The discussion has centered on the superposition of two separate load cases with the same pattern of loading for each. The argument could have been formulated with equal validity for the superposition of two separate load patterns.

2.10 STIFFNESS AND FLEXIBILITY

The concepts of force and displacement were introduced in Sections 2.1 and 2.6, respectively. It is useful to consider the interactive relationships between these two quantities, that is, to establish displacement–force or force–displacement relationships.

Consider the structure shown in Fig. 2.11a. Points 1 and 2 identify two points on the structure—the forces at these points are identified as P_1 and P_2, and the corresponding displacements are Δ_1 and Δ_2.

It is instructive to consider the two separate loading cases shown in Fig. 2.11b, where unit values of P_1 and P_2 are applied in turn. For $P_1 = 1$ and $P_2 = 0$, the following displacement quantities are noted:

$$D_{11} = \text{value of } \Delta_1 \text{ when } P_1 = 1 \text{ and } P_2 = 0$$
$$D_{21} = \text{value of } \Delta_2 \text{ when } P_1 = 1 \text{ and } P_2 = 0$$

Similarly, for $P_1 = 0$ and $P_2 = 1$, the displacements are as follows:

$$D_{12} = \text{value of } \Delta_1 \text{ when } P_1 = 0 \text{ and } P_2 = 1$$
$$D_{22} = \text{value of } \Delta_2 \text{ when } P_1 = 0 \text{ and } P_2 = 1$$

The D quantities are referred to as *flexibility coefficients*, and it is clear that D_{ij} *is the value of the ith displacement associated with a unit value of the jth force.*

If P_1 and P_2 represent specific values of the applied forces on a linearly elastic structure, then superposition leads to

$$\Delta_1 = D_{11}P_1 + D_{12}P_2$$
$$\Delta_2 = D_{21}P_1 + D_{22}P_2 \tag{2.13}$$

where Δ_1 and Δ_2 are the total displacements associated with the applied forces. These equations can be written in matrix form as

$$\begin{Bmatrix} \Delta_1 \\ \Delta_2 \end{Bmatrix} = \begin{bmatrix} D_{11} & D_{12} \\ D_{21} & D_{22} \end{bmatrix} \begin{Bmatrix} P_1 \\ P_2 \end{Bmatrix} \tag{2.14}$$

or in the abbreviated form

$$\{\Delta\} = [D]\{P\} \tag{2.15}$$

where $\{\Delta\}$ and $\{P\}$ are the displacement and force vectors, respectively, and $[D]$ is the *flexibility matrix*.

Figure 2.11c shows two displacement patterns, where unit values of Δ_1 and Δ_2 are separately induced. For $\Delta_1 = 1$ and $\Delta_2 = 0$, the following forces must be introduced at point 1 and 2:

$$K_{11} = \text{value of } P_1 \text{ for } \Delta_1 = 1 \text{ and } \Delta_2 = 0$$
$$K_{21} = \text{value of } P_2 \text{ for } \Delta_1 = 1 \text{ and } \Delta_2 = 0$$

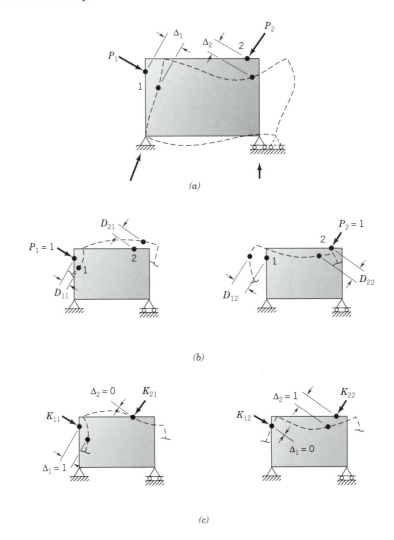

Figure 2.11 Forces and displacements. (*a*) Applied forces and corresponding displacements. (*b*) Unit loads and flexibility coefficients. (*c*) Unit displacements and stiffness coefficients.

Similarly, for $\Delta_1 = 0$ and $\Delta_2 = 1$, the required forces are

$$K_{12} = \text{value of } P_1 \text{ for } \Delta_1 = 0 \text{ and } \Delta_2 = 1$$
$$K_{22} = \text{value of } P_2 \text{ for } \Delta_1 = 0 \text{ and } \Delta_2 = 1$$

The K quantities are called *stiffness coefficients*, and they are defined such that K_{ij} is *the value of the ith force associated with a unit value of the jth displacement.*

If Δ_1 and Δ_2 are taken as the final values of an imposed deformation configuration for a linearly elastic structure, then superposition gives

$$P_1 = K_{11}\Delta_1 + K_{12}\Delta_2$$
$$P_2 = K_{21}\Delta_1 + K_{22}\Delta_2 \tag{2.16}$$

where P_1 and P_2 are the total forces associated with the imposed displacement pattern. In matrix form, Eqs. 2.16 become

$$\begin{Bmatrix} P_1 \\ P_2 \end{Bmatrix} = \begin{bmatrix} K_{11} & K_{12} \\ K_{21} & K_{22} \end{bmatrix} \begin{Bmatrix} \Delta_1 \\ \Delta_2 \end{Bmatrix} \tag{2.17}$$

or

$$\{P\} = [K]\{\Delta\} \tag{2.18}$$

where $\{P\}$ and $\{\Delta\}$ are as previously defined, and $[K]$ is the *stiffness matrix*.

The concepts of stiffness and flexibility are encountered in numerous instances throughout the book in the context of overall structure considerations, such as is illustrated in the discussions related to Fig. 2.11 and for individual member considerations. In Part Five of the book, these concepts take on increased emphasis when matrix methods are considered.

2.11 WORK

The concept of *work* is of prime importance in structure analysis. Consider the situation shown in Fig. 2.12a, where a force of magnitude P acts through the incremental displacement $d\Delta$. The increment of work, dW, done by the force is defined as

$$dW = P \cdot d\Delta \tag{2.19}$$

It is important to note that the displacement must be along the same line of action as the force in order for the force to do work.

If the magnitude of force is plotted against the displacement, then the increment of work at any load level is given by Eq. 2.19 and is represented by the crosshatched area of Fig. 2.12b. Thus, the *total work*, W, done over the full displacement Δ is given by

$$W = \int_0^{\Delta} P \cdot d\Delta \tag{2.20}$$

where P is a function of Δ, and integration is over the full range of displacement. Equation 2.20 represents the shaded area under the load–displacement curve between displacements of zero and Δ as shown in Fig. 2.12b.

One common case is when the load has a constant value over the full range of the displacement as shown in Fig. 2.12c. In this instance, the work is given by

$$W = P \cdot \Delta \tag{2.21}$$

This case corresponds to the situation in which the force is at its full magnitude on a structure and is subjected to a displacement caused by an action independent of the force. Another case of interest is when the force is proportional to the displacement. This case is represented by Fig. 2.12d and the work is

$$W = \tfrac{1}{2}P \cdot \Delta \tag{2.22}$$

This situation occurs when the force itself is causing the displacement that produces the work, and the structure is linearly elastic.

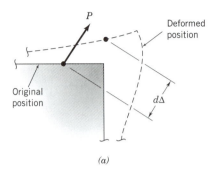

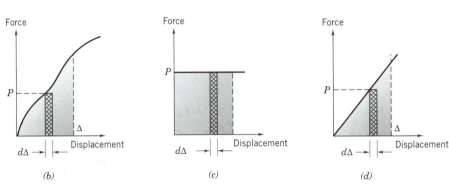

Figure 2.12 Work. *(a)* Force and increment of displacement. *(b)* General force–displacement relationship. *(c)* Constant force–displacement relationship. *(d)* Linear force–displacement relationship.

Everything that has been stated thus far is related to a single force and a single displacement. For a system of n forces and n displacements, the integration indicated by Eq. 2.20 must be carried out for each of the n components. The total work is then given by

$$W = \sum_{i=1}^{n} \int_{0}^{\Delta_i} P_i \cdot d\Delta_i \qquad (2.23)$$

For constant load values over the full range of the displacements, this becomes

$$W = P_1 \Delta_1 + P_2 \Delta_2 + \cdots + P_i \Delta_i + \cdots + P_n \Delta_n \qquad (2.24)$$

or

$$W = \sum_{i=1}^{n} P_i \Delta_i = \{P\}^T \{\Delta\} \qquad (2.25)$$

where $\{P\}^T$ is the transpose of the load vector and $\{\Delta\}$ is the corresponding displacement vector, each of which possesses n elements. For linear force–displacement relations, Eq. (2.23) leads to

$$W = \tfrac{1}{2} \sum_{i=1}^{n} P_i \Delta_i = \tfrac{1}{2} \{P\}^T \{\Delta\} \qquad (2.26)$$

Equations 2.25 and 2.26 are multidimensional matrix versions of Eqs. 2.21 and 2.22, respectively.

2.12 COMPLEMENTARY WORK

If work is done by an incremental force dP being swept through a displacement of Δ, as shown in Fig. 2.13a, then the resulting work is referred to as *complementary work*. The increment of complementary work, dW^*, is given by

$$dW^* = \Delta \cdot dP \tag{2.27}$$

Figure 2.13b is a repeat of the force–displacement relationship given in Fig. 2.12b, and an increment of complementary work is crosshatched. The total complementary work, W^*, done over the full magnitude of the load P is given by

$$W^* = \int_0^P \Delta \cdot dP \tag{2.28}$$

where Δ is a function of P, and integration is over the full range of the load. Equation 2.28 represents the shaded area above the load–displacement curve between force magnitudes of zero and P as noted in Fig. 2.13b.

When the total complementary work results from the interaction of a displacement of constant value over the full range of loading (Fig. 2.13c), then the complementary work is

$$W^* = \Delta \cdot P \tag{2.29}$$

This represents a case where the displacement is not dependent on changes in force. Another special case is when the displacement is proportional to the force (Fig. 2.13d). Here, the structure is linearly elastic, and the integration of Eq. 2.28 leads to

$$W^* = \tfrac{1}{2}\Delta \cdot P \tag{2.30}$$

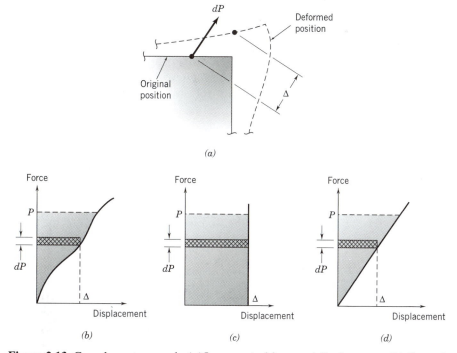

Figure 2.13 Complementary work. (*a*) Increment of force and displacement. (*b*) General displacement–force relationship. (*c*) Constant displacement–force relationship. (*d*) Linear displacement–force relationship.

For a system with n forces and n displacements, the total complementary work is given by

$$W^* = \sum_{i=1}^{n} \int_0^{P_i} \Delta_i \cdot dP_i \qquad (2.31)$$

When constant displacements exist over the full range of the forces, Eq. 2.31 gives

$$W^* = \sum_{i=1}^{n} \Delta_i P_i = \{\Delta\}^T \{P\} \qquad (2.32)$$

For the linearly elastic case, the integration of Eq. 2.31 yields

$$W^* = \frac{1}{2} \sum_{i=1}^{n} \Delta_i P_i = \frac{1}{2} \{\Delta\}^T \{P\} \qquad (2.33)$$

Equations 2.32 and 2.33 are multidimensional representations of Eqs. 2.29 and 2.30, respectively.

A comparison of Eqs. 2.32 and 2.33 with Eqs. 2.25 and 2.26, respectively, shows that $W = W^*$ for these two special cases. This is evident from comparing the corresponding portions of Figs. 2.12 and 2.13. It is clear, however, that this equality does not hold true for the general case.

2.13 PRINCIPLE OF VIRTUAL DISPLACEMENTS

The principle of virtual displacements is credited to John Bernoulli in 1717. To develop this principle, consider the planar rigid body shown in Fig. 2.14a, which is in static equilibrium under the set of P forces shown. It is now imagined that the body is subjected to a translation, without rotation, as shown in Fig. 2.14b. This imagined translation is called a *virtual displacement*, and this is a fitting term because the word virtual means "being in essence or effect, but not in fact." The fact that this is a virtual displacement is indicated by a δ prefix on the displacement components δu and δv shown in Fig. 2.14b. As the body undergoes the virtual displacement, each of the P forces does *virtual work* by an amount equal to its magnitude times the virtual displacement along the line of action of the force. It is most convenient to work with force components and the corresponding displacement components as shown for the specific force P_i in Fig. 2.14b. For this force, the virtual work would be $P_{ix}\delta u + P_{iy}\delta v$. Letting δW_T represent the total virtual work of translation for the complete set of n forces, we have

$$\delta W_T = P_{1x}\delta u + P_{1y}\delta v + \cdots + P_{ix}\delta u + P_{iy}\delta v + \cdots + P_{nx}\delta u + P_{ny}\delta v \qquad (2.34)$$

This work expression is written in a form for the general loading under consideration without anticipating the signs of the individual terms. However, in its application, care must be exercised regarding the sign of each force and displacement component. Positive work would result when the force component and its corresponding virtual displacement component are in the same direction, as shown for the specific force components P_{ix} and P_{iy} in Fig. 2.14b, whereas negative work would result when a force component and its associated virtual displacement component are in opposing directions. Continuing with the general formulation, we can rewrite Eq. 2.34 as

$$\delta W_T = (P_{1x} + \cdots + P_{ix} + \cdots + P_{nx})\delta u + (P_{1y} + \cdots + P_{iy} + \cdots + P_{ny})\delta v$$

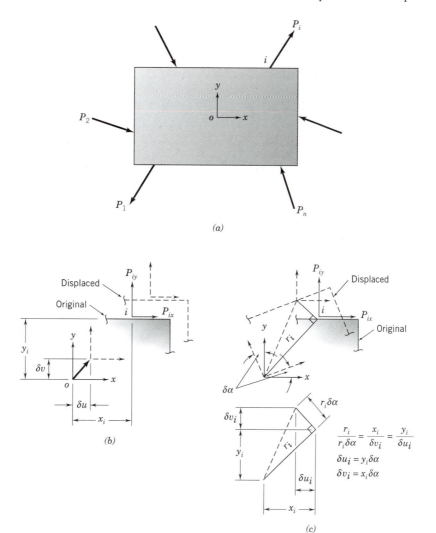

Figure 2.14 Virtual displacement for rigid body. *(a)* Rigid body in equilibrium. *(b)* Rigid body translation. *(c)* Rigid body rotation.

or

$$\delta W_T = \left[\sum_{i=1}^{n} P_{ix} \right] \delta u + \left[\sum_{i=1}^{n} P_{iy} \right] \delta v \qquad (2.35)$$

However, because the body is in equilibrium, each of the bracketed terms is zero as shown in Section 2.4. Thus,

$$\delta W_T = 0 \qquad (2.36)$$

It is next imagined that the body is subjected to a virtual rotation of $\delta\alpha$, which induces horizontal and vertical displacements of δu_i and δv_i, respectively, at point i, as shown in Fig. 2.14c. It is clear from the figure that a positive virtual rotation of $\delta\alpha$ induces a negative δu_i ($\delta u_i = -y_i\delta\alpha$) and a positive δv_i ($\delta v_i = x_i\delta\alpha$), where x_i and y_i

are the coordinates of point i. It is easily verified that these expressions are general for all x's and y's. Thus, the virtual work for the specific force P_i in Fig. 2.14c would be $P_{ix}\delta u_i + P_{iy}\delta v_i = -P_{ix}y_i\delta\alpha + P_{iy}x_i\delta\alpha$. Thus, representing the total virtual work of rotation by δW_R, we have

$$\delta W_R = (-P_{1x} \cdot y_1\delta\alpha + P_{1y} \cdot x_1\delta\alpha) + \cdots + (-P_{ix} \cdot y_i\delta\alpha$$
$$+ P_{iy} \cdot x_i\delta\alpha) + \cdots - (P_{nx} \cdot y_n\delta\alpha + P_{ny} \cdot x_n\delta\alpha) \quad (2.37)$$

Again, this work expression is for the general loading, but care must be exercised regarding the signs of the coordinates of the load point and the individual load terms. Rewriting Eq. 2.37, we have

$$\delta W_R = \sum_{i=1}^{n} [-P_{ix}y_i + P_{iy}x_i]\delta\alpha \quad (2.38)$$

Because the body is in equilibrium, the bracketed term, which represents the summation of moments about point o, is zero, and we have

$$\delta W_R = 0 \quad (2.39)$$

Because a rigid-body virtual displacement can be composed of a combination of translation and rotation, the total virtual work, δW, also must vanish. That is,

$$\delta W = \delta W_T + \delta W_R = 0 \quad (2.40)$$

Thus, the *principle of virtual displacements* can be summarized as follows: *If a rigid body is in equilibrium under a set of* P *forces and it is subjected to any virtual displacement, the virtual work done by the* P *forces is zero.*

2.14 VIRTUAL WORK FOR A DEFORMABLE BODY

The concept of virtual work was introduced in the previous section, and the principle of virtual displacements was based on virtual work considerations for a rigid body. We now extend the application of virtual work to a deformable body. As these techniques will be used throughout the text for all types of structures, the development here is of a general nature.

Consider the planar deformable body shown in Fig. 2.15a, which is in equilibrium under the P forces. Any element within the body is in equilibrium under a set of σ_P stresses that are caused by the P forces. The body is then subjected to a geometrically compatible virtual distortion. This distortion results in a corresponding set of virtual displacements, $\delta\Delta$, at each of the external load points and a set of virtual strains, $\delta\epsilon_\Delta$, of the element.

The total virtual distortion of the body is made up of two components. There is a rigid-body motion of the body as a whole and a deformation of the body as illustrated in Fig. 2.15b. Thus, the total virtual work done by the external P forces, δW_e, has two components. There is a work term that couples the P forces with the virtual displacements that result from rigid-body motion, δW_r, and a work term that couples the P forces with the virtual displacements that result from the deformation of the body, δW_d. Thus,

$$\delta W_e = \delta W_r + \delta W_d$$

or

$$\delta W_r = \delta W_e - \delta W_d \quad (2.41)$$

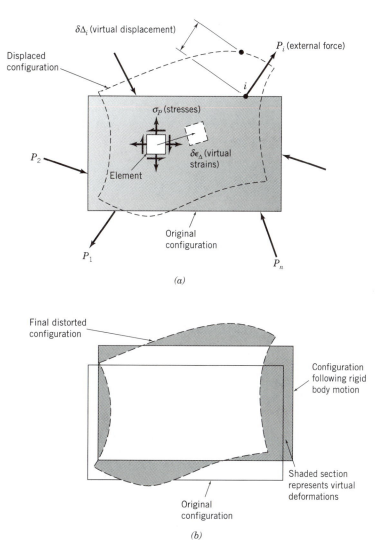

Figure 2.15 Virtual deformation of deformable body. *(a)* Deformable body in equilibrium. *(b)* Components of virtual displacements.

However, from the principle of virtual displacements, δW_r, the virtual work associated with rigid-body motion, is zero, and thus

$$\delta W_e = \delta W_d \tag{2.42}$$

It is clear from Fig. 2.15*a* and from the considerations of Section 2.11 that the external virtual work is given by

$$\delta W_e = \sum_{i=1}^{n} P_i(\delta\Delta)_i \tag{2.43}$$

With respect to the virtual work of deformation, or the internal virtual work, δW_d, consider the element shown in Fig. 2.15*a*. Each σ_P stress component acts over an elemental area dA to produce an internal force component $F = \sigma_P \cdot dA$. The accumulation of the corresponding $\delta\epsilon_\Delta$ virtual strain component over the elemental length dl yields a virtual displacement $\delta = \delta\epsilon_\Delta \cdot dl$. Thus, the elemental virtual work done by this single stress component is given by

$$d(\delta W_d) = (\sigma_P \cdot dA)(\delta\epsilon_\Delta \cdot dl) \tag{2.44}$$

Noting that the volume of the element is $d\,\mathrm{Vol} = dA \cdot dl$, and summing the virtual work of all stress components over the entire body, we obtain

$$\delta W_d = \int_{\mathrm{Vol}} \sigma_P(\delta\epsilon_\Delta)d\,\mathrm{Vol} \tag{2.45}$$

Substituting Eqs. 2.43 and 2.45 into Eq. 2.42, we obtain

$$\sum_{i=1}^{n} P_i(\delta\Delta)_i = \int_{\mathrm{Vol}} \sigma_P(\delta\epsilon_\Delta)d\,\mathrm{Vol} \tag{2.46}$$

Equation 2.46 expresses the *method of virtual work,* and it states that *if a deformable body is in equilibrium under a set of* P *forces and it is subjected to a geometrically compatible virtual distortion, then the external virtual work done by the* P *forces is equal to the internal virtual work done by the* σ_P *stresses.*

In applying the method of virtual work, one must think of two separate systems. One system is a *P system* of forces that is in equilibrium, and the other system is a Δ *system* of virtual deformations that is geometrically compatible. Equation 2.46 thus states the relationship involving the work done by the P system as it undergoes the virtual motions corresponding to the Δ system.

2.15 COMPLEMENTARY VIRTUAL WORK FOR A DEFORMABLE BODY

A review of Sections 2.11 and 2.12 reveals that the expressions for work resulted from integrating the effects of forces acting through incremental displacements, whereas the formulations for complementary work evolved from integrating the effects of displacements interacting with incremental forces. That is, complementary work relates to work through a reversal in the roles played by forces and displacements.

The virtual work concepts developed in the previous section stem from allowing a structure that is in equilibrium under a set of forces to experience a virtual deformation (incremental variations in deformation quantities) that is associated with a set of geometrically compatible virtual displacements. If the roles of forces and displacements are reversed, that is, if geometrically compatible displacements interact with a set of virtual forces (incremental variations in force quantities) in equilibrium, then *complementary virtual work* results.

Following the reasoning advanced in Sections 2.13 and 2.14, we find that

$$\delta W_e^* = \delta W_d^* \tag{2.47}$$

where δW_e^* is the external complementary virtual work that combines the external virtual forces with the corresponding external displacements, and δW_d^* is the complementary virtual work of deformation, or the internal complementary virtual work, that couples the internal virtual stresses with their respective internal strains. In an expanded form, Eq. 2.47 becomes

$$\sum_{i=1}^{n} (\delta P)_i \Delta_i = \int_{\mathrm{Vol}} (\delta\sigma_P)\epsilon_\Delta d\,\mathrm{Vol} \tag{2.48}$$

where δP represents the external virtual forces, and $\delta \sigma_P$ are the associated internal virtual stresses; and Δ identifies the external displacements at the locations of the external virtual forces, and ϵ_Δ are the corresponding internal strains.

Equation 2.48 establishes the *method of complementary virtual work,* and it states that *if a deformable body is in a geometrically compatible state of deformation with a set of Δ displacements and it is subjected to a set of virtual forces that satisfy equilibrium, then the external complementary virtual work done through the Δ displacements is equal to the internal complementary virtual work done through the ϵ_Δ strains.*

The application of the method of complementary virtual work requires the use of two separate systems. One system is Δ *system* of deformations that is geometrically compatible, and the other is a *P system* of virtual forces that satisfies equilibrium. Equation 2.48 states the relationship involving the complementary work done by the virtual *P* system as it experiences the deformation corresponding to the Δ system.

2.16 APPLICATION OF VIRTUAL METHODS

The general strategies for the application of the virtual work method and the complementary virtual work method are briefly outlined in this section. These methods are applicable over a wide range of structural types, and their uses are demonstrated in subsequent sections of the book.

2.16.1 Virtual Work Method Using Virtual Displacements

This formulation is consistent with the development in Section 2.14. It is used when the analyst is attempting to determine a selected *P* force through the introduction of a judiciously selected set of $\delta \Delta$ virtual displacements. This approach requires the following systems, which are shown in Fig. 2.16:

P System: *Actual* force system in equilibrium.

Δ System: *Virtual* deformation pattern that is geometrically compatible.

Use of Eq. 2.46 gives

$$\sum_{i=1}^{n} P_i(\delta \Delta)_i = \int_{\text{Vol}} \sigma_P(\delta \epsilon_\Delta) d\,\text{Vol} \qquad (2.49)$$

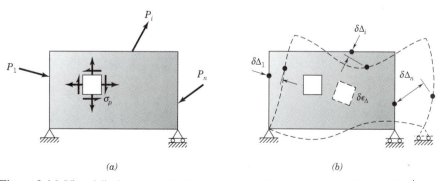

(a)   *(b)*

Figure 2.16 Virtual displacements. *(a) P* system—actual force system in equilibrium. *(b) Δ* system—virtual deformation pattern that is geometrically compatible.

By selecting a proper set of $\delta\Delta$ displacements, the desired P force can be isolated on the left-hand side of the equation and thus determined.

2.16.2 Complementary Virtual Work Method Using Virtual Forces

This formulation follows the development of Section 2.15. Here, a fictitious loading system is introduced, and the displacements of the actual system are used as displacements for this loading. This fictitious loading system is a set of virtual forces, δP, that are used to determine an actual displacement Δ. This procedure thus requires the following systems, which are illustrated in Fig. 2.17:

P System: *Virtual* force system in equilibrium.

Δ System: *Actual* deformation pattern that is geometrically compatible.

For this form, Eq. 2.48 gives

$$\sum_{i=1}^{n}(\delta P)_i\Delta_i = \int_{\text{Vol}}(\delta\sigma_P)\epsilon_\Delta d\,\text{Vol} \qquad (2.50)$$

By selecting a proper set of δP forces, the desired Δ displacement can be isolated on the left side of the equation and thus determined. Generally, the virtual force system involves a unit force at the point and in the direction of the desired displacement. For this reason, the virtual force approach is sometimes referred to as the *dummy unit load method*.

2.17 PRINCIPLE OF STATIONARY TOTAL POTENTIAL ENERGY

In Section 2.14, it was shown that for a deformable body to be in equilibrium, the virtual work quantities have to satisfy the relationship

$$\delta W_e = \delta W_d \qquad (2.51)$$

where δW_e is the external virtual work and δW_d is the virtual work of deformation, or the internal virtual work, which are associated with a geometrically compatible virtual deformation. Because the work of deforming the system is also defined as the *strain energy*, δW_d represents the virtual change in strain energy, δU, and Eq. 2.51 can be written in the form

$$\delta U - \delta W_e = 0 \qquad (2.52)$$

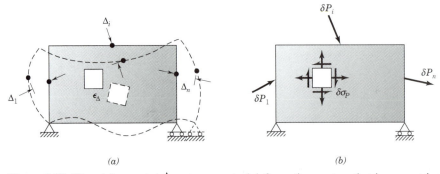

(a)   (b)

Figure 2.17 Virtual forces. *(a)* Δ system—actual deformation system that is geometrically compatible. *(b) P* system—virtual force system in equilibrium.

or

$$\delta(U - W_e) = 0 \tag{2.53}$$

In energy terms, δW_e is referred to as the change in *potential of the external forces*, and the quantity $(U - W_e)$ is called the *total potential energy* of the system and is represented by V. Therefore, Eq. 2.53 can be written as

$$\delta V = 0 \tag{2.54}$$

Equation 2.54 states that the virtual change, or variation, in the total potential energy associated with any geometrically compatible set of virtual displacements is zero. It is merely a restatement of the method of virtual work, and it is known as the *principle of stationary total potential energy*. This principle states that *for a system to be in equilibrium, the variation in the total potential energy must vanish for any geometrically compatible virtual deformation.* In other words, *the total potential energy is stationary with respect to variations in displacements.*

Note that this principle is not limited to linearly elastic structures, but is applicable to any so-called elastic structure in which the force is uniquely defined by the displacement as shown in Fig. 2.12*b*.

2.17.1 Castigliano's First Theorem

When the change in total potential energy results from a virtual change in the *i*th displacement quantity, then Eqs. 2.52, 2.54, and 2.19 combine to give

$$\delta V = \delta U - \delta W_e = \frac{\partial U}{\partial \Delta_i} \delta \Delta_i - P_i \delta \Delta_i = 0 \tag{2.55}$$

where $\delta \Delta_i$ is a virtual displacement corresponding to the force P_i, and $\partial U / \partial \Delta_i$ measures the change in the strain energy per unit change in Δ_i. Equation 2.55 can be written in the form

$$\left(\frac{\partial U}{\partial \Delta_i} - P_i \right) \delta \Delta_i = 0 \tag{2.56}$$

and, since $\delta \Delta_i \neq 0$, the bracketed term must vanish. This leads to

$$\frac{\partial U}{\partial \Delta_i} = P_i \tag{2.57}$$

This equation is valid for $i = 1, 2, \ldots, n$ and thus represents n equations of equilibrium corresponding to the n displacement degrees of freedom. The terms *force* and *displacement* are used in the generalized sense; therefore, moments and rotations are included.

Equation 2.57 is the mathematical statement of *Castigliano's first theorem,* which states that *the partial derivative of the strain energy with respect to any displacement quantity is equal to the force corresponding to that displacement.* Although only linearly elastic structures are treated in this book, Castigliano's first theorem is not limited to this class of structures.

2.18 PRINCIPLE OF STATIONARY TOTAL COMPLEMENTARY POTENTIAL ENERGY

In Section 2.15, it was demonstrated that for a body to be in a geometrically compatible state of deformation, the complementary virtual work quantities must satisfy the requirement that

$$\delta W_e^* = \delta W_d^* \tag{2.58}$$

where δW_e^* is the external complementary virtual work and δW_d^* is the internal complementary virtual work of deformation which correspond to a statically admissible set of virtual forces. The quantity δW_d^* can also be referred to as the virtual change in *complementary strain energy*, δU^*, and Eq. 2.58 can be rearranged to give

$$\delta U^* - \delta W_e^* = 0 \tag{2.59}$$

or

$$\delta(U^* - W_e^*) = 0 \tag{2.60}$$

Furthermore, δW_e^* is called the change in *complementary potential of the external forces*, and the quantity $(U^* - W_e^*)$ is termed *the total complementary potential energy*, which can be represented as V^*. Thus, Eq. 2.60 can be written in the form

$$\delta V^* = 0 \tag{2.61}$$

Equation 2.61 is an alternate expression of the method of complementary virtual work, and it is referred to as the *principle of stationary total complementary potential energy*. It states that *for a system to be in a geometrically compatible state of deformation, the variation in the total complementary potential energy must vanish for any set of virtual forces that satisfies the requirements of equilibrium*. In alternate terms, *the total complementary potential energy is stationary with respect to variations in the forces*.

Again, this principle is valid for both elastic and linearly elastic structures.

2.18.1 Complementary Strain Energy Theorem

When the change in complementary strain energy results from a virtual change in the *i*th force quantity, Eqs. 2.61, 2.59, and 2.27 collectively produce

$$\delta V^* = \delta U^* - \delta W_e^* = \frac{\partial U^*}{\partial P_i} \delta P_i - \Delta_i \delta P_i = 0 \tag{2.62}$$

where δP_i is a virtual force corresponding to the displacement Δ_i, and $\partial U^*/\partial P_i$ measures the change in complementary strain energy per unit change in P_i. Writing Eq. 2.62 in the form

$$\left(\frac{\partial U^*}{\partial P_i} - \Delta_i\right)\delta P_i = 0 \tag{2.63}$$

and recalling that $\delta P_i \neq 0$, we have

$$\frac{\partial U^*}{\partial P_i} = \Delta_i \tag{2.64}$$

This equation is valid for $i = 1, 2, \ldots, n$, and it expresses the so-called *complementary strain energy theorem*, which is credited to Engesser in 1889. It states that *the partial derivative of the complementary strain energy with respect to any force quantity is equal to the displacement corresponding to that force*. Since Eq. 2.64 follows directly from the principle of stationary total complementary energy, it too is valid for both elastic and linearly elastic structures.

2.18.2 Castigliano's Second Theorem

Actually, Castigliano's work predated Engesser's by 10 years. However, because Castigliano focused on linearly elastic structures, $U = U^*$, and Eq. 2.64 becomes

$$\frac{\partial U}{\partial P_i} = \Delta_i \qquad (2.65)$$

This equation, which is the mathematical expression of *Castigliano's second theorem*, states that *for a linearly elastic structure, the partial derivative of the strain energy with respect to any force quantity is equal to the displacement corresponding to that force.*

Equation 2.65 is valid for $i = 1, 2, \ldots, n$ and, therefore, leads to n compatibility equations. These equations can be used to determine individual displacement quantities or can be used to write compatibility equations in the analysis of statically indeterminate structures. Again, the terms *force* and *displacement* are general, and they can include moment and rotation, respectively.

2.19 MAXWELL'S AND BETTI'S LAWS

Consider again the structure shown in Fig. 2.11*a*, which is subjected to the forces P_1 and P_2 and experiences the displacements Δ_1 and Δ_2, respectively. As expressed in Eq. 2.13, the force–displacement relations in terms of flexibility coefficients are

$$\Delta_1 = D_{11}P_1 + D_{12}P_2$$
$$\Delta_2 = D_{21}P_1 + D_{22}P_2 \qquad (2.66)$$

Two different loading sequences are now considered:

1. Apply P_1 with $P_2 = 0$, and then P_2 with P_1 already on the structure.
2. Apply P_2 with $P_1 = 0$, and then P_1 with P_2 already on the structure.

The load–displacement diagrams for the two load points for both loading sequences are shown in Fig. 2.18. Recalling from Section 2.11 that work is given by the area under the load–displacement diagram, we can express the total work for the first loading sequence, W_1, as

$$W_1 = \tfrac{1}{2}D_{11}P_1^2 + D_{12}P_2P_1 + \tfrac{1}{2}D_{22}P_2^2 \qquad (2.67)$$

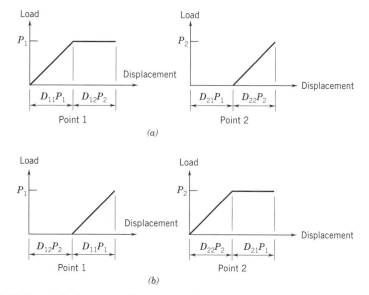

Figure 2.18 Load–displacement diagrams. (*a*) Loading sequence 1. (*b*) Loading sequence 2.

Similarly, the total work for the second loading sequence, W_2, is

$$W_2 = \tfrac{1}{2}D_{11}P_1^2 + \tfrac{1}{2}D_{22}P_2^2 + D_{21}P_1P_2 \tag{2.68}$$

Since the final displacements are the same, regardless of the loading sequence, the total work done by the forces P_1 and P_2 must be the same for either path. Thus, $W_1 = W_2$, from which

$$D_{12}P_2P_1 = D_{21}P_1P_2 \tag{2.69}$$

Letting $D_{12}P_2 = \Delta_{12}$, the displacement at point 1 due to the force at point 2, and $D_{21}P_1 = \Delta_{21}$, the displacement at point 2 due to the force at point 1, we obtain

$$P_1\Delta_{12} = P_2\Delta_{21} \tag{2.70}$$

Equation 2.70 states that the work done by the force P_1 as it experiences the displacement caused by the force P_2 is equal to the work done by the force P_2 as it experiences the displacement caused by the force P_1.

If instead of two single loads, two complete loading systems are considered, then the same considerations that led to Eq. 2.70 lead to

$$\sum_{i=1}^{n}(P_i)_1(\Delta_i)_2 = \sum_{j=1}^{m}(P_j)_2(\Delta_j)_1 \tag{2.71}$$

In this case, $(P_i)_1$ is one of the n loads of the first loading system and $(\Delta_i)_2$ is the displacement that this load experiences as a result of the second loading system. Similarly, $(P_j)_2$ is one of the m loads of the second loading system and $(\Delta_j)_1$ is the displacement that this load experiences as a result of the first loading system.

Equation 2.71 expresses *Betti's law.* That is, *for a linearly elastic structure, the work done by the n forces of loading system 1 through the displacements caused by loading system 2 is equal to the work done by the m forces of loading system 2 through the displacements caused by the loading system 1.*

Returning to the case of two single loads and canceling the load terms from Eq. 2.69, we have $D_{12} = D_{21}$. In a more general form

$$D_{ij} = D_{ji} \tag{2.72}$$

This equation expresses *Maxwell's law.* That is, *for a linearly elastic structure, the displacement at point i due to a unit load at point j is equal to the displacement at point j due to a unit load at point i.*

The development of Betti's and Maxwell's laws as given here is completely general, and they are valid for any linearly elastic structure. As such, they will be used in subsequent portions of the book when various structural systems are considered.

Part Two

Analysis of Statically Determinate Structures

Chapter 3

Reactions

Sears Tower, Chicago, Illinois (courtesy of Skidmore, Owings & Merrill, architects/ engineers, photo by Timothy Hursley).

3.1   REACTION FORCES FOR PLANAR STRUCTURES

As noted earlier, reaction forces develop at support points of a structure to equilibrate the effects of the applied forces. The focus of this chapter is on the determination of these reaction forces. Only planar structures are considered in this section; nonplanar cases are treated in Section 3.6.

It is recalled that a force is completely defined by its magnitude, direction, and point of application. Since applied forces are fully specified, these three characteristics are necessarily known for each. Reaction forces are not specified a priori, however, but rather develop as a consequence of the applied forces and in conformance with the support mechanisms.

In determining the unknown quantities associated with a reaction force, it is helpful to consider the physical constraints provided by the support point. For instance, consider the frictionless pin support at point a on the beam shown in Fig. 3.1a, first shown in its realistic form, and then in an idealized representation. This mechanism has the capacity to furnish translational restraints. Thus, the unknown force components are R_{ax} and R_{ay} as shown. Clearly, the magnitude of the resultant reaction is R_a, as given by the first of Eq. 2.2, its direction is given by the second of Eq. 2.2, and its point of application is at point a. For the frictionless roller shown in Fig. 3.1b, again shown in both realistic and idealized forms, only a vertical restraint is provided with the corresponding unknown vertical force component of R_{ay}. Here, neither horizontal nor rotational restraints are available, and the magnitude of the resultant reaction force is given directly by R_{ay}, the direction is vertical, and the point of application is at point a.

For a fixed support, such as the one shown in Fig. 3.1c, in addition to the vertical and horizontal restraints in the x and y directions, a rotational restraint about the z axis is present. Here, the unknown force components of R_{ax} and R_{ay} yield the magnitude and direction of the resultant reaction force, R_a, in accordance with Eqs. 2.2. In addition, there is an unknown moment, M_{az}, which is associated with the rotational restraint of the support. Actually, the total reaction effect can be represented by the force

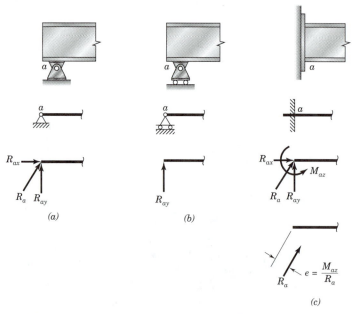

Figure 3.1   Unknown reaction force components. (a) Frictionless pin support. (b) Frictionless roller support. (c) Fixed support.

R_a, with its magnitude and direction as previously described, acting at a point whose eccentricity with respect to point a is given by $e = M_{az}/R_a$. This alternative way of representing the reaction force is also shown in Fig. 3.lc.

Table 3.1 gives a summary of the known and unknown reaction force components for a variety of support conditions. For each case considered, realistic representations are presented for both steel and concrete members along with their

Table 3.1   Reaction Forces

Type of Support	Realistic Representations	Idealized Representations	Restraints	Reaction Components
Hinge or pin	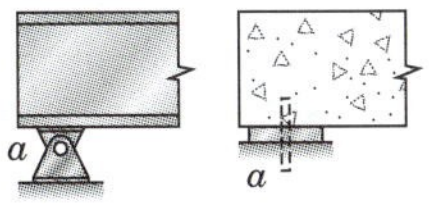	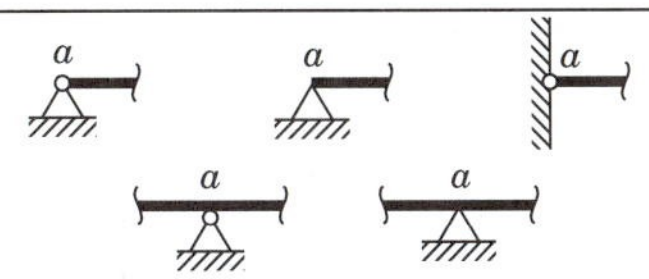	Horizontal Vertical	Unkn: R_{ax}, R_{ay} Known: $M_{az} = 0$
Roller or rocker	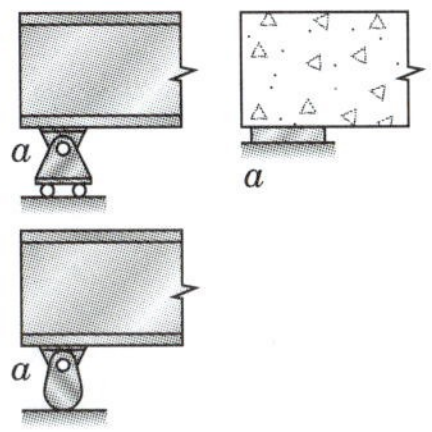	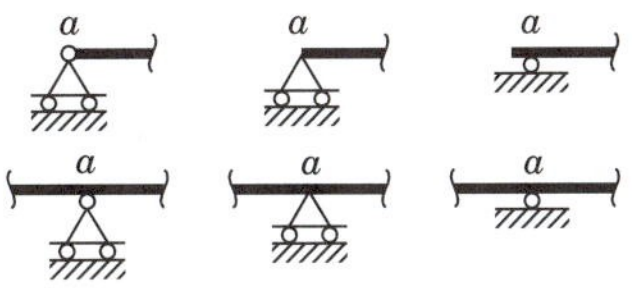	Vertical	Unkn: R_{ay} Known: $R_{ax} = M_{az} = 0$
Roller on incline	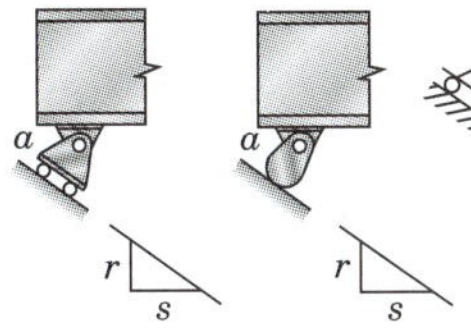	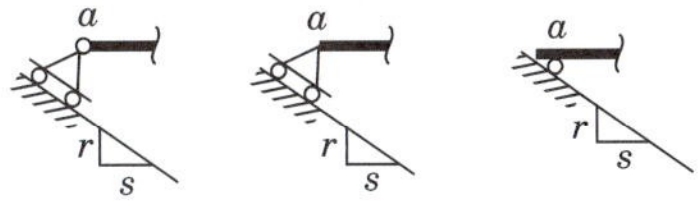	Inclined	Unkn: R_{ay} Known: $R_{ax} = R_{ay}\left(\dfrac{r}{s}\right)$ $M_{az} = 0$
Fixed or clamped	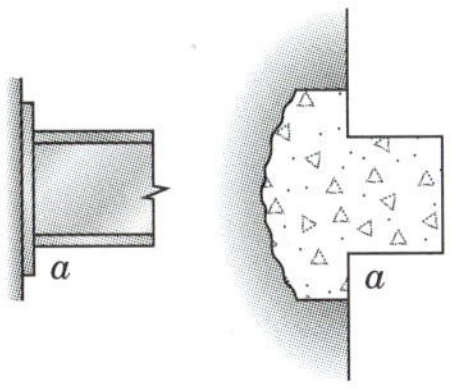		Horizontal Vertical Rotational	Unkn: R_{ax}, R_{ay}, M_{az} Known: —
Link	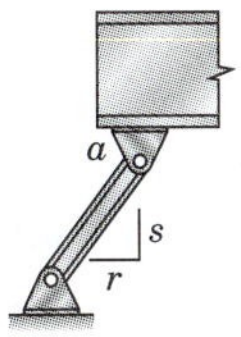	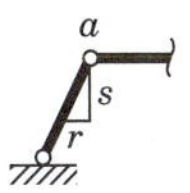	Inclined	Unkn: R_{ay} Known: $R_{ax} = R_{ay}\left(\dfrac{r}{s}\right)$ $M_{az} = 0$
Guide	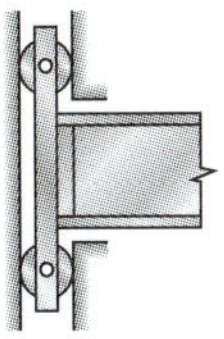	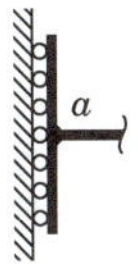	Horizontal Rotational	Unkn: R_{ax}, M_{az} Known: $R_{ay} = 0$

idealized representations, and the restraints provided for each condition are then noted. These constraints are then expressed in terms of the reaction force components.

3.2   SUPPORT CONDITIONS: REAL VERSUS IDEALIZED

The idealized representations given in Table 3.1 and the accompanying restraints are simplifications of the real conditions. In actuality, because friction-free pins and rollers are nonexistent, all supports for planar structures will have horizontal, vertical, and rotational restraints. However, special hardware is available, or a construction detail is specified, that provides support conditions that come very close to those simulated by the idealized cases. Figure 3.2 shows examples of pintle or hinge support bearings and Fig. 3.3 illustrates typical roller and rocker support bearings. For concrete beams, sometimes a simple rubber-type bearing pad is used for a roller-type support, and an embedded dowel is added for a pin support. Both of these arrangements are symbolized in Table 3.1. For all of these cases, the ideal assumptions are in order.

Even though all supports provide certain components of restraint, it should also be noted that all supports exhibit deformations in the directions corresponding to these force restraints. This fact becomes clear if we note that the supporting mechanism must undergo strains in order to develop the stresses that are associated with the reaction forces. These strains then result in support deformations. The larger the effective stiffnesses of the supports are, the smaller the deformations will be; however, theoretically such support movements will always exist. Thus, there is no such thing as complete fixity corresponding to a constraint, although in many cases the deformations are so small that the assumptions of fixity are acceptable.

3.3   EXTERNAL STATICAL DETERMINACY AND STABILITY FOR PLANAR STRUCTURES

Based on the discussion of Section 3.1, each support point for a planar structure can be evaluated regarding the number of unknown reaction components. For the structure as a whole, the actual total number of unknown reaction components, r_a, is given by the summation of the unknowns for all support points. The *statical classification* of the structure is dependent on the quantity r_a and the arrangement of the reaction components.

To determine the unknown reaction components, it is necessary to apply the equations of static equilibrium that were presented in Section 2.4. Because three equations are available for a planar structure, it is possible to solve for three independent reaction components. Thus, if $r_a = 3$, the structure is said to be *statically determinate externally*. If $r_a > 3$, there are more unknowns than there are equations for their determination and the structure is classified as *statically indeterminate externally*. Here, the equations of static equilibrium remain a vital part of the solution, but they must be augmented by equations that relate the deformation characteristics of the structure. For cases where $r_a < 3$, there are fewer unknowns than equations. Structures of this type are *statically unstable externally* for a general loading condition because there is no solution that can simultaneously satisfy the requirements of equilibrium.

Figure 3.2 Typical pintle or hinge bearings: *(a)* (courtesy M.L. Sheridan); *(b)* (courtesy Stewart C. Watson, R. J. Watson, Inc., Amherst, N.Y.); *(c)* (courtesy Stewart C. Watson, R. J. Watson, Inc., Amherst, N.Y.).

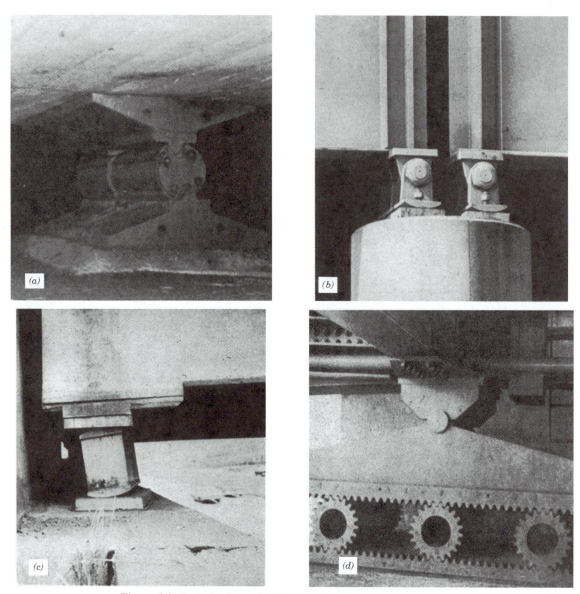

Figure 3.3 Typical roller and rocker bearings (courtesy Stewart C. Watson, R. J. Watson, Inc., Amherst, N.Y.). *(a)* Roller bearing. *(b)* Rocker bearing. *(c)* Rocker bearing. *(d)* Large-displacement roller bearing.

In summary, the following criteria hold true:

$r_a < 3$;   structure is statically unstable externally

$r_a = 3$;   structure is statically determinate externally

$r_a > 3$;   structure is statically indeterminate externally

The preceding criteria must be applied with some discretion. If $r_a < 3$, the structure is definitely unstable for general loading, but there may be unique loading conditions for which the structure is stable. For such cases, $(3 - r_a)$ equations of equilibrium are automatically satisfied, and the remaining equations are used to solve for the remaining r_a reaction components. In cases where $r_a \geq 3$, the structure is not necessarily stable. It is possible that the reaction components are not properly arranged to ensure external stability. Structures in this category are termed *geometrically unstable.*

Table 3.2 gives the statical classification of several structures. Cases *a* through *f* represent straightforward applications of the criteria. The structure given as case *g* is unstable with respect to general loading because $r_a < 3$. However, for the special case of vertical loading, the equation for equilibrium of the horizontal forces is automatically satisfied. The two remaining equations of equilibrium can be solved for the two unknown reaction components. The structure under case *h* would appear to be determinate according to the criteria; however, the three reaction components are not arranged in a way that would produce stability under all loading conditions. Only if the applied forces were vertical would the structure be stable. For this special case, the equilibrium equation dealing with horizontal forces is automatically satisfied, and only two equations are available to solve for the three unknown reaction quantities. Thus, the structure is indeterminate even if its stability is ensured by the loading. Case *i* also appears initially to be statically determinate; however, further inspection shows that all reactions pass through point *p*. Any external load that does not pass through this point cannot be equilibrated by the reactions and thus the structure is unstable under general loading. If the resultant of the applied forces passes through point *p*, the structure is stable. For this special case, the moment equilibrium equation is inherently satisfied; however, the two remaining equations are not sufficient to determine the three unknown reaction components. Thus, the structure is either geometrically unstable or indeterminate.

Cases *h* and *i* fall within a broad category of structures whose reactions form a system of parallel or concurrent forces. Such cases were discussed in Section 2.4. Situations of this type are potentially unstable even if $r_a \geq 3$ and are statically indeterminate for the special loading cases that lead to stability.

3.4   COMPUTATION OF REACTIONS USING EQUATIONS OF EQUILIBRIUM

The equations of static equilibrium play a vital role in determining the reactions for any structure. If the structure is statically determinate externally, these equations provide all that is needed for the solution, whereas for statically indeterminate cases, the equations of equilibrium still apply, but they are not sufficient for a solution. This section considers only planar structures that are statically determinate.

In determining the reactions, a free-body diagram of the entire structure is used. All of the known forces are shown to act in their prescribed directions, and each unknown reaction component is assumed to act in a specified direction. The equations of equilibrium are written in consistency with the assumed directions, and these equations are solved for the unknown reactions. If the solution yields a positive reaction component, then the assumed direction is correct, whereas a negative result indicates that the wrong direction was assumed. In the latter case, care must be exercised in subsequent calculations. If the negative sign is retained, then the originally assumed direction must also be retained. However, the direction can be reversed on the free-body diagram and a positive sign can then be used in subsequent work.

Table 3.2 External Statical Classification of Structures

Idealized Representation of Structure	Independent Reaction Components, r_a	Classification
(a)	3	$r_a = 3$; determinate, stable
(b)	3	$r_a = 3$; determinate, stable
(c)	4	$r_a > 3$; indeterminate, stable
(d)	2	$r_a < 3$; unstable
(e)	5	$r_a > 3$; indeterminate, stable
(f)	4	$r_a > 3$; indeterminate, stable
(g)	2	$r_a < 3$; unstable for general loading, stable for vertical loads only
(h)	3	$r_a = 3$; geometrically unstable
(i)	3	$r_a = 3$; geometrically unstable

In general, the three equations of static equilibrium can be solved simultaneously for the three unknown reaction components. However, in most instances, the equations are, or can be, uncoupled such that an individual equation will have but a single unknown. This uncoupling can frequently be accomplished through a judicious selection of the z axis for the moment equation of equilibrium. With one unknown determined, a second equation may have but a single unknown, and then a third.

In the translation from the realistic structure to the idealized structure, the applied and reaction force components are assumed to be applied at the centroidal axes of the members. The actual eccentricities may have to be accounted for in the detailed member designs, but they are not considered here.

EXAMPLE 3.1

Compute the reactions for the beam structure shown.

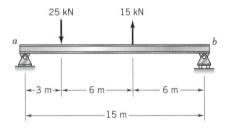

Free-Body Diagram

The known forces are specified, and the unknown reaction components are shown in their assumed directions.

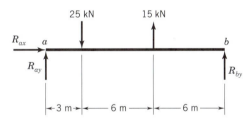

Determination of Reactions

Since $r_a = 3$, and the arrangement is stable, the structure is statically determinate. By locating the z axis at point a, the moment equation will involve R_{by} as the only unknown.

$$\sum P_x = 0 \;\xrightarrow{+}\; R_{ax} = \underline{\underline{0}}$$

$$\sum M_{za} = 0 \;\circlearrowleft+\; (25 \times 3) - (15 \times 9) - (R_{by} \times 15) = 0; \qquad R_{by} = \underline{\underline{-4 \text{ kN}}}$$

The direction of R_{by} on the free-body diagram will be retained, but the negative sign for R_{by} will be retained in all subsequent calculations.

$$\sum P_y = 0 \;\uparrow+\; R_{ay} - 25 + 15 + R_{by} = 0; \qquad R_{ay} = 25 - 15 - (-4) = \underline{\underline{14 \text{ kN}}}$$

Because R_{ay} is positive, it acts in the direction assumed on the free-body diagram; however, the negative sign for R_{by} indicates that it acts downward rather than upward as assumed.

EXAMPLE 3.2   Compute the reactions for the beam structure shown.

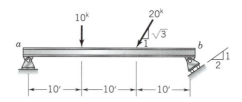

Free-Body Diagram

It is convenient to show all forces, known and unknown, in terms of their components. The reaction at point a has two unknown components, whereas the reaction at point b has only a single unknown component, which for convenience is taken as R_{by}. As noted earlier, all force components are assumed to be applied at the centroidal axis of the member.

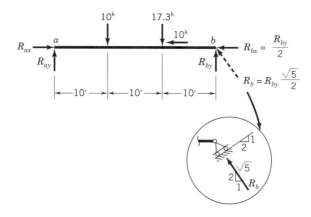

Determination of Reactions

Since $r_a = 3$, and the arrangement is stable, the structure is statically determinate. The reactions are determined by systematically applying the equations of equilibrium.

$$\sum P_x = 0 \quad \circlearrowright +$$

$$(10 \times 10) + (17.3 \times 20) - (R_{by} \times 30) = 0; \quad R_{by} = \underline{\underline{14.9^k}}$$

$$\therefore R_{bx} = \frac{R_{by}}{2} = \frac{14.9}{2} = \underline{\underline{7.5^k}}; \quad R_b = R_{by}\frac{\sqrt{5}}{2} = 14.9\frac{\sqrt{5}}{2} = \underline{\underline{16.7^k}}$$

$$\sum P_y = 0 \quad \uparrow +$$

$$R_{ay} - 10 - 17.3 + R_{by} = 0$$

$$R_{ay} = 27.3 - R_{by} = 27.3 - 14.9 = \underline{\underline{12.4^k}}$$

$$\sum P_x = 0 \quad \overset{+}{\rightarrow}$$

$$R_{ax} - 10 - R_{bx} = 0$$

$$R_{ax} = 10 + R_{bx} = 10 + 7.5 = \underline{\underline{17.5^k}}$$

EXAMPLE 3.3

Determine the reactions for the beam structure shown. The overhanging portion b–c is referred to as a cantilever beam.

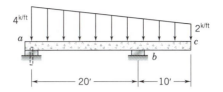

Free-Body Diagram

The trapezoidal load is broken up into rectangular and triangular loads from which equivalent concentrated loads are then determined and used.

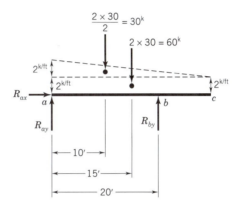

Determination of Reactions

$$\sum P_x = 0 \text{ yields } R_{ax} = \underline{\underline{0}}$$

$$\sum M_{za} = 0 \;\; \circlearrowright + \;\; (30 \times 10) + (60 \times 15) - (20 \times R_{by}) = 0 ; \qquad R_{by} = \underline{\underline{60}}^{k}$$

$$\sum P_y = 0 \;\; \uparrow + \;\; R_{ay} - 30 - 60 + R_{by} = 0; \qquad R_{ay} = \underline{\underline{30}}^{k}$$

Note: The trapezoidal load could have been replaced by two triangular loads in the following manner:

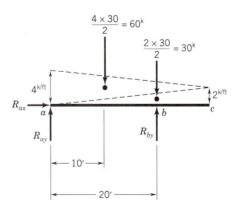

In this case,

$$\sum M_{za} = 0 \;\; \circlearrowright + \;\; (60 \times 10) + (30 \times 20) - (R_{by} \times 20) = 0 ; \qquad R_{by} = \underline{\underline{60}}^{k}$$

EXAMPLE 3.4   Determine the reactions for the structure shown in the following sketch.

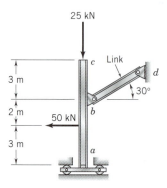

Free-Body Diagram

The known forces and the three unknown reaction components are shown on a free-body diagram. The link must carry a pure axial force as shown.

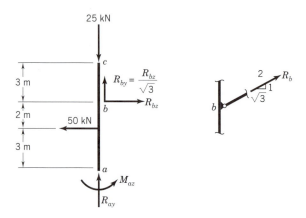

Determination of Reactions

The structure is stable, and $r_a = 3$; therefore, the structure is statically determinate. The equations of equilibrium are applied to determine the reactions.

$$\sum P_x = 0 \ \overset{+}{\rightarrow}$$

$$R_{bx} - 50 = 0 ; \qquad R_{bx} = \underline{\underline{50 \text{ kN}}}$$

$$\therefore \ R_{by} = \frac{R_{bx}}{\sqrt{3}} = 28.9 ; \qquad R_b = 2R_{by} = \underline{\underline{57.8 \text{ kN}}}$$

$$\sum P_y = 0 \ \uparrow +$$

$$R_{ay} + R_{by} - 25 = 0$$

$$R_{ay} = 25 - R_{by} = 25 - 28.9 = \underline{-3.9 \text{ kN}}$$

The negative sign indicates that the vertical component of reaction at point a acts downward, not upward, as assumed on the free-body diagram.

$$\sum M_{za} = 0 \ \curvearrowright +$$

$$(R_{bx} \times 5) - (50 \times 3) - M_{az} = 0$$

$$M_{az} = -150 + (50 \times 5) = \underline{\underline{100 \text{ kN} \cdot \text{m}}}$$

EXAMPLE 3.5

Calculate the reactions for the frame structure shown.

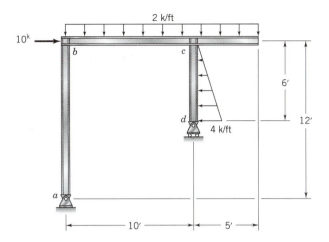

Free-Body Diagram

The distributed forces are replaced by resultant concentrated forces at their centroidal positions. The unknown reaction components are also shown.

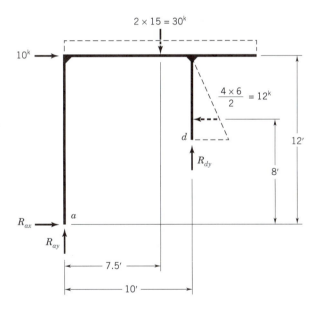

Determination of Reactions

The arrangement of the reaction components provides stability, and $r_a = 3$; thus, the structure is statically determinate. Application of the equations of equilibrium leads to the determination of the reactions.

$$\sum M_{za} = 0 \quad \circlearrowright +$$

$$(10 \times 12) + (30 \times 7.5) - (12 \times 8) - (R_{dy} \times 10) = 0$$

$$R_{dy} = \underline{\underline{24.9}}^k$$

$$\sum P_x = 0 \;\xrightarrow{+}$$

$$R_{ax} + 10 - 12 = 0; \qquad R_{ax} = \underline{\underline{2.0}}^k$$

$$\sum P_y = 0 \;\uparrow+$$

$$R_{ay} - 30 + R_{dy} = 0$$

$$R_{ay} = 30 - R_{dy} = 30 - 24.9 = \underline{\underline{5.1}}^k$$

3.5 CONDITION EQUATIONS FOR PLANAR STRUCTURES

For the structures that have been considered thus far, it has been possible to determine a maximum of three unknown reaction components for a statically determinate and stable structure. These reaction components have been determined by systematically applying the three equations of static equilibrium.

As discussed in Section 2.5, it is possible to have more than three statically determinate reaction components. This occurs when a structure is composed of two or more substructural units connected in a manner that allows one or more statical equations to be written involving some of the reaction components in addition to the overall equations of static equilibrium. These equations are referred to as *condition equations*, and each condition increases by one the number of statically determinate reaction components.

It was shown in Section 3.3 that for a planar structure without condition equations, a minimum of three reaction components are required for external stability. This number is increased by one for each condition equation that is available. Thus, if r is taken as the least number of reaction components required for external stability of a structure with n condition equations, then $r = 3 + n$. If r_a is the actual number of reaction components, then the same line of reasoning that was applied to structures without condition equations leads to the following criteria:

$$r_a < r; \quad \text{structure is statically unstable externally}$$

$$r_a = r; \quad \text{structure is statically determinate externally}$$

$$r_a > r; \quad \text{structure is statically indeterminate externally}$$

As before, when $r_a < r$, it is possible for the structure to be conditionally stable for unique loading conditions. Also, once again the criteria for $r_a \geq r$ are necessary but not sufficient conditions for statical classification. The reaction components must be arranged in a manner that ensures stability.

Table 3.3 shows the statical classification of several structures with condition equations. For example, consider structure a. In accordance with the notation introduced, $r_a = 4$, $n = 1$, and $r = 3 + n = 3 + 1 = 4$. Since $r_a = r$, and the reactions provide a stable arrangement, the structure is statically determinate. The procedure for determining the reactions is an orderly one. Using the equations of equilibrium for the entire structure, one can write three equations in terms of the four unknown reaction components. A fourth equation involving some of the unknowns can be written by taking a free-body diagram to either the left or right of the internal hinge and by applying the condition that the moment about a z axis is zero at the hinge. Thus, there are four independent equations to solve for the four unknowns. It is important to note that only one independent condition equation results from the internal hinge. The equation for the section to the left of the hinge is not independent of the equation for the section to the right of the hinge and vice versa. Each one can be formed by combining the other one with the equations for overall equilibrium of the structure.

Table 3.3   External Statical Classification of Structures with Condition Equations

Idealized Representation of Structure	Indep. React. Comps. r_a	Number of Conditions n	Req'd. React. Comps. $r = 3 + n$	Classification
(a)	4	1	$3 + 1 = 4$	$r_a = r$; determinate, stable
(b)	5	2	$3 + 2 = 5$	$r_a = r$; determinate, stable
(c)	4	1	$3 + 1 = 4$	$r_a = r$; determinate, stable
(d)	5	1	$3 + 1 = 4$	$r_a > r$; indeterminate, stable
(e)	5	2	$3 + 2 = 5$	$r_a = r$; determinate, stable
(f)	4	2	$3 + 2 = 5$	$r_a < r$; unstable
(g)	3	1	$3 + 1 = 4$	$r_a < r$; unstable
(h)	4	2	$3 + 2 = 5$	$r_a < r$; unstable for general loading, stable for vertical loads only

EXAMPLE 3.6

Compute the reactions for the cantilevered structure shown below. Here, section BC cantilevers out from span AB, section DE cantilevers out from span EF, and span CD is supported by the two cantilevers.

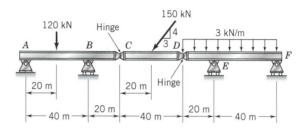

Classification

Two condition equations can be written: $M_{Cz} = M_{Dz} = 0$. Thus, $r = 3 + n = 3 + 2 = 5$. Since $r_a = 5$ also, and since these reactions are arranged so that the structure is stable, the structure is statically determinate.

It is possible to write the three equations of equilibrium for the entire structure and the two condition equations and then solve the resulting five equations for the five unknown reaction components. However, it is more convenient to disassemble the structure into several free-body diagrams and to apply statics to each part. Here, the condition equations are used as each free body is analyzed.

Free-Body Diagrams

The structure is broken into three separate free bodies, and the unknown reactions and internal forces at the hinges are labeled.

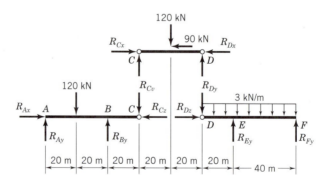

Determination of the Reactions

At first glance, there appear to be more unknowns than available equations for each of the three individual free-body diagrams. However, examination of DEF reveals that $R_{Dx} = 0$. Thus, there are only three unknowns for section CD and, therefore, the analysis will begin there.

Free-Body CD

Here we treat CD as a simply supported beam and thus inherently use the conditions that $M_{Cz} = M_{Dz} = 0$. As already noted, $R_{Dx} = 0$ from examining the free-body diagram for DEF. The reactions (internal forces transferred at hinges) are determined to be

$$R_{Cx} = +90 \text{ kN}; \qquad R_{Cy} = +60 \text{ kN}; \qquad R_{Dx} = 0; \qquad R_{Dy} = +60 \text{ kN}$$

Free-Body ABC

This segment is analyzed for the applied forces with R_{Cx} and R_{Cy} acting at point C. This leads to

$$R_{Ax} = +90 \text{ kN}; \qquad R_{Ay} = +30 \text{ kN}; \qquad R_{By} = +150 \text{ kN}$$

Free-Body DEF

This segment is analyzed for the applied forces and R_{Dy} acting at point D. This leads to

$$R_{Ey} = +225 \text{ kN}; \qquad R_{Fy} = +15 \text{ kN}$$

EXAMPLE 3.7

Compute the reactions for the three-hinged arch shown.

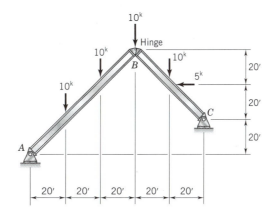

Classification

$r_a = 4$; $r = 3 + n = 3 + 1 = 4$; since $r_a = r$, structure is statically determinate, and it is stable.

Free-Body Diagrams

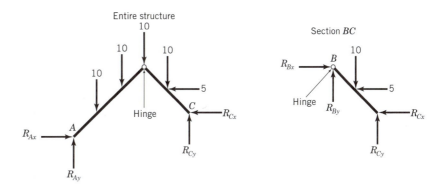

Note: The required dimensions are given in the sketch in the problem statement.

Determination of Reactions

For the entire structure:

$$\sum M_{zA} = 0 \quad \circlearrowright +$$

$$10(20 + 40 + 60 + 80) - 5(40) - R_{Cy}(100) - R_{Cx}(20) = 0$$

$$R_{Cx} + 5R_{Cy} = 90$$

$$\sum P_y = 0 \quad \uparrow + \quad R_{Ay} + R_{Cy} - 4(10) = 0$$

$$R_{Ay} + R_{Cy} = 40$$

$$\sum P_x = 0 \quad \xrightarrow{+} R_{Ax} - R_{Cx} - 5 = 0$$

$$R_{Ax} - R_{Cx} = 5$$

For section BC:

$$\underset{\text{(condition equation)}}{\sum M_{zB} = 0 \quad \circlearrowright +} \quad 10(20) + 5(20) + R_{Cx}(40) - R_{Cy}(40) = 0$$

$$4R_{Cy} - 4R_{Cx} = 30$$

In summary, the equations of equilibrium are

$$R_{Cx} + 5R_{Cy} = 90$$

$$R_{Ay} \qquad + R_{Cy} = 40$$

$$R_{Ax} \qquad - R_{Cx} \qquad = 5$$

$$-4R_{Cx} + 4R_{Cy} = 30$$

The solution is

$$R_{Ax} = 13.75^k; \qquad R_{Ay} = 23.75^k; \qquad R_{Cx} = 8.75^k; \qquad R_{Cy} = 16.25^k$$

3.6   REACTION FORCES FOR NONPLANAR STRUCTURES

For planar structures, the restraints provided by the support in the xy plane were described in detail in Sections 3.1 and 3.2. For nonplanar structures, the same considerations must be taken into account in the three mutually perpendicular xy, xz. and yz planes.

For instance, consider the restraint conditions that are provided by the support point a in Fig. 3.4. As was the case with a planar structure, the three reaction force components of R_{ax}, R_{ay}, and M_{az} define the restraints within the xy plane, where M_{az} is the moment about the z axis. Similarly, the reaction components R_{ax}, R_{az}, M_{ay} and R_{ay}, R_{az}, M_{ax} define the support restraints in the xz and yz planes, respectively. Thus, a total of three force components and three moment components completely define the resultant reaction.

In most cases, the support mechanism is such that fewer than six reaction components are developed. The considerations here are similar to those discussed in Section 3.1 for planar structures and summarized in Table 3.1. In this case, however, the restraints in three separate support planes must be considered. Table 3.4 gives some representative support situations in which the restraints are enumerated. In these cases, two members are assumed to frame into a support at point a as shown in the key sketch. One of these members is in the xz plane, and the other member is inclined to this plane, but in the same vertical plane. In each case, the support restraints are noted and the corresponding unknown reaction components are identified. Of course, many other possibilities beyond the representative cases of Table 3.4 are possible when all the possible combinations in the three planes of restraint are considered.

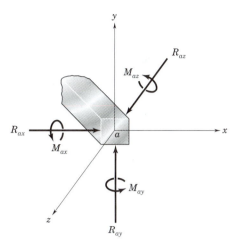

Figure 3.4 Components of reaction force at point a.

3.7 EXTERNAL STATICAL DETERMINACY AND STABILITY FOR NONPLANAR STRUCTURES

Based on the considerations of the previous section, each support point can be evaluated with regard to the number of unknown reaction components. For the entire structure, the actual total number of unknown reaction components r_a is given by the summation of the unknowns at all support points. The statical classification of a structure is dependent on the value of r_a and the arrangement of the corresponding reaction components.

The unknown reaction components are determined by the systematic application of the equations of static equilibrium, which are given as Eqs. 2.10. Since six equations are available for the solution, it is possible to solve for six independent reaction components. Thus, if $r_a = 6$, the structure is classified as *statically determinate externally.* However, if $r_a > 6$, there are more unknowns than there are equations available for their determination, and thus the structure is classified as *statically indeterminate externally.* Where $r_a < 6$, there are fewer unknowns than there are equations. Structures in this latter category are *statically unstable externally* since there is no solution that will satisfy the requirements of equilibrium for general loading conditions.

As was true with planar structures, when the structure is statically indeterminate, the equations of static equilibrium remain a necessary part of the solution. They do not form the basis for a solution in themselves, but when augmented with equations of compatibility, a solution is possible.

Summarizing the criteria, we have the following:

$$r_a < 6; \quad \text{structure is statically unstable externally}$$

$$r_a = 6; \quad \text{structure is statically determinate externally}$$

$$r_a > 6; \quad \text{structure is statically indeterminate externally}$$

As was true for planar structures, the criteria summarized above must be applied with care. If $r_a < 6$, the structure is definitely unstable for general loading, but there may be unique loading conditions for which the structure is stable. For such cases, $(6 - r_a)$ equations are automatically satisfied by virtue of the uniqueness of the loading, and the remaining equations are used to solve for the r_a reaction components. However, $r_a \geq 6$ are necessary but not sufficient criteria for structural classification. The reaction components must be arranged to resist rigid-body translation along, and rotation about, each

Table 3.4   Types of Supports

Type of Support	Realistic Representation	Idealized Representation		Restraints		Reaction Components	
		Plan[a]	Elevation[a]	Translational	Rotational	Known	Unknown
Fixed	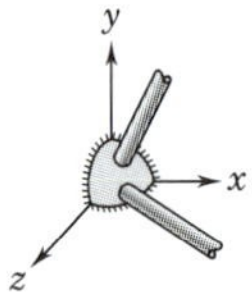	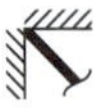	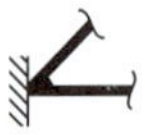	x, y, z	x, y, z	Nothing	$R_{ax}, R_{ay}, R_{az}, M_{ax}, M_{ay}, M_{az}$
Sleeve (pinned about z axis)	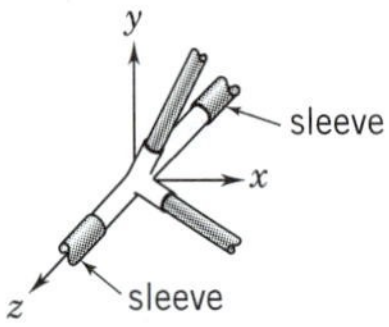	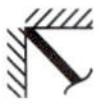	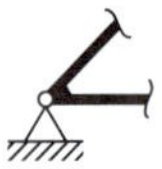	x, y, z	x, y	$M_{az} = 0$	$R_{ax}, R_{ay}, R_{az}, M_{ax}, M_{ay}$
Universal joint	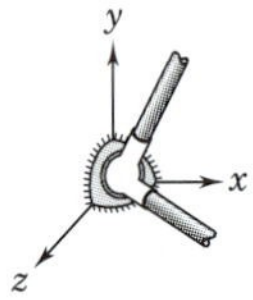	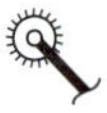	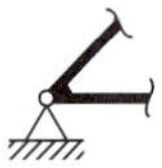	x, y, z	—	$M_{ax} = M_{ay} = M_{az} = 0$	R_{ax}, R_{ay}, R_{az}
Roller (free to roll in x direction)	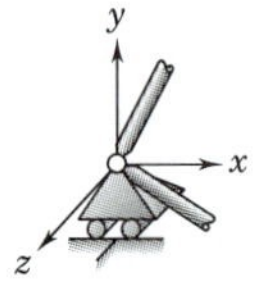	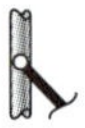	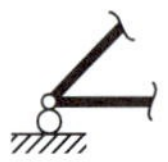	y, z	—	$R_{ax} = 0$ $M_{ax} = M_{ay} = M_{az} = 0$	R_{ay}, R_{az}
Ball	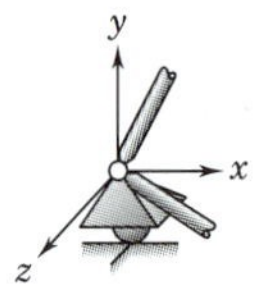		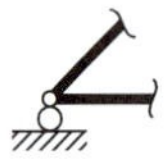	y	—	$R_{ax} = R_{az} = 0$ $M_{ax} = M_{ay} = M_{az} = 0$	R_{ay}

[a]Key Sketch
Plan: xz plane
Elevation: xy plane

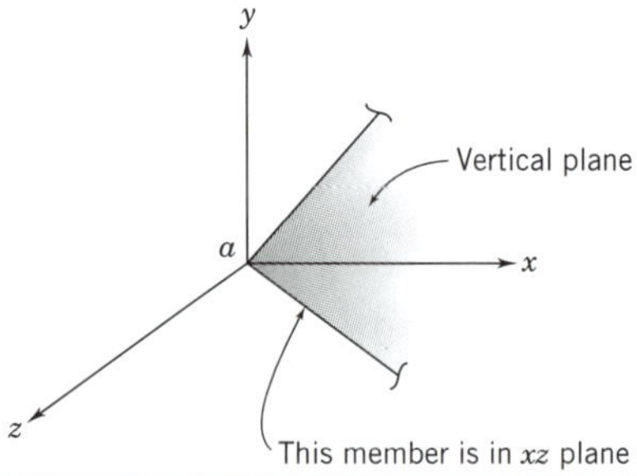

of the coordinate axes. That is, the reactions must be arranged to ensure external stability. If this is not satisfied, the structure is classified as *geometrically unstable*. As was the case for planar structures, geometrically unstable structures may be stable for unique loading conditions. Such situations are, however, statically indeterminate.

As was true for planar structures, it is possible to have condition equations that will increase the number of reactions that can be determined from statics. If there are n condition equations, then $r_a = 6 + n$ becomes the condition for statical determinacy. Again, the structure may be unstable if the reaction components are not properly arranged.

3.8   COMPUTATION OF REACTIONS FOR NONPLANAR STRUCTURES

In determining the reactions for nonplanar structures, the same general techniques are used that were used in Section 3.4 for planar structures. The initial step is to construct appropriate free-body diagrams and to show each unknown reaction component to act in a specified direction. The equations of equilibrium are then written in consistency with the arrangements shown on the free-body diagrams, and these equations are solved for the unknown reactions. When the solution produces a positive reaction component, then the assumed direction is correct; however, a negative reaction component signifies that the wrong direction was assumed on the free-body diagram.

This section will treat only those structures that are statically determinate externally. Thus, the six equations of static equilibrium given as Eqs. 2.10 provide all that is needed for the determination of the six independent reaction components, which must be arranged to ensure external stability. Additional reaction components are necessary for stability when conditions of construction allow for the expression of condition equations that augment the overall equations of equilibrium, and these additional components are determinate. For statically indeterminate structures, the equations of equilibrium continue to play a necessary role in the solution, but they are not sufficient for the determination of the reactions.

EXAMPLE 3.8

Compute the reactions for the fixed-ended cantilever structure shown below.

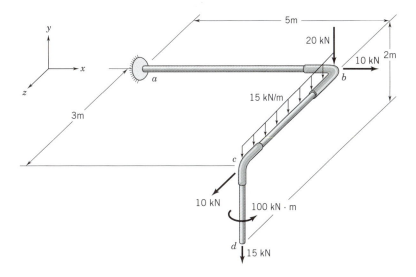

Free-Body Diagram

The six unknown reaction components at point a are identified as unknowns. The uniformly distributed load is replaced with an equivalent concentrated load.

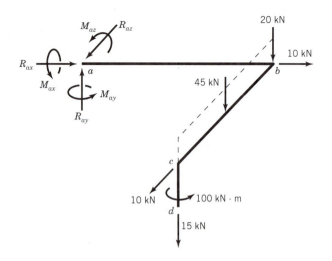

Determination of Reactions

Since $r_a = 6$ and the arrangement is stable, the structure is statically determinate. The reaction components are determined from the equations of equilibrium.

$$\sum P_x = 0 \;\xrightarrow{+}\; R_{ax} + 10 = 0\,; \qquad R_{ax} = -10 \text{ kN}$$

$$\sum P_y = 0 \;\uparrow+\; R_{ay} - 20 - 45 - 15 = 0\,; \qquad R_{ay} = +80 \text{ kN}$$

$$\sum P_z = 0 \;\nearrow\!\!+\; R_{az} + 10 = 0\,; \qquad R_{az} = -10 \text{ kN}$$

$$\sum M_{xa} = 0 \;+\curvearrowleft\; M_{ax} + (45 \times 1.5) + (15 \times 3) = 0\,; \qquad M_{ax} = -112.5 \text{ kN} \cdot \text{m}$$

$$\sum M_{ya} = 0 \;\circlearrowleft\; M_{ay} - (10 \times 5) + 100 = 0\,; \qquad M_{ay} = -50 \text{ kN} \cdot \text{m}$$

$$\sum M_{za} = 0 \;+\nearrow\!\!\!\!\times\; M_{az} - (20 \times 5) - (45 \times 5) - (15 \times 5) = 0\,; \qquad M_{az} = +400 \text{ kN} \cdot \text{m}$$

EXAMPLE 3.9

Determine the reaction components for the structure shown. The admissible reaction components are shown symbolically.

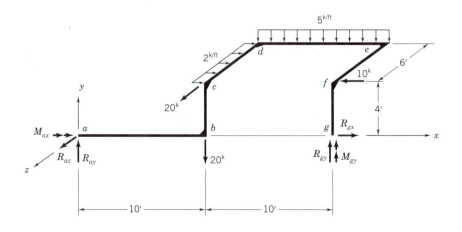

Free-Body Diagram

The uniformly distributed loads are replaced with equivalent concentrated loads.

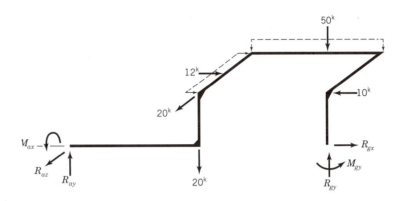

Determination of Reactions

Since $r_a = 6$ and the arrangement is stable, the structure is statically determinate. The reaction components are determined from the equations of equilibrium.

$$\sum P_x = 0 \ \xrightarrow{+} \ R_{gx} + 12 - 10 = 0;$$

$$R_{gx} = \underline{\underline{-2}}^k$$

$$\sum P_z = 0 \ \nearrow \ R_{az} + 20 = 0;$$

$$R_{az} = \underline{\underline{-20}}^k$$

$$\sum M_{xa} = 0 \ ^a\!\!\curvearrowleft^+ \ M_{ax} + (20 \times 4) - (50 \times 6) = 0;$$

$$M_{ax} = \underline{\underline{220}}^{ft\text{-}k}$$

$$\sum M_{za} = 0 \ _+\!\!\nwarrow^{a} \ -(20 \times 10) - (12 \times 4) - (50 \times 15) + (10 \times 4)$$
$$+ (R_{gy} \times 20) = 0;$$

$$R_{gy} = \underline{\underline{47.9}}^k$$

$$\sum P_y = 0 \ \uparrow + \ R_{ay} - 20 - 50 + R_{gy} = 0;$$

$$R_{ay} = 70 - R_{gy} = \underline{\underline{22.1}}^k$$

$$\sum M_{yg} = 0 \ \curvearrowright +_g \ M_{gy} - (12 \times 3) + (20 \times 10) + (R_{az} \times 20) = 0$$

$$M_{gy} = -164 - 20R_{az} = \underline{\underline{236}}^{ft\text{-}k}$$

EXAMPLE 3.10   Determine the reactions for the structure given below. Each support is a roller-type mechanism that is capable of developing the reaction components shown.

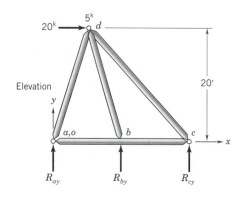

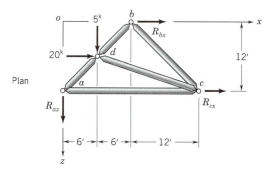

Free-Body Diagrams

The entire structure is taken as a free-body diagram and the given sketches (with the assumed directions of the reaction components) will be used to write the equations of equilibrium.

Determination of Reactions

Because $r_a = 6$, and the arrangement is stable, the structure is statically determinate. One could write the six equations of equilibrium and solve them simultaneously for the six unknowns. Indeed, this is necessary in some cases; however, in the present case, an effort is made to solve for the unknowns in a sequence that will uncouple the equations. This can be accomplished by a judicious selection of axes for the moment equations and by the order in which the equations are considered.

$$\sum M_{xa} = 0 \qquad a \xrightarrow{+} c \longrightarrow x$$

$$(R_{by} \times 12) + (5 \times 20) = 0; \qquad R_{by} = \underline{-8.33}^k$$

$$\sum M_{za} = 0 \qquad (R_{cy} \times 24) + (R_{by} \times 12) - (20 \times 20) = 0$$

$$R_{cy} = \frac{400 - (-8.33 \times 12)}{24} = \underline{20.83}^k$$

$$\sum P_y = 0 \ \uparrow + \quad R_{ay} + R_{by} + R_{cy} = 0$$
$$R_{ay} = 8.33 - 20.83 = -\underline{\underline{12.50}}^{\mathrm{k}}$$

$$\sum M_{yo} = 0 \qquad (R_{cx} \times 12) + (20 \times 6) - (5 \times 6) = 0$$

$$R_{cx} = \frac{-120 + 30}{12} = -\underline{\underline{7.5}}^{\mathrm{k}}$$

$$\sum P_x = 0 \ \xrightarrow{+} \ R_{bx} + R_{cx} + 20 = 0$$
$$R_{bx} = -20 + 7.5 = -\underline{\underline{12.5}}^{\mathrm{k}}$$

$$\sum P_z = 0 \ +\!\!\nearrow \ R_{az} + 5 = 0; \qquad R_{az} = -\underline{\underline{5}}^{\mathrm{k}}$$

Note:

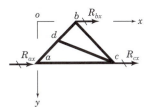

If the roller at point *a* had been oriented to resist R_{ax}, as shown here, then $r_a = 6$ would still have been satisfied. However, the structure would be unstable because the 5^{k} load in the *z* direction could not be resisted. The structure would be conditionally stable if no loads were applied in the *z* direction, but it would then be classified as geometrically unstable, and the three horizontal reaction components would be statically indeterminate.

3.9   VARIATIONS IN BOUNDARY CONDITIONS

A variety of boundary conditions can be employed to provide external stability for a given structure. Consider again the structure of Example 3.9 but with full fixity at point *a* and complete freedom at point *g* as shown in Fig. 3.5*a*. Here, the six reaction components at point *a* are determined by applying the six equilibrium equations in the fashion employed in Example 3.8.

If the restraint associated with M_{az} were to be removed, the structure would be unstable with respect to rotation about the *z* axis, but this could be restored by introducing either an M_{gz} restraint or an R_{gy} restraint at point *g*. The latter selection is shown in Fig. 3.5*b*. Similarly, if M_{ay} is removed, R_{gz} or M_{gy} is needed to stabilize the system; M_{gy} is shown in Fig. 3.5*b*. If M_{ax} were removed, M_{gx} would have to be introduced; in this case, rotational stability about the *x* axis cannot be ensured by introducing a force reaction at point *g*. If R_{ax} is removed, however, translational stability can be ensured by introducing R_{gx} as shown in Fig. 3.5*b*. The set of reactions shown in Fig. 3.5*b* corresponds to those used in Example 3.9. Other boundary restraint tradeoffs are possible; Fig. 3.5*c* provides yet another example. In each case, the reaction components must be arranged so as to preclude rigid-body instabilities.

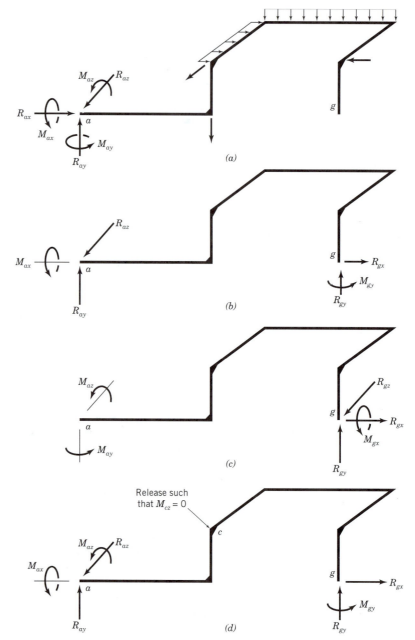

Figure 3.5 Alterations in boundary conditions.

Now, consider again the case shown in Fig. 3.5b; however, in this instance incorporate a release at point c that requires the moment about the z axis to be zero. Because this restricts the capacity of the structure to carry moment about the z axis, the reaction component M_{az} must be added to ensure rotational stability. The solution now requires the six equilibrium equations and the condition equation that arises from taking a free-body diagram of the segment ca and setting M_{cz} equal to zero.

3.10   SUGGESTED PROBLEMS

3.1 Classify each of the following structures, with respect to its external reactions, as statically determinate or indeterminate and stable or unstable.

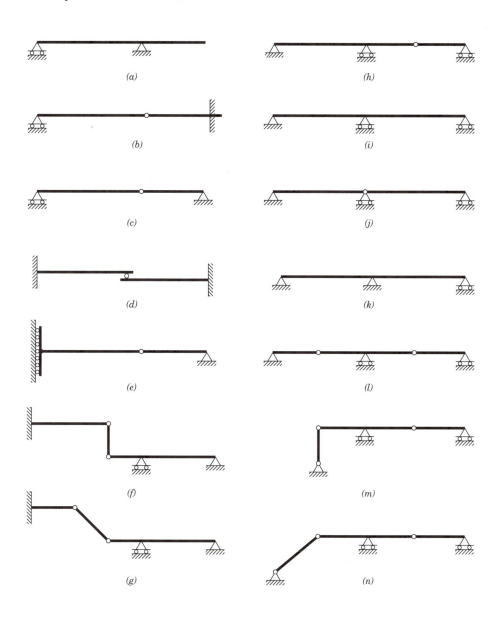

3.1 *continued*

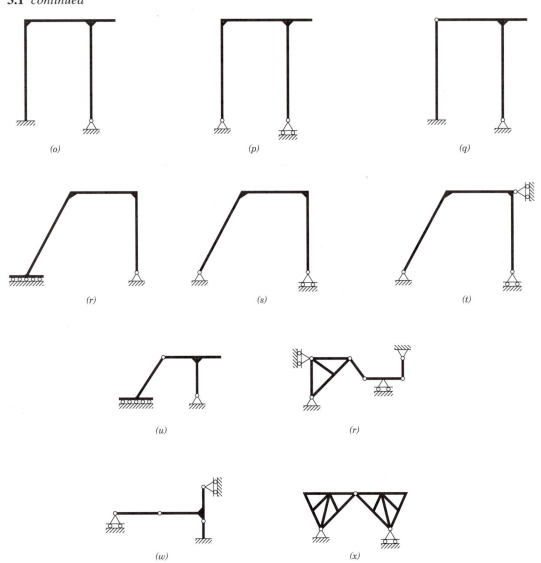

(o)

(p)

(q)

(r)

(s)

(t)

(u)

(r)

(w)

(x)

3.2 through 3.31   Verify that each structure is stable and determinate. Then, using the equations of equilibrium and any available condition equations, determine the support reactions for each structure under the prescribed loading. As illustrated in the examples, in forming the idealized structures for the analysis, assume all force components to be applied at the centroidal axes of the members.

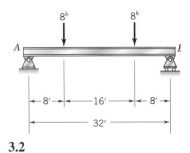

3.2

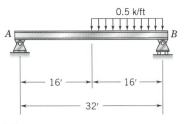

3.3

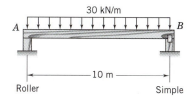

30 kN/m

A   B

|← 10 m →|

Roller   Simple

3.4

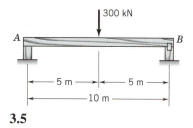

300 kN

A   B

|← 5 m →|← 5 m →|

|← 10 m →|

3.5

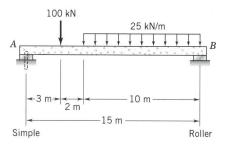

100 kN

25 kN/m

A   B

|← 3 m →|← 2 m →|← 10 m →|

|← 15 m →|

Simple   Roller

3.6

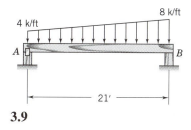

40^k

5 k/ft

A   B

|← 8' →|← 6' →|← 6' →|

|← 20' →|← 10' →|

3.7

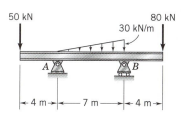

50 kN   80 kN

30 kN/m

A   B

|← 4 m →|← 7 m →|← 4 m →|

3.8

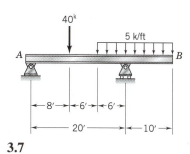

4 k/ft   8 k/ft

A   B

|← 21' →|

3.9

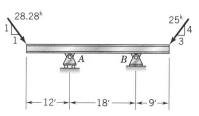

28.28^k 25^k

1   4

1   3

A   B

|← 12' →|← 18' →|← 9' →|

3.10

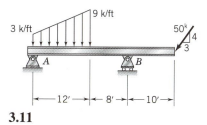

9 k/ft

3 k/ft   50^k

4

3

A   B

|← 12' →|← 8' →|← 10' →|

3.11

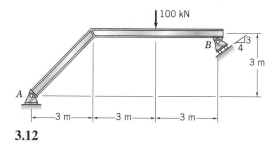

100 kN

B

3

4

3 m

A

|← 3 m →|← 3 m →|← 3 m →|

3.12

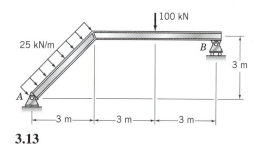

100 kN

25 kN/m   B

3 m

A

|← 3 m →|← 3 m →|← 3 m →|

3.13

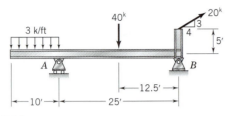

3.14

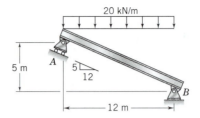

3.15

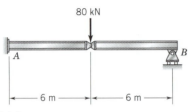

3.16

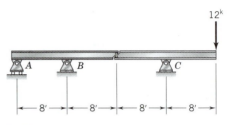

3.17

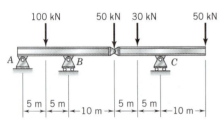

3.18

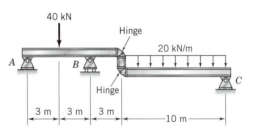

3.19

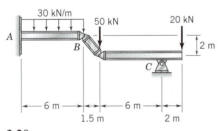

3.20

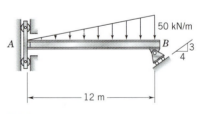

3.21

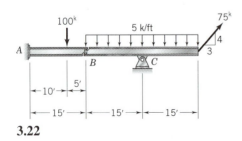

3.22

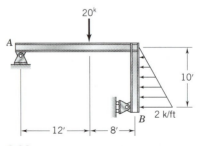

3.23

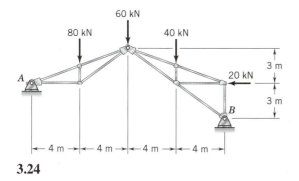

3.24

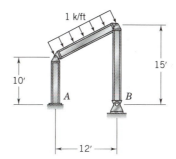

3.25

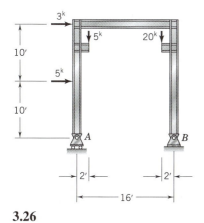

3.26

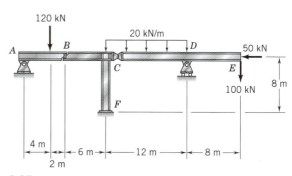

3.27

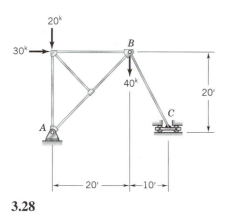

3.28

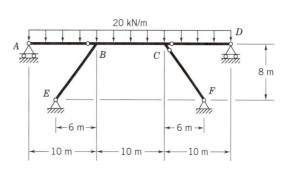

3.29

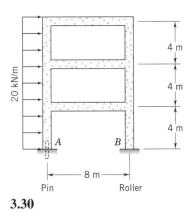

4 m

4 m

4 m

20 kN/m

A   B

8 m

Pin   Roller

3.30

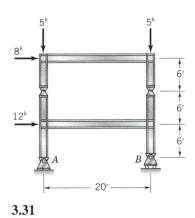

5^k 5^k

8^k

6'

6'

6'

12^k

A   B

20'

3.31

3.32 through 3.42   Verify that each three-dimensional structure is statically determinate and stable. Then, using the appropriate equations of equilibrium and condition equations, determine the support reactions for the specified loading.

⊬→ represents an unknown force reaction

⊬⇒ represents an unknown moment reaction

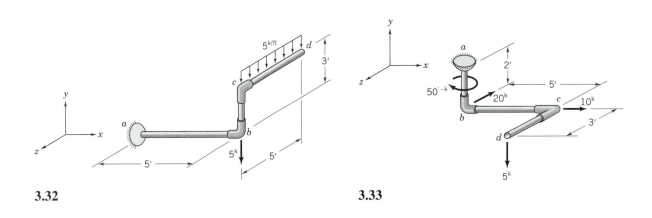

5$^{k/ft}$ d

3'

c

y

a   b

x

z   5^k

5'   5'

3.32

y

x

z

a

2'

50$^{'-k}$

20^k

b   c   10^k

5'

d   3'

5^k

3.33

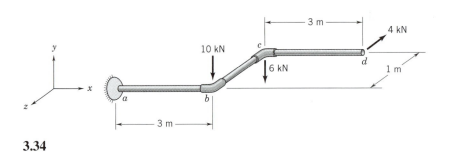

3 m

4 kN

10 kN

c

y   6 kN

d

x   a   b   1 m

z

3 m

3.34

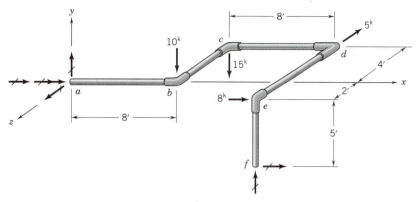

3.35

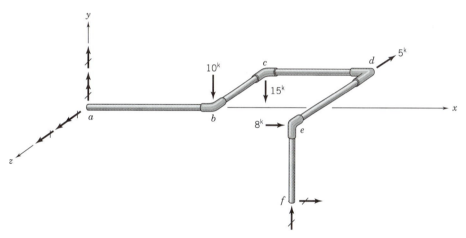

3.36 The same structure and loading as Problem 3.35, but with the altered set of boundary conditions shown.

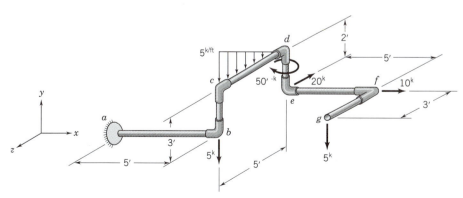

3.37

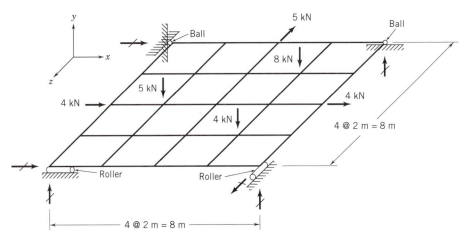

3.38

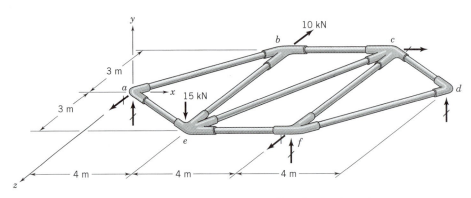

Note: Structure in xz plane

3.39

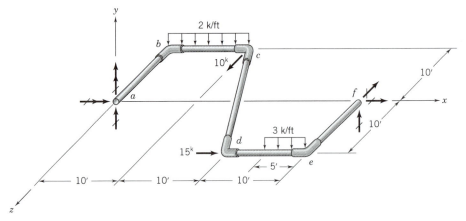

Note: Structure in xz plane

3.40

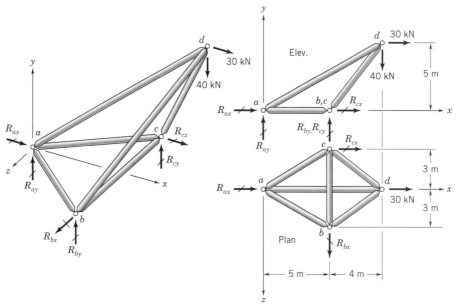

3.41

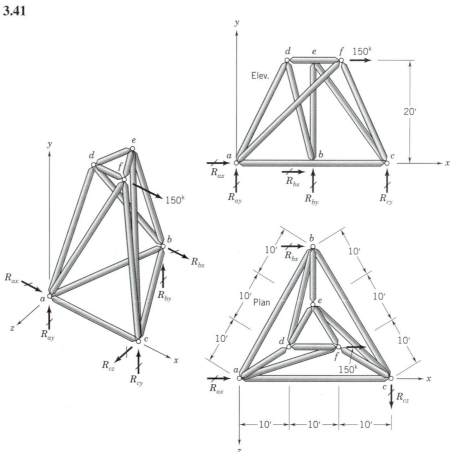

3.42

Chapter 4

Member Forces in Planar Trusses and Space Frameworks

Firth of Forth Railroad Bridge, Scotland (photo by Harry G. West).

4.1   PLANAR TRUSS STRUCTURES

A structure that is composed of a number of members that lie in a plane and are assumed to be pin connected at their ends to form a stable framework is called a *planar truss*. For such structures to be stable, there can be no relative movement between any two points in the structure beyond that which is caused by the deformation of the members of the framework. A common stable arrangement results from an assemblage of triangular units as shown in Fig. 4.1*a*, which is shown in its idealized pin connected form in Figure 4.1*b*.

In actuality, truss structures are not planar, but are three dimensional in nature. This three dimensionality exists on two levels. First, because of the possible nonplanar nature of the member cross sections, and the fact that the physical joining of the members involves a layered nonplanar assemblage of members (Figs. 4.1*c* and 4.1*d*), the structure has a dimension normal to the plane of the truss. Second, many truss-type structures actually enclose three-dimensional space.

The nonplanar nature of the first category is generally ignored. The minor eccentricities of the member axes with respect to the plane of the structure, which may be an issue in the design of the individual members, are not important in the analysis of

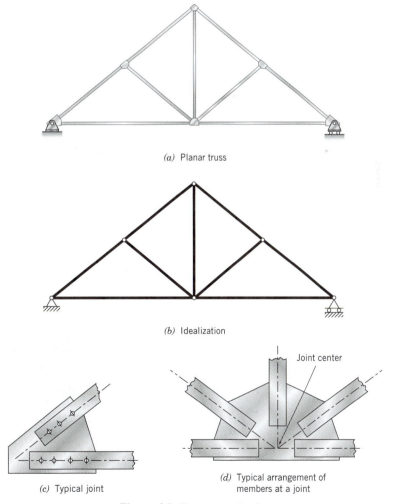

(a) Planar truss

(b) Idealization

(c) Typical joint

(d) Typical arrangement of members at a joint

Joint center

Figure 4.1 Planar truss details.

the truss structure as a whole. For the second category, it may be appropriate to consider the full spatial interconnection of the members when analyzing some structures, such as towers, microwave dishes, complicated roof systems, and aerospace structures. However, in many cases, such as bridge structures and simple roof systems, the framework can be subdivided into planar components for analysis as planar trusses without compromising the accuracy of the results.

Truss structures can be statically indeterminate in two different ways. If there are more reaction components than can be determined by consideration of statics, the structure is statically indeterminate externally. However, if there are more member forces than can be determined from statics, the structure is statically indeterminate internally. The topic of statical classification will be covered fully in Sections 4.8 and 4.14 for planar and nonplanar structures, respectively.

4.2   IDEALIZATIONS FOR PLANAR TRUSSES

As shown in Fig. 4.1*a*, those structures that are collectively called truss-type structures are not actually composed of pin-connected members. In fact, the assumption of pin-connected members is only one of several assumptions that are customarily made in the analysis of trussed frameworks.

The normal assumptions made in the analysis of planar trusses are enumerated below along with a comparison of how each of these idealizations relates to an actual structure.

1. Member ends are connected at the truss joints with frictionless pins.

   Most members are not pinned at their ends, but are welded or bolted as shown in Fig. 4.1*c*. Even if they were pinned, some friction would be present. However, this assumption gives reasonably good results for most truss-type connections.

2. Loads and reactions are applied to the truss at the joints only.

   This is realistic in terms of the applied loading because joints can be placed at points where loads are to be applied. Frequently, these loads come from other members that frame into the truss. For instance, a bridge might be composed of two parallel planar trusses as shown in Fig. 4.2. These trusses are connected by floorbeams that span between them and frame into each truss at a joint. The floor system of the deck, stringers, and floorbeams constitutes a complicated grid arrangement that is supported by the trusses at the floorbeam connection points.

   There is, of course, the dead load of the truss members, which is distributed along their lengths. These loads are generally small and are lumped at the joints for purposes of analysis.

3. The centroidal axis of each member is straight and coincides with the line connecting the joint centers at each end of the member.

   To avoid eccentricity at a joint, the centroidal axes of all members framing into a joint should intersect at a point. This is called the joint center, as shown in Fig. 4.1*d*. To ensure that a member is not loaded eccentrically, the centroidal axis of each member should pass through the joint centers at its ends. This assumption can be fulfilled by careful detailing, and it is especially important for structures that are subjected to repetitive loads.

   If all of the above assumptions are satisfied, then a free-body diagram of any truss member will show that it must carry axial load only, with no bending moment or

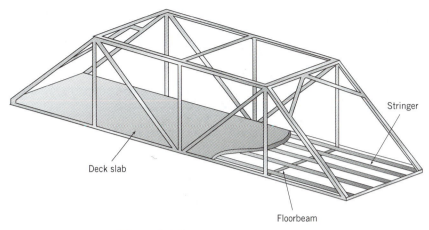

Figure 4.2 Typical truss bridge structure.

transverse force present. Because of this simple loading condition, truss members are frequently referred to as *bars* and the corresponding member forces are called *bar forces.*

An analysis based on the assumptions of an ideal truss will yield the *primary bar forces.* An actual truss, which will be in variance with the assumptions of an ideal truss, will have bar forces that differ from the primary bar forces. These differences are the *secondary bar forces.* In most cases, the secondary effects are negligible when compared with the primary effects; however, the analyst must be sensitive to the fact that they may be important in certain situations.

4.3   VARIATIONS IN PLANAR TRUSS CONFIGURATION

The simplest form of a stable planar truss is a triangular arrangement of three members as shown in Fig. 4.3*a*. Here, relative joint movements are precluded except for those that result from the member elongations that accompany the material strains. In the present context, these member deformations are ignored, and therefore the triangle forms a rigid unit. In contrast, the four-member arrangement of Fig. 4.3*b* admits relative joint movements through a realignment of the members as shown by the dashed lines.

If it is required to enlarge the structure of Fig. 4.3*a* to include an additional joint *d,* as shown in Fig. 4.3*c*, then two new members must be added to the truss. Further expansion of the truss follows the same pattern—each new joint requires the addition of two new members. The structure of Fig. 4.3*d* results from adding members *ed* and *ea* to establish joint *e* and members *fb* and *fe* to include joint *f.* A structure formed by this basic procedure is called a *simple truss.* Of course, in forming a simple truss, the original triangular unit must connect joints that are not located along a straight line, and each added joint must not lie along a line passing through the two joints to which it is being connected.

Other stable truss configurations are possible that do not evolve from the basic procedure described above. For instance, a *compound truss* results when two or more simple trusses are joined together to form a stable framework. Consider the two simple trusses shown in Fig. 4.4*a*. To join these separate trusses to form a stable compound truss, three conditions of connectivity are required to ensure that there will be no relative translation or rotation between the two sections. If the two trusses are connected at joint *a,* as shown in Fig. 4.4*b*, relative translations are precluded. The addition of member *bc* prevents relative rotation, and therefore the overall structure is stable.

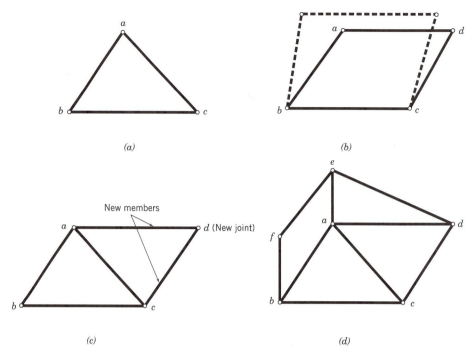

Figure 4.3 Arrangement of members for simple planar trusses. *(a)* Basic triangular arrangement. *(b)* Four-member arrangement. *(c)* Enlargement of truss to include joint at *d*. *(d)* Further expansion.

An alternate connection scheme would be to connect the two simple trusses with three nonparallel members as shown in Fig. 4.4*c*. Here, members *ad* and *ac* eliminate relative translations, and member *bd* prohibits relative rotations.

Trusses that cannot be classified as either simple or compound are called *complex trusses*. Structures of this type are discussed briefly in Section 4.10, and a complete treatment is offered by Timoshenko and Young, whose work is cited in Section 4.17.

4.4   TRUSS JOINT IDENTIFICATION AND MEMBER FORCE REPRESENTATION

There are numerous ways to identify the truss joints and to represent symbolically a given bar force. Any notation scheme is acceptable if it effectively communicates the meaning and if the analyst is consistent in its use.

In the development given here, each truss joint is labeled with a letter as shown in the portion of the truss in Fig. 4.5*a*. A bar force is represented by the letter F with a double subscript appended to designate the bar. For instance, F_{ij} represents the force in the member that connects joints i and j.

If a free-body diagram cuts through a member, then the force in that member must be shown to act on the cut face. Figure 4.5*b* shows a section through member ij and the force F_{ij} acting on the cut face of the member. Since the member must carry an axial force, the force must act in a direction that is coincident with the member direction. In the analysis process, it is frequently convenient to work with components of a bar force. Actually, a force can be resolved into any set of components that yield the given force as a resultant. However, the most commonly used components are those in the vertical and horizontal directions. In resolving a bar force into these components, the horizontal component is designated by X and the vertical component by Y. Again, a double subscript is appended to represent the bar. Figure 4.5*b* shows the bar force F_{ij} along with its components X_{ij} and Y_{ij}.

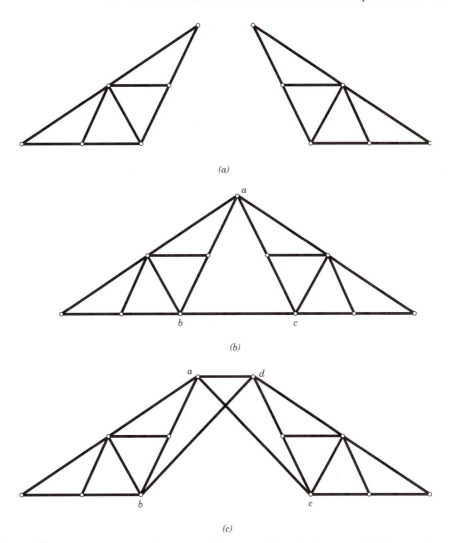

Figure 4.4 Member arrangement for compound trusses. *(a)* Two simple trusses. *(b)* Compound truss formation using common joint and one member. *(c)* Compound truss formation using three members.

Because the bar force and member direction are coincident, similar triangles exist between the member and its geometric components and the bar force and its components. These similar triangles are lightly shaded in Fig. 4.5. From these similar triangles, the following relationships exist between the bar force and its components:

$$\frac{F_{ij}}{L_{ij}} = \frac{X_{ij}}{x_{ij}} = \frac{Y_{ij}}{y_{ij}} \tag{4.1}$$

From this, any element of the force triangle can be determined from any other. That is,

$$F_{ij} = X_{ij}\left(\frac{L}{x}\right)_{ij} = Y_{ij}\left(\frac{L}{y}\right)_{ij}$$

$$X_{ij} = F_{ij}\left(\frac{x}{L}\right)_{ij} = Y_{ij}\left(\frac{x}{y}\right)_{ij} \tag{4.2}$$

$$Y_{ij} = F_{ij}\left(\frac{y}{L}\right)_{ij} = X_{ij}\left(\frac{y}{x}\right)_{ij}$$

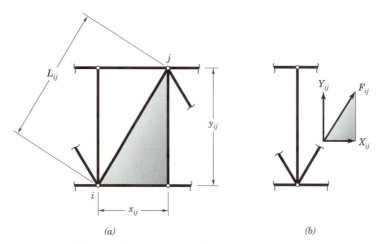

(a)   (b)

Figure 4.5 *(a)* Joint identification. *(b)* Bar force notation.

4.5 SIGN CONVENTION AND MEMBER FORCE REPRESENTATION

Along with a comprehensive and consistent notation, a sign convention must be adopted. In any analysis, there is no unique sign convention. However, whatever the analyst chooses must be used with consistency. In truss analysis, the sign convention that is almost universally employed designates a tensile force as a positive bar force and a compressive force as a negative bar force. This notation has a convenient physical interpretation in which a positive force is accompanied by a lengthening of the member, whereas a negative force is accompanied by a shortening of the member.

The process of determining an unknown bar force requires that the analyst isolate a free-body diagram in which the desired bar force is exposed as an unknown. This unknown force is assumed to act in the positive direction (tension) on the cut face as shown for member ij in Fig. 4.5b. Application of the equations of static equilibrium will yield the unknown bar force. If the solution produces a positive force, there is a dual significance in the plus sign. First, it means that the correct direction was assumed, and second, it means that the member is elongated or in tension. On the other hand, a negative force means that the assumed direction was incorrect and that the member is being shortened in compression.

Sometimes it is necessary to isolate a member and represent the forces acting on it. There are two ways in which it is convenient to represent such member forces. In the first, the forces that are exerted on the member by the joints are indicated, whereas in the second, the actions of the member on the joints are shown. These two approaches are shown in Fig. 4.6.

In the first method, a free-body diagram of the member is considered and the forces that the joints apply to its ends are shown. In the second approach, free-body diagrams of the joints are considered and the actions of the member on the joints are shown.

The two bar force representations are equivalent—each has advantages in certain situations and both will be used.

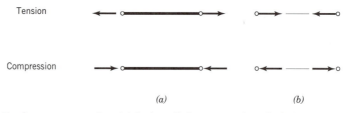

(a)   (b)

Figure 4.6 Bar force representation. *(a)* Action of joints on member. *(b)* Action of member on joints.

4.6 STRATEGY FOR PLANAR TRUSS ANALYSIS

The term *truss analysis* is used to describe the process whereby the bar forces in a truss are determined. For a statically determinate truss, these forces can be found by employing the laws of statics to ensure internal equilibrium of the structure. The process requires repeated use of free-body diagrams from which individual bar forces are determined.

In some cases, a single joint is isolated as shown in Fig. 4.7a. Here, the free body is acted on by a concurrent force system (all forces pass through the joint in the free-body diagram) and moment equilibrium is automatically satisfied. Thus, for a joint in a planar truss, two independent equilibrium equations are available, and only two unknown bar forces can be determined. The *method of joints* is a technique of truss analysis in which the bar forces are determined by the sequential isolation of joints—the unknown forces at one joint are determined and become known bar forces at subsequent joints. Simple trusses can always be analyzed by the method of joints. This is evident if it is noted that there are only two members framing into the joint that was positioned last in the formation processes. Therefore, these two bar forces can readily be determined by applying the equilibrium equations. Isolating the joints sequentially, in the reverse order from that used in forming the truss, ensures that the remaining bar forces can be determined.

In other cases, a larger portion of the structure is isolated by taking a section through the structure as shown in Fig. 4.7b. Here, since more than a single joint is included in the free-body diagram, the forces are not concurrent, and all three equations of static equilibrium are available for the planar case. Thus, three bar forces can be determined. Analysis by this method is referred to as the *method of sections*. This technique is particularly useful when only certain bar forces are required. It is

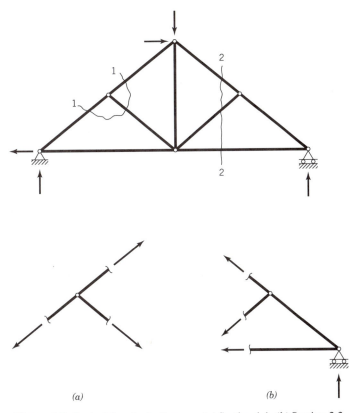

(a)   *(b)*

Figure 4.7 Typical free-body diagram. *(a)* Section 1.1. *(b)* Section 2.2.

also necessary to use this approach when the sequential consideration of joints fails because there are more than two unknowns at a joint. This is frequently the case for compound trusses.

In many cases, it is found that a combination of the method of joints and the method of sections is employed. One should determine the desired bar forces by applying the equations of equilibrium to the most convenient free-body diagrams and should not be committed to some particular method.

4.7   CONDITION EQUATIONS FOR DETERMINING REACTIONS

As discussed in Chapter 3, three independent reaction components can be determined for a planar system through the systematic application of the equations of static equilibrium. Again, as explained in Section 2.5, additional reaction components can be determined if there are internal conditions that permit the expression of special condition equations.

In the case of truss structures, the conditions might take a number of forms, some of which are described in Fig. 4.8. The structure of Fig. 4.8*a* is merely a trussed version of a three-hinged arch. For a free-body diagram either left or right of the hinge, the condition at point *b* augments the equations of equilibrium applied to the entire structure to enable the solution of the four reaction components.

The structure of Fig. 4.8*b* is a statically determinate stable truss, and the three reaction components can be determined from statics. If this structure is modified to form the *cantilever truss* of Fig. 4.8*c,* the condition at point *c* allows for the determination of the additional reaction component at point *b*. Further modification yields the structure of Fig. 4.8*d*. Here, the link (member *ce*) provides two conditions ($M_z = 0$ at each end), and therefore one more reaction component is added to maintain stability—the horizontal reaction at point *d*.

Normally, the absence of a diagonal member in a truss panel, such as in any of the panels for the truss shown in Fig. 4.3*b*, would render the truss unstable. However, under proper support conditions, this configuration merely introduces a condition. In Fig. 4.8*e*, a vertical section through panel *bc* reveals that no vertical force component can be transferred through this panel. This condition, coupled with the overall equations of equilibrium, enables one to determine the reaction components.

Other conditions are possible, and these are discussed in detail by Timoshenko and Young and by Norris, Wilbur, and Utku, as noted in Section 4.17. Special care must be exercised to make certain that trusses with conditions are stably arranged.

4.8   STATICAL DETERMINACY AND STABILITY OF PLANAR TRUSSES

The most fundamental form of a stable planar truss structure is the simple triangular arrangement of three members along with the accompanying three reaction components shown in Fig. 4.9*a*. In this case, there are six unknowns (three bar forces and three independent reaction components) and six independent equations of equilibrium (two equations at each joint), and thus the system is statically determinate. Of course, the reactions could be determined first by applying the conditions of static equilibrium to the entire structure, and then the bar forces could be determined by joint considerations. In either case, however, there are only six *independent* equations of equilibrium.

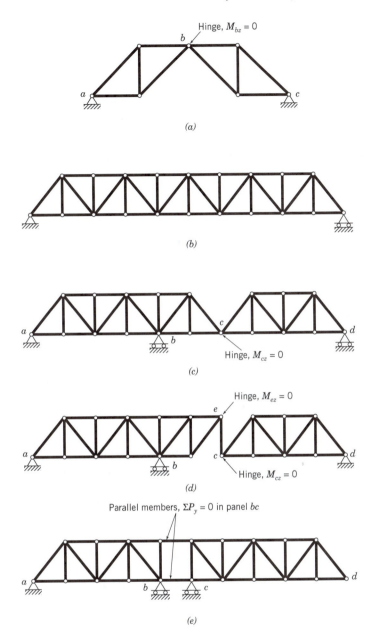

Figure 4.8 Condition equations for trusses.

If it is required to enlarge the truss to include an additional joint at *d,* as shown in Fig. 4.9*b,* then two members must be added to the truss as discussed earlier in Section 4.3. The new joint furnishes two new equations of equilibrium for the determination of the two new bar forces. If required, additional joints can be added; each additional joint requires two additional bars to include it as a stable portion of the truss, and it furnishes two additional equations of equilibrium. Expansion of the truss according to this pattern ensures that the truss remains stable and statically determinate, and the so-called simple truss results.

If a reaction component were added without including a condition that would enable its determination, the structure would be statically indeterminate externally. Or, if

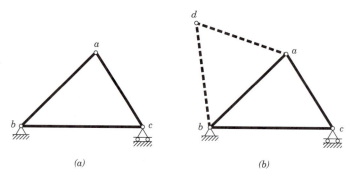

Figure 4.9 Formation of statically determinate planar truss. *(a)* Stable truss. *(b)* Enlargement of truss.

a member were added without increasing the number of joints, the structure would become statically indeterminate internally.

Based on the foregoing discussion, criteria can be developed concerning the statical determinacy of truss structures. First, the structure can be tested for external determinacy and stability according to the criteria developed in Chapter 3. That is, if r is taken as the least number of reaction components required for external stability and r_a is the actual number of reaction components, then the following criteria hold:

$$r_a < r \text{ ; structure is statically unstable externally}$$

$$r_a = r \text{ ; structure is statically determinate externally}$$

$$r_a > r \text{ ; structure is statically indeterminate externally}$$

As pointed out earlier, the conditions for $r_a \geq r$ are necessary but not sufficient conditions for statical classification. The reactions must be properly arranged to ensure stability.

For internal classification, in addition to the above definition for r, let m be the total number of members (bar forces) and j the total number of joints. Then

$$2j = m + r \tag{4.3}$$

is the relationship that must be satisfied if the truss is to be statically determinate internally. The left-hand side of the equation represents the number of equilibrium equations that are available, and the right-hand side represents the total number of unknown bar forces and unknown required reaction components. Equation 4.3 can be rewritten as

$$m = 2j - r \tag{4.4}$$

In this form, m is the number of members required to form an internally statically determinate truss that connects j joints and has r reaction components required for external stability. If m_a is the actual number of members in the truss, then the following conditions prevail:

$$m_a < m \text{ ; truss is statically unstable internally}$$

$$m_a = m \text{ ; truss is statically determinate internally}$$

$$m_a > m \text{ ; truss is statically indeterminate internally}$$

Care must be exercised in using the above conditions. If $m_a < m$, the truss is definitely unstable, but if $m_a \geq m$ it does not necessarily follow that the truss is stable. It is possible that the m members are not properly arranged to ensure internal stability. Such trusses are said to have *critical form*.

The foregoing criteria for the internal classification of a truss are clearly applicable for simple trusses because the development of Eq. 4.4 is based on principles that parallel the development of the simple truss. Although it will not be proven here, Eq. 4.4 is also applicable for compound trusses.

Table 4.1 illustrates the statical classification of several planar truss structures. In each case, both external and internal classification are considered according to the criteria that have been developed. It should be recalled that r is equal to three (corresponding to the three equations of static equilibrium) plus the number of additional condition equations available for the determination of the reactions.

Table 4.1   Statical Classification of Trusses

Structure	Structure Characteristics			External Classification		Internal Classification	
	j	m_a	r_a	r	Classification	$m = 2j - r$	Classification
(a)	8	13	3	3	$r_a = r$, deter. stable	$13 = 16 - 3$	$m_a = m$, deter. stable
(b)	8	15	3	3	$r_a = r$, deter. stable	$13 = 16 - 3$	$m_a > m$, indet. stable
(c)	8	13	3	3	$r_a = r$, deter. stable	$13 = 16 - 3$	$m_a = m$, but unstable
(d)	7	9	4	3	$r_a > r$, indet. stable	$11 = 14 - 3$	$m_a < m$, unstable
(e) $M_z = 0$	14	24	4	$3 + 1 = 4$	$r_a = r$, deter. stable	$24 = 28 - 4$	$m_a = m$, deter. stable
(f)	12	23	4	3	$r_a > r$, indet. stable	$21 = 24 - 3$	$m_a > m$, indet. stable
(g)	7	11	3	3	$r_a = r$, deter. stable	$11 = 14 - 3$	$m_a = m$, deter. stable
(h) $\Sigma P_y = 0$ $M_z = 0$	14	22	6	$3 + 3 = 6$	$r_a = r$, deter. stable	$22 = 28 - 6$	$m_a = m$, deter. stable

4.9   NUMERICAL TRUSS ANALYSIS PROBLEMS

Several numerical problems will be considered in this section. Each problem has been selected because it demonstrates a specific concept regarding truss analysis. The first example illustrates the use of the method of joints for a complete truss analysis. The other examples involve both the method of joints and the method of sections.

Regardless of the method being employed, there are certain common points. For any free-body diagram, all known bar forces and the applied forces are shown to act in their proper directions. Unknown bar forces are assumed to act in tension—if the subsequent analysis yields a positive bar force, the bar is in tension, whereas a negative bar force corresponds to compression in the bar.

In most cases, there are opportunities to perform independent checks on certain bar forces. Such opportunities should always be used to ensure that the results are consistent with all of the requirements of equilibrium.

EXAMPLE 4.1   Compute the bar forces in all members of the truss shown for the designated loading condition.

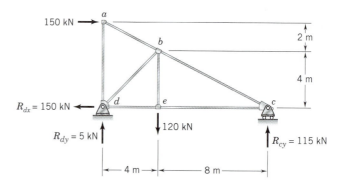

Determination of Reactions

The truss is statically determinate externally, and the reactions are determined and shown on the structure.

Determination of Bar Forces

Application of Eq. 4.4 yields $m = 2j - r = (2 \times 5) - 3 = 7$. Since $m_a = 7$, the structure is statically determinate internally.

In this problem, the method of joints will be employed. Each of the five joints will be isolated as a free-body diagram (*FBD*), and the equations of equilibrium will be applied to each. The analysis proceeds from joint to joint, with the unknown bar forces being determined at each joint.

Isolate joint a as FBD:

Summation of forces in the x direction, with the positive direction taken to the right $(\Sigma P_x = 0 \xrightarrow{+})$ yields

$$150 + X_{ab} = 0; \quad X_{ab} = -150 \text{ kN}$$

$$Y_{ab} = X_{ab}\left(\frac{y}{x}\right)_{ab} = -150\left(\frac{2}{4}\right) = -75 \text{ kN} \tag{4.2}$$

$$F_{ab} = X_{ab}\left(\frac{L}{x}\right)_{ab} = -150\left(\frac{4.47}{4}\right) = \underline{\underline{-167.6 \text{ kN}}} \tag{4.2}$$

Summation of forces in the y direction, with the positive direction taken as upward $(\Sigma P_y = 0 \uparrow+)$ yields

$$F_{ad} + Y_{ab} = 0; \quad F_{ad} = -(-75) = \underline{\underline{+75 \text{ kN}}}$$

Isolate joint d as FBD:

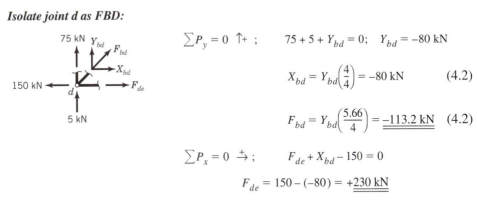

$\sum P_y = 0 \uparrow+ \; ; \qquad 75 + 5 + Y_{bd} = 0; \quad Y_{bd} = -80 \text{ kN}$

$$X_{bd} = Y_{bd}\left(\frac{4}{4}\right) = -80 \text{ kN} \tag{4.2}$$

$$F_{bd} = Y_{bd}\left(\frac{5.66}{4}\right) = \underline{\underline{-113.2 \text{ kN}}} \tag{4.2}$$

$\sum P_x = 0 \xrightarrow{+} \; ; \qquad F_{de} + X_{bd} - 150 = 0$

$$F_{de} = 150 - (-80) = \underline{\underline{+230 \text{ kN}}}$$

Isolate joint e as FBD:

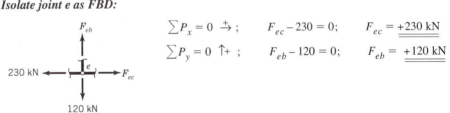

$\sum P_x = 0 \xrightarrow{+} \; ; \qquad F_{ec} - 230 = 0; \qquad F_{ec} = \underline{\underline{+230 \text{ kN}}}$

$\sum P_y = 0 \uparrow+ \; ; \qquad F_{eb} - 120 = 0; \qquad F_{eb} = \underline{\underline{+120 \text{ kN}}}$

Isolate joint c as FBD:

$\sum P_x = 0 \xleftarrow{+} \; ; \qquad X_{bc} + 230 = 0; \qquad X_{bc} = -230 \text{ kN}$

$$Y_{bc} = X_{bc}\left(\frac{4}{8}\right) = -115 \text{ kN}$$

$$F_{bc} = X_{bc}\left(\frac{8.94}{8}\right) = \underline{\underline{-257.0 \text{ kN}}}$$

$\sum P_y = 0 \uparrow+ \; ; \qquad Y_{bc} + 115 = 0; \qquad Y_{bc} = -115 \text{ kN} \quad \text{check}$

At this point, all bar forces have been determined, but joint b should be considered as check.

Isolate joint b as FBD:

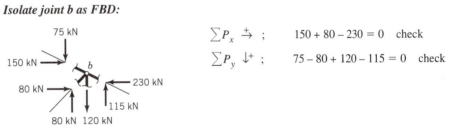

$\sum P_x \xrightarrow{+} \; ; \qquad 150 + 80 - 230 = 0 \quad \text{check}$

$\sum P_y \downarrow+ \; ; \qquad 75 - 80 + 120 - 115 = 0 \quad \text{check}$

Summary of Bar Forces

The bar forces (in kips), along with the appropriate signs, are shown on the following sketch of the structure:

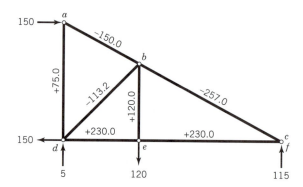

Solution by Inspection

The above treatment illustrates a complete analysis using the method of joints. Each joint has been isolated and the unknown forces have been determined by using the two equations of equilibrium. The unknown bar forces at one joint become the known bar forces of subsequent joints. This problem is so straightforward that it could actually be done by inspection.

The truss is lightly drawn in the sketch shown below with the applied loads and reactions. Starting at joint a, $\Sigma P_x = 0$ shows that bar ab must impart a horizontal component of 150 kN toward the joint. Since the bar force must act along the member, the horizontal component must be accompanied by a vertical component of 75 kN. Use of Eq. 4.2 yields the actual bar force of 167.6 kN. Figure 4.6b shows that this kind of member action on a joint corresponds to a compressive member force and there is a similar action on the joint at end b of the member. Application of $\Sigma P_y = 0$ at joint a shows that member ad must impart a force that pulls away from the joint with a vertical component of 75 kN. Since this member is vertical, the bar force itself is 75 kN, and Fig. 4.6b shows that the member is in tension and imparts a force that pulls away from joint d.

By proceeding in the same fashion to joints d, e, and c, we are able to determine the remaining bar forces. Joint b is then used as a check to see that the bar forces determined are consistent with the requirements of equilibrium at that joint.

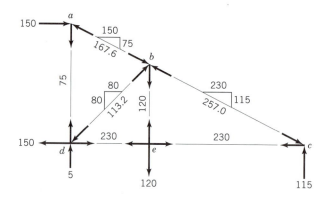

This example was simplified by the fact that at each joint one of the two unknown bar forces acted either vertically or horizontally. Thus, the application of one equation of equilibrium led to one bar force and the second equation led to the second bar force. If both bars were inclined so that each had vertical and horizontal components, then the two bar forces would result from the simultaneous solution of the two equations of equilibrium. Frequently, the solution of simultaneous equations can be avoided by employing a moment equation, as illustrated in subsequent problems.

EXAMPLE 4.2

Determine the forces in members *cd, Cd, CD, BC,* and *cC* (identified with ✕ on sketch) for the truss shown. The loading and the corresponding statically determinate reactions are given.

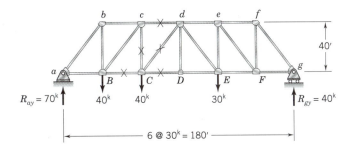

Determination of Required Bar Forces

Application of Eq. 4.4 yields $m = 2j - r = (2 \times 12) - 3 = 21$. Since $m_a = 21$, the structure is statically determinate internally.

In this problem, the method of sections will be used. Isolate the portion of the truss left of a cut through members *cd, Cd,* and *CD.* The unknown bar forces are shown (including the *x* and *y* components of F_{Cd}).

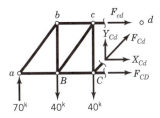

Since F_{Cd} and F_{CD} pass through point *C*, F_{cd} can be determined directly by applying moment equilibrium about point *C*. With clockwise moments taken to be positive ($\Sigma M_C = 0 \; \circlearrowright +$), we obtain

$$(70 \times 60) - (40 \times 30) + (F_{cd} \times 40) = 0$$

$$F_{cd} = \frac{-3000}{40} = \underline{\underline{-75^k}}$$

Summation of forces in the *y* direction, with the positive direction taken as upward ($\Sigma P_y = 0 \; \uparrow +$) yields

$$Y_{Cd} + 70 - 40 - 40 = 0; \qquad Y_{cd} = +10^k$$

$$X_{Cd} = Y_{Cd}\left(\frac{3}{4}\right) = +7.5^k; \qquad F_{Cd} = Y_{Cd}\left(\frac{5}{4}\right) = \underline{\underline{+12.5^k}} \tag{4.2}$$

Summation of forces in the *x* direction, with the positive direction taken to the right ($\Sigma P_x = 0 \; \overset{+}{\rightarrow}$) gives

$$F_{CD} + X_{Cd} + F_{cd} = 0; \qquad F_{CD} = -(7.5) - (-75) = \underline{\underline{+67.5^k}}$$

Isolate the portion of the truss left of a cut through members cd, cC, and BC.

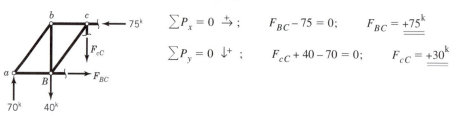

$$\sum P_x = 0 \ \overset{+}{\rightarrow} ; \qquad F_{BC} - 75 = 0; \qquad F_{BC} = +75^k$$

$$\sum P_y = 0 \ \downarrow^+ ; \qquad F_{cC} + 40 - 70 = 0; \qquad F_{cC} = +30^k$$

EXAMPLE 4.3

Determine the forces in members U_1L_2 and U_1U_2 (identified with $\times$ on sketch) for the truss and loading shown. The structure is statically determinate both internally and externally, and the reactions are given.

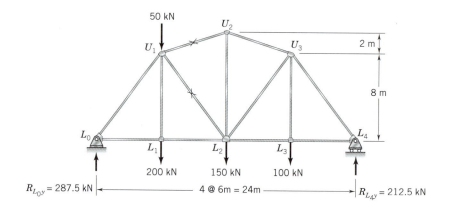

Determination of Required Bar Forces

Isolate the portion left of section through U_1U_2, U_1L_2, and L_1L_2. Locate pt. O, at intersection of $F_{U_1U_2}$ & $F_{L_1L_2}$ in order to solve directly for the force in member U_1L_2 by summing moments about pt. O. From similar triangles,

$$\frac{r}{8} = \frac{6}{2}; \quad r = \frac{6 \times 8}{2} = 24 \text{ m}$$

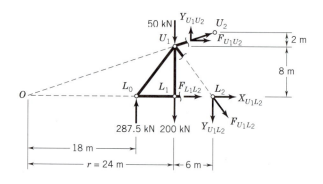

Resolve $F_{U_1L_2}$ into components at pt. L_2; then $X_{U_1L_2}$ also passes through point O.

$$\sum M_O = 0 \; \text{⟳+;} \qquad (200 + 50)24 + (Y_{U_1L_2} \times 30) - (287.5 \times 18) = 0$$

$$Y_{U_1L_2} = \frac{-6000 + 5175}{30} = \frac{-825}{30} = -27.5 \text{ kN}$$

$$F_{U_1L_2} = -27.5\left(\frac{10}{8}\right) = \underline{\underline{-34.4 \text{ kN}}} \qquad (4.2)$$

$$\sum P_y = 0 \; \text{↑+ ;} \qquad 287.5 - 250 + Y_{U_1U_2} - Y_{U_1L_2} = 0$$

$$Y_{U_1U_2} = 250 - 287.5 + (-27.5) = -65 \text{ kN}$$

$$F_{U_1U_2} = -65\left(\frac{6.32}{2}\right) = \underline{\underline{-205.4 \text{ kN}}} \qquad (4.2)$$

EXAMPLE 4.4

Determine the bar forces in members ad and bd (identified with $\times$ on truss) for the tower shown. The structure is statically determinate.

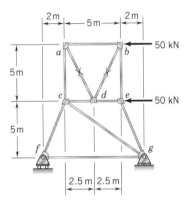

Determination of Required Bar Forces

Isolate joint d:

$$\sum P_y = 0 \; \text{↑+ ;} \qquad Y_{ad} + Y_{bd} = 0; \quad Y_{ad} = -Y_{bd}$$

Inclinations of ad and bd are the same; therefore,

$$X_{ad} = -X_{bd}; \qquad F_{ad} = -F_{bd}$$

Take section below a and b. Isolate portion above section.

$$\sum P_x = 0 \; \xrightarrow{+} ; \qquad X_{ad} - X_{bd} - 50 = 0; \qquad \text{but } X_{bd} = -X_{ad}$$

$$\therefore X_{ad} + X_{ad} = 50; \qquad X_{ad} = +25 \text{ kN}$$

$$F_{ad} = +25\left(\frac{5.59}{2.5}\right) = \underline{\underline{+55.9 \text{ kN}}} \qquad (4.2)$$

$$F_{bd} = -F_{ad} = \underline{\underline{-55.9 \text{ kN}}}$$

EXAMPLE 4.5

Determine the bar forces for the truss shown below. The structure is statically determinate both externally and internally, and the reactions are given.

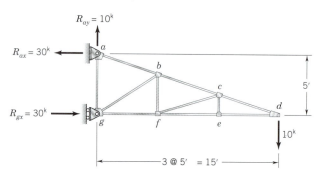

Determination of Bar Forces

The method of joints is used, with the solution being completed by inspection with the results shown on the accompanying sketch of the truss.

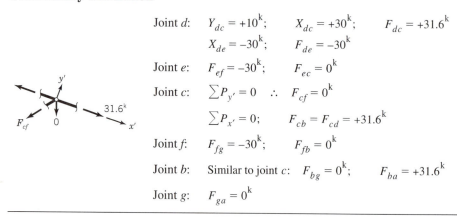

Commentary on Solution

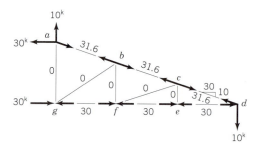

Joint d: $Y_{dc} = +10^k$; $X_{dc} = +30^k$; $F_{dc} = +31.6^k$

$X_{de} = -30^k$; $F_{de} = -30^k$

Joint e: $F_{ef} = -30^k$; $F_{ec} = 0^k$

Joint c: $\sum P_{y'} = 0$ $\therefore$ $F_{cf} = 0^k$

$\sum P_{x'} = 0$; $F_{cb} = F_{cd} = +31.6^k$

Joint f: $F_{fg} = -30^k$; $F_{fb} = 0^k$

Joint b: Similar to joint c: $F_{bg} = 0^k$; $F_{ba} = +31.6^k$

Joint g: $F_{ga} = 0^k$

4.10   COMPLEX PLANAR TRUSSES

There are planar truss configurations that are neither simple nor compound. For instance, consider the simple truss shown in Fig. 4.10a. Beginning with the triangular unit *abc*, the truss is systematically assembled by adding two members to establish each new joint. A count of the members and joints verifies that Eq. 4.4 is satisfied, and therefore the structure is statically determinate. If member *cd* is removed and a member connecting joints *e* and *f* is added, as shown in Fig. 4.10b, the structure remains statically determinate; however, its form is no longer that of a simple truss. In addition, it is not a compound truss.

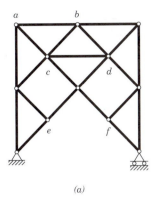
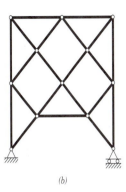

(a)   *(b)*

Figure 4.10 Formation of a complex planar truss. *(a)* Simple truss. *(b)* Complex truss.

Such a configuration is referred to as a *complex truss*. Complex trusses can be analyzed by a strategy involving a combination of the method of joints and the method of sections. Several special techniques have been developed for the analysis of complex trusses, and Timoshenko and Young have one of the most complete treatments. Their work is cited in Section 4.17. These procedures had greater importance in the precomputer age because of the need to avoid solving large numbers of simultaneous equations. However, given the current computer capabilities, more general methods are viable for the analysis of complex trusses, as are the general matrix methods of Part Five.

Complex trusses are particularly prone to possess critical forms. Although these forms are frequently difficult to detect by inspection, they possess certain mathematical characteristics that are evident in the general methods of analysis.

4.11   NONPLANAR TRUSS-TYPE SPACE FRAMEWORKS

The characteristics of a planar truss can be extended to three dimensions, with the pin connections of a planar truss being replaced by *universal joints* that are not capable of sustaining a member-end moment. For *space frameworks* of this type to be stable, there can be no relative motion between the joints beyond that which occurs as a result of member deformations. Such stability generally results from a combination of tetrahedral units that are interconnected so as to include all points that are to be embraced within the structure space. The details concerning member arrangements for stable frameworks are treated in Sections 4.12 and 4.16.

Space frameworks can be statically indeterminate either externally or internally. In the former case, there are more reaction components than can be determined from statics, whereas in the latter case, there are more member forces than can be determined by statics. It is possible for the structure to be statically indeterminate externally but have an overall classification of statically determinate.

This chapter considers only statically determinate structures, and it is limited to a presentation of the fundamental concepts and procedures needed for the analysis of a space framework. Complicated frameworks, whether statically determinate or indeterminate, would be treated by the more general matrix techniques that are developed in Part Five.

The key assumptions necessary for the analysis of space frameworks are mere extensions of those employed in the analysis of planar trusses. Since these assumptions were justified in detail in Section 4.2, they are stated here simply in the context of a three-dimensional structure as follows:

1. Members are connected at their ends with frictionless universal joints.

2. All loads and reactions are applied to the framework at the joints only.

3. The centroidal axis of each member is straight and coincides with the line connecting the joint centers at each end of the member.

When these assumptions are satisfied, each member of the structure will carry pure axial load, with no bending moment or shear present.

4.12 VARIATIONS IN FRAMEWORK CONFIGURATIONS

The simplest form of a stable space framework is the tetrahedral arrangement of members shown in Fig. 4.11*a*. In this case, to the basic planar triangular unit *abc*, point *d* is attached by the addition of three members. The resulting tetrahedron includes four separate stable triangular units, and the relative movements between points result only from the member elongations that accompany the material strains. Since these small movements are ignored in this context, the tetrahedral unit *abcd* is taken as a rigid unit.

This simple tetrahedral framework can be enlarged to include an additional point *e* by adding three members, as shown in Fig. 4.11*b*. This procedure can be repeated to enlarge the system further to any size and configuration, and the structure formed by this approach is called a *simple space framework*. For example, Fig. 4.11*c* shows an enlargement of the structure shown in Fig. 4.11*b* by the ordered addition of joints *f, g,* and *h*. The resulting structure is composed of five tetrahedra. An alternative arrangement is shown in Fig. 4.11*d*—here the original tetrahedron of Fig. 4.11*a* is enlarged by a different strategy, through the ordered addition of points *e, f, g,* and *h*. This example

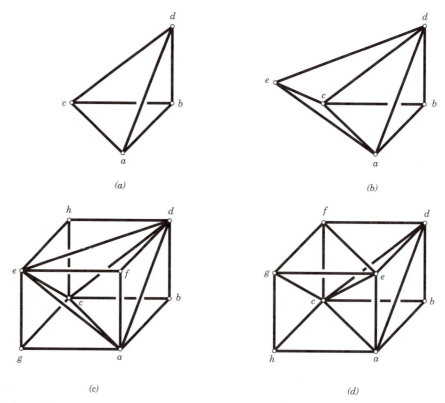

Figure 4.11 Arrangement of members for simple frameworks. *(a)* Basic tetrahedral arrangement. *(b)* Enlargement of framework to include joint at *e*. *(c)* Further enlargement to include joints at *f, g,* and *h*. *(d)* Alternative enlargement of original tetrahedron.

serves to illustrate that not all stable structures are composed of a nested array of tetra-hedra. In this case, there are four tetrahedra and a five-sided polygon.

It must be noted that in forming a simple framework, the original triangular unit must connect points that are not located along a straight line, and each added joint must not lie in the plane of the triangular unit to which it is being connected.

Other stable framework configurations are possible that do not evolve from the procedure outlined above. Consider the case where two or more simple frameworks are connected to form a *compound space framework*. For example, take the two simple frameworks shown in Fig. 4.12*a*. As a first step, these two frameworks are joined by bringing together joints *b* and *c'*. This arrangement is not stable because rotation could occur about any axis through the common point. If a member were added between points *a* and *a'*, the resulting structure would be unstable with respect to rotations about axes along members *ab* and *a'c'*. The addition of a member between joints *d*

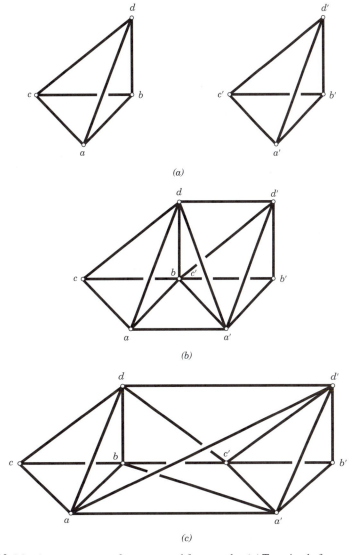

Figure 4.12 Member arrangement for compound frameworks. *(a)* Two simple frameworks. *(b)* Compound framework formation using common joint and three additional members. *(c)* Compound framework formation using six additional members.

and a' would inhibit rotation about an axis along member ab, and the addition of a member between joints d and d' would remove the remaining instability associated with rotation about an axis along member $a'c'$. The compound framework is now stable as shown in Fig. 4.12b—the attachment at the common points prevents relative translations, and the addition of three members restricts relative rotations.

An alternative connection would be the addition of six members between the original simple frameworks as shown in Fig. 4.12c. Here, member bc' prevents overall translation between the original two tetrahedra, but it takes five additional members to restrict all possible relative rotations.

It should be noted that although the structure of Fig. 4.12b was formed as a compound framework, it is actually a simple framework that could have been formed by the sequential addition of joints a', d', and b' to the original tetrahedron $abcd$. However, the structure of Fig. 4.12c is truly a compound framework in that it could not be formed from the procedure used to form a simple framework.

In this discussion, only simple cases are considered for the purposes of definition. When more complicated simple frameworks are connected, such as those shown in Figs. 4.11c and d, the connecting sections may contain polyhedra. For additional information on more complicated frameworks, the reader is directed to the references in Section 4.17.

4.13 JOINT IDENTIFICATION, MEMBER FORCE NOTATION, AND SIGN CONVENTION

The techniques for joint identification, bar force notation, and sign convention adopted here are direct extensions of what was done for planar trusses in Sections 4.4 and 4.5.

Each joint of the structure is labeled with a letter as shown in Fig. 4.13a. A given member force is again symbolically represented by the letter F with an appended double subscript indicating the terminal points of the member. For instance, F_{ij} represents the force in the member that connects joints i and j of the structure. This force, which is an internal force, becomes an external force for any free-body diagram that cuts through member ij, as shown in Fig. 4.13b. Since the member must carry an axial force, the force acts in the same direction as the member itself, and the relationships between the member length and its geometric components are similar to those between the member force and its force components. Thus, if member ij is of length L_{ij}

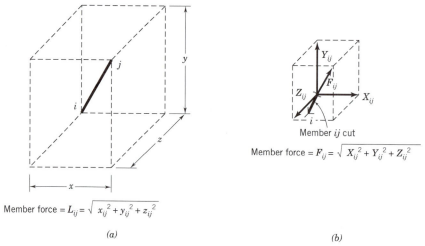

Member force $= L_{ij} = \sqrt{x_{ij}^2 + y_{ij}^2 + z_{ij}^2}$

Member force $= F_{ij} = \sqrt{X_{ij}^2 + Y_{ij}^2 + Z_{ij}^2}$

(a) (b)

Figure 4.13 Member and force identification. (a) Member. (b) Force.

and it has projections of x_{ij}, y_{ij}, and z_{ij} along the three coordinate axes, and if the bar force F_{ij} has components X_{ij}, Y_{ij}, and Z_{ij} along the same axes, then

$$\frac{F_{ij}}{L_{ij}} = \frac{X_{ij}}{x_{ij}} = \frac{Y_{ij}}{y_{ij}} = \frac{Z_{ij}}{z_{ij}} \tag{4.5}$$

From this relationship, any bar force component can be determined from any other. That is,

$$F_{ij} = X_{ij}\left(\frac{L}{x}\right)_{ij} = Y_{ij}\left(\frac{L}{y}\right)_{ij} = Z_{ij}\left(\frac{L}{z}\right)_{ij}$$

$$X_{ij} = F_{ij}\left(\frac{x}{L}\right)_{ij} = Y_{ij}\left(\frac{x}{y}\right)_{ij} = Z_{ij}\left(\frac{x}{z}\right)_{ij}$$

$$Y_{ij} = F_{ij}\left(\frac{y}{L}\right)_{ij} = X_{ij}\left(\frac{y}{x}\right)_{ij} = Z_{ij}\left(\frac{y}{z}\right)_{ij}$$

$$Z_{ij} = F_{ij}\left(\frac{z}{L}\right)_{ij} = X_{ij}\left(\frac{z}{x}\right)_{ij} = Y_{ij}\left(\frac{z}{y}\right)_{ij}$$

$$\tag{4.6}$$

As was done in planar trusses, a tensile member force is taken to be positive and a compressive member force, negative. In the analysis process, the analyst must isolate a free-body diagram in which the desired member force is exposed as an unknown. This unknown force should be assumed to act in the positive direction (tension) on the cut face of the member. The subsequent analysis leads to the desired member force; a positive result indicates that the assumed direction is correct and that the member is in tension, whereas a negative result means that the assumed direction is incorrect and that the member is in compression.

4.14   OVERALL STATICAL DETERMINACY AND STABILITY OF SPACE FRAMEWORKS

The discussion of determinacy and stability in Section 3.7 was based solely on the application of the six equations of static equilibrium to the entire structure. In some cases, it turns out that the reaction components are not statically determinate based on external considerations. However, the total analysis of the structure (reaction components and member forces) is statically determinate when both external and internal conditions are considered.

Each joint of a space framework forms a concurrent force system in which the three moment equations of Eqs. 2.10 are automatically satisfied. Thus, only three independent equations of static equilibrium can be written at each joint. It must be noted that the static relations written externally for the entire structure will not provide further independent equations when this approach is used. Therefore, if both external reaction components and internal member forces are considered, a necessary, but not sufficient, condition for the overall determinacy of the structure is that the total number of unknown member forces m, plus the actual number of unknown reaction components r_a, must not exceed three times the number of joints j. That is,

$$m + r_a = 3j$$

or

$$m = 3j - r_a \tag{4.7}$$

In this form, m is the number of members required to form a statically determinate structure that will connect j joints and possess r_a reaction components. If m_a is the actual number of members in the structure, then the following criteria hold:

$$m_a < m; \qquad \text{structure is statically unstable}$$

$$m_a = m; \qquad \text{structure is statically determinate}$$

$$m_a > m; \qquad \text{structure is statically indeterminate}$$

It must be emphasized that $m_a \geq m$ are necessary but not sufficient conditions. Also, if $m_a > m$, it is not clear from this test whether the indeterminacy stems from redundant reaction components or redundant members. Of course, the criteria of Section 3.7 could be used to classify the structure externally, and those results could be coupled with the overall classification of this section to determine the degrees of external and internal redundancies.

The criteria enumerated above clearly apply to simple frameworks, but it is not immediately evident that they also apply to compound frameworks. The proof of this, however, will not be addressed here.

The treatment given here is somewhat brief, and it does not address some subtle points regarding structure classification for complicated structures with unusual arrangements of members. Section 4.16 briefly treats the topic of *complex frameworks*, but for a more complete treatment of advanced topics, the reader is directed to the references given at the end of this chapter.

4.15   ANALYSIS OF TRUSS-TYPE SPACE FRAMEWORKS

Two slightly different techniques can be employed in the analysis of statically determinate space frameworks. The first approach is used for structures that are statically determinate externally. Here, all the reaction components can be determined by applying the six equations of static equilibrium to the entire structure. For these cases, the reactions are determined first, and then the individual member forces are determined by taking a sequence of free-body diagrams that permit the systematic determination of all member forces. The second technique involves situations where the reactions are statically indeterminate externally and cannot be determined by the external application of the equations of static equilibrium. In these cases, once it has been determined that the overall structure is statically determinate, one proceeds directly to the free-body diagrams of the various parts of the structure that lead to the requisite number of equations for the combined solution of the reactions and the member forces.

By either approach, a free-body diagram may involve the isolation of a joint of the structure, which is a simple extension of the method of joints that is used in planar truss analysis, or it may treat any convenient portion of the structure that allows the analyst to isolate and solve for the desired unknown forces. There are several special theorems that result from considering a free-body diagram of a joint under prescribed conditions of member arrangement and loading. The first theorem follows from Fig. 4.14 and is given as follows:

1. If a joint is arranged so that all members except one lie in the same plane, then the force component normal to the plane in the one member not in the plane is equal to the resultant force component normal to the plane of the external loads applied at the joint.

The remaining theorems are actually corollaries to the first:

2. For the same arrangement of bars, if the resultant normal force component of the external loads at the joint is zero, then the force in the member not in the plane is zero.

3. At a joint where three noncoplanar members meet, and at which no external loads act, all members have zero force unless two of the members are colinear.

4. If all but two of the members framing into a joint have zero force, and these two members are not colinear, and if, further, there are no external forces acting at the joint, then the force in each of these two bars is zero.

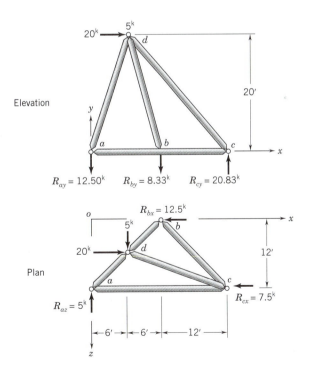

Figure 4.14 Special arrangement of members at a joint.

It is not necessary that these theorems be employed in the analysis process; however, their use frequently simplifies the analysis by permitting certain bar forces to be determined by inspection.

EXAMPLE 4.6

Determine the complete set of bar forces for the structure and loading shown. The reactions that were previously determined in Example 3.10 are given.

Structure Classification

$$j = 4; \quad r_a = 6; \quad m_a = 6$$

$$m = 3j - r_a = 12 - 6 = 6$$

$m_a = m$ ∴ Structure is statically determinate. Arrangement is stable.

Determination of Member Forces

It is convenient to arrange the geometric components and the force components in a table that permits easy access to the data as they are needed. The geometric data are entered in the table immediately with the aid of Fig. 4.13a, whereas the force data are added as the analysis proceeds. At the conclusion of the analysis, the complete solution is within the table.

Member	Geometry (ft)				Force (kips)			
	x	y	z	L	X	Y	Z	F
ab	12.0	0	12.0	17.0	−8.75	0	−8.75	−12.40
bc	12.0	0	12.0	17.0	+6.25	0	+6.25	+8.85
ac	24.0	0	0	24.0	+5.00	0	0	+5.00
ad	6.0	20.0	6.0	21.7	+3.75	+12.50	+3.75	+13.56
bd	6.0	20.0	6.0	21.7	+2.50	+8.33	+2.50	+9.04
cd	18.0	20.0	6.0	27.6	−18.75	−20.83	−6.25	−28.75

Isolate joint a as FBD:

$$\sum P_y = 0 \;\uparrow+ \;;\qquad Y_{ad} - 12.50 = 0; \quad Y_{ad} = +12.50^k$$

$$X_{ad} = Y_{ad}\left(\frac{x}{y}\right)_{ad} = (+12.50)\left(\frac{6}{20}\right) = +3.75^k$$

$$Z_{ad} = Y_{ad}\left(\frac{z}{y}\right)_{ad} = (+12.50)\left(\frac{6}{20}\right) = +3.75^k \Bigg\} \;(4.6)$$

$$F_{ad} = Y_{ad}\left(\frac{L}{y}\right)_{ad} = (+12.50)\left(\frac{21.7}{20}\right) = +13.56^k$$

These results are added to the table for member ad.

$$\sum P_z = 0 \;\uparrow+ \;;\qquad 5 + Z_{ad} + Z_{ab} = 0$$

$$Z_{ab} = -5 - 3.75 = -8.75^k$$

$$X_{ab} = Z_{ab}\left(\frac{x}{z}\right)_{ab} = (-8.75)\left(\frac{12}{12}\right) = -8.75^k$$

$$Y_{ab} = Z_{ab}\left(\frac{y}{z}\right)_{ab} = (-8.75)\left(\frac{0}{12}\right) = 0 \Bigg\} \;(4.6)$$

$$F_{ab} = Z_{ab}\left(\frac{L}{z}\right)_{ab} = (-8.75)\left(\frac{17.0}{12}\right) = -12.40^k$$

$$\sum P_x = 0 \;\rightarrow \;;\qquad X_{ad} + X_{ab} + X_{ac} = 0$$

$$X_{ac} = -3.75 + 8.75 = 5.00^k$$

$$Y_{ac} = X_{ac}\left(\frac{y}{x}\right)_{ac} = (+5.00)\left(\frac{0}{24}\right) = 0$$

$$Z_{ac} = X_{ac}\left(\frac{z}{x}\right)_{ac} = (+5.00)\left(\frac{0}{24}\right) = 0 \Bigg\} \;(4.6)$$

$$F_{ac} = X_{ac}\left(\frac{L}{x}\right)_{ac} = (+5.00)\left(\frac{24}{24}\right) = +5.00^k$$

These results are added to the table for members ab and ac.

Isolate joint b as FBD:

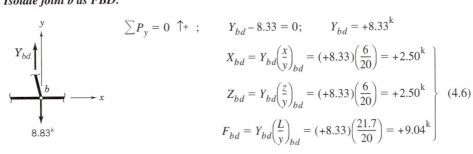

$\sum P_y = 0 \ \uparrow +$; $\qquad Y_{bd} - 8.33 = 0; \qquad Y_{bd} = +8.33^k$

$$X_{bd} = Y_{bd}\left(\frac{x}{y}\right)_{bd} = (+8.33)\left(\frac{6}{20}\right) = +2.50^k$$

$$Z_{bd} = Y_{bd}\left(\frac{z}{y}\right)_{bd} = (+8.33)\left(\frac{6}{20}\right) = +2.50^k$$

$$F_{bd} = Y_{bd}\left(\frac{L}{y}\right)_{bd} = (+8.33)\left(\frac{21.7}{20}\right) = +9.04^k$$

(4.6)

$\sum P_x = 0 \ \xrightarrow{+}$; $\qquad X_{bc} - X_{ba} - X_{bd} - 12.50 = 0$

$$X_{bc} = 12.50 - 8.75 + 2.50 = +6.25^k$$

$$Y_{bc} = X_{bc}\left(\frac{y}{x}\right)_{bc} = (+6.25)\left(\frac{0}{12}\right) = 0$$

$$Z_{bc} = X_{bc}\left(\frac{z}{x}\right)_{bc} = (+6.25)\left(\frac{12}{12}\right) = +6.25^k$$

$$F_{bc} = X_{bc}\left(\frac{L}{x}\right)_{bc} = (+6.25)\left(\frac{17.0}{12}\right) = +8.85^k$$

(4.6)

These results are added to the table for members *bd* and *bc*. As a check, consider $\Sigma P_z = 0 \ \downarrow +$

$$Z_{ba} + Z_{bd} + Z_{bc} = 0$$

$$-8.75 + 2.50 + 6.25 = 0$$

Isolate joint c as FBD:

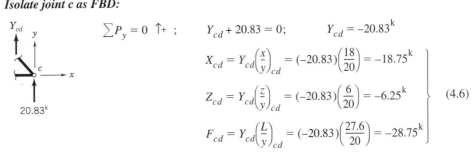

$\sum P_y = 0 \ \uparrow +$; $\qquad Y_{cd} + 20.83 = 0; \qquad Y_{cd} = -20.83^k$

$$X_{cd} = Y_{cd}\left(\frac{x}{y}\right)_{cd} = (-20.83)\left(\frac{18}{20}\right) = -18.75^k$$

$$Z_{cd} = Y_{cd}\left(\frac{z}{y}\right)_{cd} = (-20.83)\left(\frac{6}{20}\right) = -6.25^k$$

$$F_{cd} = Y_{cd}\left(\frac{L}{y}\right)_{cd} = (-20.83)\left(\frac{27.6}{20}\right) = -28.75^k$$

(4.6)

These results are added to the table for member *cd*. As a check, consider $\Sigma P_x = 0 \ \xrightarrow{+}$

$$-X_{ca} - X_{cd} - X_{cb} - 7.5 = 0$$

$$-5.00 + 18.75 - 6.25 - 7.5 = 0$$

Also, as a check, consider $\Sigma P_z = 0 \ \uparrow +$

$$Z_{cd} + Z_{cb} = 0$$

$$-6.25 + 6.25 = 0$$

Isolate joint d as FBD:

Here, we check, $\Sigma P_x = 0, \Sigma P_y = 0, \Sigma P_z = 0$ and find that these conditions are satisfied.

EXAMPLE 4.7

Determine the reactions and member forces for the structure and loading shown. The support mechanisms include ball supports at c and d, a roller at a, and a universal joint at b. The unknown reaction components associated with these support conditions are shown on the sketch.

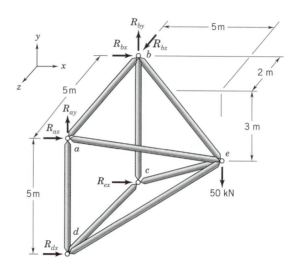

Structure Classification

External: $r_a = 7 > 6$; thus, structure is statically indeterminate externally.

Overall: $j = 5$; $r_a = 7$; $m_a = 8$; $m = 3j - r_a = 15 - 7 = 8$.
$m_a = m$; thus, structure is statically determinate. Arrangement is stable.

Analysis

One could write three equations of equilibrium at each joint and solve the resulting 15 equations for the 15 unknowns (7 reactions and 8 member forces).

Reactions

Proceed as if R_{ax} were known, and solve for all other reaction components in terms of R_{ax}.

$$\sum M_{zd} = 0 \quad \overset{c}{\underset{d}{\curvearrowleft}}{}_{+} \quad -(R_{ax} \times 5) - (R_{bx} \times 5) - (50 \times 5) = 0$$
$$R_{bx} = -(50 + R_{ax})$$

$$\sum M_{yd} = 0 \quad \overset{a}{\underset{d\,|+}{\curvearrowright}} \quad -(R_{bx} \times 5) - (R_{cx} \times 5) = 0$$
$$R_{cx} = -R_{bx} = 50 + R_{ax}$$

$$\sum P_x = 0 \quad \overset{+}{\rightarrow} \quad R_{ax} + R_{bx} + R_{cx} + R_{dx} = 0$$
$$R_{ax} - (50 + R_{ax}) + (50 + R_{ax}) + R_{dx} = 0$$
$$R_{dx} = -R_{ax}$$

$$\sum M_{xb} = 0 \quad \overset{+}{\underset{b}{\curvearrowleft}} \quad (50 \times 2) - (R_{ay} \times 5) = 0; \quad R_{ay} = \underline{\underline{20 \text{ kN}}}$$

$$\sum P_z = 0 \quad \diagup \quad R_{bz} = \underline{\underline{0}}$$

$$\sum P_y = 0 \quad \uparrow + \quad R_{ay} + R_{by} - 50 = 0; \quad R_{by} = 50 - 20 = \underline{\underline{30 \text{ kN}}}$$

The actual value of R_{ax} is determined by establishing the force in member ab at joints a and b and equating the results (see calculations immediately following table).

Member Forces

Member	Geometry (m)				Force (kN)			
	x	y	z	L	X	Y	Z	F
ab	0	0	5.0	5.0	0	0	−12.0	−12.0
bc	0	5.0	0	5.0	0	+12.0	0	+12.0
cd	0	0	5.0	5.0	0	0	+12.0	+12.0
da	0	5.0	0	5.0	0	+8.0	0	+8.0
ae	5.0	3.0	3.0	6.56	+20.0	+12.0	+12.0	+26.24
be	5.0	3.0	2.0	6.16	+30.0	+18.0	+12.0	+36.96
ce	5.0	2.0	2.0	5.74	−30.0	−12.0	−12.0	−34.44
de	5.0	2.0	3.0	6.16	−20.0	−8.0	−12.0	−24.64

Joint a: $X_{ae} = -R_{ax};$ $Z_{ae} = X_{ae}\left(\dfrac{z}{x}\right)_{ae} = (-R_{ax})\left(\dfrac{3}{5}\right) = -0.6R_{ax}$

$$Z_{ab} = -Z_{ae} = 0.6R_{ax}$$

Joint b: $X_{be} = -R_{bx} = (50 + R_{ax})$

$$Z_{be} = X_{be}\left(\dfrac{z}{x}\right)_{be} = (50 + R_{ax})\left(\dfrac{2}{5}\right) = 20 + 0.4R_{ax}$$

$$Z_{ba} = -Z_{be} = -(20 + 0.4R_{ax})$$

but

$$Z_{ab} = Z_{ba};\qquad 0.6R_{ax} = -20 + 0.4R_{ax};\qquad R_{ax} = \underline{\underline{-20 \text{ kN}}}$$

The reactions can now be summarized:

$$R_{ax} = -20 \text{ kN}$$
$$R_{bx} = -(50 + R_{ax}) = -30 \text{ kN}$$
$$R_{cx} = (50 + R_{ax}) = +30 \text{ kN}$$
$$R_{dx} = -R_{ax} = +20 \text{ kN}$$
$$R_{ay} = +20 \text{ kN}$$
$$R_{bz} = 0$$
$$R_{by} = +30 \text{ kN}$$

Continuing with the member forces:

$Z_{ab} = 0.6R_{ax} = -12 \text{ kN } (X_{ab}, Y_{ab}, F_{ab} \text{ in table from Eq. 4.6})$

$X_{ae} = -R_{ax} = -(-20) = +20 \text{ kN } (Y_{ae}, Z_{ae}, F_{ae} \text{ in table from Eq. 4.6})$

$X_{be} = (50 + R_{ax}) = (50 - 20) = +30 \text{ kN } (Y_{be}, Z_{be}, F_{be} \text{ in table from Eq. 4.6})$

Joint d: $X_{de} = -R_{dx} = -20$ kN (Y_{de}, Z_{de}, F_{de} in table from Eq. 4.6)

$Z_{dc} = -Z_{de} = -(-12) = +12$ kN (X_{dc}, Y_{dc}, F_{dc} in table from Eq. 4.6)

$Y_{da} = -Y_{de} = -(-8) = +8$ kN (X_{da}, Z_{da}, F_{da} in table from Eq. 4.6)

Joint c: $X_{ce} = -R_{cx} = -30$ kN (Y_{ce}, Z_{ce}, F_{ce} in table from Eq. 4.6)

$Y_{cb} = -Y_{ce} = -(-12) = +12$ kN (X_{cb}, Z_{cb}, F_{cb} in table from Eq. 4.6)

4.16 COMPLEX FRAMEWORKS

As was the case with planar trusses, there are space frameworks that are neither simple nor compound. For instance, consider the structure of Fig. 4.11c, which is redrawn in Fig. 4.15a along with the six reaction components that are needed for external stability. This structure is a simple framework that was constructed by starting with the basic tetrahedral unit *abcd* and systematically adding joints *e, f, g,* and *h* through the addition of three members to secure each new joint. A count of the members, joints, and reaction components ($m_a = 18, j = 8, r_a = 6$) and the application of Eq. 4.7 ($m = 3 \times 8 - 6 = 18$) reveals that $m_a = m$, and therefore the structure is statically determinate.

An examination of the framework shows that the diagonals slope in opposite directions on opposing faces of the cube. If, however, members *ae* and *cd* are removed and are replaced by members *gf* and *hb*, the framework shown in Fig. 4.15b results. This structure remains stable and statically determinate, but it is neither simple nor compound. Such a structure is referred to as a *complex framework*.

It is noted again that the treatment given here is brief. Much more could be stated regarding special cases and more complicated structures. Such topics are beyond the scope of this treatment, and the reader is encouraged to examine the works cited in Section 4.17.

4.17 REFERENCES

Norris, C. H., Wilbur, J. B., and Utku, S., *Elementary Structural Analysis,* 4th Ed., Chapter 4, McGraw-Hill, New York, 1991.

Timoshenko, S. P., and Young, D. H., *Theory of Structures,* 2nd Ed., Chapters 2 and 4, McGraw-Hill, New York, 1965.

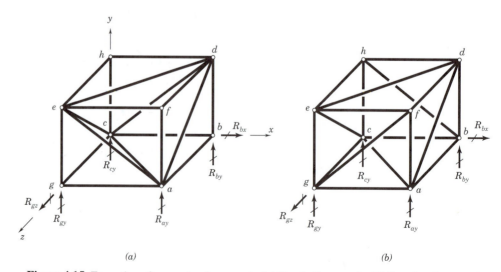

(a)   *(b)*

Figure 4.15 Formation of a complex framework. *(a)* Simple framework. *(b)* Complex framework.

4.18   SUGGESTED PROBLEMS

4.1 Classify each of the following structures with respect to statical determinacy and stability. Consider both external and internal criteria. In addition, for each stable truss, note whether it is simple, compound, or complex.

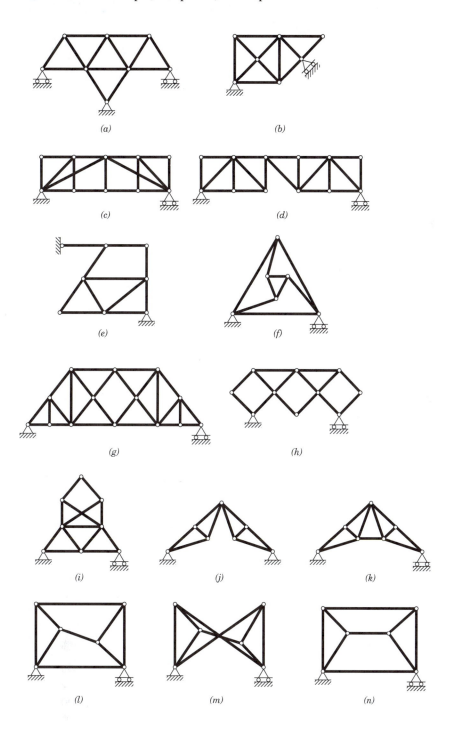

4.1 (*continued*)

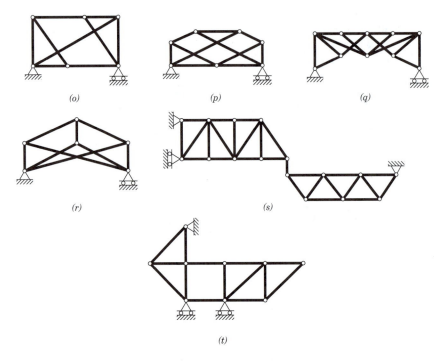

(o) (p) (q)

(r) (s)

(t)

4.2 through 4.26 Determine the bar forces for all members in the given trusses.

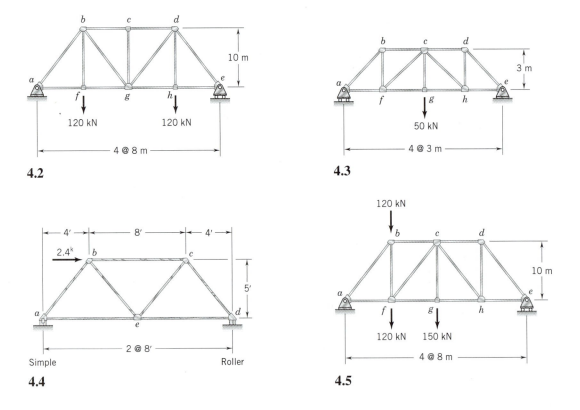

4.2

4.3

4.4

4.5

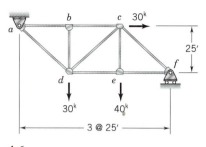

4.6

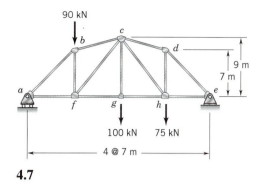

4.7

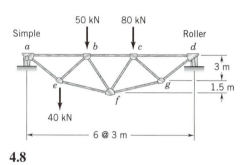

4.8

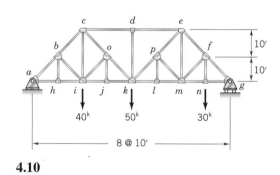

4.9

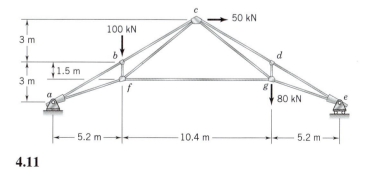

4.10

4.11

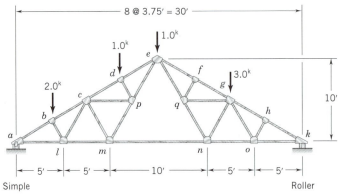

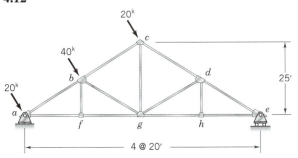

4.12

4.13

4.14

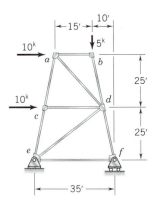

4.15

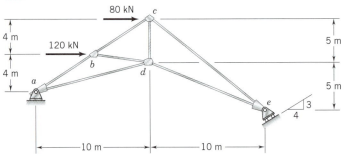

4.16

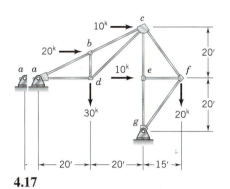

4.17

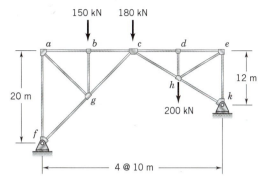

4.18

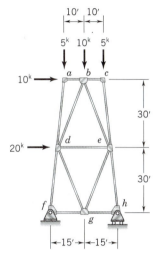

4.19

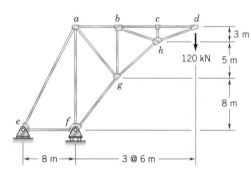

4.20

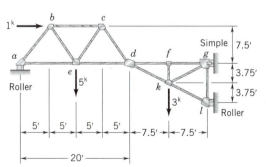

4.21

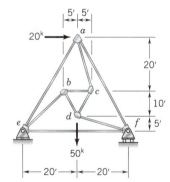

4.22

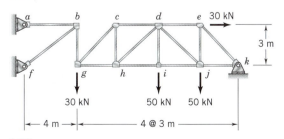

4.23

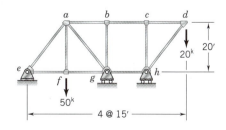

4.24

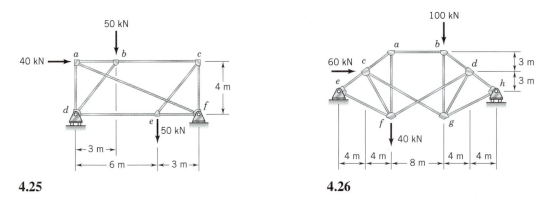

4.25

4.26

4.27 through 4.29 Determine the member forces in all members for each of the structures for the indicated loads.

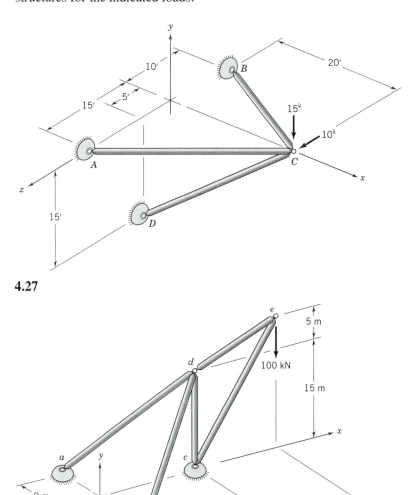

4.27

4.28

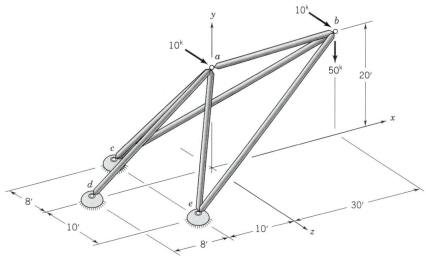

4.29

4.30 through 4.32   Determine the reactions and all member forces for the given structure with the indicated loading. Classify the structure with respect to statical determinacy and stability before proceeding with the analysis. The symbol ⊷ indicates an admissible reaction component.

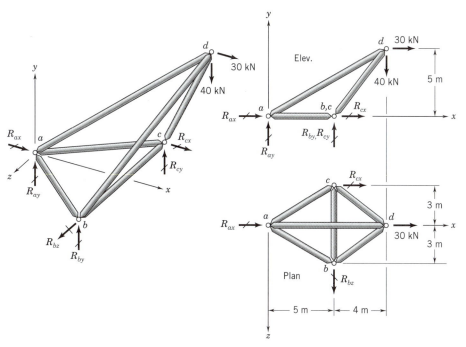

4.30

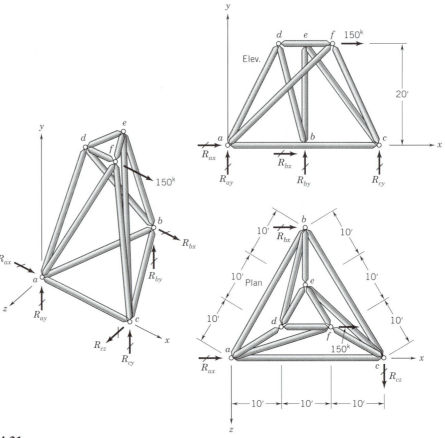

4.31

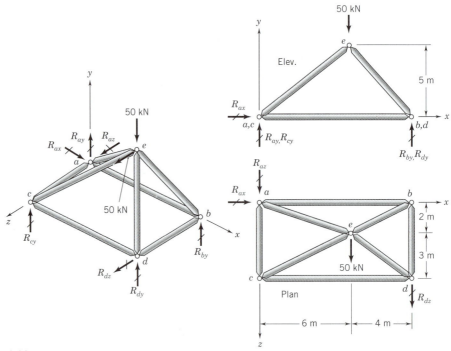

4.32

Chapter **5**

Member Forces
in Beams and Frames

John Hancock Center, Chicago, Illinois (courtesy of Skidmore, Owings & Merrill, architects/
engineers, photo by Ezra Stoller Esto).

5.1   BEAM AND FRAME STRUCTURES

In Chapter 4, we studied the internal equilibrium of pin-connected or truss structures in which all members were subjected to a pure axial force. The assumptions that had to be satisfied for a structure to behave as a truss were enumerated in Section 4.2. If any of these restrictive assumptions are relaxed, the constituent members are subjected to flexure in addition to axial force, and the structure collectively supports the applied loads by a mechanism that is more complicated than that of a truss. Structures of this type are classified as beam and frame structures, and the members are flexural members. As was the case with trusses, we will limit our consideration to stable structures in which relative motion within the structure is limited to that which results from the deformation of members.

Most beam and frame structures are three-dimensional, and a complete analysis should recognize this fact. In many cases, however, it is appropriate to consider portions of the structural system as planar substructures. In this chapter, the emphasis is on the analysis of plane structures or planar substructures of more complicated systems. Nonplanar beam and frame structures are treated briefly at the end of the chapter and in Part Five of the text.

Planar beam and frame structures may be statically classified according to the usual practice. That is, if all the reaction and member-end forces can be determined by employing the laws of statics, the structure is statically determinate; however, if it is necessary to employ conditions of compatibility, the structure is statically indeterminate. As was true with trusses, an indeterminate structure can result from redundant reaction components, redundant internal force components, or both. Similarly, instabilities can result from an inadequate number of independent reaction components or internal force components. The topics of determinacy and stability are treated in Section 5.4.

This chapter focuses on the determination of the internal forces to which a member is subjected at any point along its length. In most cases, the structures treated here are statically determinate, but a few indeterminate structures are considered. In either case, once the member-end forces are determined, the technique for studying the internal forces is basically the same.

5.2   INTERNAL FORCES FOR FLEXURAL MEMBERS

With flexural members, it is not sufficient to know only the member-end forces because the member forces may vary along the length of the member. If the internal member forces are desired at some specified point, it is necessary to cut the structure at that point and determine the forces that must be applied at the cut surface to equilibrate the resulting free-body diagrams. Consider the beam ab in Fig. 5.1 and assume that it is desired to determine the internal forces at point c. If the member is cut at point c and the equations of static equilibrium are applied to portion ac, the internal forces at point c are found. That is, the summation of vertical forces reveals that a transverse force V is required, the summation of horizontal forces leads to the axial force F, and the summation of moments shows that a bending moment M is required— all at point c. If portion cb is treated as a free body, similar forces, $V, F,$ and $M,$ are found to be required at point c. For segment $ac,$ the equilibrating forces at point c are provided by portion $cb,$ and vice versa. Thus, the forces on each side of point c must be equal and opposite. As the two portions are merged, the internal forces cancel and we are left with the original structure and loading.

Directing attention again to the internal forces, we define the force F as the *axial force,* the transverse force V as the *shear force,* and the moment M as the *bending moment.*

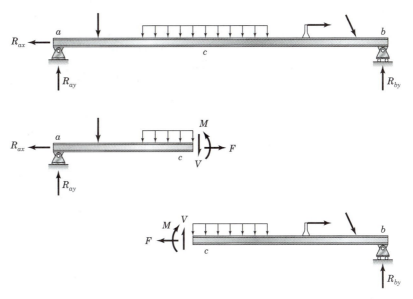

Figure 5.1 Internal member forces.

As has been noted, our present concern is with planar structures, and therefore the internal forces are limited to those shown in Fig. 5.1. For nonplanar structures, one must include a torsional moment about the longitudinal member axis and a bending action in the plane normal to the page. Figure 5.2 shows the torsion as T and the shear and bending moment as V' and M', respectively, for flexural action in the xz plane. These effects are considered more fully in subsequent sections.

5.3 NOTATION AND SIGN CONVENTION

The notation that will be used to represent the internal forces in flexural members was introduced in the previous section. That is, the axial force is represented by F, the transverse shear force by V, and the bending moment by M. In certain instances, these symbols carry a subscript to identify the point on the structure to which they correspond.

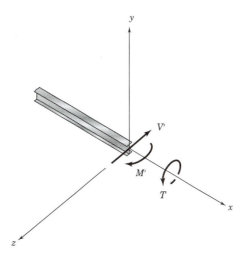

Figure 5.2 Torsion and out-of-plane bending

A sign convention must be adopted regarding these internal forces. Two different approaches are used; they are totally equivalent. In one case, the forces acting on each side of a cut section at a given point are considered, whereas in the other case, the forces acting on an infinitesimal element at the point in question are observed. The two methods of isolating point *a* are shown in Fig. 5.3*a*.

For axial force, tension is taken to be positive, as was the case with trusses. This condition is illustrated in Fig. 5.3*b* for both the section and the element representations. In either case, it is seen that positive axial force is represented by a vector directed away from the face on which it acts. For the element, positive axial force tends to elongate the element.

Positive shear force is shown in Fig. 5.3*c* by both representations. In either case, it is seen that positive shear acts downward on an exposed right-hand face, whereas it acts upward on an exposed left-hand face. Positive shear tends to deform an element by "shearing" the right side downward relative to the left side.

Positive moment is illustrated in Fig. 5.3*d*. As shown by either representation, positive moment places the top portion of a cut face in compression and the bottom portion in tension. Under positive moment, the element tends to deform so as to be concave upward.

For members that are not oriented horizontally, the member is viewed normal to its longitudinal axis, and the same conventions are applied.

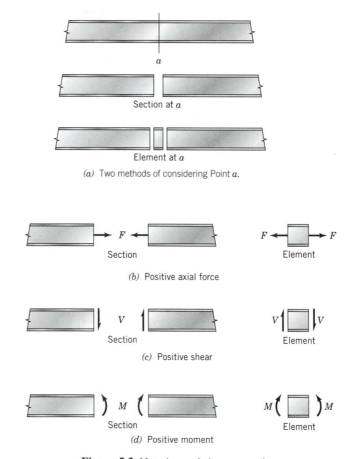

(*a*) Two methods of considering Point *a*.

(*b*) Positive axial force

(*c*) Positive shear

(*d*) Positive moment

Figure 5.3 Notation and sign convention.

5.4   STATICAL DETERMINACY AND STABILITY OF BEAM AND FRAME STRUCTURES

Before the analysis of a beam or frame structure can proceed, it is necessary to establish criteria for its statical classification. This section considers planar structures only, and the treatment is an extension of the procedures presented in Section 4.8 for planar trusses.

First, the structure can be tested for external determinacy and stability according to the criteria developed in Chapter 3. According to that approach, if r is taken as the least number of independent reaction components that are required for external stability and r_a is the actual number of independent reaction components, then the following criteria hold:

$r_a < r$; structure is statically unstable externally

$r_a = r$; structure is statically determinate externally

$r_a > r$; structure is statically indeterminate externally

The reader is reminded that the conditions $r_a \geq r$ are necessary but not sufficient conditions for the statical classification of the structure. The reaction components must be properly arranged to ensure stability. However, just as was true for trusses, statically determinate reaction components do not ensure that all of the member forces can be determined from statics.

It is not particularly useful to classify beam-type structures internally, as was done with trusses; however, an overall statical classification, which includes both internal and external considerations, is useful. Here, it is necessary to note that for any beam-type member, the internal forces are completely defined throughout the member once the axial force, shear force, and bending moment are determined at any point along the member. Thus, three member actions must be determined for each member of the structure. Thus, if m_a is taken to represent the actual number of members, and r_a continues to represent the actual number of independent reaction components, then there are $(3m_a + r_a)$ unknown quantities that must be determined.

To determine these unknown quantities, Eq. 2.5 can be used to generate three equations of equilibrium at each joint of the structure. In addition, there may be equilibrium equations that can be written by virtue of conditions of construction. Thus, if there are j joints in the structure and n condition equations, then there are $(3j + n)$ equations available for the solution. Thus,

$$(3m_a + r_a) = (3j + n) \tag{5.1}$$

is the relationship that must be satisfied if a planar beam-type structure is to be statically determinate. This leads to the following criteria:

$(3m_a + r_a) < (3j + n)$; structure is statically unstable

$(3m_a + r_a) = (3j + n)$; structure is statically determinate

$(3m_a + r_a) > (3j + n)$; structure is statically indeterminate

Again, care must be used in applying these criteria. If $(3m_a + r_a) < (3j + n)$, the structure is definitely unstable; however, $(3m_a + r_a) \geq (3j + n)$ are necessary but not sufficient conditions for the classification of the structure. The individual members, the construction conditions, and the reaction components must be properly arranged to ensure stability of the structure.

In applying these criteria, it is recalled from Section 3.5 that for external classification, $r = 3 + n$. Here, n is the number of condition equations that can be written

Table 5.1 Statical Classification of Beam and Frame Structures

Structure	Structure Characteristics				External Classification		Overall Classification		
	j	n	m_a	r_a	$r = 3 + n$	Classification	$(3m_a + r_a)$	$(3j + n)$	Classification
(a)	4	1	3	4	$3 + 1 = 4$	$4 = 4$ Determinate, stable	$(9 + 4) = 13$	$(12 + 1) = 13$	$13 = 13$ Determinate, Stable
(b)	5	2	4	6	$3 + 2 = 5$	$6 > 5$ Indet., 1st degree, stable	$(12 + 6) = 18$	$(15 + 2) = 17$	$18 > 17$ Indet., 1st degree, stable
(c)	4	1	3	3	$3 + 1 = 4$	$3 < 4$ Unstable	$(9 + 3) = 12$	$(12 + 1) = 13$	$12 < 13$ Unstable
(d)	8	0	10	3	3	$3 = 3$ Determinate	$(30 + 3) = 33$	$(24 + 0) = 24$	$33 > 24$ Indet., 9th degree, stable
(e)	8	3 2 (Ext)	8	9	$3 + 2 = 5$	$9 > 5$ Indet., 4th degree, stable	$(24 + 9) = 33$	$(24 + 3) = 27$	$33 > 27$ Indet., 6th degree, stable
(f)	5	1 0 (Ext)	5	5	3	$5 > 3$ Indet., 2nd degree, stable	$(15 + 5) = 20$	$(15 + 1) = 16$	$20 > 16$ Indet., 4th degree, stable

that will involve the external reaction components only at the exclusion of any member forces. However, the n value used for the overall classification in Eq. 5.1 includes all the condition equations that can be written throughout the structure, some of which will be related to internal quantities.

Several structures are classified in Table 5.1 with regard to both the external criteria and the overall criteria. Here, if n for the external classification is different from n for the overall classification, it is so noted. A comparison of the external classification and the overall classification will indicate the degree of internal indeterminacy. For instance, case e is statically indeterminate externally to the fourth degree, and it has an overall indeterminacy to the sixth degree. Thus, the structure is statically indeterminate internally to the second degree.

The construction detail at the center support of case f introduces one condition equation. The pin ensures that there is no support moment; however, a condition of zero moment must be specified at one member end in order to ensure the intended pin end for each member. In fact, if there are p members that frame into a common pin support, $(p-1)$ member-end conditions must be introduced. Also, note that the free end of a cantilever beam must be treated as a joint. This point is illustrated in the classification of case a.

5.5   DETERMINATION OF INTERNAL FORCES

When the internal forces are to be determined at a given point in a beam or frame structure, the same general technique is employed that was used in trusses. That is, a free-body diagram is taken that cuts through the structure at the desired location, and each unknown internal force is assumed to act in the positive direction. The analysis proceeds, and if it produces a positive value for the unknown, then the assumed direction is correct and the internal force is indeed positive. If, however, the analysis yields a negative value for the unknown, then the assumed direction is incorrect and the internal force is negative.

In this approach a separate free-body diagram must be considered for each point on the structure where the internal forces are desired. The point may be explicitly defined, or it may be given a variable location. The analysis for each case yields the desired internal forces at that cut section.

EXAMPLE 5.1

Determine the internal forces (axial force, shear, and bending moment) at points c and f for the beam structure shown. The structure is statically determinate and stable, and the reactions are given.

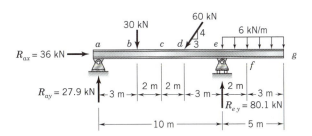

Point c: Isolate segment ac as a FBD.

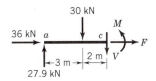

$$\sum P_x = 0 \xrightarrow{+} ; \qquad 36 + F = 0; \qquad F = \underline{-36\,\text{kN}}$$

$$\sum P_y = 0 \uparrow+ ; \qquad 27.9 - 30 - V = 0; \qquad V = \underline{-2.1\,\text{kN}}$$

$$\sum M_{zc} = 0 \;+\circlearrowright; \quad M + (30 \times 2) - (27.9 \times 5) = 0; \qquad M = \underline{79.5\,\text{kN}\cdot\text{m}}$$

The negative signs associated with F and V indicate that these forces are directed in the opposite senses from those assumed and are thus negative by our sign convention.

Point f: The segment fg is taken as a FBD.

$$\sum P_x = 0 \xrightarrow{+} ; \qquad F = \underline{0\,\text{kN}}$$

$$\sum P_y = 0 \uparrow+ ; \qquad V - (6 \times 3) = 0; \qquad V = \underline{18\,\text{kN}}$$

$$\sum M_{zf} = 0; \;\circlearrowleft+; \qquad M + (6 \times 3) \times 1.5 = 0; \qquad M = \underline{-27\,\text{kN}\cdot\text{m}}$$

Again, the negative sign for M indicates that the moment is opposite from that assumed and thus negative for our sign convention.

It should be noted that segment af could have been taken as a free-body diagram, and the same internal forces at f would have resulted.

EXAMPLE 5.2

Use the x coordinate shown to write general expressions for the axial force, shear, and bending moment for regions bd and eg for the beam structure of Example 5.1.

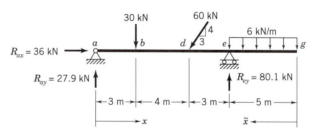

Region bd: In this region, x varies from 3 m to 7 m.

$$\sum P_x = 0 \xrightarrow{+} ; \qquad F(x) + 36 = 0; \qquad \underline{F(x) = -36\,\text{kN}}$$

$$\sum P_y = 0 \uparrow+ ; \qquad 27.9 - 30 - V(x) = 0; \qquad \underline{V(x) = -2.1\,\text{kN}}$$

$$\sum M_z = 0 \;\circlearrowright+; \qquad 27.9x - 30(x-3) - M(x) = 0;$$

$$\underline{M(x) = -2.1x + 90\,\text{kN}\cdot\text{m}}$$

Substitution for $x = 5$ m yields the specific values for point c as determined in Example 5.1.

Region *eg*: In this region, x varies from 10 m to 15 m.

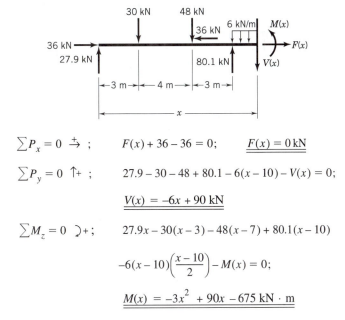

$$\sum P_x = 0 \ \xrightarrow{+} \ ; \qquad F(x) + 36 - 36 = 0; \qquad \underline{F(x) = 0 \, kN}$$

$$\sum P_y = 0 \ \uparrow+ \ ; \qquad 27.9 - 30 - 48 + 80.1 - 6(x-10) - V(x) = 0;$$

$$\underline{V(x) = -6x + 90 \, kN}$$

$$\sum M_z = 0 \ )+ ; \qquad 27.9x - 30(x-3) - 48(x-7) + 80.1(x-10)$$

$$-6(x-10)\left(\frac{x-10}{2}\right) - M(x) = 0;$$

$$\underline{M(x) = -3x^2 + 90x - 675 \, kN \cdot m}$$

Substitution for $x = 12$ m produces the specific values for point f that were found in Example Problem 5.1.

As an alternate procedure for region *eg,* use an $\tilde{x}$ coordinate measured to the left from point g (see original figure), where $\tilde{x}$ ranges from 0 m to 5 m.

$$\sum P_x = 0 \ \xrightarrow{+} \ ; \qquad \underline{F(\tilde{x}) = 0 \, kN}$$

$$\sum P_y = 0 \ \uparrow+ \ ; \qquad V(\tilde{x}) - 6\tilde{x} = 0; \qquad \underline{V(\tilde{x}) = 6\tilde{x} \, kN}$$

$$\sum M_z = 0 \ )+ ; \qquad M(\tilde{x}) + 6\tilde{x}\left(\frac{\tilde{x}}{2}\right) = 0; \qquad \underline{M(\tilde{x}) = -3\tilde{x}^2 \, kN \cdot m}$$

Substitution for $\tilde{x} = 3$ m gives the values determined for point f in Example Problem 5.1.

If $\tilde{x} = 15 - x$ is substituted into these equations, the original equations for region *eg* in terms of x are determined.

5.6   RELATIONSHIPS BETWEEN LOAD, SHEAR, AND BENDING MOMENT

The procedure outlined in Section 5.5 can be used to determine the internal forces at as many points as are desired. However, there is a more efficient technique for determining the shears and bending moments associated with transverse loading.

To develop the necessary relationships, consider the beam shown in Fig. 5.4*a*, which is subjected to a distributed transverse load of varying intensity $p(x)$. We have already introduced a sign convention for shear and moment, and we now add to this the convention that upward load is positive and that x increases from left to right. An element of the beam, Δx in length, is isolated as a free-body diagram in Fig. 5.4*b*. On the left-hand side of the element, the load intensity is $p(x)$ and the corresponding shear and moment are $V(x)$ and $M(x)$, respectively. These functions are dependent on x as indicated; however, for convenience they are represented by p, V, and M as shown in Fig. 5.4b. Since these functions vary with x, there may be incremental changes over the distance Δx—hence, on the right-hand side of the element, the respective values of load, shear, and moment are $(p + \Delta p)$, $(V + \Delta V)$, and $(M + \Delta M)$. The load, which varies from p to $(p + \Delta p)$ over the length Δx, has an average value of p_a, and the resultant load of $(p_a \cdot \Delta x)$ acts at a distance of $(\lambda \cdot \Delta x)$ from the right-hand face of the element.

The desired relationships result from the application of the equations of static equilibrium to the element of Fig. 5.4b. Horizontal equilibrium is automatically satisfied. The summation of forces in the vertical direction yields

$$V + (p_a \cdot \Delta x) - (V + \Delta V) = 0$$

from which

$$\frac{\Delta V}{\Delta x} = p_a$$

In the limiting process, as Δx approaches zero, the load intensity p_a approaches $p(x) = p$, and the resulting differential equation is

$$\frac{dV}{dx} = p \tag{5.2}$$

The summation of moments about a z axis through point O gives

$$M + (V \cdot \Delta x) + (p_a \cdot \Delta x)(\lambda \cdot \Delta x) - (M + \Delta M) = 0$$

or

$$\frac{\Delta M}{\Delta x} = V + (p_a \cdot \lambda \cdot \Delta x)$$

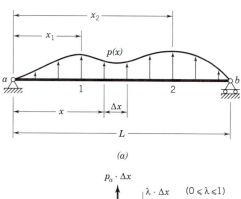

(a)

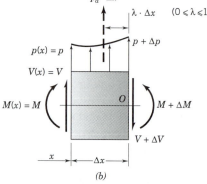

(b)

Figure 5.4 Statics of beam element.

As Δx approaches zero, the term $(p_a \cdot \lambda \cdot \Delta x)$ vanishes, and the resulting differential equation is

$$\frac{dM}{dx} = V \tag{5.3}$$

Equations 5.2 and 5.3 are differential equations of equilibrium that must be satisfied at any point along the beam length. Substitution of Eq. 5.3 into Eq. 5.2 gives

$$\frac{d^2 M}{dx^2} = p \tag{5.4}$$

which relates the bending moment at any section to the load intensity at that section.

Equation 5.2 can be written in the form

$$dV = p \, dx \tag{5.5}$$

which upon integration gives

$$V = \int p \, dx + C_1 \tag{5.6}$$

Equation 5.6 gives the expression for shear at any point along the beam. The constant of integration, C_1, that results from the indicated integration is evaluated by using the boundary conditions on shear. If Eq. 5.5 is integrated between sections 1 and 2, as shown in Fig. 5.4a, then we have

$$V_2 - V_1 = \Delta V_{1-2} = \int_{x_1}^{x_2} p \, dx \tag{5.7}$$

where ΔV_{1-2} is the *change* in shear between sections 1 and 2.

If Eq. 5.3 is written in the form

$$dM = V \, dx \tag{5.8}$$

then integration will yield

$$M = \int V \, dx + C_2 \tag{5.9}$$

This gives an expression for bending moment throughout the beam, and the resulting constant of integration, C_2, is determined by using the boundary conditions on moment. Integrating Eq. 5.8 between the limits of sections 1 and 2 gives

$$M_2 - M_1 = \Delta M_{1-2} = \int_{x_1}^{x_2} V \, dx \tag{5.10}$$

where ΔM_{1-2} is the *change* in moment between sections 1 and 2.

5.7   SHEAR AND BENDING MOMENT DIAGRAMS

For any flexural member, once the load is specified, it is possible to use the relationships developed in Section 5.6 to construct graphs in which the shear and moment are plotted as ordinates against x as abscissa. The resulting graphs are called *shear and moment diagrams*. These diagrams are extremely useful because they simultaneously give the values for shear and moment at any point along the member, thus obviating the need to determine these functions point by point as was demonstrated in Section 5.5.

Two methods are customarily used to construct shear and moment diagrams. Each approach is illustrated with a simple example problem in this section, and further examples are provided in greater detail in Section 5.10.

The first method is based on Eqs. 5.6 and 5.9, in which formal integration leads to mathematical expressions for the shear and moment diagrams. This method is most appropriate when the load intensity is represented by a continuous mathematical function that is readily integrated.

The second method is based on the systematic application of Eqs. 5.2, 5.3, 5.7, and 5.10. Each of these equations contributes to the construction of the shear and moment diagrams; the role of each is as follows:

Eq. 5.2: The slope of the shear diagram at any point is given by the load intensity at that point.

Eq. 5.3: The slope of the moment diagram at any point is given by the value of the shear at that point.

Eq. 5.7: The change in shear between two points is equal to the area under the load intensity diagram between these two points.

Eq. 5.10: The change in moment between two points is equal to the area under the shear diagram between these two points.

In this method, changes in shear and moment are determined by incrementally integrating the load and shear diagrams. These changes are accumulated to determine key points on the shear and moment diagrams.

EXAMPLE 5.3

Construct the load, shear, and moment diagrams by the formal integration method for the simply supported beam under the triangular load shown. The reactions are given.

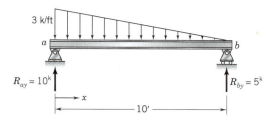

Transverse Load Diagram

The load intensity, p (k/ft), is taken directly from the given problem. The equation for a straight line is employed ($p = mx + b$, where m is the slope and b is the intercept). The end reactions P (k) are shown as concentrated loads.

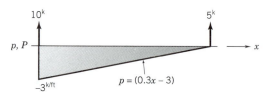

Shear Diagram

The shear diagram is determined from Eq. 5.6, where

$$V = \int p \, dx + C_1 = \int (0.3x - 3) \, dx + C_1 = 0.15x^2 - 3x + C_1$$

The constant of integration, C_1, is determined from the boundary conditions. In this case, for a section an infinitesimal distance right of point a, $V(x = 0) = +10$ (see figure below). Therefore,

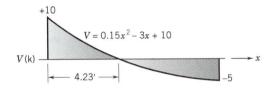

$$V(x = 0) = 0.15(0)^2 - 3(0) + C_1 = + 10$$
$$\therefore C_1 = + 10$$

and thus

$$V = 0.15x^2 - 3x + 10$$

The shear diagram is given by

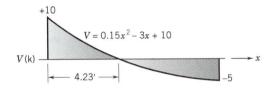

Substitution for $x = 10$ gives $V(x = 10) = -5$, which verifies the boundary condition on the shear at point b.

The point of zero shear is determined from

$$V = 0.15x^2 - 3x + 10 = 0$$

from which $x = 4.23$ ft.

Moment Diagram

The moment diagram is determined from Eq. 5.9, where

$$M = \int V\,dx + C_2 = \int (0.15x^2 - 3x + 10)dx + C_2$$
$$= 0.05x^3 - 1.5x^2 + 10x + C_2$$

The constant of integration, C_2, is determined from the boundary conditions. In this case, at point a, $M(x = 0) = 0$ (see figure below), and thus

$$M(x = 0) = 0.05(0)^3 - 1.5(0)^2 + 10(0) + C_2 = 0$$
$$\therefore C_2 = 0$$

and therefore,

$$M = 0.05x^3 - 1.5x^2 + 10x$$

The moment diagram is given by

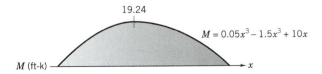

Substitution for $x = 10$ gives $M(x = 0) = 0$, which verifies the boundary condition on the moment at point b.

Since $dM/dx = V$, the point of zero shear corresponds to the point of maximum moment. Thus, the maximum moment occurs at $x = 4.23$ ft and is given by

$$M_{max} = 0.05(4.23)^3 - 1.5(4.23)^2 + 10(4.23) = +19.24 \text{ ft-k}$$

The shear and moment can be determined at any x by substituting the value of x in the given formulae.

EXAMPLE 5.4

Construct the load, shear, and moment diagrams for the beam shown under the given loading using the incremental change method. The reactions are given.

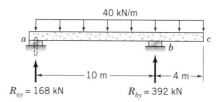

Transverse Load Diagram

The applied distributed load intensity, p (kN/m), is taken directly from the problem statement. The reactions are shown as concentrated loads, P (kN), and these produce load intensities of positive infinity.

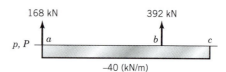

Shear Diagram

The shear diagram is determined by considering the boundary conditions on shear for each span and applying the changes in shear from Eq. 5.7. The slopes of the shear diagram are controlled by Eq. 5.2.

Point a:

$$V_a = +168 \text{ kN}$$

$R_{ay} = 168$ kN

a to b:

$$\Delta V_{a-b} = \left\{ \begin{array}{c} \text{Area under load diagr.} \\ \text{between } a \text{ \& } b \end{array} \right\} = \left\{ -40\frac{\text{kN}}{\text{m}} \times 10 \text{ m} \right\} = -400 \text{ kN}$$

$$\frac{dV}{dx} = p = -40\frac{\text{kN}}{\text{m}} \text{ throughout}$$

Just left of point b:

$$V_{b,\,\text{left}} = V_a + \Delta V_{a-b} = 168 - 400 = -232 \text{ kN}$$

Just right of point b:

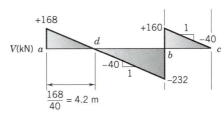

$$V_{b,\,\text{right}} = 392 - 232 = 160 \text{ k}$$

b to c:

$$\Delta V_{b-c} = \left\{ \begin{array}{c} \text{Area under load diagr.} \\ \text{between } b \ \& \ c \end{array} \right\} = \{-40 \times 4\} = -160 \text{ kN}$$

$$\frac{dV}{dx} = p = -40 \ \frac{\text{kN}}{\text{m}} \text{ throughout}$$

Point c:

$$V_c = V_{b,\,\text{right}} + \Delta V_{b-c} = 160 - 160 = 0$$

Thus, the shear diagram takes the following form:

Note: The slope of the shear diagram is positive infinity (vertical) at points a and b where the concentrated reactions occur (infinite load intensity).

Moment Diagram

The moment diagram is determined by considering the boundary conditions on moment for each span and applying the changes in moment from Eq. 5.10. The slopes of the moment diagram are controlled by Eq. 5.3.

Point a: $M_a = 0$ (simple support)

a to d: (see shear diagram for point d)

$$\Delta M_{a-d} = \left\{ \begin{array}{c} \text{Area under shear diagr.} \\ \text{between } a \ \& \ d \end{array} \right\} = \tfrac{1}{2}[168 \text{ kN} \times 4.2 \text{ m}]$$

$$= 352.8 \text{ kN} \cdot \text{m}$$

$$\frac{dM}{dx} = V \ \therefore \ \text{Slope varies from} + 168 \text{ kN to zero}$$

Point d:

$$M_d = M_a + \Delta M_{a-d} = 0 + 352.8 = +352.8 \text{ kN} \cdot \text{m}$$

d to b:

$$\Delta M_{d-b} = \left\{ \begin{array}{c} \text{Area under shear diagr.} \\ \text{between } d \text{ \& } b \end{array} \right\} = \tfrac{1}{2}[-232 \times 5.8]$$

$$= -672.8 \text{ kN} \cdot \text{m}$$

$$\frac{dM}{dx} = V \quad \therefore \text{ Slope varies from zero to } -232 \text{ kN}$$

Just left of point b:

$$M_{b,\,\text{left}} = M_d + \Delta M_{d-b} = +352.8 - 672.8$$

$$= -320 \text{ kN} \cdot \text{m}$$

Just right of point b:

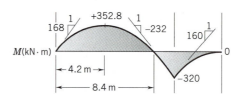

$$M_{b,\text{right}} = -320 \text{ kN} \cdot \text{m}$$

b to c:

$$\Delta M_{b-c} = \left\{ \begin{array}{c} \text{Area under shear diagr.} \\ \text{between } b \text{ and } c \end{array} \right\} = \tfrac{1}{2}[160 \times 4]$$

$$= +320 \text{ kN} \cdot \text{m}$$

$$\frac{dM}{dx} = V \quad \therefore \text{ Slope varies from } +160 \text{ kN to zero}$$

Point c:

$$M_c = M_{b,\text{right}} + \Delta M_{b-c} = -320 + 320 = 0 \text{ kN} \cdot \text{m}$$

Therefore, the moment diagram is as follows:

5.8 QUALITATIVE DEFLECTED STRUCTURES

The determination of deflections for flexural members is very important in structural analysis, and this subject will be developed quantitatively in Chapter 8. However, it is very useful to be able to sketch the general deflected shape of a structure in a qualitative fashion. This is easily accomplished by using the moment diagram in conjunction with the sign convention for moment.

Recalling that positive moment is associated with member curvature that is concave from above whereas negative moment corresponds to convex curvature, we can sketch the deflected structure from the moment diagram. Points of zero moment are referred to as *points of inflection* and the curvature changes from concave to convex, or vice versa, at these points. Of course, in sketching the deflected structure, adherence to the boundary conditions and the requirements of continuity must be observed. Figure 5.5 shows the qualitative deflected shape of the structure analyzed in Section 5.4, along with the shear and moment diagrams. The same procedure can be followed for sketching the deflected shape of any structure once the moment diagram has been determined.

It is strongly suggested that the deflected shape of the structure be sketched at the conclusion of each shear and moment diagram problem. This will help the student to develop an appreciation for the way structures respond to loading. Errors can frequently be detected by alert analysts who have sharpened their intuition regarding structural behavior by developing the practice of sketching deflected structures based upon the load and moment diagrams. Also, being able to sketch a qualitative deflected shape will be of great assistance when the quantitative deflection problem is studied in Chapter 8.

5.9   DETAILED CONSTRUCTION OF *V* AND *M* DIAGRAMS

The manner in which the key equations of Section 5.6 are used to construct shear and moment diagrams was described in Section 5.7. Two methods were described in that section—one used formal integration and the other accumulated changes in shear and moment. The second method has the greater practical value, and this section will focus attention on its application. The roles of Eqs. 5.2, 5.3, 5.7, and 5.10 were described briefly in Section 5.7. We will now expand on the use of these equations and offer some detailed guidance for the construction of shear and moment diagrams through the following summary points:

1. Construct a load diagram reflecting the loading on the structure in adherence with the sign convention for load. The diagram should have units of load intensity. Concentrated loads, which actually result from the integration of loads of very high intensities over very short segments of the member, are idealistically taken to have infinite load intensities and are plotted as force spikes on the load diagram with their magnitudes indicated.

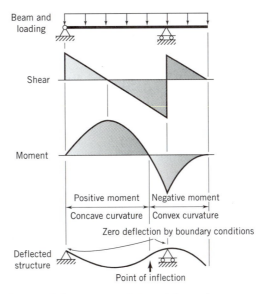

Figure 5.5 Qualitative deflected shape of structure.

2. Construct the shear diagram, noting the following:

   a. Based on Eq. 5.2, the slope of the shear diagram at any point is equal to the load intensity at that point. Specifically, this means that
   - for uniform load, there is a constant slope of the shear diagram which is equal to the load intensity.
   - for varying load intensity, the slope of the shear diagram will vary accordingly.
   - at a point of concentrated load, the load intensity is infinite, and thus there is an abrupt change in the ordinate of the shear diagram.

   b. Based on Eq. 5.7, the change in shear between two points is equal to the area under the load intensity diagram between these two points. Thus,
   - for a distributed load, the change in shear over a segment of the beam is given by the area under the load intensity diagram over the corresponding segment.
   - for a concentrated load, the change in shear is given by the magnitude of the load. This is verified in Fig. 5.6.
   - for general loading, the change in shear between two points is given by the total applied load between these two points.

3. Construct the moment diagram, noting the following:

   a. Based on Eq. 5.3, the slope of the moment diagram at any point is given by the value of the shear at that point. Specifically, this means that
   - in regions of constant shear, there is a constant slope of the moment diagram which is equal to the shear.
   - for varying shear, the slope of the moment diagram will vary accordingly.
   - at points of zero shear, there is a stationary point (zero slope) on the moment diagram.
   - at a point of concentrated load, there is an abrupt change in the ordinate of the shear diagram. Thus there is a corresponding change in the slope of the moment diagram—that is, there is a *cusp* in the moment diagram.

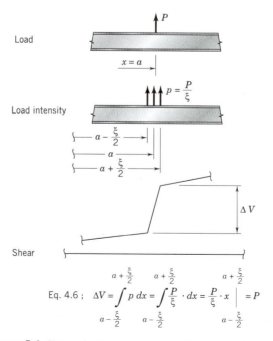

$$\text{Eq. 4.6}; \quad \Delta V = \int_{a-\frac{\xi}{2}}^{a+\frac{\xi}{2}} p \, dx = \int_{a-\frac{\xi}{2}}^{a+\frac{\xi}{2}} \frac{P}{\xi} \cdot dx = \frac{P}{\xi} \cdot x \, \Big|_{a-\frac{\xi}{2}}^{a+\frac{\xi}{2}} = P$$

Figure 5.6 Change in shear associated with concentrated load.

b. Based on Eq. 5.10, the change in moment between two points is equal to the area under the shear diagram between these two points. Thus,

- for a typical variation in shear, the change in moment over a segment of the beam is given by the area under the shear diagram over the corresponding segment.
- for a concentrated moment, there is an abrupt change in the ordinate of the moment diagram at the point where the moment is applied. This change is not evident from the transverse load diagram or the shear diagram.

In applying the techniques enumerated above, it is generally best to start at an end point of the structure. Based on the boundary conditions, the end values of shear and moment are determined, and the diagrams are then constructed by accumulating the incremental changes in shear and moment.

Some convenient area formulae for a variety of segments are given in the figure located inside the back cover. These are helpful in evaluating the incremental changes needed in the construction of shear and moment diagrams, and they are used in the example problems that follow. The centroidal positions that are shown are of no interest at this point, but they will be used in Chapter 8 when beam deflections are studied. The area formulae are presented for figures with a horizontal base. They are also valid for skewed-base situations when the base dimension, l, is taken as the horizontal projection of the base and the altitude is measured perpendicular to the horizontal. Two skewed-base situations are shown in Fig. 5.7. In Fig. 5.7a, the total area cannot be taken as $\frac{2}{3}lh$ because the tangent at B is not parallel to the base. However, by constructing AC parallel to this tangent, we can find the areas of the two subsections. The area of the upper section is $\frac{2}{3}lh_2$, where l is the horizontal projection of the skewed base and h_2 is the altitude of this subsection measured perpendicular to the horizontal. The area of the lower subsection is simply that of a triangle. Similarly, in Fig. 5.7b, if AC is constructed tangent to the curve AB at A, the area of the upper segment is $\frac{1}{3}lh_2$, and, again, the area of the lower subsection is that of a triangle.

5.10 EXAMPLE SHEAR AND MOMENT DIAGRAM PROBLEMS

This section gives several example problems involving the construction of shear and moment diagrams. No commentary is given where a straightforward application of the principles of Section 5.9 is involved. However, commentary is provided where an extension of these principles is used. It is important to note that the locations of the points of zero shear and moment are shown on the shear and moment diagrams. Also, a qualitative sketch of the deflected structure is included for each problem.

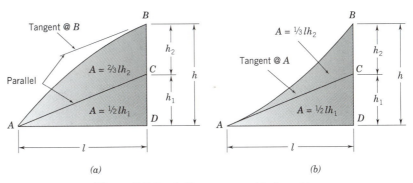

Figure 5.7 Parabolic segments with skewed base.

It should be noted that all of the examples treated in this section are statically de-terminate. One would normally ascertain, as a first step in the analysis procedure, the statical classification according to the criteria of Section 5.4. Statically indeterminate cases are treated in Section 5.11.

EXAMPLE 5.5

Construct the complete shear and moment diagrams and sketch the deflected structure for the beam and loading considered in Example 5.1

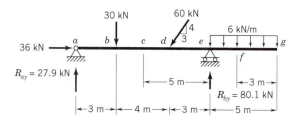

Transverse Load Diagram

The applied distributed load, p (kN/m), and both the applied transverse concentrated loads and reactions, P (kN), are taken from the problem statement. The axial (horizontal) loads are not in-cluded because they do not affect the shear and moment diagrams.

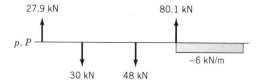

Shear Diagram

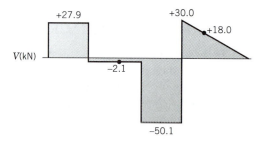

Moment Diagram

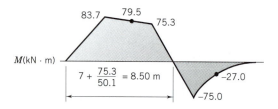

The values of shear and moment for points c and f, which were determined in Example 5.1, are verified as shown on the shear and moment diagrams.

Deflected Structure

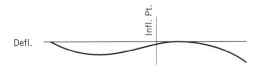

EXAMPLE 5.6

Construct the complete shear and moment diagrams for the beam and loading shown.

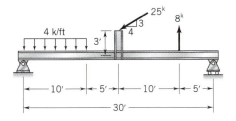

Transverse Load Diagram

The reactions are determined and all of the transverse loads are shown. The axial (horizontal) loads are not required, but the moment caused by the 20-kip horizontal component interacting with the 3-ft moment arm is included as 60 ft-kip concentrated moment.

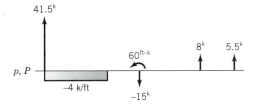

Shear Diagram

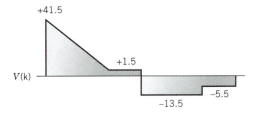

Moment Diagram

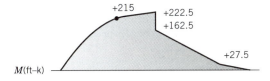

Deflected Structure

There is positive moment throughout, with sharper curvature in the areas of largest bending moment.

A problem arises regarding how the concentrated moment of 60 ft-k affects the moment diagram. This can be resolved by considering the moment as acting alone. It then becomes clear that there must be a 60 ft-k decrease in moment as one proceeds from left to right in order for the boundary conditions to be satisfied.

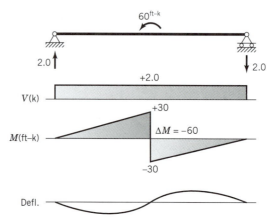

The deflected structure, which could be sketched intuitively from the given loading, confirms that there is a change from positive moment to negative moment as one passes from left to right over the point of concentrated moment.

EXAMPLE 5.7

Construct the complete shear and moment diagrams for the beam and loading shown.

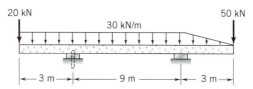

Transverse Load Diagram

The reactions are determined and included with the applied loadings.

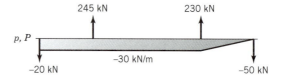

Shear Diagram

Areas under the individual portions of the shear diagram are encircled. These details were not shown in earlier examples because of their simplicity; however, it is done here to show how the areas are subdivided.

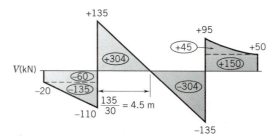

Moment Diagram

The shear diagram areas (shown on shear diagram) are used as moment changes to construct the moment diagram. The total moment change on the cantilever spans results from superimposing the individual areas as shown. The broken lines show the contributions to the moment changes from the rectangular areas of the shear diagram.

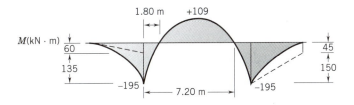

The points of zero moment are determined as follows:

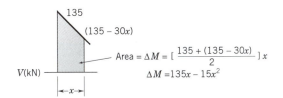

$$\text{Area} = \Delta M = \left[\frac{135 + (135 - 30x)}{2} \right] x$$

$$\Delta M = 135x - 15x^2$$

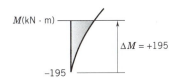

$$\Delta M = +195$$

Thus,

$$135x - 15x^2 = +195$$

or

$$x^2 - 9x + 13 = 0$$

from which

$$x = 1.80, 7.20 \text{ m}$$

Both of these roots are valid because they fall within the range of x for which the equation is applicable $(0 \leq x \leq 9)$.

Deflected Structure

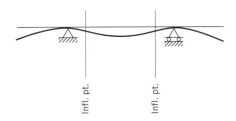

EXAMPLE 5.8

Construct the shear and moment diagrams for the frame structure shown. The column will be viewed from the left, and the normal sign conventions will be employed from that orientation; however, the diagrams will be constructed on horizontal axes. The reactions are given.

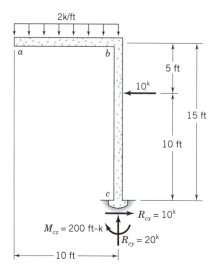

Member Free-Body Diagrams

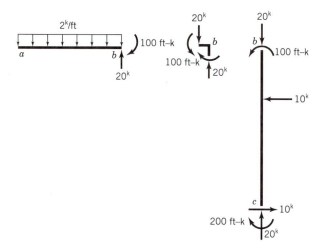

Transverse Load Diagrams

The member-end forces and reactions are included with the applied transverse loads.

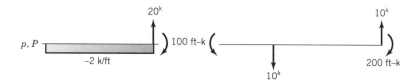

Shear Diagrams

Moment Diagrams

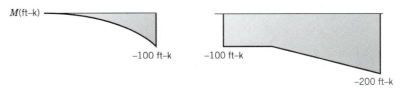

M(ft–k)

−100 ft–k

−100 ft–k

−200 ft–k

Deflected Structure

EXAMPLE 5.9   Construct the shear and moment diagrams for the beam structure and loading shown.

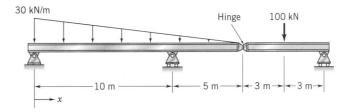

30 kN/m

Hinge   100 kN

10 m   5 m   3 m   3 m

x

Transverse Load Diagram

The reactions are determined and included with the applied transverse loads.

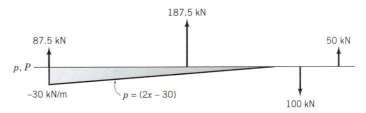

187.5 kN

87.5 kN   50 kN

p, P

−30 kN/m $p = (2x − 30)$

100 kN

Shear Diagram

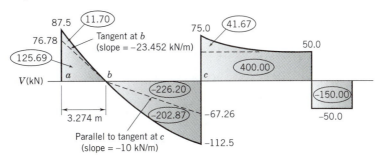

87.5   11.70   75.0   41.67

76.78   Tangent at b 50.0
(slope = −23.452 kN/m)

125.69

V(kN) a b 400.00 c

−226.20   −150.00

3.274 m   −202.87   −67.26   −50.0

Parallel to tangent at c
(slope = −10 kN/m)   −112.5

Point of zero shear (free-body diagram of section left of x, set $V_x = 0$):

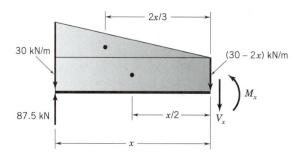

$$V_x = 87.5 - \frac{1}{2}[30 + (30 - 2x)]x = 0$$

$$x^2 - 30x + 87.5 = 0$$

$$x = 3.274, \ 26.726 \text{ m}$$

Here, only the first root is valid because the second root falls beyond the range of applicability $(x > 10 \text{ m})$.

The load intensity at the point of zero shear is $p(x = 3.274 \text{ m}) = (2 \times 3.274 - 30) = -23.452$ (kN/m), which is the slope of the shear diagram at point b. The load intensity, or the slope of the shear diagram at point c, is $p\ (x = 10 \text{ m}) = (2 \times 10 - 30) = -10$ (kN/m) .

These slopes are used to establish convenient subsections under the shear diagram. Standard area formulae and the principles of Fig. 5.7 are then used to determine the areas (encircled) under the shear diagram.

Moment Diagram

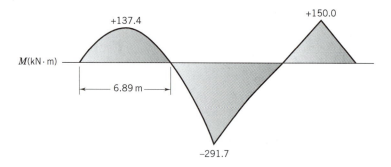

Point of zero moment (FBD of section left of x; set $M_x = 0$):

Use the same free-body diagram that was employed to establish the point of zero shear.

$$87.5x - (30 - 2x)x \cdot \frac{x}{2} - (2x)\frac{x}{2} \cdot \frac{2x}{3} = 0$$

$$x^3 - 45x^2 + 262.5x = 0; \qquad x = 0, 6.89, 38.11 \text{ m}$$

Deflected Structure

EXAMPLE 5.10

Construct shear and moment diagrams for the given frame and the designated loading. The vertical members are viewed from within the frame, and the normal sign conventions are used from that orientation. The diagrams are developed along horizontal axes.

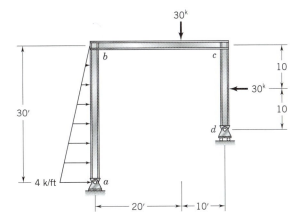

Free-Body Diagrams

The reactions are determined, and each member is isolated as a free-body diagram with its applied loading and member-end forces shown.

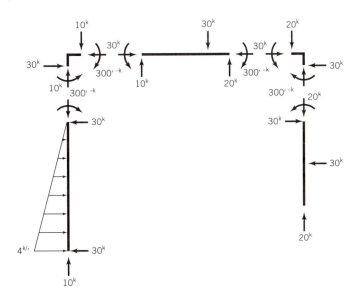

Transverse Load Diagrams

All of the applied transverse loads and member-end forces are included.

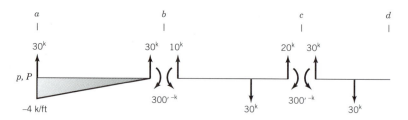

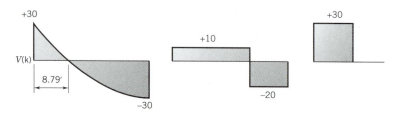

Shear Diagrams

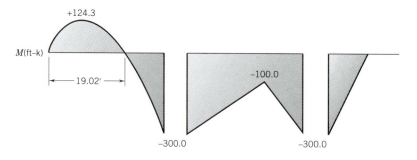

Moment Diagrams

Deflected Structure

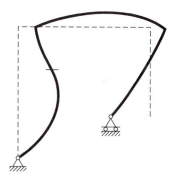

5.11 SHEAR AND MOMENT DIAGRAMS FOR STATICALLY INDETERMINATE STRUCTURES

Although we have limited our treatment thus far to statically determinate structures, it is also necessary to construct shear and moment diagrams for statically indeterminate structures. As will be seen later, most methods of indeterminate analysis yield member-end moments. Once these moments are available, the analyst can proceed to construct the shear and moment diagrams by statics.

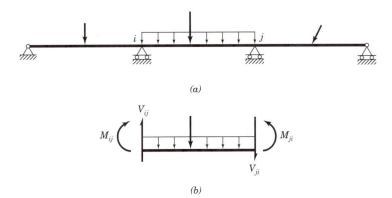

(a)

(b)

Figure 5.8 Statically indeterminate structures.

For instance, consider span ij of the continuous beam given in Fig. 5.8a. A statically indeterminate analysis results in the member-end moments M_{ij} and M_{ji}, which are shown according to our positive sign convention in Fig. 5.8b. The first subscript on each M term identifies the member end on which the moment acts, whereas the second subscript indicates the distant end of the member. Using these moments and the loading on span ij, we proceed to obtain the member-end shears, V_{ij} and V_{ji}, by statics. These shears are shown in Fig. 5.8b in their positive directions, and the same subscripting convention is used as was employed for the moment terms.

The specific approach requires that one sums moments about point i for span ij. The shear V_{ij} passes through the moment center and, therefore, because the loading and the end moments are known, V_{ji} can be determined. The shear V_{ij} can then be determined by either summing moments about point j or by summing forces in the direction normal to the member axis. With the end shears having been determined, the shear and moment diagrams are readily constructed.

When the structure is composed of more than a single member, each individual member is treated in the manner just described. For the interior supports of continuous beams, the support reactions are given by the abrupt changes that occur in the shear at these points. For frame structures, joint equilibrium considerations will yield the member-end shears and member axial forces.

EXAMPLE 5.11

Construct the shear and moment diagrams for the continuous beam given. The member-end moments for each span have been determined by a statically indeterminate analysis and are given as shown.

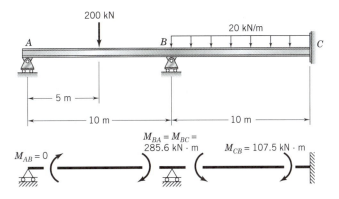

Determination of Member-End Shears

Span AB

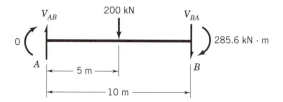

$$\sum M_{Az} = 0 \;\circlearrowright+; \qquad (200 \times 5) + 285.6 + (V_{BA} \times 10) = 0; \qquad V_{BA} = -128.56 \text{ kN}$$

$$\sum P_y = 0 \;\uparrow+; \qquad V_{AB} - 200 - V_{BA} = 0; \qquad V_{AB} = 200 + (-128.56) = 71.44 \text{ kN}$$

Span BC

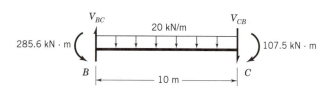

$$\sum M_{Bz} = 0 \;\circlearrowright+; \qquad -285.6 + (20 \times 10)5 + 107.5 + (V_{CB} \times 10) = 0; \qquad V_{CB} = -82.19 \text{ kN}$$

$$\sum P_y = 0 \;\uparrow+; \qquad V_{BC} - (20 \times 10) - V_{CB} = 0; \qquad V_{BC} = 200 + (-82.19) = 117.81 \text{ kN}$$

Shear Diagram

Knowing the member loading and having established the member-end shears, we can proceed directly to the shear diagram. Of course, the member-end shears must be consistent with the loading.

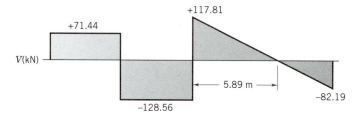

Moment Diagram

With the given member-end moments and the moment changes based on the areas under the shear diagram, the moment diagram is readily constructed. As a check, the moment changes must be compatible with the member-end moments.

Note: Positive moment causes compression on the top fiber of the beam. Therefore, the given end moments are negative.

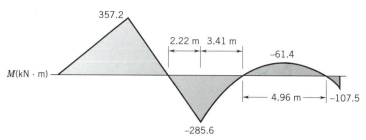

Deflected Structure

Defl. Str.

EXAMPLE 5.12

Construct the shear and moment diagrams for the frame shown. The member-end moments have been determined by a statically indeterminate analysis and are given as shown.

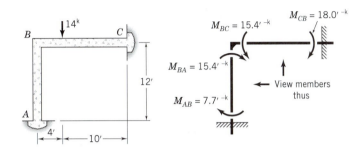

Determination of Member-End Shears

Column AB (view member from the right)

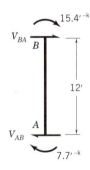

$$\sum M_{Az} = 0 \;\circlearrowright+; \qquad 7.7 + 15.4 + (V_{BA} \times 12) = 0; \qquad V_{BA} = -1.93^{k}$$

$$\sum P_{y} = 0 \;\overset{+}{\leftarrow}; \qquad V_{AB} - V_{BA} = 0; \qquad V_{AB} = -1.93^{k}$$

Beam BC

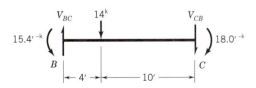

$$\sum M_{Bz} = 0 \;\circlearrowright+; \qquad -15.4 + (14 \times 4) + 18 + (V_{CB} \times 14) = 0; \qquad V_{CB} = -4.19^{k}$$

$$\sum P_{y} = 0 \;\uparrow+; \qquad V_{BC} - 14 - V_{CB} = 0; \qquad V_{BC} = 14 + (-4.19) = 9.81^{k}$$

Shear Diagram

The shear diagram for each member can be established from the member-end shears and the member loading. The diagram for the column is constructed along a horizontal axis.

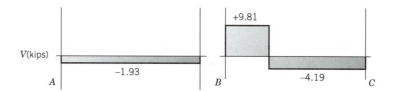

Moment Diagram

The moment diagram for each member can be developed from the given member-end moments and the moment changes from the shear diagram. Again, the diagram for the column is constructed along a horizontal axis.

Note: The column is viewed from the right and the beam from below. Based on these orientations, positive moments cause compression on the top fiber.

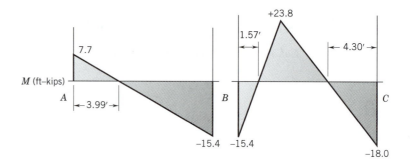

Deflected Structure

5.12 INTERNAL FORCES FOR NONPLANAR STRUCTURES

The emphasis in this chapter has been on planar structures; however, most structures are in fact three-dimensional structures. In this section, a few observations are made concerning nonplanar beam and frame structures. The analysis concepts are direct extensions of those employed in planar systems, but the member actions are more complicated, as will be explained in the next section.

We will limit discussion to statically determinate structures. Therefore, the determination of the reactions is accomplished through a systematic application of the equations of equilibrium, as expressed by Eqs. 2.10. It is thus clear that a statically determinate system has six independent reaction components, which must be arranged to ensure external stability. Additional reaction components are necessary for stability, and they are determinate, when conditions of construction allow for the expression of condition equations that would augment the equations of equilibrium.

5.13 MEMBER FORCE NOTATION AND SIGN CONVENTION

The member force notation for nonplanar beam and frame structures is a bit more complicated than the one employed for planar structures, which was described in Section 5.3; but it is a mere extension of that simpler case.

Consider the element shown in Fig. 5.9a. For planar structures, axial force was considered, as was flexure in the xy plane. Figure 5.9b shows the member forces associated with these modes of behavior. The member actions and notation are as they were introduced in Section 5.3, with the addition of some clarifying subscripts: F_x is the axial

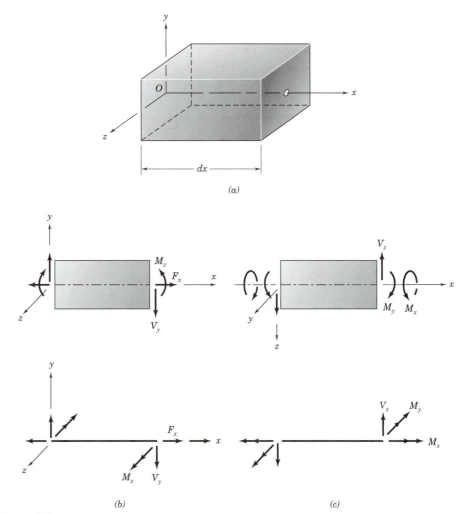

(a)

(b)   (c)

Figure 5.9 Notation and sign convention. *(a)* Member element and member axes. *(b)* Member forces for planar case—*xy* plane. *(c)* Additional member forces for nonplanar case—*xz* plane.

force along the x axis, V_y is the shear directed along the y axis, and M_z is the bending moment about the z axis. For the nonplanar case, flexure must be added in the xz plane and torsion about the x axis must be included. The member forces associated with bending, the shear along the z axis, V_z, and the moment about the y axis, M_y, are shown in Fig. 5.9c, as is the moment about the x axis, M_x, which is the torsional moment.

With respect to sign convention, the general concepts considered in planar structures are carried forward. Positive member forces are as follows: Positive F_x produces tension on the element; positive M_z causes compression on the positive y fibers in the xy plane; positive V_y acts in the negative y direction on the right-exposed face and in the positive y direction of a left-exposed face; positive M_y causes compression on the positive z fibers in the xz plane; positive V_z acts in the negative z direction on a right-exposed face and in the positive z direction on the left-exposed face; and positive M_x acts according to the right-hand rule on the right-exposed face and in the opposite sense on the left-exposed face. All of the member forces shown on the element of Fig. 5.9 are positive, with vector representations for the moments shown according to the right-hand rule.

The sign convention described above lacks elegance and generality, but it will suffice for our present considerations. In the formal development of matrix methods in Part Five, a superior sign convention will be employed that has greater generality for three-dimensional structures.

5.14 INTERNAL FORCE DIAGRAMS FOR NONPLANAR STRUCTURES

In addition to shear and moment, some members in beam and frame structures are subjected to axial forces. The structures considered in Examples 5.5, 5.6, 5.8, and 5.10 show portions of members or entire members that have axial forces present. Diagrams can be plotted to show the variation in axial force along the members.

For three-dimensional structures, axial forces and twisting moments routinely occur, and bending occurs in two planes. This information can be conveyed through a set of internal force diagrams, which are merely generalizations of the shear and moment diagrams that were considered for planar structures.

The procedure follows the same general approach that was used for planar structures. The reactions are first determined. A free-body diagram is then established for each member showing the member loading and the complete set of member-end actions. The internal force diagrams are then constructed for each member using the sign convention introduced in the previous section.

EXAMPLE 5.13   Determine the internal force diagrams for the structure shown.

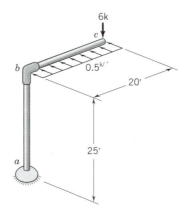

The reactions at point a are determined from the application of Eqs. 2.10. The free-body diagrams of the individual members (superimposed on member coordinate systems), including all member-end forces, are as follows:

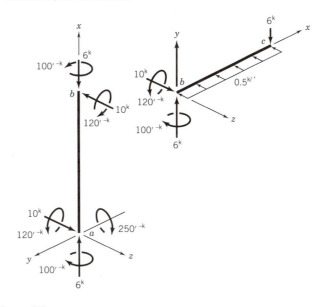

Internal Force Diagrams

Based on the sign convention described in Section 5.13, we have the following:

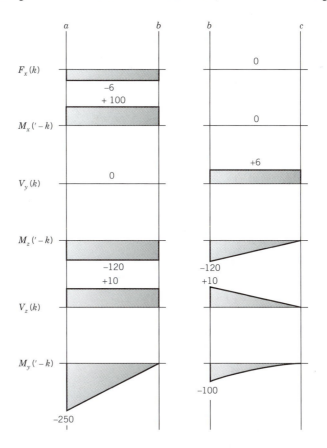

5.15 SUGGESTED PROBLEMS

5.1 Classify each of the following structures with respect to statical determinacy and stability. Consider both external and overall criteria.

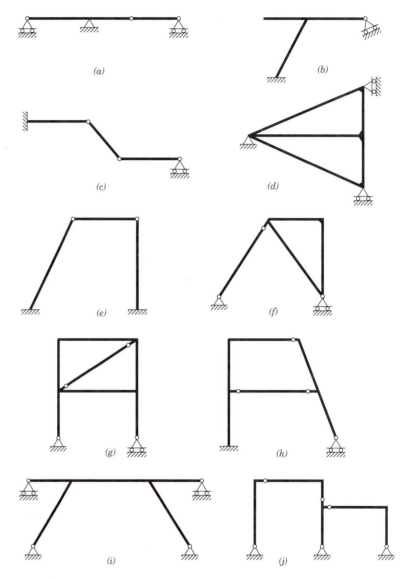

5.2 through 5.6   Use the x coordinate shown to write general expressions for shear and moment for each labeled region shown. Evaluate the shear and moment at each end of each region

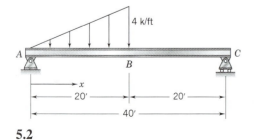

5.2

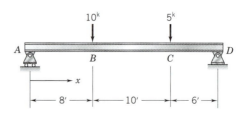

5.3

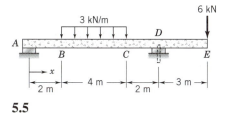

5.4

5.5

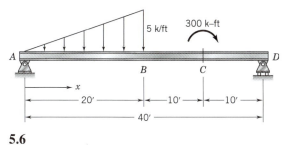

5.6

5.7 Using the given x coordinate, write general expressions for axial load, shear, and moment for regions BC and CD. Use the resulting expressions to evaluate these quantities at point C for each of the specified regions. *Note:* Shear acts normal to the member axis.

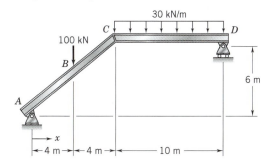

5.8 through 5.11   Construct the shear and moment diagrams by the formal integration method for beams and loading conditions shown.

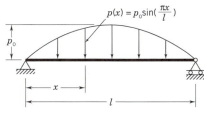

5.8

5.9

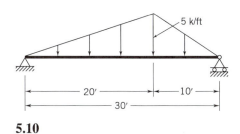

5.10

5.11

5.12 through 5.39   Use the incremental change method to construct the shear and moment diagrams for all members for each of the structures and loadings shown. In each case, *sketch* the deflected structure.

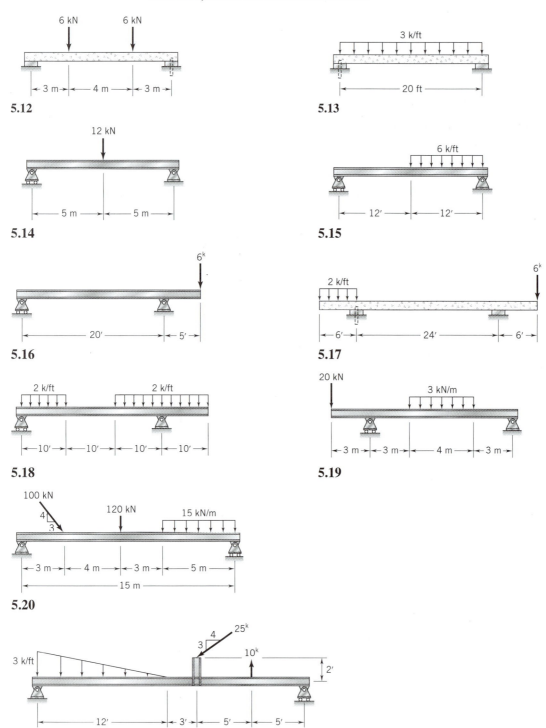

5.12

5.13

5.14

5.15

5.16

5.17

5.18

5.19

5.20

5.21

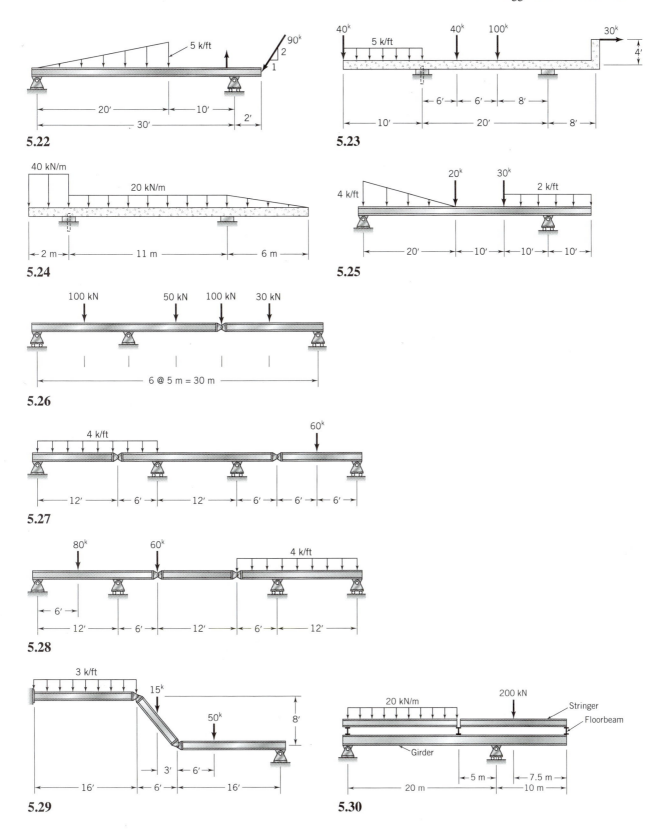

5.22

5.23

5.24

5.25

5.26

5.27

5.28

5.29

5.30

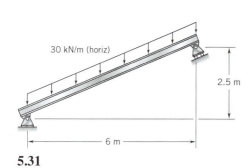

5.31

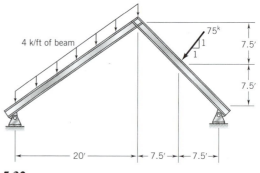

5.32

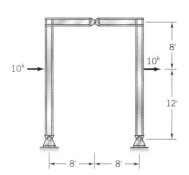

5.33

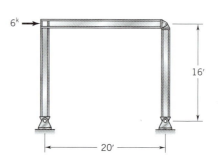

5.34

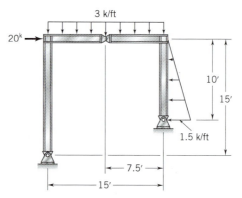

5.35

5.36

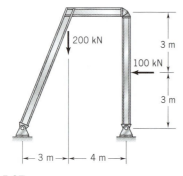

5.37

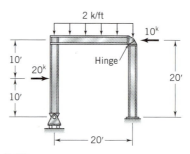

5.38

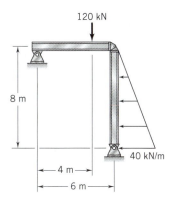

5.39

5.40 through 5.41   Construct the shear and moment diagrams of the statically indeterminate structures shown, and *sketch* the deflected structure for each. The end moments for each span are given.

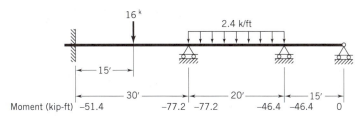

5.40

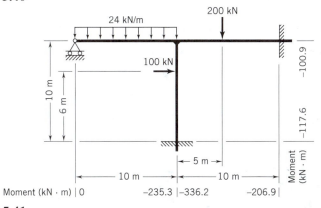

5.41

5.42 through 5.47   Construct the internal force diagrams for all members for each of
the structures and loadings designated. In each case, *sketch* the deflected structure.
5.42 The structure and loading of Problem 3.32
5.43 The structure and loading of Problem 3.33
5.44 The structure and loading of Problem 3.35
5.45 The structure and loading of Problem 3.36
5.46 The structure and loading of Problem 3.37
5.47 The structure and loading of Problem 3.40

Chapter 6

Influence Lines and Maximum Load Effects

Golden Gate Bridge, San Francisco, Calif. (photo by Louis F. Geschwindner).

6.1   VARIABLE LOADINGS

The design of any structural component requires knowledge of the maximum load effects to which the component will be subjected in fulfilling its functional requirements. These load effects are determined from various response functions, such as the reactions, bar forces, shears, moments, and so forth that result from detailed structural analyses. However, for most structures, there is an array of possible loading conditions, and there are unique loading conditions associated with the maximum response functions and the associated maximum load effects for each member.

There are, therefore, two major concerns that motivate the study of variable loading conditions. First, of all possible loading conditions, which ones will maximize the load effects in a given structural component? And second, what specific load effects correspond to these loading conditions?

In this chapter, only statically determinate structures are considered. The problem of variable loading conditions for statically indeterminate structures is not treated in this book.

6.2   VARIATION IN RESPONSE FUNCTION WITH POSITION OF LOAD: INFLUENCE LINE

A useful technique for the study of variable loading problems focuses on consideration of the variation in a given response function with the position of load. For instance, consider the simply supported beam of Fig. 6.1a. Selecting the reaction at A as the response function of interest, we will determine its value for various positions of a unit downward load. Figure 6.1b shows the value of R_A for each of four different positions of the unit load. If the value of R_A is plotted as ordinate along an abscissa that represents the position of the unit load on the structure, we obtain the diagram shown in Fig. 6.1c. A diagram of this type is called an *influence line*—it records the influence that a unit load has on a particular response function as the load traverses the structure. In this case, the response function is R_A, and thus Fig. 6.1c is the influence line for R_A (I.L. R_A).

Each ordinate of Fig. 6.1c gives the value of R_A that results when a unit downward load is applied at a point on the structure corresponding to the abscissa. In this case, it is clear that a positive (upward) value of R_A results for any position of the unit load on the structure and that R_A diminishes linearly as the unit downward load traverses the structure. In fact, in this case, the equation for the influence line is $1(l - x)/l$, which gives the expression for R_A in terms of the position of the unit load.

In this example, an influence line was developed for a reaction. Actually, an influence line can be constructed for any response function—reaction, bar force, shear, moment, stress, and so forth. Thus, an *influence line is defined as a curve, the ordinate to which at any abscissa equals the value of some particular response function attributable to a unit downward load acting at the point on the structure corresponding to that abscissa.*

The units for an influence line reflect the units of the response function and the unit load. For the example of Fig. 6.1c, the units are force per force (kN/kN). Thus, the ordinate of 0.75 kN/kN indicates that R_A has a value of 0.75 kN per 1 kN load acting downward on the structure at that point.

Influence lines are very useful in design for determining the response of a structure to a specific loading condition. By definition, an influence line indicates the response of a structure for a unit load. However, it is clear from the principle of superposition that the response for a load larger than unity would merely be amplified accordingly. Likewise, if there is more than one load on the structure, the appropriate influence line, coupled with the principle of superposition, will enable one to determine the total response for the combined

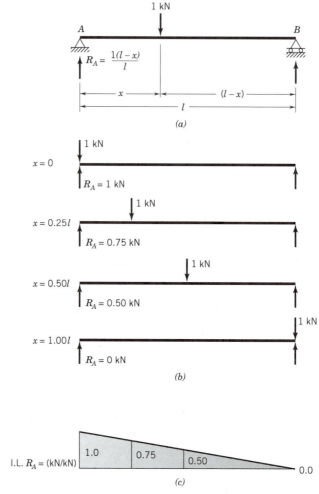

Figure 6.1 Development of influence line. *(a)* Structure and loading. *(b)* Selected load positions. *(c)* Influence line for R_A.

loading. And even for distributed loads, superposition can be employed to integrate the individual effects to get the total response. These concepts will be treated in detail in Section 6.5.

6.3   INFLUENCE LINES BY EQUILIBRIUM METHODS

The most fundamental approach to the construction of influence lines is to allow the unit load to occupy sequentially several positions on the structure and to compute the value of the desired response function for each load position. The ordinates are plotted to form the influence line for the given response quantity.

The following suggestions are offered for use of this technique in constructing influence lines:

1. Apply the usual analysis procedure to solve for the response function for which the influence line is to be constructed. That is, begin by indicating the positive symbolic representation of the desired response function on the appropriate free-body diagrams.

2. For each position of the unit load, the most convenient free-body diagram is used in conjunction with the equations of static equilibrium to solve for the response

function. The sign of the response function will be consistent with the established sign convention and the corresponding ordinate of the influence line will reflect the appropriate sign at the abscissa corresponding to the load point.

3. This procedure is carried out for as many positions of the unit load as are needed to establish the shape of the influence line.

4. In solving for an internal response function, the solution generally includes terms involving the reactions. Thus, it is convenient to construct the influence lines for reactions first. These influence lines are then used in the construction of other influence lines.

EXAMPLE 6.1

Construct the influence lines for the shear and moment at point C for the structure shown.

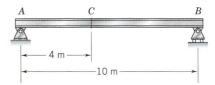

Influence Lines for Reactions

As an initial step, the influence lines for the reactions are constructed. Even though these are not specifically requested, they are invaluable for the construction of the desired influence lines. Taking a unit downward load of 1 kN, whose position varies over the range identified by the horizontal two-headed arrow, and assuming positive R_A and R_B to act upward, we obtain the following influence lines:

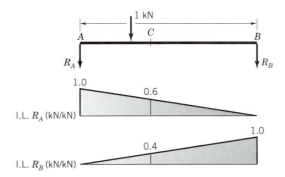

As a check on these influence lines, it is noted that the sum of the ordinates at any point must be 1 kN/kN; that is, the unit downward load must be equilibrated vertically by the reaction components. The influence line ordinates at point C are identified for subsequent use.

Influence Line for Shear at C

To determine the shear at point C, the structure is cut at point C, V_C is shown in the positive sense, and statics is applied to determine the value of V_C. When the unit load acts anywhere to the left of point C, it is most convenient to consider the free-body diagram to the right of point C. For this section, which is enclosed in a box, vertical equilibrium requires that $V_C = -R_B$.

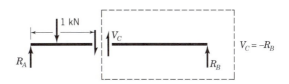

When the unit load acts to the right of point C, it is most convenient to consider the free-body diagram to the left of point C. For this portion, vertical equilibrium leads to $V_C = R_A$.

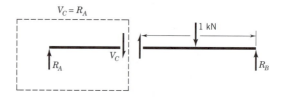

Thus, the influence line is constructed from these two relationships for V_C and the influence lines for reactions.

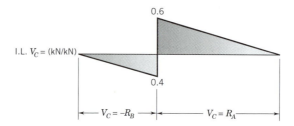

Influence Line for Moment at C

To determine the moment at point C, the structure is cut at point C, M_C is shown in the positive sense, and statics is applied to determine the value of M_C. Again, when the unit load acts to the left of point C, it is most convenient to consider the free-body diagram to the right of point C. For this region, moment equilibrium at point C requires that $M_C = 6R_B$.

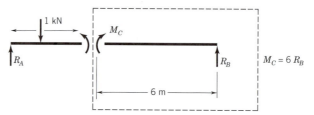

When the unit load is right of point C, the free-body diagram to the left of point C is used. Here, moment equilibrium imposes the condition that $M_C = 4R_A$.

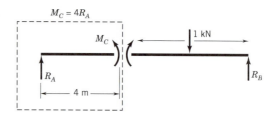

The final influence line is constructed from these two relationships for M_C and the influence lines for reactions.

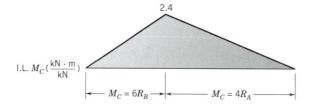

EXAMPLE 6.2

Construct the influence lines for the shear at the hinge at B and for the moment at D for the structure shown below.

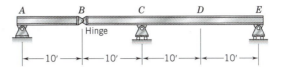

Influence Lines for Reactions

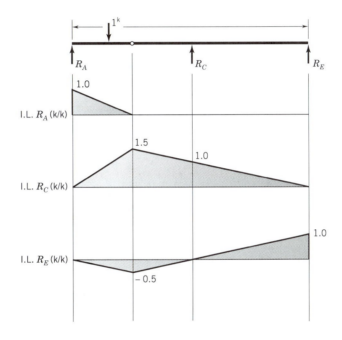

Influence Line for Shear at B

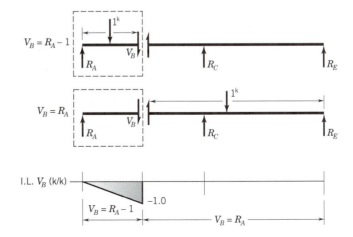

Influence Line for Moment at D

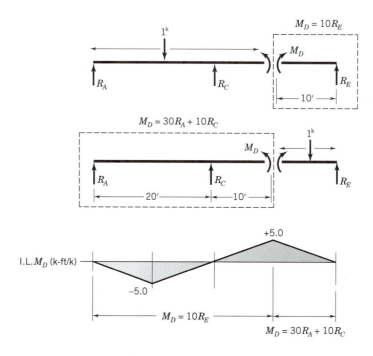

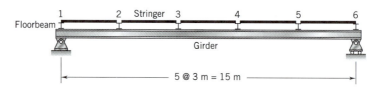

EXAMPLE 6.3

Construct the influence lines for the shear in panel 2–3 and the moment at point 5 for the girder shown below. The unit load is applied to the stringers.

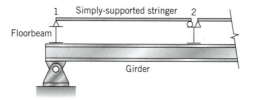

Girder–Floorbeam–Stringer Detail

Note: The stringers transmit loads to the girder at the floorbeam connection points only. Therefore, for any given loading condition, the girder shear is constant between load points, and the girder moment varies linearly between the load points. For these reasons, it is appropriate to construct the influence lines for girder shear in a panel and for girder moment at a panel point.

Influence Lines for Girder Reactions

Here, we take a free-body diagram of the entire structure with the unit load applied to the stringers.

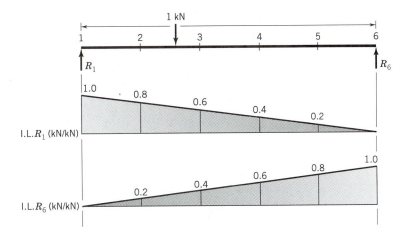

I.L.R_1 (kN/kN)

I.L.R_6 (kN/kN)

Influence Line for the Shear in Panel 2–3 of the Girder, $V_{2\text{-}3}$

In this case, the structure is transversely cut through panel 2–3.

First, consider the unit load to be on stringer 1–2, left of panel 2–3. The portion of the structure to the right of the cut is taken as a free-body diagram.

Next, consider the unit load to be to the right of panel 2–3, moving along the stringers from point 3 to point 6. The portion of the structure left of the cut is now taken as the free-body diagram.

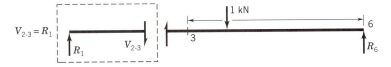

Thus, with the unit load either to the left or to the right of panel 2–3, we can construct the following partial influence line.

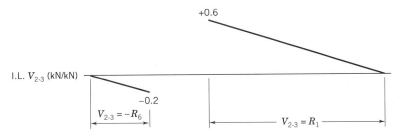

I.L. $V_{2\text{-}3}$ (kN/kN)

When the unit load is on the stringer between points 2 and 3, it is useful to consider the influence lines for the stringer reactions for stringer 2–3, which, in turn, are the influence lines for the girder loads at points 2 and 3.

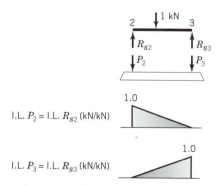

I.L. P_2 = I.L. R_{s2} (kN/kN)

I.L. P_3 = I.L. R_{s3} (kN/kN)

The partial influence line for V_{2-3} shows that the portion of the unit load that enters the girder as P_2 contributes to V_{2-3} with an influence of -0.2 kN/kN, whereas the portion of the unit load that enters the girder as P_3 contributes to V_{2-3} with an influence of $+0.6$ kN/kN. Since the girder loads P_2 and P_3 vary linearly between the limits of zero and unity as the unit load moves from 2 to 3, their effects on V_{2-3} must also vary linearly between the limits of -0.2 and $+0.6$ as shown.

This leads to the following complete influence line.

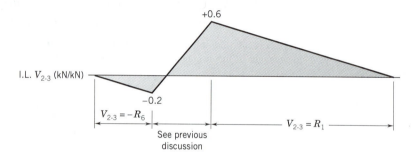

Influence Line for Moment in the Girder at Point 5, M_5

Here, the structure is cut at point 5, and the unit load is first placed to the left of point 5 with a free-body diagram considered to the right of the cut. Then the unit load is placed to the right of point 5, and the portion of the structure left of the cut is taken as the free-body diagram.

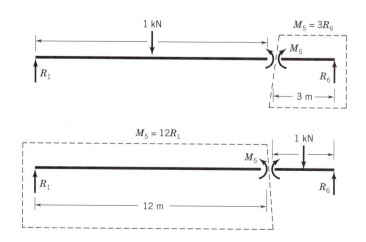

The complete influence line is as follows:

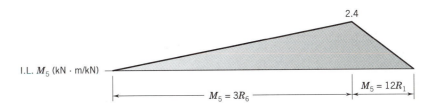

$$\text{I.L. } M_5 \text{ (kN} \cdot \text{m/kN)}$$

2.4

$$M_5 = 3R_6$$

$$M_5 = 12R_1$$

EXAMPLE 6.4

Construct the influence lines for the bar forces in members U_1L_1, U_2L_3, and U_3U_4 (identified by X on sketch). Consider the unit load to pass along the top chord of the truss.

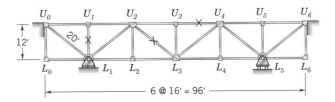

Note: A truss can receive loads only at the joints. When the unit load is between top chord joints it will enter the truss through adjacent joints in a manner similar to that explained in Example 6.3 for panel point loads for girders.

Influence Lines for Truss Reactions

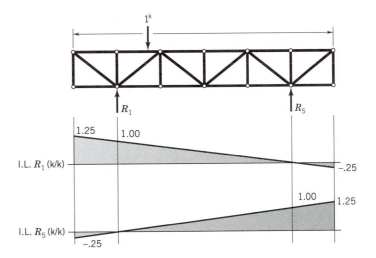

Influence Line for the Bar Force in Member U_1L_1, $F_{U_1L_1}$

$F_{U_1L_1}$ is directly related to the load P_1, which enters the truss at joint U_1. An influence line for P_1 leads directly to the influence line for $F_{U_1L_1}$.

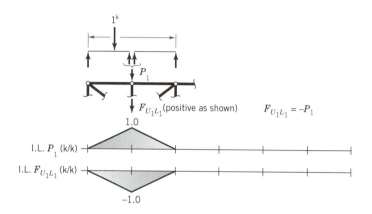

Influence Line for the Bar Force in Member U_2L_3, $F_{U_2L_3}$

The vertical component of $F_{U_2L_3}$ must take the shear in panel 2–3. Thus, we proceed in the same fashion as we did for the girder shear in Example 6.3.

For unit load left of joint U_2, with all quantities shown in the positive direction, we have

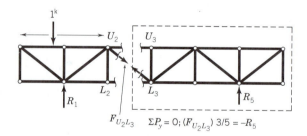

$$\Sigma P_y = 0; (F_{U_2L_3}) \, 3/5 = -R_5$$

For unit load right of joint U_3

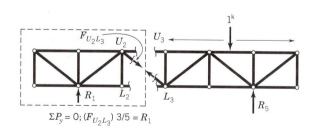

$$\Sigma P_y = 0; (F_{U_2L_3}) \, 3/5 = R_1$$

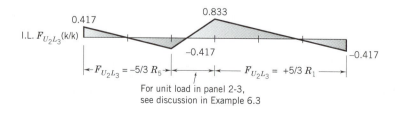

For unit load in panel 2-3,
see discussion in Example 6.3

Influence Line for the Bar Force in Member U_3U_4, $F_{U_3U_4}$

For unit load left of joint U_3, with all quantities again shown in the positive direction, we have

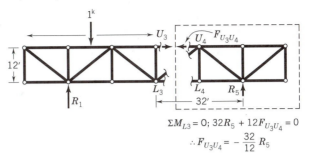

$$\Sigma M_{L3} = 0; \quad 32R_5 + 12F_{U_3U_4} = 0$$

$$\therefore F_{U_3U_4} = -\frac{32}{12}R_5$$

For unit load right of joint U_3

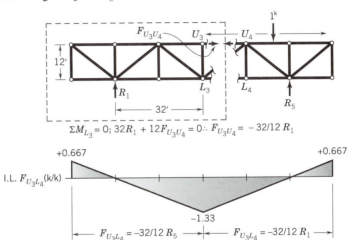

$$\Sigma M_{L_3} = 0; \quad 32R_1 + 12F_{U_3U_4} = 0 \quad \therefore F_{U_3U_4} = -32/12 \, R_1$$

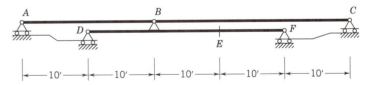

I.L. $F_{U_3L_4}$(k/k)

$+0.667$ ... $+0.667$

-1.33

$F_{U_3L_4} = -32/12 \, R_5$ ⟶⟵ $F_{U_3L_4} = -32/12 \, R_1$

EXAMPLE 6.5

Determine the influence lines for the vertical force transferred at point B, the reactions at D and F, and the shear and moment at point E.

The unit load traverses structure from A to C:

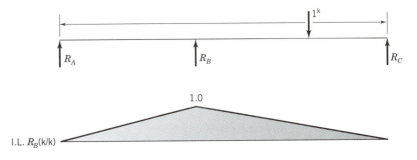

Influence Line for Force at B

Point B is a simple support for spans AB and BC. The vertical reaction at that point, R_B, provides the load at point B for span DF. The influence line for R_B is easily determined.

R_A R_B R_C

1.0

I.L. R_B(k/k)

Influence Lines for Span *DF*

As the unit load moves from A to C, the load transmitted to span DF at point B is given by the influence line for R_B.

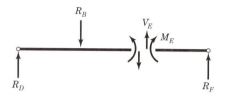

From statics,

$$R_D = \frac{2}{3}R_B, \ R_F = \frac{1}{3}R_B, \ V_E = -R_F, \ M_E = 10R_F$$

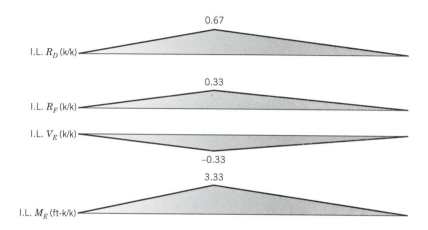

0.67

I.L. R_D (k/k)

0.33

I.L. R_F (k/k)

I.L. V_E (k/k)

−0.33

3.33

I.L. M_E (ft-k/k)

EXAMPLE 6.6

Determine the influence lines for all reaction components and for the moment at the top of column *BD*. The unit load moves from *A* to *C*.

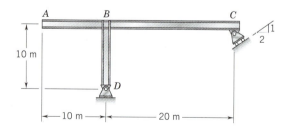

Influence Line Construction

The free-body diagram shows the response quantities of interest, and the application of statics yields the expressions shown.

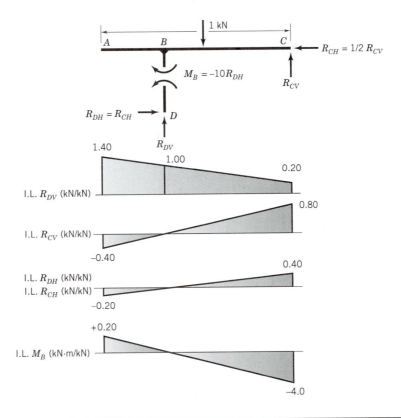

6.4   INFLUENCE LINES BY VIRTUAL WORK: MÜLLER–BRESLAU PRINCIPLE

Betti's law, which was presented in Section 2.19, can be used to develop a very useful concept for the construction of influence lines. Consider the structure shown in Fig. 6.2*a*. Assume that this structure is subjected to a unit downward load at point *x*. In the process, the structure deforms and develops reactions R_A and R_B as shown in Fig. 6.2*b*. This situation is referred to as "System 1." Next, assume that the given structure is deformed by removing the restraint associated with the reaction R_A and that a unit displacement is introduced at point *A* in the positive direction of R_A. This situation is shown in Fig. 6.2*c* and is referred to as "System 2."

Recalling from Eq. 2.71 that the virtual work done by the forces of System 1 through the displacements of System 2 is equal to the virtual work done by the forces of System 2 through the displacements of System 1, we can write

$$\sum_{i=1}^{n} (P_i)_1 (\Delta_i)_2 = \sum_{j=1}^{m} (P_j)_2 (\Delta_j)_1 \qquad (6.1)$$

where $(P_i)_1$ and $(\Delta_i)_2$ represent the force at point *i* for System 1 and the displacement at point *i* for System 2, respectively. Similarly, $(P_j)_2$ and $(\Delta_j)_1$ give the force at point *j* for System 2 and the displacement at point *j* for System 1, respectively. The above work quantities are virtual in that they exist in essence or effect, but not in fact—that is, they are imagined for the purposes of analysis.

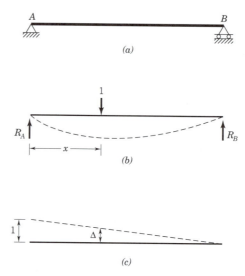

Figure 6.2 Müller–Breslau principle. *(a)* Given structure. *(b)* System 1: structure in equilibrium under unit load. *(c)* System 2: structure subjected to unit displacement corresponding to R_A.

Thus, for the situation in Fig. 6.2, application at Eq. 6.1 gives

$$
\overset{(\Delta)_2}{(R_A)(1) + (1)(-\Delta)} = 0 \tag{6.2}
$$
$$
(P)_1
$$

The negative sign on the Δ quantity simply means that it is in a direction opposite the unit load; thus, negative work results. In the present problem, there are no forces for System 2 since the entire member is merely rotated as a rigid body about point B. However, even if a force were required at point A to impose the deformation pattern of System 2, it would do no work, because there is no displacement at point A in System 1.

Rewriting Eq. 6.2, we have

$$
R_A = \Delta \tag{6.3}
$$

Equation 6.3 states that the value of R_A for a unit load at point x is given by the displacement at point x when the structure is subjected to a unit displacement corresponding to the positive direction of R_A. Because x can represent any point along the span, the deflection pattern of System 2 traces the influence line for R_A.

Influence lines for other response functions can be determined by the virtual work method in the same general fashion. For instance, consider the influence lines for the shear and moment at point C of the structure shown in Fig. 6.3a. System 1 is given in Fig. 6.3b, in which the unit load is shown at point x and the associated shear and moment are symbolically represented in their positive directions on the cut section at point C.

To determine the influence line for V_C, System 2 requires a unit displacement that will interact with V_C of System 1 to do work, as shown in Fig. 6.3c. This deformation pattern must retain the same slope on each side of the cut section so that the rotational displacements of System 2 do not produce any net work as they interact with M_C of System 1.

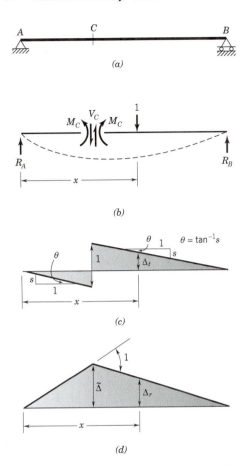

Figure 6.3 Müller–Breslau principle for shear and moment influence lines. *(a)* Given structure. *(b)* System 1. *(c)* System 2 used to obtain influence line for V_C. *(d)* System 2 used to obtain influence line for M_C.

Thus, applying Eq. 6.1, we have

$$(\Delta)_2$$

$$(V_c)(1) + (M_c)(-\theta) + (M_c)(\theta) + (1)(-\Delta_t) = 0 \tag{6.4}$$

$$(P)_1$$

or,

$$V_C = \Delta_t \tag{6.5}$$

The influence line for the shear at point C is thus given by the displacement pattern that results when the structure is broken at point C and a unit displacement is introduced corresponding to positive V_C.

To construct the influence line for M_C, System 2 must reflect a unit rotational displacement that will interact with M_C of System 1 to do work. This situation is shown in Fig. 6.3*d*. This deformation pattern must retain the same transverse displacement on each side of the cut section so that the displacements of System 2 do not produce any net work as they interact with V_C of System 1.

The application of Eq. 6.1 then leads to

$$(M_c)(1) + (V_c)(-\tilde{\Delta}) + (V_c)(\tilde{\Delta}) + (1)(-\Delta_r) = 0 \tag{6.6}$$

$(\Delta)_2$

$(P)_1$

from which

$$M_C = \Delta_r \tag{6.7}$$

Thus, the influence line for the moment at point C is given by the displacement pattern that occurs when the structure is broken at point C and a unit rotational displacement is introduced corresponding to positive M_C.

For all three cases cited, that is, for the construction of influence lines for R_A, V_C, and M_C, the same general pattern was followed. This technique is known as the *Müller–Breslau principle*, and it can be formally stated as follows: *The influence line for any response function is given by the deflection curve that results when the restraint corresponding to that response function is removed and a unit displacement is introduced in its place.*

In applying the Müller–Breslau principle, care must be taken to make certain that the deformation pattern of System 2 interacts with the forces of System 1 such that only the unit load and the response function for which the influence line is being constructed do work. A careful examination of the procedures used to obtain Eqs. 6.3, 6.5, and 6.7 will verify the importance of this fact.

In this context, the Müller–Breslau principle is most helpful in a qualitative sense for verifying the shape of influence lines. This principle can be used both qualitatively and quantitatively for statically indeterminate structures, but those topics are not considered here. In fact, it is in this context that the principle is of greatest value.

EXAMPLE 6.7   Use the Müller–Breslau principle to construct the influence lines for the reaction at point C, the shear at point B, and the moment at point D for the structure studied in Example 6.2.

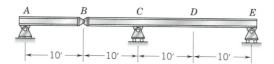

Influence Line for R_C

The reaction for which the influence line is desired is shown in its positive direction.

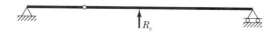

The restraint associated with R_C is removed, and a unit vertical displacement is introduced at point C.

1.5

1.0 (induced)

I.L. R_C (k/k)

Influence Line for V_B

The structure is cut at point B and the shear quantity for which the influence line is desired is shown to act in its positive sense.

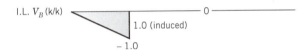

V_B

The restraint associated with V_B is removed, and a unit transverse displacement is introduced at point B. In this case, because there is no resistance to transverse displacement afforded by the segment AB (it merely rotates as a rigid body), whereas, the section BCE remains a stable structure which would resist transverse displacement, the entire unit transverse displacement occurs to the left of the cut at point B.

I.L. V_B (k/k)

0

1.0 (induced)

– 1.0

Influence Line for M_D

The structure is cut at point D, and the moment quantity for which the influence line is desired is shown to act in its positive direction.

A M_D B

The rotational restraint associated with M_D is removed, and a unit rotation is introduced at point D.

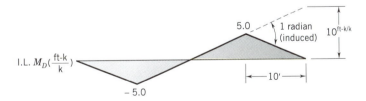

5.0   1 radian $10^{ft\text{-}k/k}$
(induced)

I.L. $M_D(\frac{ft\text{-}k}{k})$

|← 10' →|

– 5.0

EXAMPLE 6.8

Use the Müller–Breslau principle to construct the influence line for the girder shear in panel 2–3 for the structure shown below. The unit load is applied to the stringers.

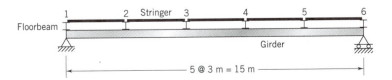

1   2   Stringer   3   4   5   6

Floorbeam

Girder

5 @ 3 m = 15 m

Influence Line for V_{2-3}

The response function for which the influence line is to be constructed is shown in a positive sense on the figure given below.

Remove the restraint corresponding to V_{2-3} and introduce a unit transverse displacement, while maintaining the same slope on each side of the cut. Although the displacement is introduced in the girder, the influence line is traced by the stringers, since the loads are applied along this line.

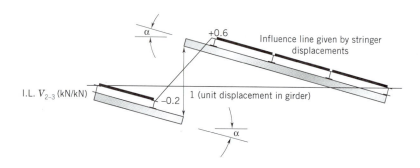

Note: The unit displacement could be introduced anywhere in panel 2–3, and the same influence line would result.

6.5   USE OF INFLUENCE LINES

Influence lines are useful to the structural analyst in two specific ways. These can be summarized as follows:

1. An influence line for a given response function can be used *qualitatively* to determine how a structure should be loaded in order to maximize a particular response function.

2. An influence line for a given response function can be used *quantitatively* to determine the value of the corresponding response function for any given loading arrangement.

The first point is useful in establishing the pattern of live loading that will give the maximum value of the response function, whereas the second point is useful for both dead and live loading to establish the value of the response function for a designated loading condition.

It is helpful to consider separately the case of a single concentrated load and that of a uniformly distributed load.

6.5.1   Concentrated Loads

Recalling that each ordinate of an influence line at a given abscissa gives the value of a response function caused by the placement of a unit load on the structure corresponding to that abscissa, and noting that superposition is valid, we can state the following principles for the use of influence lines with concentrated loads:

1. To determine the maximum signed value of a given response function due to a single concentrated load, the load should be placed on the structure at the point where the corresponding signed ordinate to the influence line for that response function is at a maximum.

2. The value of a response function attributable to the action of a single concentrated load equals the product of the magnitude of the load and the ordinate to the influence line for that response function measured at the abscissa corresponding to the point of application of the load.

3. The value of a response function due to a series of concentrated loads equals the summation of the products of the magnitudes of the loads and the respective ordinates to the influence line for the response function at the abscissas corresponding to the points of application of the loads.

6.5.2   Uniformly Distributed Load

The principles stated for concentrated loads in Section 6.5.1 are now used to develop two additional principles for uniformly distributed loads.

Consider the structural segment shown in Fig. 6.4, which is subjected to a uniformly distributed downward load of intensity p between points A and B. Also given in this figure is a segment of an influence line for the response function F_R. At section x, an element of the structure dx in length is taken, and an elemental load of $dP = p\,dx$ can be taken as a concentrated load at point x. Using the second principle of Section 6.5.1, we can express the increment of the response function that results from the load $dP = p\,dx$ as

$$dF_R = p\,dx\,[\text{I.L. } F_R(x)] \tag{6.8}$$

where $[\text{I.L. } F_R(x)]$ is the ordinate to the influence line at point x.

In accordance with the third principle of Section 6.5.1, the elemental contributions of Eq. 6.8 can be summed between A and B to give the total response F_R as

$$F_R = p\int_{x_A}^{x_B} [\text{I.L. } F_R(x)]\,dx \tag{6.9}$$

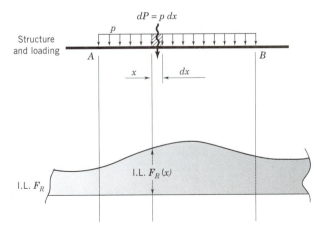

Figure 6.4 Use of influence lines for uniformly distributed loads.

In the above, p has been moved outside the integral since it is a constant. In the form of Eq. 6.9, it is clear that the integral represents the area under the influence line between the limits of points A and B on the structure.

Thus, the following principles can be stated for the use of influence lines with uniformly distributed loads:

1. To determine the maximum signed value of a given response function attributable to a uniformly distributed load, the load should be placed over all sections of the structure where the influence line for that response function has the corresponding signed ordinates.

2. The value of a response function attributable to the action of a uniformly distributed load over a portion of the structure equals the product of the load intensity and the net area under the influence line for that response function between the abscissas corresponding to the limits of the portion of the structure loaded.

EXAMPLE 6.9

Consider the simple beam structure shown below. For the loading specified, determine the maximum reaction at point A, the maximum positive shear at point C, and the maximum positive moment at point C. The required influence lines were developed in Example 6.1.

Dead load: Uniform load of 10 kN/m.

Live load: Uniform load of 25 kN/m; two concentrated loads of 100 kN that are separated by 3.

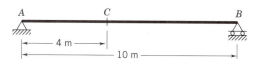

Note: The dead load acts along the entire structure in accordance with the prescribed magnitude, therefore, there is no judgment needed on the part of the analyst regarding where to position this load. The live load, however, should be positioned so as to maximize the signed response quantity under consideration. That is, concentrated live loads should be placed so as to interact with the maximum signed values on the influence line to produce the maximum effect, and distributed live loads should be spread over those regions of the structure for which the ordinates of the influence line have the desired signed values.

Reaction at Point A

The influence line for R_A is taken from Example 6.1. The ordinates are labeled for positioning the concentrated loads for maximum effect, and area under the influence line is given.

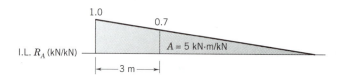

Dead load:	$10 \text{ kN/m} \times 5 \text{ kN·m/kN} =$	50 kN
Live load:	$25 \text{ kN/m} \times 5 \text{ kN·m/kN} =$	125 kN
	$100 \text{ kN} \times (1.0 + 0.7) \text{ kN/kN} =$	170 kN
Total R_A		= 345 kN

Positive Shear at Point C

The influence line for V_C is taken from Example 6.1. The appropriate ordinates and areas are noted.

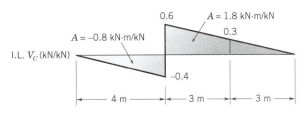

Dead load:	$10 \text{ kN/m} \times (1.8 - 0.8) \text{ kN·m/kN} =$	10 kN
Live load:	$25 \text{ kN/m} \times 1.8 \text{ kN·m/kN} =$	45 kN
	$100 \text{ kN} \times (0.6 + 0.3) \text{ kN/kN} =$	90 kN
Total Positive V_C		= 145 kN

Positive Moment at Point C

The influence line for M_C is taken from Example 6.1. The required ordinates and area are specified.

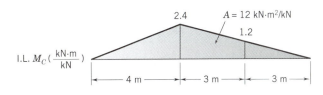

Dead load:	$10 \text{ kN/m} \times 12 \text{ kN·m}^2/\text{kN} = 120 \text{ kN·m}$
Live load:	$25 \text{ kN/m} \times 12 \text{ kN·m}^2/\text{kN} = 300 \text{ kN·m}$
	$100 \text{ kN} \times (2.4 + 1.2) \text{ kN·m/kN} = 360 \text{ kN·m}$
Total Positive M_C	= 780 kN·m

EXAMPLE 6.10

Consider the structure shown below. For the designated loading, determine the maximum positive (upward) reaction at point E and the maximum negative moment at point D. The required influence lines were developed in Example 6.2.

Dead load: Uniform load of 1 k/ft.

Live load: Uniform load of 3 k/ft; two concentrated loads of 5^k and 10^k that are separated by 5 ft.

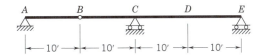

Upward Reaction at Point E

The influence line for R_E is taken from Example 6.2. The required ordinates and areas are designated.

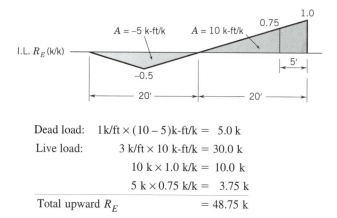

Dead load: $1\text{k/ft} \times (10 - 5)\text{k-ft/k} = 5.0 \text{ k}$

Live load: $3 \text{ k/ft} \times 10 \text{ k-ft/k} = 30.0 \text{ k}$

$10 \text{ k} \times 1.0 \text{ k/k} = 10.0 \text{ k}$

$5 \text{ k} \times 0.75 \text{ k/k} = 3.75 \text{ k}$

Total upward R_E $ = 48.75 \text{ k}$

Negative Moment at Point D

The influence line for M_D is taken from Example 6.2. The ordinates and areas that are needed are identified.

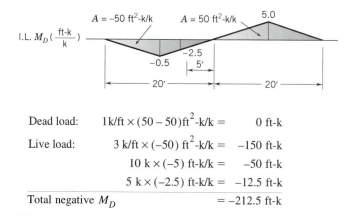

Dead load: $1\text{k/ft} \times (50 - 50)\text{ft}^2\text{-k/k} = 0 \text{ ft-k}$

Live load: $3 \text{ k/ft} \times (-50) \text{ ft}^2\text{-k/k} = -150 \text{ ft-k}$

$10 \text{ k} \times (-5) \text{ ft-k/k} = -50 \text{ ft-k}$

$5 \text{ k} \times (-2.5) \text{ ft-k/k} = -12.5 \text{ ft-k}$

Total negative M_D $ = -212.5 \text{ ft-k}$

6.6   MAXIMUM RESPONSE FUNCTIONS IN BEAMS

The sole purpose for considering variable loading conditions is to determine the loading arrangements that will maximize the response functions and to determine the corresponding values of these maximized response functions. Several topics related to the general area of maximum effects are discussed in this section.

6.6.1   Maximum Absolute Moment

In the example problems of Section 6.5 the maximum moments were computed at designated points on the structure. However, moments computed in this fashion may not include the maximum possible moment on the structure.

On a simply supported beam, the maximum moment occurs at midspan for a uniformly distributed load or at the point of loading for a single concentrated load. One need merely construct the influence line for the moment at the appropriate point and use it in conjunction with the specified loading to determine the maximum moment.

Another case of interest occurs when a simply supported beam is subjected to a series of concentrated loads. For any position of these loads, the moment diagram will consist of a series of straight-line segments between the load points; thus, the maximum moment must occur at a load point. However, it is not apparent which load will occupy the position of maximum moment or where the loads should be positioned to create the maximum possible moment.

Consider the simply supported beam shown in Fig. 6.5, which supports a series of n concentrated loads. These loads can be represented by the resultant, $\overline{P}$, which is located at a distance d from the load P_l. The maximum moment will generally occur at one of the loads that is adjacent to $\overline{P}$, and it is assumed that this condition occurs at the load P_l when it is located at a position x from the midspan of the beam. This condition is illustrated in Fig. 6.5.

Summing moments about point B, we determine the reaction at A to be

$$R_{Ay} = \frac{\overline{P}(L/2 + x - d)}{L} \tag{6.10}$$

Thus, the moment at the load P_l is

$$M_{P_l} = R_{Ay}\left(\frac{L}{2} - x\right) - \sum_{i=1}^{k} P_i x_i \tag{6.11}$$

where x_i gives the position of P_i with respect to P_l. Substituting Eq. 6.10 into Eq. 6.11, we obtain

$$M_{P_l} = \overline{P}\left(\frac{L}{4} - \frac{d}{2} - \frac{x^2}{L} + \frac{xd}{L}\right) - \sum_{i=1}^{k} P_i x_i \tag{6.12}$$

For the maximum value of M_{P_l}

$$\frac{dM_{P_l}}{dx} = \overline{P}\left(-\frac{2x}{L} + \frac{d}{L}\right) = 0 \tag{6.13}$$

from which $x = d/2$.

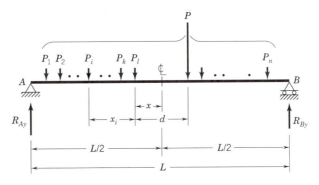

Figure 6.5 Loading for absolute maximum moment.

We thus conclude that *the maximum moment at one of a series of concentrated loads that act on a simply supported beam occurs when the midspan of the structure is located midway between that particular load and the resultant of all loads acting on the span.* This criterion gives the position of the loads for maximum moment to occur at a given load. To obtain the absolute maximum moment, the above location criterion would have to be applied to all loads, and the largest of all maximum moments would then be selected. In general, the largest of the two loads adjacent to the resultant is the load at which the maximum moment occurs.

In many statically determinate beams, the section where the absolute maximum moment occurs cannot be established by inspection, nor can convenient criteria be established. For such cases, it is necessary to compute the maximum moment for sections where the absolute maximum value is likely to occur and then select the controlling maximum value.

6.6.2   Critical Loading Conditions from Influence Lines

As was illustrated in Section 6.5, the loading conditions that cause the maximum value for any response function can be determined from the influence line for that function. For complicated loading conditions, however, it is not always apparent how the loads should be arranged for maximum effects. It is possible to develop detailed criteria for the determination of the critical load position for maximizing any response function; however, such criteria will not be presented here. The problems of Section 6.5 illustrate the essential features for the utilization of influence lines.

With the computer capabilities now available, it is frequently easier to enter those loading patterns that are feasible critical loading conditions and then analyze the structure for each loading case. A comparison of the results then yields the critical loading case and gives the desired maximum effects.

6.6.3   Highway Loadings

Design specifications and building codes usually dictate the loading conditions that must be considered in the design of any particular structure. As an example of a typical situation, we will consider the bridge loadings that are specified in the *LRFD Bridge Design Specifications* of the American Association of State Highway and Transportation Officials (AASHTO). Section 1.8.2.2 outlines the loadings specified by AASHTO. According to this specification, a critical condition can result from one of two combinations, whichever produces the maximum values of the desired response functions. These combinations are:

  1. The effect of one design truck (Fig. 1.23), with the variable axle spacing specified, combined with the effect of the design lane load.
  2. The effect of the design tandem combined with the effect of the design lane load.

The design lanes and the 10-foot, loaded width in each lane shall be positioned so as to produce extreme force effects. For the design truck and tandem, they shall be positioned transversely such that the center of any wheel load is no closer than 2 feet from the edge of the design lane for the design of most members. This dimension, coupled with the 6-foot spacing between wheels, places the axle transversely, as if it were centered in a 10-foot-wide lane.

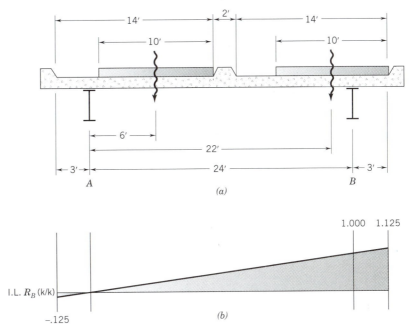

Figure 6.6 Position of lane loadings for maximum girder load. *(a)* Bridge cross section. *(b)* Influence line for R_B.

Each of these loading arrangements is for a single lane of traffic, which occupies a 10-foot width of the roadway. The number of lanes of traffic that must be supported by each supporting bridge member is determined from a transverse analysis of the structure. For instance, consider a two-lane bridge, of which a cross section is shown in Fig. 6.6. Each traffic lane is 14 feet wide, and the two lanes are separated by a 2-foot median strip. If the transverse section of the bridge is assumed to be simply supported by girders A and B, then the influence line for the force between the bridge deck and girder B, R_B, is given in Fig. 6.6b. Thus, to maximize R_B, each lane loading should be placed to the extreme edge of the traffic lane, as shown in Fig. 6.6a.

Taking moments about girder A for this position of lane loading, we have

$$(\text{Lane})\ 6 + (\text{Lane})\ 22 = R_B \cdot 24$$

$$R_B = \frac{28\ (\text{Lane})}{24} = 1.17\ \text{Lanes}$$

Normally, one would proceed with the analysis of girder B with the normal considerations of maximum effects, using the single lane loadings of Fig. 1.23 in conjunction with the required influence lines. Then, the final results would be multiplied by a factor of 1.17 to account for the transverse positioning of the lane loading.

When the bridge deck is transversely supported by more than two longitudinal members, the lane loading taken by each girder would result from a statically indeterminate analysis of a continuous beam on deflecting supports. For this case, the AASHTO specifications provide factors that can be used to determine the loading to be applied to each girder.

6.6.4 Envelope of Maximum Effects

In the design process, each member must be proportioned to sustain the largest possible values of *all* response quantities at every point along the structure. Thus, the placement of load for the maximization of a response quantity and the corresponding maximum value of that quantity must be determined for a number of points along the structure. The collection of these results for any one response function forms an *envelope of maximum effects*.

Consider the beam structure shown in Fig. 6.7a along with the prescribed intensities for dead load and live load, p_D and p_L, respectively. The dead load extends over the entire structure, whereas the live load can be applied over any desired section of

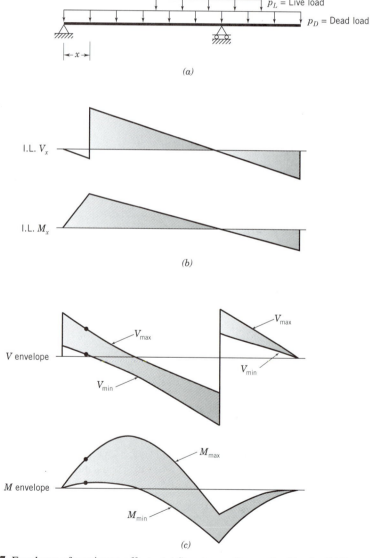

Figure 6.7 Envelopes of maximum effects. (*a*) Structure and prescribed loads. (*b*) Representative influence lines. (*c*) Envelopes for shear and moment.

the structure. Figure 6.7*b* shows representative influence lines for shear and moment at point *x* along the span. These influence lines are used in conjunction with the pre-scribed loads to determine the extreme values for both shear and moment. These values are plotted as dots in Fig. 6.7*c* corresponding to the *x* position. If this procedure is followed for several values of *x*, the full envelope of extreme values is produced as shown in Fig. 6.7*c*.

Problems of this type are best carried out through the systemization provided by tabular calculations. This is illustrated by Example 6.11.

Of course, an envelope of maximum effects can be generated for any response function coupled with any loading situation. Example 6.12 involves the determination of the envelope of maximum live load moments associated with the HS 20-44 truck loading.

EXAMPLE 6.11

Construct the envelopes for maximum shears and moments for the structure shown. Consider 2-meter intervals along the structure.

$$\text{Dead load: Uniform load} = p_D = 10 \text{ kN/m}$$

$$\text{Live load: Uniform load} = p_L = 15 \text{ kN/m}$$

Influence Lines

The influence lines for shear and moment are required at 2-meter intervals along the structure. These are portrayed below in general form.

Shear:

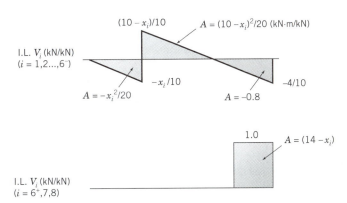

Moment:

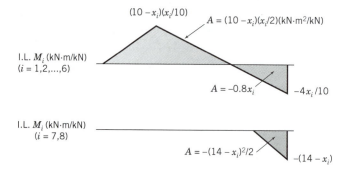

Response Values

The extreme values of the shears and moments are computed in the following tables:

Shear:

		Infl. Line Areas			Shear = $p \times$ Area (kN)			Envelope Extremes	
Pt.		(kN · m/kN)			Dead Load	Live Load		$V_{max} =$	$V_{min} =$
i	x_i	Pos.	Neg.	Net	V_D	V_L^+	V_L^-	$V_D + V_L^+$	$V_D + V_L^-$
1	0.0	5.0	−0.8	4.2	42.0	75.0	−12.0	117.0	30.0
2	2.0	3.2	−1.0	2.2	22.0	48.0	−15.0	70.0	7.0
3	4.0	1.8	−1.6	0.2	2.0	27.0	−24.0	29.0	−22.0
4	6.0	0.8	−2.6	−1.8	−18.0	12.0	−39.0	−6.0	−57.0
5	8.0	0.2	−4.0	−3.8	−38.0	3.0	−60.0	−35.0	−98.0
6^-	10.0	0.0	−5.8	−5.8	−58.0	0.0	−87.0	−58.0	−145.0
6^+	10.0	4.0	0.0	4.0	40.0	60.0	0.0	100.0	40.0
7	12.0	2.0	0.0	2.0	20.0	30.0	0.0	50.0	20.0
8	14.0	0.0	0.0	0.0	0.0	0.0	0.0	0.0	0.0

Moment:

		Infl. Line Areas			Moment = $p \times$ Area (kN · m)			Envelope Extremes	
Pt.		(kN · m²/kN)			Dead Load	Live Load		$M_{max} =$	$M_{min} =$
i	x_i	Pos.	Neg.	Net	M_D	M_L^+	M_L^-	$M_D + M_L^+$	$M_D + M_L^-$
1	0.0	0.0	0.0	0.0	0.0	0.0	0.0	0.0	0.0
2	2.0	8.0	−1.6	6.4	64.0	120.0	−24.0	184.0	40.0
3	4.0	12.0	−3.2	8.8	88.0	180.0	−48.0	268.0	40.0
4	6.0	12.0	−4.8	7.2	72.0	180.0	−72.0	252.0	0.0
5	8.0	8.0	−6.4	1.6	16.0	120.0	−96.0	136.0	−80.0
6	10.0	0.0	−8.0	−8.0	−80.0	0.0	−120.0	−80.0	−200.0
7	12.0	0.0	−2.0	−2.0	−20.0	0.0	−30.0	−20.0	−50.0
8	14.0	0.0	0.0	0.0	0.0	0.0	0.0	0.0	0.0

Envelopes for Shear and Moment

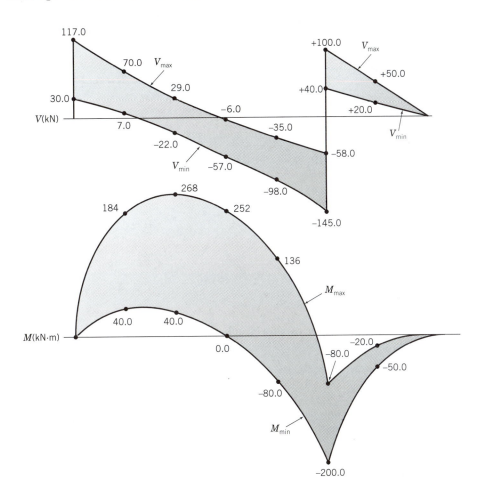

Note: The shaded portions identify the range of possible shears and moments for the prescribed loads.

EXAMPLE 6.12

Compute the envelope of maximum live load moments for one of the supporting girders for the single-span bridge shown below. Assume that the structure is a two-lane bridge with the arrangement shown in Fig. 6.6 and that it supports an AASHTO Design Truck.

 The envelope ordinates should be determined at 5-foot intervals along the structure.

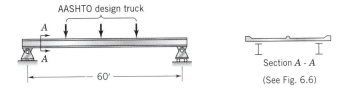

Influence Line Construction

Moment influence lines are required for points at 5-foot intervals along the structure.

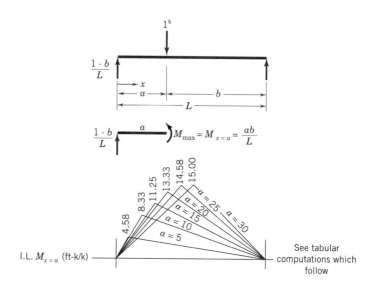

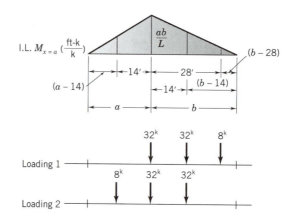

I.L. $M_{x=a}$ (ft-k/k)   See tabular computations which follow

Moments for Single Lane of AASHTO Design Truck

For Loading 1

$$M_{x=a} = 32\left(\frac{ab}{L}\right) + 32\left(\frac{ab}{L}\right)\left(\frac{b-14}{b}\right) + 8\left(\frac{ab}{L}\right)\left(\frac{b-28}{b}\right)$$

$$= \frac{ab}{L}\left[32 + 32\left(\frac{b-14}{b}\right) + 8\left(\frac{b-28}{b}\right)\right] = \frac{ab}{L}\text{[Coeff. 1]}$$

For Loading 2

$$M_{x=a} = 32\left(\frac{ab}{L}\right) + 32\left(\frac{ab}{L}\right)\left(\frac{b-14}{b}\right) + 8\left(\frac{ab}{L}\right)\left(\frac{a-14}{a}\right)$$

$$= \frac{ab}{L}\left[32 + 32\left(\frac{b-14}{b}\right) + 8\left(\frac{a-14}{a}\right)\right] = \frac{ab}{L}\text{[Coeff. 2]}$$

a	b	(ab/L)	[Coeff. 1]	[Coeff. 2]	$M_{x=a} = \dfrac{ab}{L} \cdot (\text{Coeff.})_{\max}$
(ft)	(ft)	(ft-k/k)	(k)	(k)	(ft-k)
5	55	4.58	59.78	*	273.8
10	50	8.33	58.56	*	487.8
15	45	11.25	57.06	54.58	641.9
20	40	13.33	55.20	55.20	735.8
25	35	14.58	52.80	54.72	797.8
30	30	15.00	49.60	53.33	800.0

*Loading inappropriate since $a < 14$ ft.

Absolute Maximum Moment

The procedure of Section 6.6.1 is used.

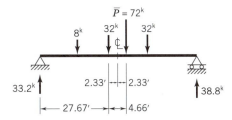

The maximum moment occurs under the wheel positioned at $x = 27.67$ ft.

$$M_{\max} = (33.2 \times 27.67) - (8 \times 14) = 806.6 \text{ ft-k}$$

Envelope of Maximum Live Load Moments

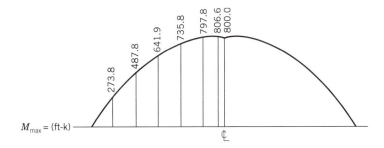

Notes:

1. The above maximum moments are for one lane of loading. These should be multiplied by 1.17 to get the maximum girder moments, as was described in Section 6.6.3, for the arrangement of Fig. 6.6.
2. The moments caused by the design lane loading must also be determined, and they too should be multiplied by 1.17 to get the girder moments.
3. The maximum live load moments (truck or lane) must be combined with the dead load moments to determine the total moments for design.

6.7 SUGGESTED PROBLEMS

6.1 through 6.20 Use the conventional equilibrium approach to construct influence lines for the indicated response functions for each of the structures shown below. In each case, use the Müller–Breslau principle to check the influence lines qualitatively.

6.1 Reactions at *A* and *B*, shear at *C*, moment at *C*.

6.2 Vertical reaction at *B*, moment at *B*, shear at *C*, moment at *C*.

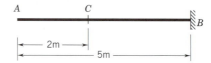

6.3 Reactions at *A* and *B*, shear at *C*, moment at *C*, shear at *D*, moment at *D*, shear at *E*, moment at *E*. *Note:* Point *C* is just right of support at *A*; point *E* is just right of support at *B*.

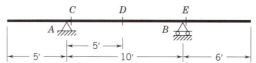

6.4 Vertical reaction at *A*, horizontal reaction at *A*, moment at *A*, shear at *C*, moment at *C*, reaction at *B*.

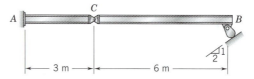

6.5 All vertical reactions, shear at *E*, moment at *E*.

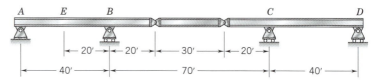

6.6 Vertical reactions at *A* and *D*, horizontal reactions at *A* and *D*, shear at *F*, moment at *F*. *Note:* Unit load moves from *A* to *C*.

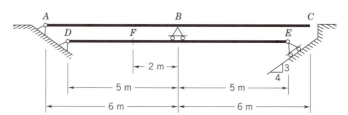

6.7 Vertical reaction at D, horizontal reaction at D, moment at D, reaction at B, moment at B, shear at C.

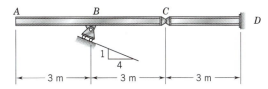

6.8 Vertical and horizontal reactions at C, shear and moment at B on member CD, shear and moment at E. Unit load moves along $A–B–D$.

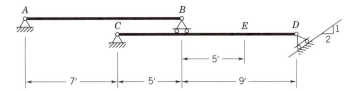

6.9 Vertical and horizontal reactions at A, reaction at E, shear and moment at G, shear and moment at D. Unit load moves from A to E.

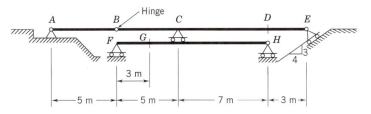

6.10 Reaction at A, reaction at B, shear in girder at C, moment in girder at C. Unit load moves along stringers.

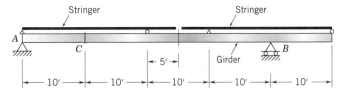

6.11 Vertical and horizontal reaction at A, vertical reaction at F, shear at C, moment at D. Unit load moves from B to E. *Note:* Point C is just right of beam-to-column connection.

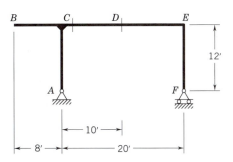

6.12 All reaction components, shear at C, moment at top of columns BA and DF. Unit load moves from B to E.

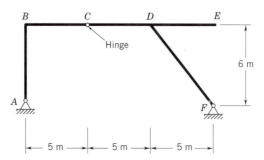

6.13 All reaction components, shear and moment at E, moment at top of column BD. Unit load moves from A to C.

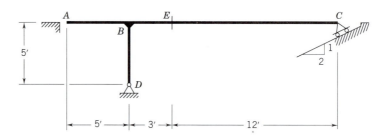

6.14 All reaction components, shear at e, moment at f, moment at top of column ed. Unit load moves from a to c.

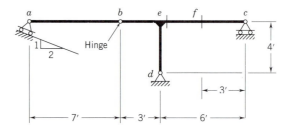

6.15 Horizontal and vertical reactions at A and B, shear and moment at E. Unit load moves from C to G.

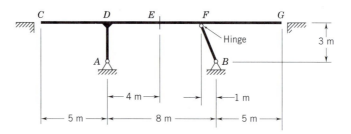

6.16 Horizontal and vertical reactions at D, member forces in AB and FG. Unit load moves from A to E.

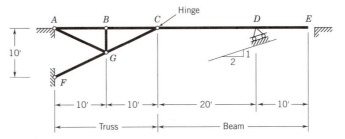

6.17 Forces in members ab, bc, cg, fg, and bf. Unit load moves from a to e.

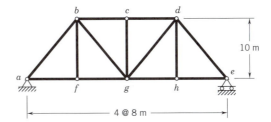

6.18 Reaction at L_0 and L_3, forces in members U_1L_1, U_2U_3, U_2L_3, L_1L_2, U_4L_4, and U_4L_5.

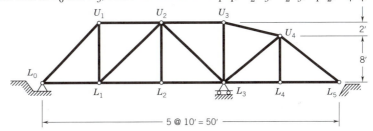

6.19 Vertical reactions at A, B, and U_4; forces in members U_1U_2, L_1U_2, U_1L_1, L_1L_2, U_2L_2, and L_3U_4. Unit load moves from A to U_4.

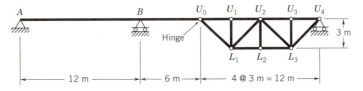

6.20 All reaction components; forces in members ab, ag, af, bg, fg, and gc. Unit load moves from a to e.

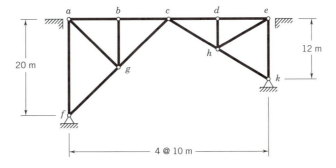

6.21 Consider the structure of Problem 3. For the loading specified, determine the maximum vertical reaction at point A, the maximum positive moment at point D, and the maximum positive shear at point D.

Dead load: Uniform load of 2 k/ft.
Live load: Uniform load of 4 k/ft and a roving concentrated load of 25 kips.

6.22 Consider the structure of Problem 5. For the loading specified, determine the maximum vertical reaction at B, the maximum negative shear at E, and the maximum negative moment at E.

Dead load: Uniform load of 2.5 k/ft.
Live load: Uniform load of 4.0 k/ft and a roving concentrated load of 30 kips.

6.23 Consider the structure of Problem 13. For the loading specified, determine the maximum horizontal reaction at D, the maximum positive and negative shear at E, and the maximum moment at the top of column BD.

Dead load: Uniform load of 1.5 k/ft.
Live load: Uniform load of 4.0 k/ft and two roving concentrated loads of 20 kips each that are spaced 6 feet apart.

6.24 Consider the structure of Problem 15. For the loading specified, determine the maximum horizontal reaction at point A, the maximum positive moment at point E, and the maximum positive shear at point E.

Dead load: Uniform load of 14 kN/m.
Live load: Uniform load of 24 kN/m and two concentrated loads of 72 kN that are separated by 2 m.

6.25 Consider the structure of Problem 18. For the loading specified, determine the maximum positive and negative axial forces in members U_1L_1, U_2U_3, U_2L_3, and U_4L_5.

Dead load: Uniform load of 4.0 k/ft.
Live load: Uniform load of 2.5 k/ft and two concentrated loads of 20 kips and 10 kips, respectively, that are separated by 8 feet.

6.26 Construct the envelopes for maximum shears and moments for the structure of Problem 3. Consider 2.5-ft intervals along the structure.

Dead load: Uniform load $= p_D = 2.0$ k/ft.
Live load: Uniform load $= p_L = 3.5$ k/ft.

6.27 Construct the envelopes for maximum shears and moments for the structure of Problem 7. Consider 1-meter intervals along the structure.

Dead load: Uniform load $= p_D = 5.0$ kN/m.
Live load: Uniform load $= p_L = 8.0$ kN/m.

6.28 Construct the envelopes for maximum shears and moments in the beam elements of the structure of Problem 13. Consider 2.5-ft intervals along the structure and include points both left and right of point B.

Dead load: Uniform load $= p_D = 1.5$ k/ft.
Live load: Uniform load $= p_L = 2.5$ k/ft.

6.29 Develop the envelope of maximum live-load moments for one of the supporting girders for the single-span bridge structure shown. The structure can carry two lanes of traffic, and the loading for a single lane is given. The envelope ordinates should be determined at 2-meter intervals.

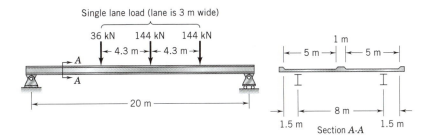

Part Three

Elastic Deflections of Structures

Chapter 7

Elastic Deflections of Trusses and Frameworks

The World Trade Center, New York City (1970–2001) (courtesy NYC & Company—the Convention and Visitors Bureau).

7.1   DESCRIPTION OF TRUSS DEFLECTION PROBLEM

When a pin-connected truss or framework is subjected to a system of applied forces, it develops a set of reactions and member forces that is consistent with the requirements of equilibrium. In response to the member forces, each bar within the truss experiences an axial deformation; however, the assemblage of deformed bars must continue to fit together in accordance with the requirements of compatibility. As a result, the truss joints that are not constrained by the support conditions will experience displacements as the structure adjusts to the deformed member lengths. These joint displacements define the deflected configuration of the truss as a whole.

There are two important reasons for determining deformation quantities. The first has to do with serviceability requirements. Structures generally are required to function within specified limitations regarding deformations if they are to perform in the desired fashion. These serviceability requirements might be related to any stage of erection or to any loading condition. The second reason for determining deformations has to do with the analysis process itself. For statically indeterminate analysis, the equations of equilibrium must be augmented by conditions of compatibility. These conditions result in equations in which deformation quantities are matched, and thus it is necessary to compute specific deformations.

Deformations may be recoverable when the loads are removed or they may be permanent or nonrecoverable in nature. The former type is referred to as an *elastic deformation,* the latter, a *plastic deformation.* Here, we limit our treatment to elastic deformations.

7.2   AXIAL FORCE–DEFORMATION RELATIONSHIPS

For any member, there are definite relationships between the member-end forces and the corresponding member-end deformations. In the case of a truss member, we seek the relationship between the axially applied force and the axial deformation.

Consider the member shown in Fig. 7.1a, which is of length l, has a cross-sectional area A, and is subjected to the axial force F. An element dx in length is shown in Fig. 7.1b. Applying the principles of elementary mechanics to this element, we recall that the stress is defined as the force per unit area. Representing this stress by σ, we thus have

$$\sigma = \frac{F}{A} \tag{7.1}$$

The strain is given by the change in length per unit length. If we represent strain by ϵ, we then have

$$\epsilon = \frac{dl}{dx} \tag{7.2}$$

where dl is the change in length or the deformation that accompanies the force F.

For a linearly elastic material, if stress is plotted as ordinate against strain as abscissa, we have the linear relationship shown in Fig. 7.1c. For any given strain, there is a unique stress given by the relationship

$$\sigma = E\epsilon \tag{7.3}$$

where E is Young's modulus or the modulus of elasticity. This modulus is a material property that must be determined from an appropriate test procedure.

Solving Eq. 7.2 for the deformation dl and then substituting Eq. 7.3 and Eq. 7.1, we obtain

$$dl = \frac{F}{EA} \, dx \tag{7.4}$$

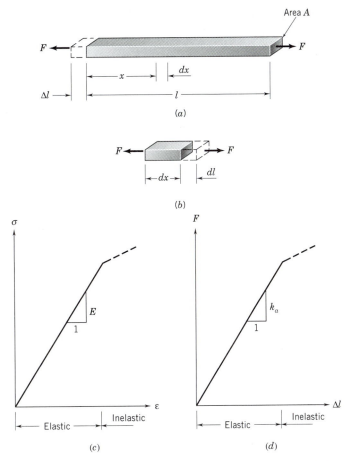

Figure 7.1 Axial force–deformation behavior. (a) Axially loaded member. (b) Element. (c) Stress–strain diagram. (d) Force–deformation relationship.

The total elongation Δl is given by integrating Eq. 7.4 over the length of the member, and because the axial force and the area are constant throughout the member length, we obtain

$$\Delta l = \int_0^l dl = \frac{F}{EA}\int_0^l dx = \frac{l}{EA}F \qquad (7.5)$$

Equation 7.5 is a *deformation–force relationship*, and it can be written in the form

$$\Delta l = d_a F \qquad (7.6)$$

where d_a, the *axial flexibility,* is given by

$$d_a = \frac{l}{EA} \qquad (7.7)$$

In an inverse form, or a *force–deformation relationship,* Eq. 7.5 can be rewritten as

$$F = \frac{EA}{l}\Delta l \qquad (7.8)$$

or

$$F = k_a \Delta l \qquad (7.9)$$

where k_a, the *axial stiffness*, is defined as

$$k_a = \frac{EA}{l} \tag{7.10}$$

The relationship between F and Δl is shown in Fig. 7.1d. The shape of this curve is shown to mirror the shape of the stress–strain curve of the constituent material.

In a more general form, consider the member shown in Fig. 7.2. Here the member-end forces and displacements are separately identified—F_1 and δ_1 at end 1, and F_2 and δ_2 at end 2.

Comparing Figs. 7.1a and 7.2, it is clear that $F = F_1 = F_2$ and $\Delta l = \delta_1 + \delta_2$. Substitution of these relationships into Eq. 7.8, and a subsequent translation into matrix form, leads to

$$\begin{Bmatrix} F_1 \\ F_2 \end{Bmatrix} = \begin{bmatrix} \dfrac{EA}{l} & \dfrac{EA}{l} \\ \dfrac{EA}{l} & \dfrac{EA}{l} \end{bmatrix} \begin{Bmatrix} \delta_1 \\ \delta_2 \end{Bmatrix} \tag{7.11}$$

In an abbreviated form, this equation can be expressed as

$$\{F\} = [k]_a\{\delta\} \tag{7.12}$$

where $\{F\}$ and $\{\delta\}$ are the member-end force and displacement vectors, respectively, and $[k]_a$ is the *member axial stiffness matrix*. Equation 7.12 is the two-dimensional equivalent of Eq. 7.9, and the individual elements of $[k]_a$ are in terms of EA/l, which reflects the axial stiffness of the member as it is defined by Eq. 7.10.

As noted earlier, Eqs. 7.6 and 7.9 are inverse expressions—the former gives a deformation–force relationship, whereas the latter states a force–deformation relationship. In the two-dimensional form, Eq. 7.12 gives the member-end forces that are associated with prescribed member-end displacements. These are clearly force–deformation relationships. Here, too, it is possible to develop inverse expressions, or displacement–force relationships. These would allow for the determination of certain member-end displacements from prescribed member-end forces, provided that the member is adequately restrained against rigid-body motions. This will be discussed fully in Chapter 13. Our purpose here is merely to emphasize that there are definite relationships between the forces that a member transmits and the deformations that it experiences and that these member relationships are central to determining the deflections of a structure as a whole.

The foregoing discussion is limited to the elastic response of the member. If the member is stressed beyond the elastic range, as shown by the dashed portion of Fig. 7.1c, the resulting force–deformation relationships will reflect this behavior as indicated in Fig. 7.1d. Of course, in the inelastic range, Eqs. 7.6, 7.9, and 7.12 are no longer valid.

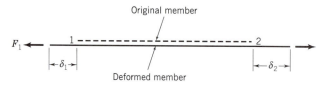

Figure 7.2 Member-end forces and displacements.

7.3   GEOMETRY OF TRUSS DEFLECTIONS

Consider the simple two-bar truss shown in Fig. 7.3*a*. Determination of the bar forces enables one to use Eq. 7.5 to compute the corresponding axial deformations Δl_{AB} and Δl_{BC} for members *AB* and *BC*, respectively, as shown in Fig. 7.3*b*. However, if the two bars are to remain connected at point *B,* the bars must each rotate until they terminate at the common point *B'*. Under this arrangement, equilibrium and compatibility are both satisfied. Of course, it is assumed that the displacements are small enough that the overall geometry changes are small and thus the bar forces are not altered by the displacements. Also, since the displacements are small, it is assumed that as each bar rotates about its support point, its terminal point moves along the tangent to the arc, or normal to the bar itself. This arrangement is shown in Fig. 7.3*c*, where the vertical and horizontal displacements are denoted as v_B and u_B, respectively. This enlarged sketch of the member elongations and the resulting joint displacements is referred to as a *Williot diagram.*

This procedure is essentially a process of triangulation in which one uses the location of two points and the deformation of two bars to establish the location of a third point. This technique can readily be applied to a larger truss by working from joint to joint throughout the truss. This, in fact, was done in earlier times when graphic methods were a useful tool in the hands of the structural analyst. It is no longer a viable method, but it does illustrate how truss deflections occur as a result of the geometric adjustments that a structure must undergo while satisfying the dual requirements of force-induced member elongations and joint compatibility.

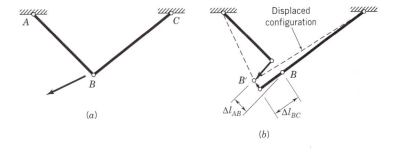

(a)

(b)

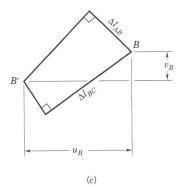

(c)

Figure 7.3 Displacements of two-bar truss. *(a)* Truss and loading. *(b)* Bar elongations. *(c)* Williot diagram.

7.4   TRUSS DEFLECTIONS BY COMPLEMETARY VIRTUAL WORK

Truss deflections can be determined by using the complementary virtual work method that uses virtual forces. Each application of this method enables the analyst to determine a single displacement quantity for a designated loading condition. In accordance with the procedure outlined in Section 2.16.2, the following systems are employed:

P System: Virtual force system in equilibrium.

Δ System: Actual deformation pattern that is geometrically compatible.

Using these systems, we equate the external complementary virtual work to the internal complementary virtual work of deformation. Thus, as the P system undergoes the displacements of the Δ system, we have

Virtual P system

$$\sum_{i=1}^{n}(\delta P)_i \Delta_i = \int_{\text{Vol}} (\delta \sigma_P) \epsilon_\Delta d\,\text{Vol} \tag{7.13}$$

Actual Δ system

Because the right-hand side of this equation represents the internal work of deformation, its form depends on the nature of the deformation. For a truss that is composed of m members, this integral represents the summation over all member lengths of the virtual forces of the P system that act through the axial deformations of the Δ system. Reference to Fig. 7.4 shows that this can be written in the form

$$\int_{\text{Vol}} (\delta \sigma_P) \epsilon_\Delta d\,\text{Vol} = \sum_{j=1}^{m} \left(\int_l (\delta F_P) dl_\Delta \right)_j \tag{7.14}$$

where l is the length of the jth member, δF_P is the virtual bar force at some point x on member j corresponding to the P system, and dl_Δ is the associated member deformation at the same point on member j that occurs over the differential element dx of the Δ system. Substitution of Eq. 7.14 into Eq. 7.13 gives

$$\sum_{i=1}^{n}(\delta P)_i \Delta_i = \sum_{j=1}^{m} \left(\int_l (\delta F_P) \cdot dl_\Delta \right)_j \tag{7.15}$$

If dl_Δ results from a Δ system of loads on the truss, then Eq. 7.4 gives

$$dl_\Delta = \frac{F_\Delta}{EA} dx \tag{7.16}$$

where F_Δ is the bar force at point x on member j corresponding to the Δ system. Substitution of Eq. 7.16 into Eq. 7.15 gives

$$\sum_{i=1}^{n}(\delta P)_i \Delta_i = \sum_{j=1}^{m} \left(\int_l (\delta F_P) \frac{F_\Delta dx}{EA} \right)_j \tag{7.17}$$

In this case, because the quantity EA and both δF_P and F_Δ are constant throughout the member lengths, Eq. 7.17 simplifies to

$$\sum_{i=1}^{n}(\delta P)_i \Delta_i = \sum_{j=1}^{m} \left(\delta F_P \frac{F_\Delta l}{EA} \right)_j \tag{7.18}$$

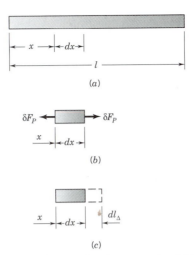

Figure 7.4 (a)

(b)

(c)

Figure 7.4 Virtual P system and actual Δ system for jth truss member. (a) jth member. (b) Virtual bar force at point x caused by P system of loads. (c) Elongation of element dx caused by Δ system of definitions.

When dl_Δ is caused by something other than loads, such as temperature change or fabrication error, Eq. 7.15 is still valid. Again, δF_P is constant throughout the member, and $\int_l dl_\Delta = \Delta l_\Delta$, the total change in length of member j. Thus, Eq. 7.15 takes the form

$$\sum_{i=1}^{n} (\delta P)_i \Delta_i = \sum_{j=1}^{m} [(\delta F_P) \cdot \Delta l_\Delta]_j \tag{7.19}$$

7.5   APPLICATION OF THE COMPLEMENTARY VIRTUAL WORK METHOD

When the complementary virtual work method is used to obtain the displacements associated with some actual Δ system, the P system is a virtual force system that must be selected with care. Examination of Eq. 7.15 shows that for the determination of any particular displacement quantity Δ_r, the P system must include a virtual force $(\delta P)_r$ that acts at the point and in the direction of the desired displacement Δ_r. There can be no other forces in the P system that couple with displacements of the Δ system. Generally, $(\delta P)_r$ is taken as a unit load, and Eq. 7.15 then yields the desired displacement Δ_r. Thus, a different virtual force system must be used for each displacement quantity that is desired for a given Δ system.

EXAMPLE 7.1

Consider the truss structure and loading shown, and determine the horizontal and vertical components of the displacement at point c. All members have a cross-sectional area of $0.002\,5\ \mathrm{m}^2$ and a modulus of elasticity of 200×10^9 Pa. The reactions are given.

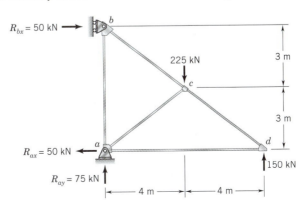

The following complementary virtual work expression is used:

$$\sum_{i=1}^{n} (\delta P)_i \Delta_i = \sum_{j=1}^{m} \left(\delta F_P \cdot \frac{F_\Delta l}{EA} \right)_j \qquad (7.18)$$

Determination of Horizontal Displacement u_c

Δ *System*

Actual system, which includes the applied loads, the resulting reactions and bar forces, and the desired displacement (identified by the vector u_c).

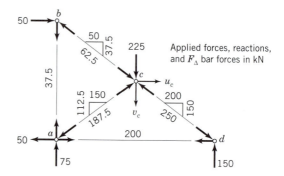

P System

Virtual system, which includes an applied unit force at the point and in the direction of the desired displacement, u_c, and the resulting reactions and bar forces.

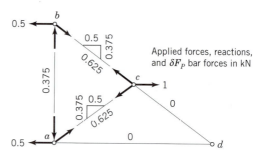

Because the reactions of the P system correspond to nondeflecting points of the Δ system, they do not contribute to the left-hand side of Eq. 7.18, and therefore,

$$\sum_{i=1}^{n} (\delta P)_i \Delta_i = (1)(u_c) \, \text{kN} \cdot \text{m}$$

And, because A and E are the same for all members, the right-hand side of Eq. 7.18 takes the following form:

$$\sum_{j=1}^{m} \left(\delta F_P \frac{F_\Delta l}{EA} \right)_j = \frac{1}{EA} \sum_{j=1}^{m} (\delta F_P \cdot F_\Delta \cdot l)_j$$

This summation can best be determined by use of the following tabulation:

Bar	l	F_Δ	δF_P	$\delta F_P \cdot F_\Delta \cdot l$
—	m	kN	kN	$(kN)^2 \cdot m$
ab	6.0	+37.5	−0.375	−84.38
bc	5.0	−62.5	+0.625	−195.31
cd	5.0	−250.0	0	0
da	8.0	+200.0	0	0
ac	5.0	−187.5	+0.625	−585.94

$$\sum (\delta F_P \cdot F_\Delta \cdot l) = -865.63$$

Thus, the right-hand side of Eq. 7.18 is

$$\frac{1}{EA} \sum_{j=1}^{m} (\delta F_P \cdot F_\Delta \cdot l) = \frac{-865.63}{200 \times 10^6 \times 0.002\ 5} = -0.001\ 73\ \text{kN} \cdot \text{m}$$

Applying Eq. 7.18, we have

$$(1)(u_c) = -0.001\ 73\ \text{kN} \cdot \text{m}$$

from which, noting that the unit force carries the unit of kN, we obtain

$$u_c = -0.001\ 73\ \text{m} = -1.73\ \text{mm}$$

The negative sign simply means that u_c is in a direction opposite to that which was assumed, that is, to the left and not to the right.

Determination of Vertical Displacement v_c

Δ System

Actual system, which is the same system as used u_c. The desired displacement is identified by the vector v_c.

P System

Virtual system whose applied loading is a unit force at the point and in the direction of the desired displacement, v_c.

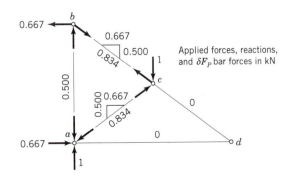

Applied forces, reactions, and δF_P bar forces in kN

Bar	l	F_Δ	δF_P	$\delta F_P \cdot F_\Delta \cdot l$
—	m	kN	kN	$(kN)^2 \cdot m$
ab	6.0	$+37.5$	-0.500	-112.50
bc	5.0	-62.5	$+0.834$	-260.63
cd	5.0	-250.0	0	0
da	8.0	$+200.0$	0	0
ac	5.0	-187.5	-0.834	$+781.88$

$$\sum (\delta F_P \cdot F_\Delta \cdot l) = - + 408.75$$

Application of Eq. 7.18 gives

$$(1)(v_c) = \frac{408.75}{200 \times 10^6 \times 0.002\ 5} = 0.000\ 82\ kN \cdot m$$

from which

$$v_c = 0.000\ 82\ m = 0.82\ mm$$

Here, the positive sign means that the displacement is downward, as assumed.

To avoid unnecessary repetition, the last two columns of the preceding table could have been appended to the first table.

EXAMPLE 7.2   Determine the vertical component of the displacement at point d for the truss and loading shown. The areas are given in parentheses (mm^2) along each member, and the modulus of elasticity is 200 GPa. The reactions are given.

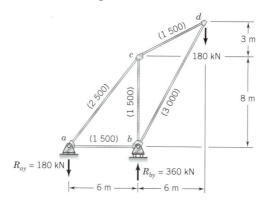

Determination of Vertical Displacement v_d

Δ *System*

Actual system that includes the desired displacement, v_d (downward vector at d).

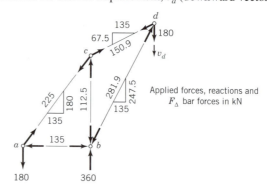

P System

Virtual system with δP loading selected to obtain v_d.

In this case a unit vertical load of 1 kN is placed at point d. Therefore, the P system forces can be determined by multiplying the Δ system forces by $\frac{1}{180}$.

The summation on the right-hand side of Eq. 7.18 is determined by use of a tabular method. Since member cross-sectional areas differ, they must be included in the tabulation.

Bar	l	A	F_Δ	δF_P	$\delta F_P F_\Delta l / A$
—	mm	mm^2	kN	kN	kN^2/mm
ab	6 000	1 500	−135	−0.75	405.0
ac	10 000	2 500	+225	+1.25	1 125.0
bc	8 000	1 500	−112.5	−0.625	375.0
cd	6 708	1 500	+150.9	+0.838	565.5
bd	12 530	3 000	−281.9	−1.566	1 843.8

$$\sum \left(\delta F_P \cdot \frac{F_\Delta l}{A} \right) = 4\ 314.3$$

Application of Eq. 7.18 yields

$$(1)(v_d) = \frac{1}{E} \sum_{j=1}^{m} \left(\frac{\delta F_P F_\Delta l}{A} \right) = \frac{4\ 314.3}{200} \text{ kN} \cdot \text{mm}$$

from which

$$v_c = 21.6 \text{ mm}$$

The positive sign indicates that the deflection is downward, as assumed.

EXAMPLE 7.3

(a) The truss shown below is distorted during fabrication because member bc is 0.5 in. short. What vertical deflection is introduced at point e because of this misfit?

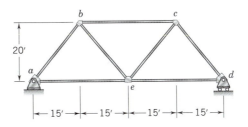

In this application, Eq. 7.19 is used.

$$\sum_{i=1}^{n} (\delta P)_i \Delta_i = \sum_{j=1}^{m} [(\delta F_P) \cdot \Delta l_\Delta]_j \tag{7.19}$$

Determination of Vertical Displacement v_e

Δ System

The actual system, which includes the imposed member length change and the desired displacement, v_e.

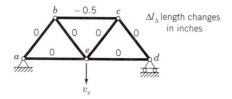

P System

Virtual system with δP loading selected to obtain v_e.

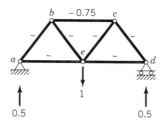

Application of Eq. 7.19 gives

$$(1)(v_e) = (-0.75)(-0.5)$$

from which, $v_e = +0.375$ in.

(b) What length change in member bc must be induced to create a camber (upward displacement) at point e of 1.0 inch?

Determination of Δl_{bc}

Δ System

The actual system, which includes the specified upward displacement, v_e, and the desired member length change in member bc.

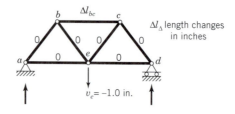

P System

Same as used in Part a.

Application of Eq. 7.19 gives,

$$(1)(v_e) = (1)(-1.0) = (-0.75)(\Delta l_{bc})$$

from which, the required member length change is $\Delta l_{bc} = +1.33$ inches

Note: The quantities A and E are not required in this problem as Δl_Δ is given. Here, v_e is not a result of loading but instead a consequence of a geometric adjustment of the structure.

EXAMPLE 7.4

Determine the vertical displacement that occurs at point b as a result of a temperature change of $+50\ °F$ in members ad and dc. The coefficient of thermal expansion is $\alpha_t = 0.000,006,5$ per $°F$.

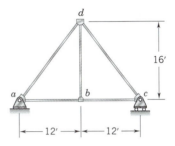

Δ System

The actual system, which includes the temperature-induced member length changes and the desired displacement, v_b.

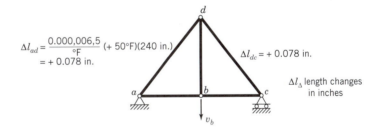

$$\Delta l_{ad} = \frac{0.000,006,5}{°F} (+50°F)(240\ in.)$$
$$= +0.078\ in.$$

$\Delta l_{dc} = +0.078\ in.$

Δl_Δ length changes in inches

P System

Virtual system with δP loading selected to obtain v_b.

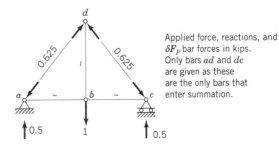

Applied force, reactions, and δF_p bar forces in kips. Only bars ad and dc are given as these are the only bars that enter summation.

The summation on the right-hand side of Eq. 7.19 is determined by use of the following table:

Bar	Δl_Δ	δF_P	$\delta F_P \cdot \Delta l_\Delta$
—	in.	kip	kip · in.
ad	+0.078	−0.625	−0.048,8
dc	+0.078	−0.625	−0.048,8
		$\sum (\delta F_P)\Delta l_\Delta = -0.097,6$	

Applying Eq. 7.19, we have

$$(1)(v_b) = -0.097,6$$

$$v_b = -0.097,6 \text{ in.}$$

The minus sign indicates that point b displaces upward, since v_b of Δ system was assumed downward.

7.6   DEFLECTIONS OF SPACE FRAMEWORKS

The complementary virtual work method can also be employed to determine space framework deflections. The method evolves from a direct application of the procedures that were employed in determining planar truss deflections.

The key relationship for a joint loaded framework is again Eq. 7.18, where Δ_i represents a desired deflection at point i corresponding to the actual loading on the framework, which causes the set of F_Δ member forces, and $(\delta P)_i$ is a virtual force that is introduced at the point and in the direction of the desired deflection, which introduces the virtual set of δF_p member forces. Of course, in the present context, the F_Δ and δF_P member forces are determined through two separate framework analyses. In the summation, j ranges over the m members of the framework.

For cases where the deflection is caused by something other than joint loading, the required formula is Eq. 7.19, where Δl_Δ represents the imposed length changes of the members, such as might result from fabrication errors or temperature changes.

Again, in computing framework deflections, it is computationally convenient to designate $(\delta P)_i$ as a unit load in determining Δ_i. Of course, a separate virtual loading must be employed for each deflection quantity that is desired.

EXAMPLE 7.5   Determine the horizontal displacement in the x direction at point d for the structure and loading of Example 4.6. In addition to the data presented earlier, the following material and member properties are specified.

Cross-sectional areas:

$$\text{Members } ab, bc, \text{ and } ac: A = 5.0 \text{ in.}^2$$

$$\text{Members } ad, bd, \text{ and } cd: A = 8.0 \text{ in.}^2$$

$$E = 29 \times 10^3 \text{ ksi}$$

The following complementary virtual work expression is used:

$$\sum_{i=1}^{n} (\delta P)_i \Delta_i = \sum_{j=1}^{m} \left(\delta F_P \frac{F_\Delta l}{EA} \right)_j \tag{7.18}$$

Determination of Horizontal Displacement u_d

Δ *System*

Actual system that includes the desired displacement, u_d. The corresponding F_Δ member forces were determined in Example 4.6 and are shown in the summary table that follows.

P System

Virtual system with δP loading selected to obtain u_d. An analysis procedure similar to that outlined in Examples 3.10 and 4.6 yields the reactions and member forces given below, and the member forces are entered in the summary table.

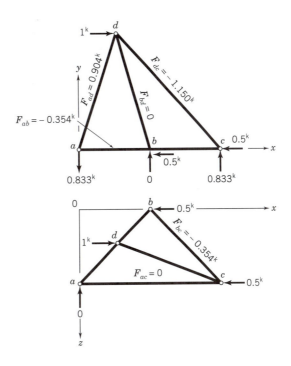

Summary Table

Member	l in.	A in.^{2}	F_Δ kips	δF_P kips	$\delta F_P F_\Delta l/A$ kips^2/in.
ab	204.0	5.0	−12.40	−0.354	179.10
bc	204.0	5.0	8.85	0.354	127.82
ac	288.0	5.0	5.00	0	0
ad	260.4	8.0	13.56	0.904	399.01
bd	260.4	8.0	9.04	0	0
cd	331.2	8.0	−28.75	−1.150	1,368.79

$$\sum\left(\delta F_P \frac{F_\Delta l}{A}\right) = 2{,}074.72$$

Application of Eq. 7.18 yields

$$(1)u_d = \frac{1}{E}\left(\delta F_P \frac{F_\Delta l}{A}\right) = \frac{2{,}074.72}{29 \times 10^3} = 0.072 \text{ kip-in.}$$

from which

$$u_d = 0.072 \text{ in.}$$

The positive sign indicates that the deflection is in the positive x direction—the direction of the applied unit load.

7.7   TRUSS DEFLECTIONS BY ENERGY METHODS

Castigliano's second theorem, which is given in Section 2.18.2, is not only well suited for obtaining individual displacement quantities but is also useful for generating the complete set of displacement–force equations for the structure as a whole. Both applications are illustrated in the example problems at the end of this section.

In the application of Castigliano's second theorem, it is essential that the total strain energy for the structure, U, be written in terms of the applied structure forces. For each structure force component P_i the corresponding displacement quantity Δ_i can be obtained from the equation

$$\frac{\partial U}{\partial P_i} = \Delta_i \tag{7.20}$$

which is derived in Section 2.18.2 as Eq. 2.65.

Since the strain energy is equal to the work of deformation, Eq. 2.22 can be used to express the strain energy associated with the elongation of the element shown in Fig. 7.1b. Thus, we have

$$dU = \frac{1}{2}F \, dl \tag{7.21}$$

where the factor $\frac{1}{2}$ results from the linear relationship between the force, F, and the elongation, dl, as was previously explained through Fig. 2.12. Substituting the expression for dl from Eq. 7.4 into Eq. 7.21, we obtain

$$dU = \frac{F^2}{2EA} dx \tag{7.22}$$

where F is the axial force on the element, A is the cross-sectional area, and E is the modulus of elasticity. In this case, since the element force is constant throughout the member length l, the total strain energy is

$$U = \int_{x=0}^{x=l} \frac{F^2 dx}{2EA} = \frac{F^2 l}{2EA} \tag{7.23}$$

Thus, the total strain energy stored in a truss that is composed of m members is given by

$$U = \sum_{i=1}^{m} \frac{F_i^2 l_i}{2E_i A_i} \tag{7.24}$$

It is stressed that F_i must be expressed in terms of the externally applied structure forces in order that the required differentiation of Eq. 7.20 can be performed.

EXAMPLE 7.6

Determine the horizontal component of displacement at point c. All members have a cross-sectional area of 2 500 mm^2 and $E = 200$ kN/mm^2. This problem was previously worked as Example 7.1. The reactions are given.

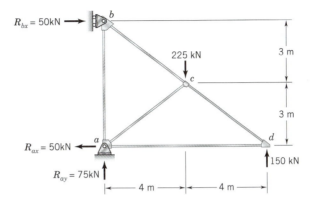

Structure Coordinate System and Member Notation

The vectors in the sketch denote the admissible force and displacement components. The circled numbers identify the individual members.

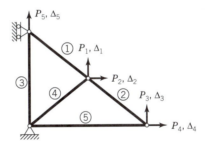

Bar Forces

Determine all bar forces in terms of the applied force components P_1, P_2, P_3, P_4, and P_5 using standard truss analysis.

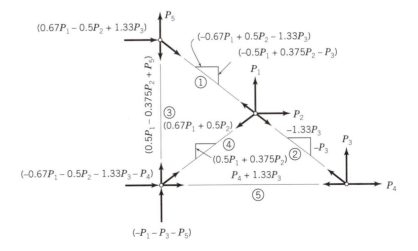

Strain Energy

The total strain energy results from summing the individual member strain energies.

$$U = \sum_{i=1}^{5} \frac{F_i^2 l_i}{2E_i A_i} = \frac{1}{2EA} \sum_{i=1}^{5} F_i^2 l_i \tag{7.24}$$

$$U = \frac{1}{2EA}\{[(-0.67P_1 + 0.5P_2 - 1.33P_3)^2 + (-0.5P_1 + 0.375P_2 - P_3)^2]l_1$$

$$+ [(-1.33P_3)^2 + (-P_3)^2]l_2 + [(0.5P_1 - 0.375P_2 + P_3 + P_5)^2]l_3$$

$$+ [(0.67P_1 + 0.5P_2)^2 + (0.5P_1 + 0.375P_2)^2]l_4$$

$$+ [(P_4 + 1.33P_3)^2]l_5\}$$

Desired Displacement

$$\Delta_i = \frac{\partial U}{\partial P_i} \tag{7.20}$$

In this case, $i = 2$ (horizontal displacement at point c)

$$\Delta_2 = \frac{\partial U}{\partial P_2} = \frac{1}{EA}\{[(-0.67P_1 + 0.5P_2 - 1.33P_3)(0.5)$$

$$+ (-0.5P_1 + 0.375P_2 - P_3)(0.375)]l_1$$

$$+ [(0.5P_1 - 0.375P_2 + P_3 + P_5)(-0.375)]l_3$$

$$+ [(0.67P_1 + 0.5P_2)(0.5) + (0.5P_1 + 0.375P_2)(0.375)]l_4\}$$

but

$$P_1 = -225, \quad P_2 = 0, \quad P_3 = 150, \quad P_4 = 0, \quad P_5 = 0$$

$$l_1 = 5, \quad l_2 = 5, \quad l_3 = 6, \quad l_4 = 5, \quad l_5 = 8$$

from which,

$$\Delta_2 = \frac{1}{EA}\{-192.05 - 84.38 - 588.00\} = \frac{-864.43 \text{ kN} \cdot \text{m}}{EA}$$

Thus

$$\Delta_2 = u_c = \frac{-864.43}{200 \times 10^6 \times 0.002\,5} = -0.001\,73 \text{ m} = -1.73 \text{ mm}$$

EXAMPLE 7.7   Use Castigliano's second theorem to generate the displacement–force equations for the given structure. From these equations, extract the structure flexibility matrix. The structure coordinate system and the member notation are shown.

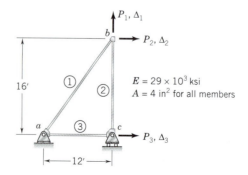

Bar Forces

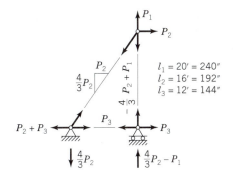

Strain Energy

$$U = \sum_{i=1}^{3} \frac{F_i^2 l_i}{2E_i A_i} = \frac{1}{2EA} \sum_{i=1}^{3} F_i^2 l_i \qquad (7.24)$$

$$U = \frac{1}{2EA}\left\{ [(\tfrac{4}{3}P_2)^2 + (P_2)^2]l_1 + (-\tfrac{4}{3}P_2 + P_1)^2 l_2 + (P_3)^2 l_3 \right\}$$

Displacement–Force Equations and Flexibility Matrix

$$\frac{\partial U}{\partial P_i} = \Delta_i \,(i = 1, 2, 3) \qquad (7.20)$$

$$\frac{\partial U}{\partial P_1} = \Delta_1 \Rightarrow \frac{1}{EA}\left\{ (-\tfrac{4}{3}P_2 + P_1)(+1)l_2 \right\} = \frac{1}{EA}\{ 192P_1 - 256P_2 \} = \Delta_1$$

$$\frac{\partial U}{\partial P_2} = \Delta_2 \Rightarrow \frac{1}{EA}\left\{ [(\tfrac{4}{3}P_2)\tfrac{4}{3} + (P_2)(1)]l_1 + (-\tfrac{4}{3}P_2 + P_1)(-\tfrac{4}{3})l_2 \right\}$$

$$= \frac{1}{EA}\{ -256P_1 + 1008P_2 \} = \Delta_2$$

$$\frac{\partial U}{\partial P_3} = \Delta_3 \Rightarrow \frac{1}{EA}\{ (P_3)(1)l_3 \} = \frac{1}{EA}\{ 144P_3 \} = \Delta_3$$

Or, in matrix form,

$$\frac{1}{EA}\begin{bmatrix} 192 & -256 & 0 \\ -256 & +1008 & 0 \\ 0 & 0 & +144 \end{bmatrix} \begin{Bmatrix} P_1 \\ P_2 \\ P_3 \end{Bmatrix} = \begin{Bmatrix} \Delta_1 \\ \Delta_2 \\ \Delta_3 \end{Bmatrix}$$

From this and in accordance with Eq. 2.15, it is clear that,

$$[D] = \frac{1}{EA}\begin{bmatrix} 192 & -256 & 0 \\ -256 & 1008 & 0 \\ 0 & 0 & 144 \end{bmatrix} \qquad \left(\frac{\text{k} - \text{in.}}{EA} \right)$$

7.8   SUGGESTED PROBLEMS

7.1 through 7.16   Use the complementary virtual work method to determine the horizontal and vertical displacements at point p for each truss and loading shown.

7.1   Cross-sectional area is 2 in.^{2} for each member; $E = 29 \times 10^3$ ksi.

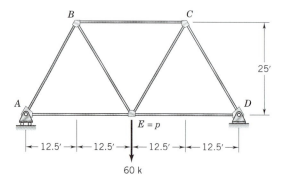

7.2   Cross-sectional area is 1000 mm^2 for each member; $E = 250$ GPa.

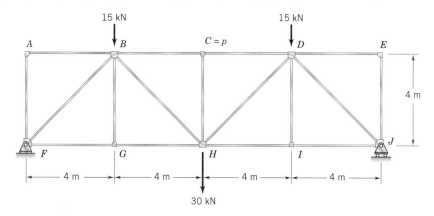

7.3   Cross-sectional areas are given in parentheses (in.^{2}); $E = 29 \times 10^3$ ksi.

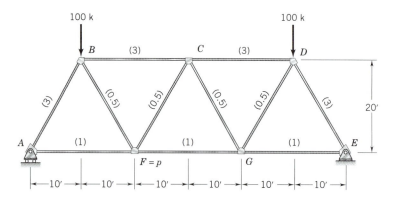

7.4 Cross-sectional area is 5 in.^{2} for each member; $E = 29 \times 10^3$ ksi.

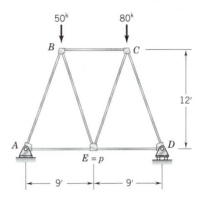

7.5 Cross-sectional areas are given in parentheses (mm^2); $E = 200$ GPa.

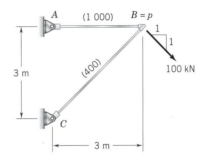

7.6 Cross-sectional areas are given in parentheses (in.^{2}); $E = 29 \times 10^3$ ksi.

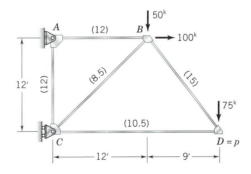

7.7 Area in parentheses (mm^2) on each member; $E = 200$ GPa.

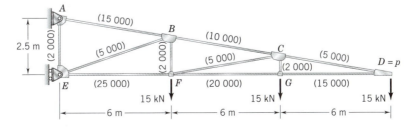

7.8 Area $= 2.5$ in.^{2} for each member; $E = 29 \times 10^3$ ksi.

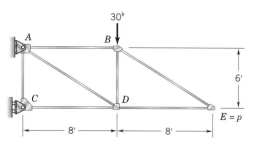

7.9 Cross-sectional areas in parentheses (mm^2); $E = 200$ GPa.

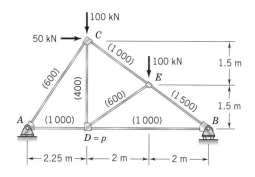

7.10 Area in parentheses (in.^{2}) on each member; $E = 29 \times 10^3$ ksi.

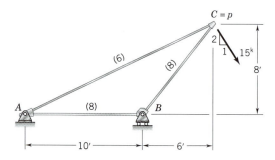

7.11 Area in parenthesis (mm^2) on each member; $E = 200$ GPa.

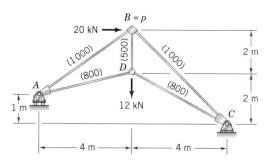

7.12 For each member, $\dfrac{L}{A} = 5\,000\ \text{m}^{-1}$; $E = 200 \times 10^6\ \text{kN/m}^2 = 200$ GPa.

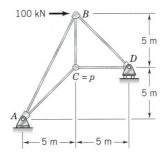

7.13 Area in parentheses (in.^{2}) on each member; $E = 29 \times 10^3$ ksi.

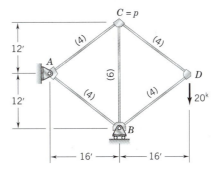

7.14 Area = $1\,500\ \text{mm}^2$ for each member; $E = 200$ GPa.

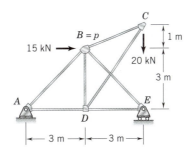

7.15 Area in parentheses (in.^{2}) on each member; $E = 29 \times 10^3$ ksi.

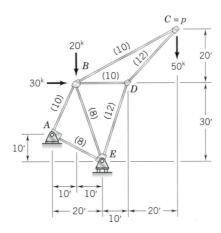

7.16 Area in parentheses (in.^{2}) on each member; $E = 29 \times 10^3$ ksi.

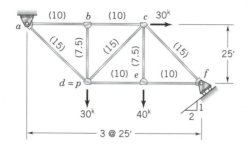

7.17 Determine the vertical displacement at joint D on the truss of Problem 7 if members AB, BC, and CD are each subjected to a temperature change of $+30°C$. Consider the loads of Problem 7 to be removed, and take $\alpha_t = 1.2 \times 10^{-5}$ per °C.

7.18 Determine the vertical displacement of point E on the truss of Problem 8 if members AB and BE are each subjected to a temperature change of $+50$ °F. Consider the loads of Problem 8 to be removed, and take $\alpha_t = 6.5 \times 10^{-6}$ per °F.

7.19 Determine the horizontal displacement of point A on the truss of Problem 11 if members AB and BC are each subjected to a temperature change of $-20°C$ and members AD and DC to a change of -50 °C. The loads of Problem 11 are to be removed, and $\alpha_t = 1.2 \times 10^{-5}$ per °C.

7.20 Determine the displacement parallel to the support surface at point f for the structure of Problem 16 if members ad, de, and ef are each subjected to a temperature decrease of 50 °F. Consider the loads of Problem 16 to be removed, and take $\alpha_t = 6.5 \times 10^{-6}$ per °F.

7.21 Compute the vertical and horizontal displacements of point E of Problem 8 if the member length changes shown below are induced in the respective members. The loads of Problem 8 are to be removed.

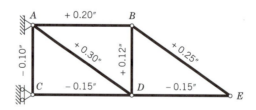

7.22 Determine the temperature change that must be induced in member AC of Problem 10 to produce an upward deflection of 0.10 inch at point C. $\alpha_t = 6.5 \times 10^{-6}$ per °F.

7.23 Determine the vertical displacement of joint E on the truss of Problem 4 if member BC is 0.5 inch too long but is still used. Consider the loads of Problem 4 to be removed.

7.24 Determine the horizontal displacement of joint D on the truss of Problem 12 if member BC is 20 mm short but is still used. Assume that the loads of Problem 12 are removed.

7.25 Determine the change in member length for member BD of the truss of Problem 11 that, when combined with the prescribed loading, will produce a net vertical displacement of zero at point D.

7.26 Calculate the change in member length that must be induced in member BC of the truss of Problem 13 to produce a horizontal displacement of 0.5 inch to the right at point D.

7.27 Use Castigliano's second theorem to determine the vertical component of displacement at joint C on the truss of Problem 10. Use the structure coordinates shown below.

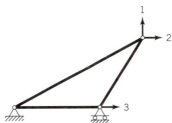

7.28 For the truss of Problem 10, use Castigliano's second theorem to generate the displacement–force equations for the structure coordinate system specified in Problem 27. From these equations, extract the structure flexibility matrix, and determine the complete set of joint displacements for the loading of Problem 10.

7.29 Use Castigliano's second theorem to determine the vertical component of displacement at joint C on the truss of Problem 12. Use the structure coordinates shown below.

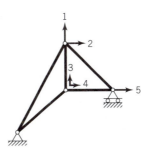

7.30 For the truss of Problem 12, use Castigliano's second theorem to generate the displacement–force equations for the structure coordinate system specified in Problem 29. From these equations, extract the structure flexibility matrix, and determine the complete set of joint displacements for the loading of Problem 12.

7.31 Determine the vertical and horizontal deflections at point C for the structure and loading of Problem 4.27. The cross-sectional area for all members is 4 in.^{2} and $E = 29 \times 10^3$ ksi.

7.32 Determine the vertical deflection under the applied load for Problem 4.28. The cross-sectional areas are given below, and $E = 200 \times 10^6$ kN/m^2.

Members *ad, bd, de:* $A = 1\,500$ mm^2
Members *dc, ec:* $A = 3\,000$ mm^2

7.33 Determine the deflections in the x and y directions at point e for the structure and loading of Problem 4.32. The cross-sectional areas are given below, and $E = 200 \times 10^6$ kN/m^2.

Members *ae, be, ce, de*: $A = 2\,000$ mm^2
Members *ab, bd, dc, ca*: $A = 1\,500$ mm^2

Chapter **8**

Elastic Deflections of Beam and Frame Structures

The Brooklyn Bridge, New York (courtesy Parsons Transportation Group).

8.1   DESCRIPTION OF FLEXURAL DEFORMATION PROBLEM

As a flexural structure responds to loading, it assumes an equilibrium configuration under the combined action of the loads and reactions. Corresponding to this external state of equilibrium, there is a distribution of internal shears and bending moments throughout the structure. These internal actions would normally be shown in the form of shear and moment diagrams for each member. At any given point within the structure, there is a curvature that is consistent with the moment at that point. These curvatures accumulate as angle changes along the member lengths, causing each member to deflect into a flexed or bent configuration. The individual members of the deformed structure must fit together in a compatible fashion, and all the displacement boundary conditions must be satisfied.

As was discussed in Section 7.1, it is necessary to compute deflections for evaluating serviceability criteria and for generating compatibility equations for statically indeterminate structures. As was the case in Chapter 7, we will again limit our discussion to elastic deformations.

8.2   FLEXURAL FORCE–DEFORMATION RELATIONSHIPS

For a flexural member, the force–deformation relationships must relate member-end moments to the corresponding member-end rotations. In a more general formulation, member-end shears and transverse deflections would be included, but they are omitted in our present discussion.

Consider a member under flexural action as shown in Fig. 8.1a, and from it isolate the beam element ab as shown in Fig. 8.1b, which is subjected to the positive moment M. As the element bends, the top fibers are compressed while the bottom fibers are elongated. In between, there is a longitudinal fiber whose length remains unchanged: this fiber is the so-called *neutral fiber* of the member.

It is assumed that as the beam deflects, plane cross sections before bending remain plane after bending. For the element ab, extensions of lines through cross sections at a and b intersect at point o, the *center of curvature*, forming the angle $d\theta$. If tangents to the deflected neutral fiber are constructed at points a and b, it is evident that $d\theta$ also measures the angular deformation over the length of the beam element. The line eb is constructed parallel to the deflected cross section at a creating triangle bde. Then, for small angles, comparing triangles bde and oab, we have

$$d\theta = \frac{dx}{\rho} = \frac{dl}{c} \tag{8.1}$$

where ρ is the *radius of curvature* of the element, c is the distance from the neutral fiber to the topmost fiber, and dl is the shortening of that top fiber. Equation 8.1 can be rewritten in the form

$$\frac{dl}{dx} = \frac{c}{\rho} \tag{8.2}$$

Because the topmost fiber is essentially an axially loaded element, the application of Eqs. 7.2, 7.3, and 8.2 leads to

$$\sigma = -E\frac{c}{\rho} \tag{8.3}$$

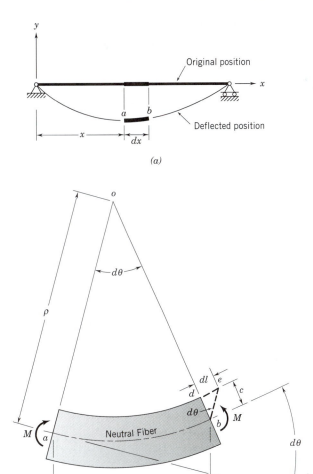

Figure 8.1 Flexural deformations of beam element. *(a)* Deflected beam. *(b)* Beam element subjected to moment.

where σ is the stress in the top fiber and E is Young's modulus. The minus sign is introduced to indicate that the element is being compressed (negative stress). This fiber stress could also be expressed by the familiar expression from basic mechanics that

$$\sigma = -\frac{Mc}{I} \tag{8.4}$$

where M is the moment acting on the element and I is the moment of inertia. Here, the negative sign indicates that a positive moment produces a negative (compressive) stress on the top fiber. Equating Eqs. 8.3 and 8.4, we obtain

$$\frac{1}{\rho} = \frac{M}{EI} \tag{8.5}$$

The transverse displacement y of the deflected structure, which is shown to be positive in the upward direction in Fig. 8.1a, can be related to the radius of the curvature according to the elementary calculus relationship

$$\kappa = \frac{1}{\rho} = \frac{\dfrac{d^2 y}{dx^2}}{\left[1 + \left(\dfrac{dy}{dx}\right)^2\right]^{3/2}} \tag{8.6}$$

where κ is defined as the *curvature*.

When the slope of the deflected structure, $\dfrac{dy}{dx}$, is small compared to unity, Eq. 8.6 reduces to

$$\kappa = \frac{1}{\rho} = \frac{d^2 y}{dx^2} \tag{8.7}$$

and Eq. 8.5 then becomes

$$\frac{d^2 y}{dx^2} = \frac{M}{EI} \tag{8.8}$$

Representing the slope of the deflected structure by θ, we can write

$$\frac{dy}{dx} = \theta \tag{8.9}$$

and then Eq. 8.8 can be expressed as

$$\frac{d}{dx}\left[\frac{dy}{dx}\right] = \frac{d\theta}{dx} = \frac{M}{EI} \tag{8.10}$$

Solving for $d\theta$ in the preceding expression, we have

$$d\theta = \frac{M}{EI}\,dx \tag{8.11}$$

For a member of length l, the total angle change, ϕ, that would accumulate over the member length is given through the integration of Eq. 8.11. This leads to

$$\phi = \int_0^l d\theta = \int_0^l \frac{M}{EI}\,dx \tag{8.12}$$

This equation is equivalent to Eq. 7.5 for axially loaded members; however, in this case M may vary along the member length. For instance, for the situation shown in Fig. 8.2a, the moment acting on an element at x is given by $M = M_1(1 - x/l)$, as shown on the moment diagram of Fig. 8.2b. Therefore, since the member is *prismatic* (EI = constant), Eq. 8.12 takes the form

$$\phi = \frac{1}{EI}\int_0^l M_1\left(1 - \frac{x}{l}\right)dx = \frac{l}{2EI}M_1 \tag{8.13}$$

Equation 8.13 is a deformation–force relationship. In the form of a force–displacement relationship, this equation can be rewritten as

$$M_1 = \frac{2EI}{l}\phi \tag{8.14}$$

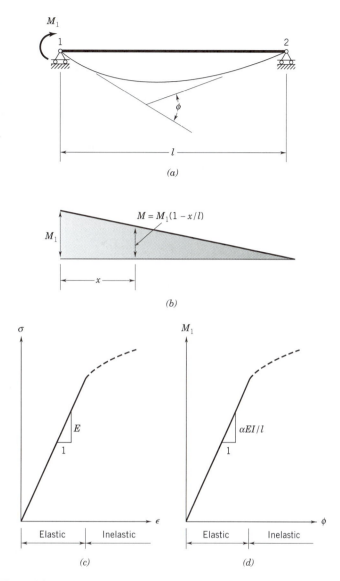

Figure 8.2 Flexural force–deformation behavior. *(a)* Flexurally loaded member. *(b)* Moment diagram. *(c)* Stress–strain diagram. *(d)* Moment–rotation relationship.

In the general form, Eq. 8.13 can be written

$$\phi = d_b M_1 \tag{8.15}$$

where d_b, the *flexural flexibility*, is given by

$$d_b = \frac{l}{\alpha EI} \tag{8.16}$$

The quantity α, which equals 2 in this case, depends on the manner in which M varies along the member length, which, in turn, depends on the end support conditions.

Likewise, Eq. 8.14 can be expressed more generally as

$$M_1 = k_b \phi \tag{8.17}$$

where k_b is a measure of the *flexural stiffness* and is given by

$$k_b = \frac{\alpha EI}{l} \tag{8.18}$$

Since a linear stress–strain diagram was assumed in deriving Eq. 8.11, the resulting moment–rotation diagram mirrors this linearity, as shown in Figs. 8.2c and 8.2d.

In a more general form, consider the member shown in Fig. 8.3. Here, the end moments and rotations are shown—M_1 and θ_1 at end 1 and M_2 and θ_2 at end 2. The specific relationships between these quantities are subsequently derived in Chapter 11 and are found to be given by

$$\begin{Bmatrix} M_1 \\ \\ M_2 \end{Bmatrix} = \begin{bmatrix} \dfrac{4EI}{l} & \dfrac{2EI}{l} \\ \\ \dfrac{2EI}{l} & \dfrac{4EI}{l} \end{bmatrix} \begin{Bmatrix} \theta_1 \\ \\ \theta_2 \end{Bmatrix} \tag{8.19}$$

In a simple form, this equation can be expressed as

$$\{M\} = [k]_b \{\theta\} \tag{8.20}$$

where $\{M\}$ and $\{\theta\}$ are the member-end moment and rotation vectors, respectively, and $[k]_b$ is the *member flexural stiffness matrix*.

Equation 8.17 gives a force–deformation relationship for the member as a whole, whereas Eq. 8.20 gives the full interrelationships between individual member-end forces and displacements. The individual elements of $[k]_b$ are in terms of the quantity EI / l, which is consistent with the flexural stiffness of the member as it is given by Eq. 8.18. As noted earlier, Eq. 8.19 could be expanded to include member-end shears and transverse deflections. This is done in Chapter 13.

Just as Eqs. 8.15 and 8.17 are inverse relationships in a one-dimensional form, Eq. 8.20 has an inverse counterpart. That is, it is possible to develop two-dimensional displacement–force relationships that would provide for the determination of member-end rotations from specified member-end moments. This will be developed in detail in Chapter 13. As was the case earlier for axially loaded members, our purpose here is to stress that there are unique relationships between member forces and deformations, and it is these relationships that form the basis for computing the deflections of structures.

The preceding treatment is based on an elastic response of the member. If the member were stressed beyond the elastic range, as indicated by the dashed line of Fig. 8.2c, the resulting moment–rotation curve would reflect this inelastic behavior as shown in Fig. 8.2d. Equations 8.15, 8.17, and 8.20 would be negated if the member were loaded beyond the elastic range, and they would also be invalid for nonprismatic members in which EI varied along the member length.

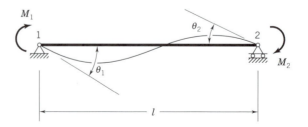

Figure 8.3 Member-end moments and rotations.

8.3   INTEGRATION METHOD

In Section 5.6, relationships were developed between load, shear, and bending moment. In this section, that family of relationships is extended to include the deformation quantities of slope and deflection.

Equation 8.11 states that

$$d\theta = \frac{M}{EI} dx \tag{8.21}$$

which, upon integration, yields

$$\theta = \int \frac{M}{EI} dx + C_3 \tag{8.22}$$

where θ is the slope of the deflected structure. From Eq. 8.9, we have

$$dy = \theta\, dx \tag{8.23}$$

from which

$$y = \int \theta\, dx + C_4 \tag{8.24}$$

where y is the transverse beam deflection.

Figure 8.4 serves to summarize the family of relationships spanning from the load intensity p through the displacement y. In this figure, a simply supported beam is subjected to a general loading and the member responses are shown through the plotted functions for load, shear, moment, slope, and deflection. The sign convention

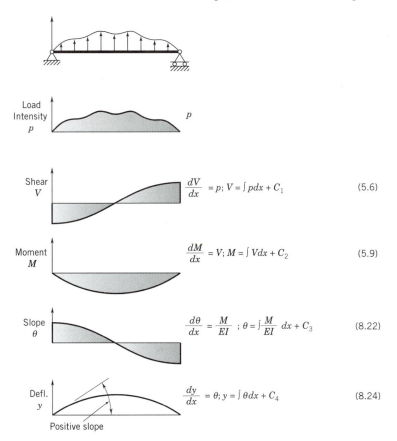

Figure 8.4 Family of beam response relationships.

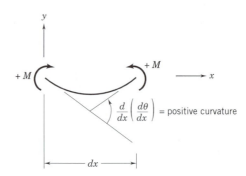

Figure 8.5 Relationship between positive moment and positive curvature.

established in Section 5.6 is applicable for load, shear, and moment—positive (upward) load produces negative moment (tension on top fiber). The convention is extended to include positive upward deflection as was suggested in Fig. 8.1a, and this then, in turn, establishes positive slope as being associated with counterclockwise rotation. This convention is consistent with Eq. 8.10 in which a positive curvature (rate of change in slope) relates to a positive moment. This is further illustrated in Fig. 8.5.

Substitution of Eq. 8.8 into Eq. 5.4 gives

$$\frac{d^2}{dx^2}\left(EI\frac{d^2y}{dx^2}\right) = p \tag{8.25}$$

For a prismatic member, in which EI is constant throughout the member length, we have

$$EI\frac{d^4y}{dx^4} = p \tag{8.26}$$

which relates the deflection to the load intensity at any section of the beam. The systematic solution of this fourth-order differential equation was shown in Fig. 8.4, where the constants of integration are determined from the boundary and continuity conditions on V, M, θ, and y.

EXAMPLE 8.1   Determine the expressions for member slope and deflection for the beam and loading of Example 5.3. Assume that the beam is prismatic, that is, EI = constant.

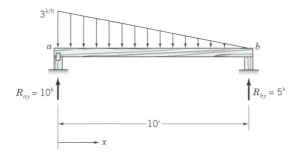

Expressions for Load, Shear, and Moment

From Example 5.3, we have

$$p = 0.3x - 3$$

$$V = 0.15x^2 - 3x + 10$$

$$M = 0.05x^3 - 1.5x^2 + 10x$$

Expressions for Slope and Deflection

The expression for slope is given by Eq. 8.22, where

$$\theta = \int \frac{M}{EI}dx + C_3 = \frac{1}{EI}\int (0.05x^3 - 1.5x^2 + 10x)dx + C_3$$

Therefore,

$$\theta = \frac{1}{EI}(0.0125x^4 - 0.5x^3 + 5x^2) + C_3$$

The constant C_3 cannot be determined at this point because there are no known boundary conditions on the slope θ. The expression for deflection is then given by Eq. 8.24, where

$$y = \int \theta\, dx + C_4 = \int \left[\frac{1}{EI}(0.0125x^4 - 0.5x^3 + 5x^2) + C_3 \right] dx + C_4$$

Integration gives

$$y = \frac{1}{EI}(0.0025x^5 - 0.125x^4 + 1.667x^3) + C_3x + C_4$$

Application of boundary conditions on deflection yields the constants of integration.

$$y(x = 0) = 0 \text{ produces } C_4 = 0$$

$$y(x = 10) = 0 \text{ produces } C_3 = -\frac{66.7}{EI}$$

Therefore, the final expressions for slope and deflection are

$$\theta = \frac{1}{EI}(0.0125x^4 - 0.5x^3 + 5x^2 - 66.7)$$

$$y = \frac{1}{EI}(0.0025x^5 - 0.125x^4 + 1.667x^3 - 66.7x)$$

EXAMPLE 8.2

Determine the expressions for slope and deflection for the beam and loading shown. The moment diagram is given, which includes the expressions for moment for regions ab and bc. The member is prismatic, therefore, EI = constant.

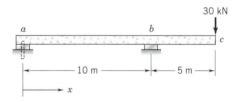

Region ab ($0 \le x \le 10$ m)

The application of Eqs. 8.22 and 8.24 yields the following expressions:

$$\theta = \int \frac{M}{EI}dx + C_3 = \int (-15x)dx + C_3$$

$$= \frac{-7.5x^2}{EI} + C_3$$

$$y = \int \theta \, dx + C_4 = \int \left(\left(\frac{-7.5x^2}{EI} + C_3 \right) + C_4 \right)$$

$$= \frac{-2.5x^3}{EI} + C_3 x + C_4$$

Region bc ($10 \text{ m} \leq x \leq 15 \text{ m}$)

Equations 8.22 and 8.24 are applied with C_3 and C_4 changed to $\overline{C}_3$ and $\overline{C}_4$, respectively, to distinguish them from the constants in region ab.

$$\theta = \int \frac{M}{EI} \, dx + \overline{C}_3 = \int \frac{1}{EI}[30x - 450] \, dx + \overline{C}_3$$

$$= \frac{15x^2}{EI} - \frac{450x}{EI} + \overline{C}_3$$

$$y = \int \theta \, dx + \overline{C}_4 = \int \left(\frac{15x^2}{EI} - \frac{450x}{EI} + \overline{C}_3 \right) dx + \overline{C}_4$$

$$= \frac{5x^3}{EI} - \frac{225x^2}{EI} + \overline{C}_3 x + \overline{C}_4$$

Boundary Conditions and Continuity Conditions

The four constants of integration (C_3, C_4, $\overline{C}_3$, $\overline{C}_4$) are determined from the governing boundary and continuity conditions.

$$y(x = 0)_{ab} = 0 \text{ produces } C_4 = 0$$

$$y(x = 10)_{ab} = 0 \text{ produces } C_3 = \frac{250}{EI}$$

$$y(x = 10)_{bc} = 0 \text{ produces } 10\overline{C}_3 + \overline{C}_4 = \frac{17\,500}{EI}$$

The slope at point b must be continuous. Thus,

$$\theta(x = 10)_{ab} = \theta(x = 10)_{bc} \text{ produces } \overline{C}_3 = \frac{2\,500}{EI}.$$

The last two conditions yield $\overline{C}_4 = \frac{-7\,500}{EI}$.

Region ab

$$\theta = \frac{1}{EI}(-7.5x^2 + 250)$$

$$y = \frac{1}{EI}(-2.5x^3 + 250x)$$

Region bc

$$\theta = \frac{1}{EI}(15x^2 - 450x + 2\,500)$$

$$y = \frac{1}{EI}(5x^3 - 225x^2 + 2\,500x - 7\,500)$$

8.3.1   Closing Remarks

The integration method is of greatest value when the loading is such as to produce a moment diagram that is a continuous function over the entire length of the beam. Example 8.1 illustrates this type of problem. Here, two constants of integration are determined through the application of the appropriate boundary conditions.

When concentrated loads occur along the span, or internal reaction points exist, such as was presented in Example 8.2, then the moment diagram has discontinuities. This leads to additional constants of integration that are evaluated by applying continuity conditions at the points of moment discontinuity. Problems of this class are treated with greater elegance when singularity functions are employed. This approach is not considered here, but the reader is referred to the work of Au and Christiano, which is cited in Section 8.10 However, the moment–area method and the conjugate beam method are considered to be superior for general loading cases, and these methods are presented in the sections that follow.

8.4   MOMENT–AREA METHOD

Consider the beam structure of Fig. 8.6a, which is shown in a deflected configuration under the action of the applied loads. An enlarged view of a portion of the deflected structure between points A and B is isolated in Fig. 8.6b. Within region AB, an element

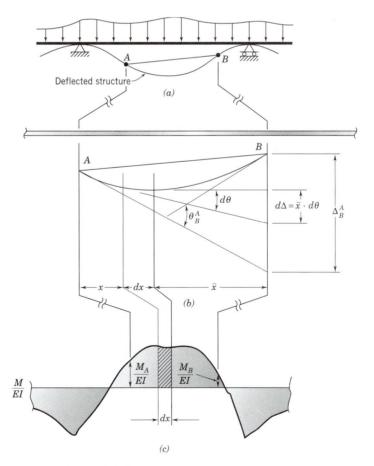

Figure 8.6 Development of moment–area theorems.

of length dx is shown with tangents to the deflected member constructed at each end of the element. The angle between these end tangents, which represents the angle change that occurs over the length dx, is denoted $d\theta$. According to Eq. 8.11, this angle change is given by

$$d\theta = \frac{M}{EI}dx \tag{8.27}$$

where M and I are the bending moment and moment of inertia at point x, respectively, and E is the modulus of elasticity of the material. If the M/EI values are plotted as shown in Fig. 8.6c, it is clear that $d\theta$ is given by the shaded area.

The total angle change that occurs between tangents constructed at points A and B is labeled θ_B^A in Fig. 8.6b. This angle is the slope at B relative to the slope at A; it results from the summation of the incremental angle changes between A and B and is given by

$$\theta_B^A = \int_A^B d\theta \tag{8.28}$$

Substitution of Eq. 8.27 into Eq. 8.28 gives

$$\theta_B^A = \int_A^B \frac{M}{EI}dx \tag{8.29}$$

Equation 8.29 is the basis for the *first moment–area theorem,* which can be stated as follows:

> *The angle change between points* A *and* B *on the deflected structure, or the slope at point* B *relative to the slope at point* A, *is given by the area under the* M/EI *diagram between these two points.*

Examination of Fig. 8.6b shows that if the tangents to the element of length dx are extended, they embrace an intercept of $d\Delta$ on a vertical line through point B. For small angles, this intercept is given by

$$d\Delta = \bar{x}d\theta \tag{8.30}$$

Substituting Eq. 8.27 into Eq. 8.30, we obtain

$$d\Delta = \bar{x}\frac{M}{EI}dx \tag{8.31}$$

which shows that the intercept $d\Delta$ is given by the static moment of the shaded area of the M/EI diagram taken about an axis through point B. The accumulation of these intercepts for all increments between points A and B gives

$$\Delta_B^A = \int_A^B d\Delta \tag{8.32}$$

where Δ_B^A is the vertical displacement of point B on the deflected structure with respect to a line drawn tangent to the structure at point A. Substitution of Eq. 8.31 into Eq. 8.32 yields

$$\Delta_B^A = \int_A^B \frac{M}{EI}\bar{x}\,dx \tag{8.33}$$

This equation expresses the essence of the *second moment–area theorem*, which can be stated as follows:

> *The deflection of point* B *on the deflected structure with respect to a line drawn tangent to point* A *on the structure is given by the static moment of the area under the* M/EI *diagram between points* A *and* B *taken about an axis through point* B.

It is emphasized that the deflection quantities that are determined by using the second moment–area theorem are normal to the original orientation of the member.

It is to be noted that the deformation quantities determined by using the moment–area theorems are relative quantities, and this fact is emphasized by the notation that is used. The first moment–area theorem is used to obtain θ_i^j, the slope at i (subscript) relative to the slope at j (superscript), whereas the second moment–area theorem is used to determine Δ_i^j, the displacement at i (subscript) with respect to a line that is tangent at point j (superscript). The relative nature of these quantities must be carefully considered when the moment–area theorems are applied. It is further emphasized that the moment–area theorems give deformations that result from flexural action only. Thus, these theorems cannot be applied over sections of a member where discontinuities are present. Discontinuities, such as a hinge, allow deformations to occur that are not related to the curvatures given by the M/EI diagram and are thus not detected by the moment–area theorems.

8.5 APPLICATION OF MOMENT–AREA METHOD

The application of the moment–area theorems to beam and frame problems is fairly routine; however, because these theorems produce relative deformation quantities, some care has to be exercised. In general, there are two categories of problems. In the first, a displacement boundary condition prescribes a known slope at a support point. This support slope serves as a reference from which the relative deformation quantities of each moment–area theorem can be applied to determine the actual slope or displacement at any point on the member. In the second situation, there is no point on the structure where the slope is initially known, and thus there is no reference from which to apply the quantities that are established by the moment–area theorems. Both problem types are among the example problems that follow.

The areas and centroidal locations for a variety of area segments are given in a table inside the back cover for easy reference. These are helpful in problem solving and are used in the example problems that follow. These formulae are presented for figures with a horizontal base. They are also valid for skewed-base situations when the dimension l is taken as the horizontal projection of the base and the altitude is measured perpendicular to the horizontal. This approach was previously demonstrated in Fig. 5.7.

For computational simplicity in determining the deflection quantities, it is generally most convenient to carry the quantity EI throughout the problem in symbolic form and then substitute the numerical value at the end.

EXAMPLE 8.3

Determine the expressions for the vertical displacement and slope at point B on the cantilever beam shown.

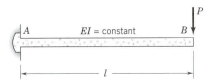

Note: This problem is of the first type—the slope at point A is specified as zero by the boundary conditions.

M/EI Diagram

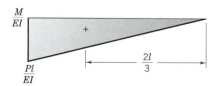

Deflected Structure

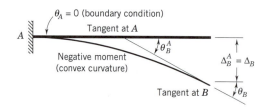

Note: All displacement quantities are taken as positive for this problem, as shown on the above sketch.

Deflection Calculations

From first moment–area theorem,

$$\theta_B^A = \frac{1}{2}\left(\frac{Pl}{EI}\right)l = \frac{Pl^2}{2EI}$$

Then,

$$\theta_B = \theta_A + \theta_B^A = 0 + \frac{Pl^2}{2EI} = \frac{Pl^2}{2EI}$$

From the second moment–area theorem,

$$\Delta_B^A = \left[\frac{1}{2}\left(\frac{Pl}{EI}\right)l\right] \times \frac{2l}{3} = \frac{Pl^3}{3EI}$$

and

$$\Delta_B = \Delta_B^A = \frac{Pl^3}{3EI}$$

EXAMPLE 8.4

Determine the slope and deflection at point C on the simply supported beam given. EI = constant: $E = 200$ GPa $= 200 \times 10^6$ kN/m^2; $I = 500 \times 10^{-6}$ m^4; $EI = 100\,000$ kN · m^2.

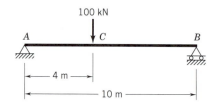

Note: This problem is of the second type; no slope is specified by the boundary conditions.

M/EI Diagram

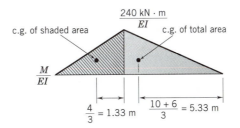

Deflected Structure

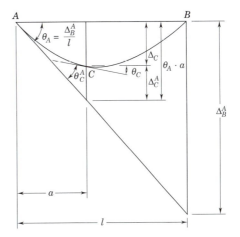

Note: All displacement quantities are taken as positive for this problem, as shown on the above sketch.

Deflection Calculations

Initially, the slope must be established at some point; point A is selected.
    From second moment–area theorem,

$$\Delta_B^A = \frac{1}{2}\left(\frac{240}{EI}\right)10 \times 5.33 = \frac{6\,396\ \text{kN} \cdot \text{m}^3}{EI}$$

and, for small displacements,

$$\theta_A = \frac{\Delta_B^A}{l} = \frac{6\,396/EI}{10} = \frac{639.6\ \text{kN} \cdot \text{m}^2}{EI}$$

From first moment–area theorem,

$$\theta_C^A = \frac{1}{2}\left(\frac{240}{EI}\right)4 = \frac{480\ \text{kN} \cdot \text{m}^2}{EI}$$

and thus,

$$\theta_C = \theta_A - \theta_C^A = \left(\frac{639.6}{EI} - \frac{480}{EI}\right) = \frac{159.6\ \text{kN} \cdot \text{m}^2}{EI}$$

From second moment–area theorem,

$$\Delta_C^A = \frac{1}{2}\left(\frac{240}{EI}\right)4 \times 1.33 = \frac{638.4 \text{ kN} \cdot \text{m}^3}{EI}$$

from which

$$\Delta_C = \theta_A \cdot a - \Delta_C^A = \left(\frac{639.6}{EI} \times 4\right) - \left(\frac{638.4}{EI}\right) = \frac{1\,920 \text{ kN} \cdot \text{m}^3}{EI}$$

For the given EI,

$$\theta_C = \frac{159.6}{100\,000} = 0.001\,596 \text{ radian}$$

$$\Delta_C = \frac{1\,920}{100\,000} = 0.019\,2 \text{ m} = 19.2 \text{ mm}$$

The positive results for θ_C and Δ_C confirm that these deflection quantities are as shown on the deflection sketch.

8.5.1   Note on Sign Convention

It is possible to develop a detailed sign convention for the moment–area method, but this will not be done here. Instead, as demonstrated in the previous two examples, an intuitive approach is used. A slightly more complicated situation is shown in Fig. 8.7a, which presents a sketch of the deflected structure that satisfies the boundary conditions and possesses member curvatures that are consistent with the moment diagram. As was done

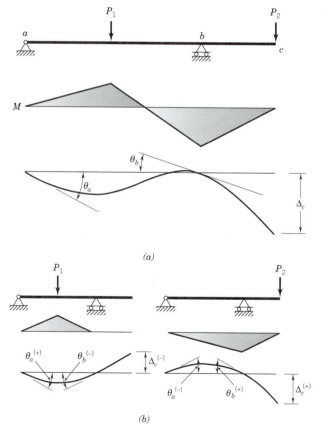

Figure 8.7 Beam deflections by superposition. (*a*) Loading and positive displacements. (*b*) Displacement contributions from individual loads.

in Examples 8.3 and 8.4, key slopes and deflections are identified, and these are shown in the positively assumed directions as noted in Fig. 8.7a. These "positive" directions do not necessarily conform to the general sign convention introduced in Section 8.3, but are rather to be taken as positive within the context of this example problem. In applying the moment–area principles, it is convenient to consider two separate cases and superimpose the results as illustrated in Fig. 8.7b. This superposition procedure leads to simpler moment diagram areas with which to work, and the signs of the individual displacement components (shown parenthetically in Fig. 8.7b) are readily ascertained by comparing their directions with those that were assigned as positive in Fig. 8.7a. For instance, for the positive direction assigned to Δ_c in Fig. 8.7a, the load P_1 produces a negative contribution while the load P_2 induces a positive contribution. After the proper superposition, a positive result for any displacement quantity confirms that the assumed direction is correct.

EXAMPLE 8.5

Determine the vertical deflection and the slope at point d for the structure shown. Here, $E = 200 \text{ GPa};$ $I = 3\,000 \times 10^{-6} \text{m}^4;$ $EI = 600\,000 \text{ kN} \cdot \text{m}^2.$

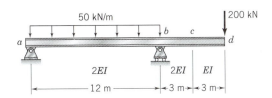

M/EI Diagram

The moment diagram would first be constructed. In this case, the separate effects of the uniform load on span ab and the concentrated load at point d are superimposed. The resulting moment diagram ordinates are then divided by the appropriate EI values to obtain the M/EI diagram.

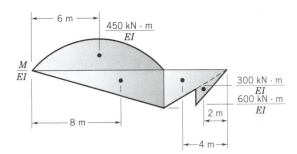

Deflected Structure

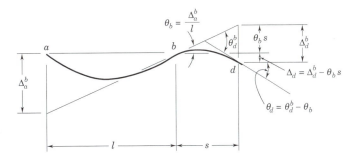

Deflection Calculations

The direction of each displacement quantity shown in this sketch is taken as positive.

Initially, the slope is established at point b. From second moment–area theorem,

$$\Delta_a^b = \underbrace{\frac{2}{3}\left(\frac{450}{EI}\right)12 \times 6}_{} - \frac{1}{2}\left(\frac{600}{EI}\right)12 \times 8 = \underbrace{\frac{-7\ 200\ \text{kN} \cdot \text{m}^3}{EI}}_{}$$

This term is taken
as positive because
it contributes positively
to Δ_a^b as shown
in the sketch.

Negative simply
indicates that Δ_a^b
is opposite to the
direction shown
in the sketch.

For small displacements,

$$\theta_b = \frac{\Delta_a^b}{l} = \frac{-7\ 200/EI}{12} = \frac{-600\ \text{kN} \cdot \text{m}^2}{EI}$$

From first moment–area theorem,

$$\theta_d^b = \frac{1}{2}\left(\frac{600}{EI}\right)6 + \frac{1}{2}\left(\frac{300}{EI}\right)3 = \frac{2\ 250\ \text{kN} \cdot \text{m}^2}{EI}$$

and therefore,

$$\theta_d = \theta_d^b - \theta_b = \frac{2\ 250}{EI} - \left(-\frac{600}{EI}\right) = \frac{2\ 850\ \text{kN} \cdot \text{m}^2}{EI}$$

From the second moment–area theorem,

$$\Delta_d^b = \frac{1}{2}\left(\frac{600}{EI}\right)6 \times 4 + \frac{1}{2}\left(\frac{300}{EI}\right)3 \times 2$$

$$= \frac{7\ 200}{EI} + \frac{900}{EI} = \frac{8\ 100\ \text{kN} \cdot \text{m}^3}{EI}$$

and then

$$\Delta_d = \Delta_d^b - \theta_b \cdot s = \frac{8\ 100}{EI} - \left(-\frac{600}{EI}\right)6 = \frac{11\ 700\ \text{kN} \cdot \text{m}^3}{EI}$$

For the specified EI,

$$\theta_d = \frac{2\ 850}{600\ 000} = 0.004\ 75\ \text{radian}$$

$$\Delta_d = \frac{11\ 700}{600\ 000} = 0.019\ 5\ \text{m}$$

The positive results for θ_d and Δ_d confirm that these deflection quantities are as shown on the deflection sketch.

EXAMPLE 8.6

Determine the slope and deflection under the concentrated load for the structure shown. Here, $E = 30 \times 10^3$ ksi; $I = 1,000$ in.^{4}; $EI = 30 \times 10^6$ k-in.^{2}

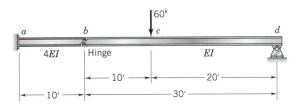

M/EI Diagram

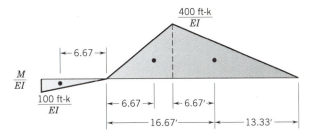

Note: The moment diagram would possess the same slope (constant shear) from point a to point c; however, the slope of the M/EI diagram changes at point b because of the abrupt change in EI.

Deflected Structure

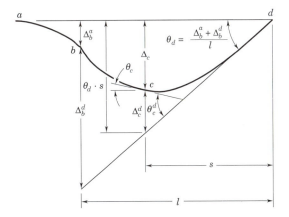

Note: Solution strategy must treat beam segments on each side of hinge separately since the moment–area theorems cannot be applied over a beam section that includes a hinge. All displacement quantities are assumed to be positive as shown.

Deflection Calculations

From second moment–area theorem,

$$\Delta_b^a = \frac{1}{2}\left(\frac{100}{EI}\right)10 \times 6.67 = \frac{3,335 \text{ ft}^3\text{-k}}{EI}$$

$$\Delta_b^d = \frac{1}{2}\left(\frac{400}{EI}\right)10 \times 6.67 + \frac{1}{2}\left(\frac{400}{EI}\right)20 \times 16.67$$

$$= \frac{13,340}{EI} + \frac{66,680}{EI} = \frac{80,020 \text{ ft}^3\text{-k}}{EI}$$

and

$$\theta_d = \frac{\Delta_b^a + \Delta_b^d}{l} = \frac{3,335 + 80,020}{30EI} = \frac{2,778.5 \text{ ft}^2\text{-k}}{EI}$$

From first moment–area theorem,

$$\theta_c^d = \frac{1}{2}\left(\frac{400}{EI}\right)20 = \frac{4,000 \text{ ft}^2\text{-k}}{EI}$$

Thus,

$$\theta_c = \theta_c^d - \theta_d = \frac{4000}{EI} - \frac{2,778.5}{EI} = \frac{1,221.5 \text{ ft}^2\text{-k}}{EI}$$

Using second moment–area theorem,

$$\Delta_c^d = \frac{1}{2}\left(\frac{400}{EI}\right)20 \times 6.67 = \frac{26,680 \text{ ft}^3\text{-k}}{EI}$$

for which

$$\Delta_c = \theta_d \cdot s - \Delta_c^d = \left(\frac{2,778.5}{EI}\right)20 - \frac{26,680}{EI} = \frac{28,890 \text{ ft}^3\text{-k}}{EI}$$

For the given EI,

$$\theta_c = \frac{1,221.5 \times 144}{30 \times 10^6} = 0.005,86 \text{ radian}$$

$$\Delta_c = \frac{28,890 \times 1,728}{30 \times 10^6} = 1.66 \text{ inches}$$

The positive results for θ_c and Δ_c indicate that the deflection quantities are in the direction shown on the deflection sketch.

EXAMPLE 8.7

Determine the vertical deflection under the load for the structure shown. Here, $E = 10 \times 10^3$ ksi; $I = 10,000$ in.^{4}; $EI = 100 \times 10^6$ k-in.^{2}

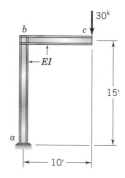

M/EI Diagrams

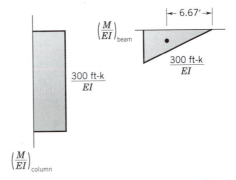

Deflected Structure

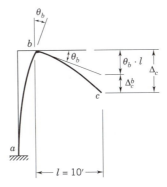

Note: All displacement quantities are assumed to be positive as shown.

Deflection Calculations

From first moment–area theorem,

$$\theta_b = \theta_b^a = \left(\frac{300}{EI}\right)15 = \frac{4{,}500 \text{ ft}^2\text{-k}}{EI}$$

This is the common rotation for the top of column *ab* and the left end of beam *bc*. From second moment–area theorem,

$$\Delta_c^b = \frac{1}{2}\left(\frac{300}{EI}\right)10 \times 6.67 = \frac{10{,}000 \text{ ft}^3\text{-k}}{EI}$$

and thus,

$$\Delta_c = \theta_b \cdot l + \Delta_c^b = \left(\frac{4{,}500}{EI}\right)10 + \frac{10{,}000}{EI} = 55{,}000 \text{ ft}^3\text{-k}$$

For the given *EI*,

$$\Delta_c = \frac{55{,}000 \times 1{,}728}{100 \times 10^6} = 0.950 \text{ inch}$$

EXAMPLE 8.8

Determine the horizontal deflection and the rotation at point b for the frame structure shown, where $E = 200$ GPa; $I = 4\,500 \times 10^{-6}$ m^4; $EI = 900\,000$ kN $\cdot$ m^2.

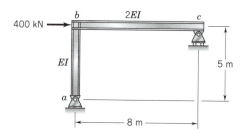

M/EI Diagrams

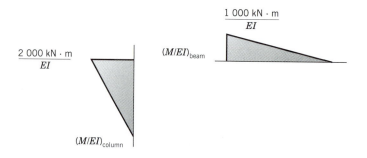

Deflected Structure

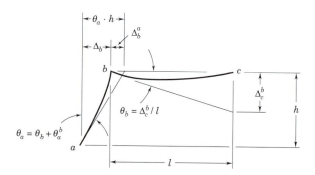

Note: All displacement quantities are assumed to be positive as shown, and θ_b is common to the left end of member bc and the top of member ba.

Deflection Calculations

From second moment–area theorem,

$$\Delta_c^b = \frac{1}{2}\left(\frac{1\,000}{EI}\right)8 \times 5.34 = \frac{21\,344\ \text{kN} \cdot \text{m}^3}{EI}$$

from which

$$\theta_b = \frac{\Delta_c^b}{l} = \frac{21\,344}{8} = \frac{2\,668\ \text{kN} \cdot \text{m}^2}{EI}$$

Then, from first moment–area theorem,

$$\theta_a^b = \frac{1}{2}\left(\frac{2\,000}{EI}\right)5 = \frac{5\,000 \text{ kN} \cdot \text{m}^2}{EI}$$

and, therefore,

$$\theta_a = \theta_b + \theta_a^b = \frac{2\,668}{EI} + \frac{5\,000}{EI} = \frac{7\,668 \text{ kN} \cdot \text{m}^2}{EI}$$

Again, second moment–area theorem gives

$$\Delta_b^a = \frac{1}{2}\left(\frac{2\,000}{EI}\right)5 \times 1.67 = \frac{8\,350 \text{ kN} \cdot \text{m}^3}{EI}$$

and

$$\Delta_b = \theta_a \cdot h - \Delta_b^a = \left(\frac{7\,668}{EI}\right)5 - \left(\frac{8\,350}{EI}\right) = \frac{29\,990 \text{ kN} \cdot \text{m}^3}{EI}$$

For the given EI,

$$\theta_b = \frac{2\,668}{900\,000} = 0.002\,964 \text{ radian}$$

$$\Delta_b = \frac{29\,990}{900\,000} = 0.033\,322 \text{ m} = 33.3 \text{ mm}$$

The positive signs for θ_b and Δ_b confirm that the deflections are in the directions shown on the sketch.

8.6   THE CONJUGATE BEAM METHOD

The problem of beam statics is governed by Eqs. 5.2 and 5.4, which state

$$\frac{d^2M}{dx^2} = \frac{dV}{dx} = p \tag{8.34}$$

This is a second-order linear differential equation, and the solution is the familiar shear and moment diagram problem: starting with the load, the first integration gives the shear and the second integration gives the moment.

Similarly, the beam deflection problem is governed by Eq. 8.10 in which

$$\frac{d^2y}{dx^2} = \frac{d\theta}{dx} = \frac{M}{EI} \tag{8.35}$$

This is also a second-order linear differential equation. Here, we start with the curvature, M/EI; the first integration yields the slope, and the second integration gives the deflection.

Comparison of Eqs. 8.34 and 8.35 forms the basis of an analogy. If M/EI is taken as loading on an imaginary beam, the resulting shears and moments will, in fact, be the slopes and deflections on the real beam. This is shown in Table 8.1 within the box of broken lines.

Of course, for the analogy to be complete the boundary conditions on shear and moment on the imaginary beam must match the boundary conditions on slope and deflection, respectively, for the real beam. This is highlighted in Table 8.1, and it forms the basis of the *conjugate beam method*

To use the conjugate beam method, an imaginary beam (conjugate beam) is conceived that has the same length as the real beam and has a set of boundary conditions and internal continuity conditions on shear and moment that match the corresponding real beam boundary conditions and internal continuity conditions on slope and deflection. This transformation procedure is outlined in Table 8.2. For example, a clamped end has geometric boundary conditions of zero slope and deflection. Thus,

Table 8.1 Basis of Conjugate Beam Method

Real Beam Parameters	Imaginary Beam Parameters	Solution
Load p Shear V Moment M		Solve $\dfrac{d^2 M}{dx^2} = \dfrac{dV}{dx} = p$ Start with p and integrate to obtain V and M.
Curvature M/EI Slope θ Deflection y	Load P Shear V Moment M	Solve $\dfrac{d^2 y}{dx^2} = \dfrac{d\theta}{dx} = \dfrac{M}{EI}$ Start with M/EI and integrate to obtain θ and y.

If the boundary conditions on V and M of the imaginary beam are forced to match the boundary conditions on θ and y of the real beam, respectively, then the analogy is complete.

Table 8.2 Transformation of Real Beam to Conjugate Beam

the conjugate beam must have force boundary conditions of zero shear and moment. These force boundary condition can occur only at a free end. After all of the necessary transformations are made, the final conjugate beam is then subjected to a loading that corresponds to the *M/EI* diagram of the real beam. The resulting shear and moment diagrams of the conjugate beam give the slope and deflection diagrams, respectively, of the real beam.

Unlike the moment–area method, an orderly sign convention can be employed with the conjugate beam method. If positive curvature (*M/EI*) is applied as positive (upward) load intensity on the conjugate beam, then the signs of the resulting shears and moments on the conjugate beam will correspond to the correct signs of the slope and deflection, respectively, on the real beam, in accordance with the convention given in Fig. 8.4.

EXAMPLE 8.9

Construct the complete slope and deflection diagrams for the structure shown, and evaluate the slope and deflection at the free end. Here, EI = Constant; E = 200 GPa; $I = 1\,000 \times 10^{-6}\,\text{m}^4$; $EI = 200\,000\,\text{kN} \cdot \text{m}^2$.

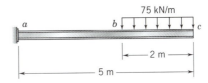

Moment Diagram

Conjugate Beam and Loading

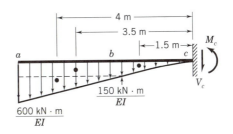

$$\sum P_y = 0: \quad V_c + \left(\frac{150}{EI}\right)3 + \frac{1}{2}\left(\frac{450}{EI}\right)3 + \frac{1}{3}\left(\frac{150}{EI}\right)2 = 0$$

$$V_c = \frac{-1\,225\,\text{kN} \cdot \text{m}^2}{EI} = \theta_c \text{ for a real beam}$$

$$\sum M_{zc} = 0: \quad M_c + \left(\frac{150}{EI}\right)3 \times 3.5 + \frac{1}{2}\left(\frac{450}{EI}\right)3 \times 4 + \frac{1}{3}\left(\frac{150}{EI}\right)2 \times 1.5 = 0$$

$$M_c = \frac{-4\,425\,\text{kN} \cdot \text{m}^3}{EI} = \Delta_c \text{ for real beam}$$

Shear Diagram for Conjugate Beam (Slope Diagram for Real Beam)

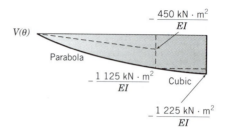

Moment Diagram for Conjugate Beam (Deflection Diagram for Real Beam)

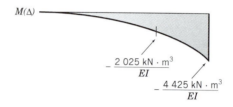

Maximum Slope and Deflection

$$\theta_c = \frac{-1\ 225}{200\ 000} = -0.006\ 13 \text{ radian (negative sign indicates } \diagdown \text{)}$$

$$\Delta_c = \frac{-4\ 425}{200\ 000} = -0.022\ 1 \text{ m} = -22.1 \text{ mm (negative sign indicates } \downarrow \text{)}$$

EXAMPLE 8.10

Determine the slope and deflection under the concentrated load for the structure shown, which is the same structure that was considered in Example 8.6. Here, $E = 30 \times 10^3$ ksi; $I = 1,000$ in.^{4}; $EI = 30 \times 10^6$ k–in.^{2}

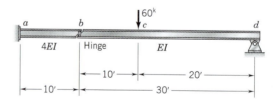

Moment Diagram

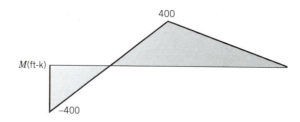

Conjugate Beam and Loading

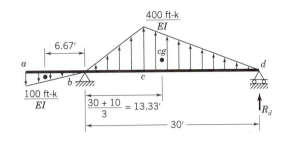

$$\sum M_{zb} = 0: \quad \frac{1}{2}\left(\frac{100}{EI}\right)10 \times 6.67 + \frac{1}{2}\left(\frac{400}{EI}\right)30 \times 13.33 + R_d \times 30 = 0$$

$$R_d = -\left(\frac{3,335 + 79,998}{30EI}\right) = \frac{-2,777.8 \text{ ft}^2\text{-k}}{EI} \quad \begin{array}{l}\text{(negative sign indicates}\\ \text{that } R_d \text{ acts downward)}\end{array}$$

Deflection Calculations

Consider section cd:

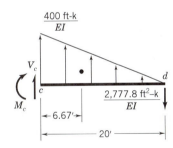

$$\sum P_y = 0: \uparrow + \quad V_c + \frac{1}{2}\left(\frac{400}{EI}\right)20 - \frac{2,777.8}{EI} = 0$$

$$V_c = \frac{-1,222.2 \text{ ft}^2\text{-k}}{EI} \quad \therefore \theta_c = \frac{-1,222.2 \times 144}{30 \times 10^6} = -0.005,867 \text{ rad}$$
$$\text{(negative sign indicates} \quad \text{)}$$

$$\sum M_{zc} = 0: \quad + \quad M_c - \frac{1}{2}\left(\frac{400}{EI}\right)20 \times 6.67 + \left(\frac{2,777.8}{EI}\right)20 = 0$$

$$M_c = \frac{-28,876 \text{ ft}^3\text{-k}}{EI} \quad \therefore \Delta_c = \frac{-28,876 \times 1,728}{30 \times 10^6} = -1.66 \text{ inches}$$
$$\text{(negative sign indicates} \downarrow)$$

Complete Diagrams

The complete loading, shear, and moment diagrams for the conjugate beam correspond to the M/EI, slope, and deflection diagrams, respectively, for the real beam.

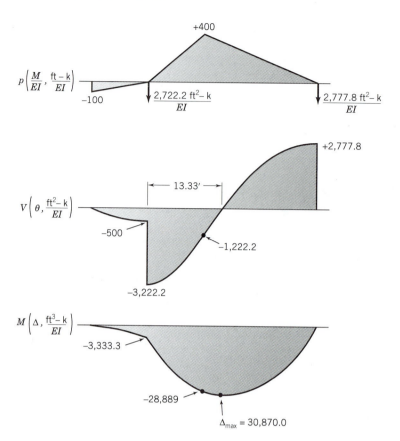

EXAMPLE 8.11

Determine the vertical deflection at point d for the structure shown under the indicated loading. $E = 200$ GPa; $I = 500 \times 10^{-6}$ m^4; $EI = 100\,000$ kN · m^2.

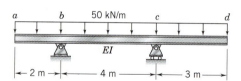

Moment Diagram

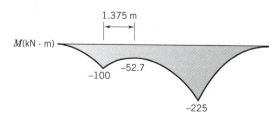

An alternative form of the moment diagram results from superimposing the effect of the loading on span bc on the effects of the loading on the cantilever spans of ab and cd.

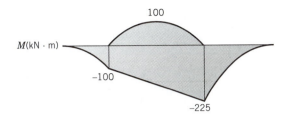

Conjugate Beam and Loading

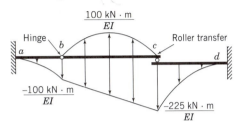

Deflection Calculations

Isolate section bc:

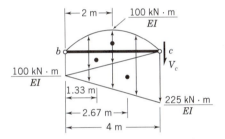

$$\sum M_{zb} = 0: \ \circlearrowright + \ \ V_c \times 4 + \frac{1}{2}\left(\frac{100}{EI}\right)4 \times 1.33 + \frac{1}{2}\left(\frac{225}{EI}\right)4 \times 2.67 - \frac{2}{3}\left(\frac{100}{EI}\right)4 \times 2 = 0$$

$$V_c = \frac{-233.55 \ \text{kN} \cdot \text{m}^2}{EI}$$

Now isolate section cd and apply V_c in the appropriate direction.

$$\sum M_{zd} = 0: + \circlearrowleft \ \ \ M_d + \left(\frac{233.55}{EI}\right)3 + \frac{1}{3}\left(\frac{225}{EI}\right)3 \times 2.25 = 0$$

$$M_d = \frac{-1 \ 206.9 \ \text{kN} \cdot \text{m}^3}{EI}$$

$$\therefore \Delta_d = \frac{-1 \ 206.9}{100 \ 000} = -0.012 \ 1 \ \text{m} = -12.1 \ \text{mm}$$
$$\text{(negative sign indicates } \downarrow \text{)}$$

8.7   BEAM AND FRAME DEFLECTIONS BY COMPLEMENTARY VIRTUAL WORK

Deformation quantities for beam and frame structures can also be determined by using the complementary virtual work method. This approach is introduced in Chapter 2 and was employed in Sections 7.4 through 7.6 for computing truss deflections. Each application yields a single displacement quantity. It is recalled from Section 2.16.2 that two separate systems must be considered. These systems are the following:

P System: *Virtual* force system in equilibrium.
Δ System: *Actual* deformation configuration which is geometrically compatible.

The P system is subjected to the deformations of the Δ system, and the external complementary virtual work is equated to the internal complementary virtual work of deformation. This leads to

Virtual P system

$$\sum_{i=1}^{n} (\delta P)_i \Delta_i = \int_{\text{Vol}} (\delta \sigma_P) \epsilon_\Delta d\text{Vol} \tag{8.36}$$

Actual Δ system

For beam and frame structures, the deformations result from flexural behavior. Thus, the right-hand side of Eq. 8.36, which represents the internal complementary virtual work of deformation, can be expressed as

$$\int_{\text{Vol}} (\delta \sigma_P) \epsilon_\Delta d\text{Vol} = \sum_{j=1}^{m} \left(\int_l (\delta M_P) d\theta_\Delta \right)_j \tag{8.37}$$

where l is the length of the jth member, (δM_P) is the virtual moment at some point x along the jth member, and $d\theta_\Delta$ is the angular deformation of the member, at the same point on member j, that occurs over the differential element of length dx for the Δ system. The summation is over the m members that make up the structure. Substitution of Eq. 8.37 into Eq. 8.36 gives

$$\sum_{i=1}^{n} (\delta P)_i \Delta_i = \sum_{j=1}^{m} \left(\int_l (\delta M_P) d\theta_\Delta \right)_j \tag{8.38}$$

If $d\theta_\Delta$ results from a Δ system of loads on the structure, then Eq. 8.11 gives

$$d\theta_\Delta = \frac{M_\Delta dx}{EI} \tag{8.39}$$

where M_Δ is the moment at point x on member j that results from the loading of the Δ system. Substitution of Eq. 8.39 into Eq. 8.38 yields

$$\sum_{i=1}^{n} (\delta P)_i \Delta_i = \sum_{j=1}^{m} \left(\int_l (\delta M_P) \frac{M_\Delta dx}{EI} \right)_j \tag{8.40}$$

When $d\theta_\Delta$ is caused by something other than direct loading, such as a differential temperature change through the member depth or a fabrication "kink," Eq. 8.37 is still valid with the proper expression being substituted for $d\theta_\Delta$.

EXAMPLE 8.12

Determine the slope and vertical deflection at point C for the simply supported beam of Example 8.4. EI = constant = 100 000 kN $\cdot$ m^2.

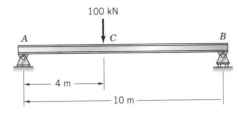

The governing complementary virtual work expression is

$$\sum_{i=1}^{n} (\delta P)_i \Delta_i = \sum_{j=1}^{m} \left(\int_l (\delta M_P) \frac{M_\Delta dx}{EI} \right)_j \tag{8.40}$$

Vertical Deflection at Point C

Δ System (Actual Load System)

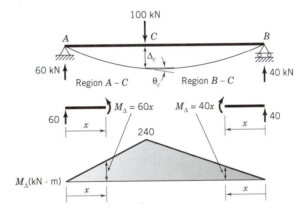

P System (Virtual Load System)

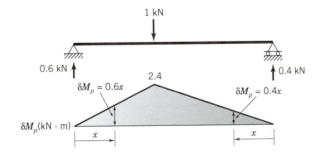

Region Function	A – C $0 \le x \le 4$	B – C $0 \le x \le 6$
M_Δ	$60x$	$40x$
δM_P	$0.6x$	$0.4x$
EI	EI	EI

Applying Eq. 8.40, we have

$$(1)(\Delta_c) = \frac{1}{EI}\left[\int_0^4 (60x)(0.6x)dx + \int_0^6 (40x)(0.4x)dx\right]$$

$$= \frac{1}{EI}[768 + 1\ 152]$$

$$\Delta_c = \frac{1\ 920}{(1)(100\ 000)} = 0.019\ 2\ \text{m} = 19.2\ \text{mm}$$

(the positive sign indicates
downward deflection as assumed)

Slope at Point *C*:

Δ *System* (*Actual Load System*)

Same system used to obtain Δ_c

P System (*Virtual Load System*)

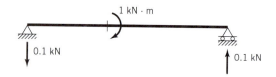

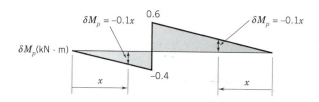

Region Function	A – C $0 \leq x \leq 4$	B – C $0 \leq x \leq 6$
M_Δ	$60x$	$40x$
δM_P	$-0.1x$	$0.1x$
EI	EI	EI

Applying Eq. 8.40, we have

$$(1)(\theta_c) = \frac{1}{EI}\left[\int_0^4 (60x)(-0.1x)dx + \int_0^6 (40x)(0.1x)dx\right]$$

$$= \frac{1}{EI}[-128 + 288]$$

$$\theta_c = \frac{160}{(1)(100\ 000)} = 0.001\ 60\ \text{radian}$$

(the positive sign indicates
clockwise rotation, as assumed)

EXAMPLE 8.13   Determine the vertical deflection and slope at point d for the structure of Example 8.5.

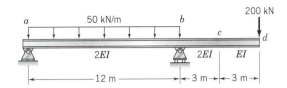

Vertical Deflection at Point d

Δ *System (Actual Load System)*

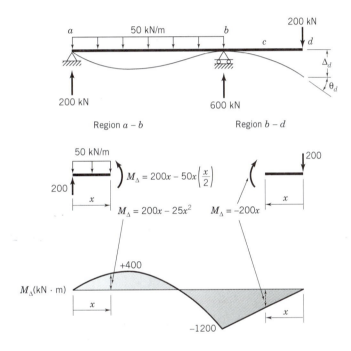

P System (Virtual Load System)

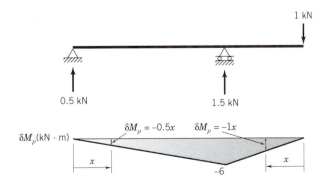

Region Function	$a - b$ $0 \le x \le 12$	$c - b$ $3 \le x \le 6$	$d - c$ $0 \le x \le 3$
M_Δ	$200x - 25x^2$	$-200x$	$-200x$
δM_P	$-0.5x$	$-1x$	$-1x$
EI	$2EI$	$2EI$	EI

Applying Eq. 8.40, we have

$$(1)(\Delta_d) = \frac{1}{2EI}\int_0^{12}(-0.5x)(200x - 25x^2)\,dx + \frac{1}{2EI}\int_3^6(-1x)(-200x)\,dx$$

$$+ \frac{1}{EI}\int_0^3(-1x)(-200x)\,dx = \frac{11\,785\ \text{kN}^2 \cdot \text{m}^3}{EI}$$

$$D_d = \frac{11\,785}{(1)(600\,000)} = 0.019\,6 \text{ m} \quad \text{(the positive sign indicates downward}$$
$$\text{deflection, as assumed)}$$

Slope at Point d

Δ System (Actual Load System)

Same system used to get Δ_d

P System (Virtual Load System)

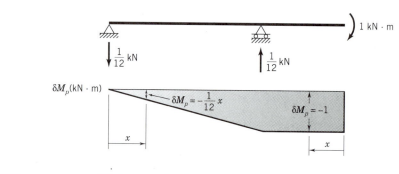

Region Function	$a - b$ $0 \le x \le 12$	$c - b$ $3 \le x \le 6$	$d - c$ $0 \le x \le 3$
M_Δ	$200x - 25x^2$	$-200x$	$-200x$
δM_P	$-\frac{1}{12}x$	-1	-1
EI	$2EI$	$2EI$	EI

Applying Eq. 8.40, we obtain

$$(1)(\theta_d) = \frac{1}{2EI}\int_0^{12}(-\tfrac{1}{12}x)(200x - 25x^2)\,dx + \frac{1}{2EI}\int_3^6(-1)(-200x)\,dx$$

$$+ \frac{1}{EI}\int_0^3(-1)(-200x)\,dx = \frac{2\,856\ \text{kN}^2 \cdot \text{m}^3}{EI}$$

$$\theta_d = \frac{2\,856}{(1)(600\,000)} = 0.004\,76 \text{ radian} \quad \text{(the positive sign indicates}$$
$$\text{clockwise rotation, as assumed)}$$

EXAMPLE 8.14

Determine the horizontal displacement at point b for the frame structure shown. Here, $E = 29 \times 10^3$ ksi, $I = 2,000$ in.^{4}

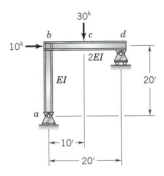

Horizontal Deflection at Point b

Δ System (Actual Load System)

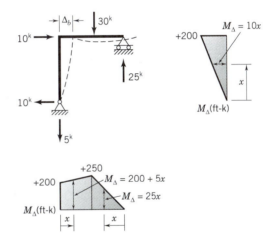

P System (Virtual Load System)

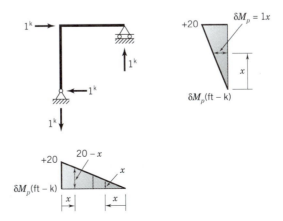

Region Function	$a-b$ $0 \le x \le 20$	$b-c$ $0 \le x \le 10$	$d-c$ $0 \le x \le 10$
M_Δ	$10x$	$200 + 5x$	$25x$
δM_P	$1x$	$20 - x$	x
EI	EI	$2EI$	$2EI$

Using Eq. 8.40, we obtain

$$(1)(\Delta_b) = \frac{1}{EI}\int_0^{20} (1x)(10x)dx + \frac{1}{2EI}\int_0^{10} (200 + 5x)(20 - x)dx$$

$$+ \frac{1}{2EI}\int_0^{10} (25x)(x)dx = \frac{47{,}500 \text{ k}^2 \cdot \text{ft}^3}{EI}$$

$$\Delta_b = \frac{47{,}500 \times 1{,}728}{1(58 \times 10^6)} = 1.415 \text{ in. (positive } \therefore \text{ to right)}$$

8.7.1 Inclusion of Shear Deformations

In addition to the deformations that result from the curvatures induced by the bending moment, there are additional curvatures induced by the shear forces. In most cases, the contribution from the shear is a small percentage of the total deformation and it can be ignored; however, for deep beams, it may be significant.

The equality of the external and internal complementary virtual work quantities is again expressed by Eq. 8.36. In the case of shear, the right-hand side of this equation must represent the internal complementary virtual work associated with the shearing deformation, which is

$$\left(\int_{\text{Vol}} (\delta\sigma_P)\epsilon_\Delta d\text{Vol}\right)_{\text{Shear}} = \sum_{j=1}^{m}\left(\int_l (\delta V_P)dy_\Delta\right)_j \tag{8.41}$$

where, as shown in Fig. 8.8, δV_p is the virtual shear at some point x along the jth member, dy_Δ is the actual transverse shearing deformation over an element dx in length at the same point along the member, and l, j, and m are as defined earlier in Section 8.7.

The shearing strain, γ, is defined as the angle change that the element experiences under the action of the shear forces, as shown in Fig. 8.8b. For the axes employed, both γ and its associated dy_Δ are negative and, therefore, assuming small angles, we may write

$$dy_\Delta = \gamma dx \tag{8.42}$$

Based on linear, elastic behavior, the shear strain is related to the shear stress, τ, by

$$\gamma = \frac{\tau}{G} \tag{8.43}$$

where G is the shear modulus, which is given by

$$G = \frac{E}{2(1 + \nu)} \tag{8.44}$$

where E is the modulus of elasticity and ν is Poisson's ratio.

The shear stress caused by the actual load system can be expressed as

$$\tau = \lambda\left(\frac{V_\Delta}{A}\right) \tag{8.45}$$

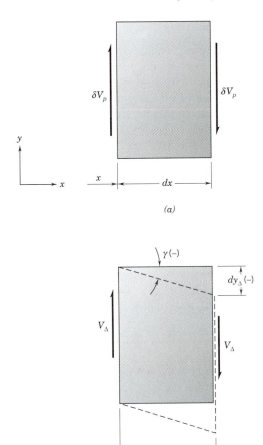

Figure 8.8 Virtual P system and actual Δ system for typical jth member under shear. *(a)* Virtual shear at point x caused by P system of loads. *(b)* Actual shear deformation over element dx caused by Δ system of loads.

where, with A taken as the cross-sectional area, (V_Δ/A) is the average transverse shear stress and λ is a shape factor to account for the variation in the shearing stress over the depth of the cross section ($\lambda = 1.2$ for a rectangular cross section, $\lambda = 10/9$ for a circular cross section, and $\lambda \cong 1$ for a wide-flange or I-beam, in which case A is taken as the area of the web only).

Therefore, substituting Eq. 8.45 into Eq. 8.43 and the resulting expression into Eq. 8.42, we obtain

$$dy_\Delta = \lambda\left(\frac{V_\Delta}{GA}\right)dx \tag{8.46}$$

Substitution of Eq. 8.46 into Eq. 8.41 then enables one to write Eq. 8.36 in the form

$$\sum_{i=1}^{n}(\delta P)_i\Delta_i = \sum_{j=1}^{m}\left(\int_l \lambda(\delta V_P)\frac{V_\Delta}{GA}dx\right)_j \tag{8.47}$$

This expression allows for the determination of a deflection Δ_i caused only by shear deformations. If the combined deflection caused by both the curvatures associated with the bending moment and the deformations from the shear forces is desired, then the right-hand side of Eq. 8.40 should be augmented by the inclusion of the right-hand side of Eq. 8.47.

8.7.2   Inclusion of Torsional Deformations

In certain situations, such as three-dimensional frameworks, torsional deformations can be of significance. Again, the equality of external and internal complementary virtual work is expressed by Eq. 8.36. The torsional internal complementary virtual work is developed using the same general approach that was employed for axial, flexural, and shear effects. Reference to Fig. 8.9 shows that the contribution to the right-hand side of Eq. 8.36 that is related to torsion would take the form

$$\left(\int_{\text{Vol}}(\delta\sigma_P)\epsilon_\Delta d\text{Vol}\right)_{\text{Torsion}} = \sum_{j=1}^{m}\left(\int_l(\delta M_{xP})d\theta_{x\Delta}\right)_j \qquad (8.48)$$

where, as shown in Fig. 8.9a, δM_{xP} is the virtual torsion about the x axis at some point x along the jth member, $d\theta_{x\Delta}$ is the associated actual twisting rotation that occurs over the differential element dx in length at the same point along the member, and l, j, and m are as defined earlier.

If the member has a circular cross-sectional area, no warping of the cross section will occur when it is loaded—the differential element shown in Fig. 8.9 represents this case. For small angles of rotation, the shearing strain, γ, on the outer surface of the element of Fig. 8.9b, which is caused by the actual twisting moment, $M_{x\Delta}$, can be expressed as

$$\gamma = \frac{r d\theta_{x\Delta}}{dx} \qquad (8.49)$$

where r is the radius of the cross section. Therefore, the actual angle of twist over the length of the element, dx, is

$$d\theta_{x\Delta} = \frac{\gamma}{r}dx \qquad (8.50)$$

From elementary mechanics, the shear stress is given by

$$\tau = \frac{M_{x\Delta}r}{J} \qquad (8.51)$$

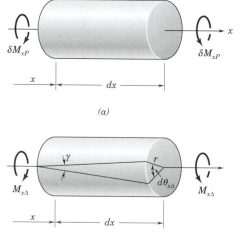

(a)

(b)

Figure 8.9 Virtual P system and actual Δ system for typical jth member in torsion. (a) Virtual torsion at point x caused by P system of loads. (b) Actual twisting deformation over element dx caused by Δ system of loads.

where J is the polar moment of inertia. Substitution of Eq. 8.51 into Eq. 8.43, which relates the linear, elastic relationship between shearing strain and stress, and then substitution of the resulting expression into Eq. 8.50, yields

$$d\theta_{x\Delta} = \frac{M_{x\Delta}dx}{GJ} \tag{8.52}$$

Substitution of Eq. 8.52 into 8.48 leads to the following form for Eq. 8.36:

$$\sum_{i=1}^{n}(\delta P)_i\Delta_i = \sum_{j=1}^{m}\left(\int_l(\delta M_{xP})\frac{M_{x\Delta}}{GJ}dx\right)_j \tag{8.53}$$

This expression allows for the determination of the deflection Δ_i caused only by the torsional deformations. For noncircular members, J is replaced by a torsional section property denoted by I_x, which depends on the nature of the cross section and is discussed in strength of materials textbooks.

8.7.3   Inclusion of Axial Effects

In addition to the deflection caused by bending, shear, and torsion, certain types of beam and frame structures may experience a contribution to the total deflection that results from axial effects. For such cases, the internal complementary virtual work associated with this effect, as developed in Section 7.4, must be taken into account. The inclusion of axial effects is incorporated in the general treatment of beam and frame deflections given in Section 8.8

8.8   DEFLECTIONS FOR NONPLANAR BEAM AND FRAME STRUCTURES

Several methods have been introduced for computing the deflections for planar beam and frame structures. The example problems used to demonstrate the use of these methods have focused on the deflections associated with flexural effects only. However, for nonplanar frame structures, axial and torsional effects are normally included, along with the flexural actions in two mutually perpendicular planes. In some cases, the shearing deformations are included in the two planes of bending. The most versatile method for the inclusion of all these effects is the complementary virtual work method. Equation 8.36 is again applicable, where the right-hand side is composed of the summation of the individual internal complementary virtual work components associated with axial effects, flexural effects, shear effects, and torsional effects. Thus, we have

$$\sum_{i=1}^{n}(\delta P)_i\Delta_i = \sum_{j=1}^{m}\left(\int_l \delta F_{xP}\frac{F_{x\Delta}dx}{EA} + \int_l \delta M_{xP}\frac{M_{x\Delta}dx}{GJ}\right.$$
$$+ \int_l \delta M_{yP}\frac{M_{y\Delta}dx}{EI_y} + \int_l \delta M_{zP}\frac{M_{z\Delta}dx}{EI_z} \tag{8.54}$$
$$\left. + \int_l \lambda\delta V_{zP}\frac{V_{z\Delta}dx}{GA} + \int_l \lambda\delta V_{yP}\frac{V_{y\Delta}dx}{GA}\right)_j$$

In Eq. 8.54, the individual terms on the right-hand side reflect the six separate deformational contributions—axial along the x axis, torsion about the x axis, bending about the y axis, bending about the z axis, shear in the xz plane, and shear in the xy plane.

The notation for the axial and flexural components is more complicated than was evidenced in Eqs. 7.17 and 8.40 because of the need for axis subscripts.

Of course, as was pointed out earlier in the application of the complementary virtual work method, Eq. 8.54 is used to find one displacement quantity at a time. For the displacement Δ_r, a single virtual load of $\delta P_r = 1$ is used. Since there are no other forces in the virtual system that couple with displacements of the Δ system, the desired deflection is isolated on the left-hand side of Eq. 8.54.

EXAMPLE 8.15

Determine the vertical deflection at point d for the structure shown. Include axial, flexural, and torsional contributions, but ignore shear effects.

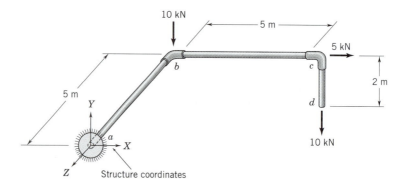

Member and Material Properties

$$A = 5\,000 \text{ mm}^2 = 5\,000 \times 10^{-6} \text{ m}^2$$

$$I_x = J = 1\,000 \times 10^6 \text{ mm}^4 = 1\,000 \times 10^{-6} \text{ m}^4$$

$$I_y = I_z = 1\,500 \times 10^6 \text{ mm}^4 = 1\,500 \times 10^{-6} \text{ m}^4$$

$$E = 200 \text{ GPa} = 200 \times 10^6 \text{ kN/m}^2$$

$$v = 0.3$$

$$G = E/2(1 + v) = 77 \times 10^6 \text{ kN/m}^2$$

Vertical Deflection at Point d

Δ System (Actual Load System)

Application of Eqs. 2.10 yields the reactions, and the member forces are shown on appropriate free-body diagrams. The downward deflection at point d is taken as v_d. Here, each member has its own set of x, y, z axes.

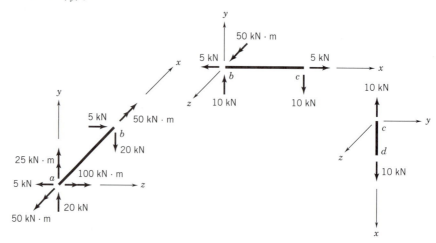

The member force diagrams for the Δ system are as follows (see Section 5.13 for sign convention):

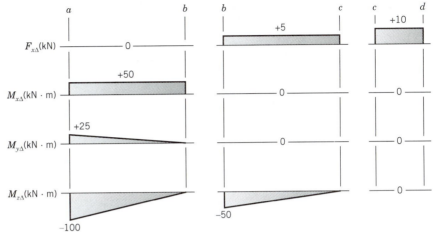

P System (Virtual Load System)

The virtual load system is composed of a unit downward load at point d. The resulting member forces are shown on the individual free-body diagrams.

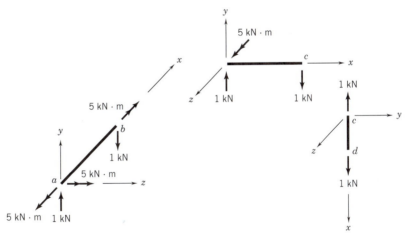

The member force diagrams for the P system are as follows:

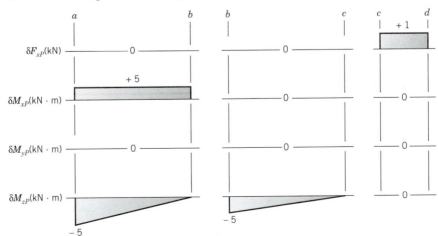

Deflection Calculation

Equation 8.54 will be used, and the following table gives a summary of the required member force expressions:

Region / Function	$a - b$ $0 \le x \le 5$	$b - c$ $0 \le x \le 5$	$c - d$ $0 \le x \le 2$
$F_{x\Delta}$	0	+5	+10
$M_{x\Delta}$	+50	0	0
$M_{y\Delta}$	$+25 - 5x$	0	0
$M_{z\Delta}$	$-100 + 20x$	$-50 + 10x$	0
δF_{xP}	0	0	+1
δM_{xP}	+5	0	0
δM_{yP}	0	0	0
δM_{zP}	$-5 + 1x$	$-5 + 1x$	0

The governing virtual expression is

$$\sum_{i=1}^{n} (\delta P)_i D_i = \sum_{j=1}^{n} \left(\int_l \delta F_{xP} \frac{F_{x\mathrm{D}} dx}{EA} + \int_l \delta M_{xP} \frac{M_{x\mathrm{D}} dx}{GJ} \right.$$

$$\left. + \int_l \delta M_{yP} \frac{M_{y\mathrm{D}} dx}{EI_y} + \int_l \delta M_{zP} \frac{M_{z\mathrm{D}} dx}{EI_z} \right)_j \tag{8.54}$$

Substitution from the table of member force quantities gives

$$(1)(v_d) = \frac{1}{EA} \int_0^2 (1)(10) dx + \frac{1}{GJ} \int_0^5 (5)(50) dx + \frac{1}{EI_z} \int_0^5 (-5 + x)(-100 + 20x) dx$$

$$+ \frac{1}{EI_z} \int_0^5 (-5 + x)(-50 + 10x) dx$$

$$= \frac{20}{EA} + \frac{1\,250}{GJ} + \frac{1\,250}{EI_z}$$

$$= \frac{20}{200 \times 10^6 \times 5\,000 \times 10^{-6}} + \frac{1\,250}{77 \times 10^6 \times 1\,000 \times 10^{-6}}$$

$$+ \frac{1\,250}{200 \times 10^6 \times 1\,500 \times 10^{-6}}$$

$$= 0.000\,020 + 0.016\,234 + 0.004\,167 \ (\mathrm{kN} \cdot \mathrm{m})$$

$$v_d = 0.020\,421 \text{ m} \quad \text{(positive sign confirms that deflection is downward)}$$

EXAMPLE 8.16

Determine the vertical deflection at point b for the structure and loading shown. Include flexural and shear deformations. Assume a rectangular cross section for which $\lambda = 1.2$.

Δ System (Actual Load System)

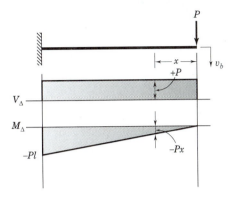

P System (Virtual Load System)

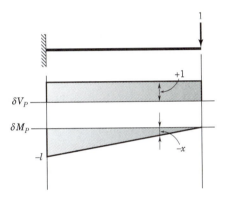

The governing virtual expression is extracted from Eq. 8.54 to be

$$\sum_{i=1}^{n} (\delta P)_i \Delta_i = \sum_{j=1}^{m} \left(\int_0^l \delta M_{zP} \frac{M_{zD} dx}{EI_y} + \int_0^l \lambda \delta V_{zP} \frac{V_{z\Delta} dy}{GA} \right)_j \qquad (8.54)$$

Substitution into this equation gives the following:

$$(1)(v_b) = \int_0^l (-x)\frac{(-Px)dx}{EI_y} + \int_0^l \lambda(1)\frac{(P)dx}{GA}$$

from which

$$v_b = \frac{Pl^3}{3EI_y} + \frac{\lambda Pl}{GA} = \frac{Pl^3}{3EI_y}\left(1 + \frac{3\lambda}{l^2} \cdot \frac{EI_y}{GA}\right)$$

For a rectangular section (width $= b$, height $= h$) for which $\dfrac{I_y}{A} = \dfrac{bh^3/12}{bh} = \dfrac{bh^2}{12}$, and $\lambda = 1.2$, we obtain

$$V_b = \frac{Pl^3}{3EI_y}\left(1 + \frac{3}{10}\left(\frac{h}{l}\right)^2 \cdot \frac{E}{G}\right)$$

For most beams, the length (l) greatly exceeds the depth (h) and, therefore, the second term of the above expression is small when compared to unity. For such cases, the total deflection is essentially equal to the contribution from flexure (first term). Only for very short, deep beams will the second term (contribution from shear) be significant.

For a wide flange steel beam, $\lambda \approx 1.0$, which would not significantly alter the above conclusions.

8.9   BEAM AND FRAME DEFLECTIONS BY ENERGY METHODS

The virtual work method is the foundation upon which all energy methods are based; however, other energy-related techniques can be employed to obtain structural deflections. Castigliano's second theorem, which is developed in Section 2.18.2, is particularly well suited for the deflection problem.

In applying Castigliano's second theorem, we must express the total strain energy for the structure U in terms of the applied loads. Then, corresponding to each structure force P_i, the corresponding displacement quantity Δ_i can be determined from the relationship

$$\frac{\partial U}{\partial P_i} = \Delta_i \tag{8.55}$$

which is presented as Eq. 2.65. This equation can be used to determine a single displacement quantity, or it can be used to establish the complete set of displacement–force equations for the structure.

Since the strain energy is equal to the work of deformation, Eq. 2.22 can be used to express the strain energy associated with flexural deformation of the element shown in Fig. 8.1b. This leads to

$$dU = \tfrac{1}{2}M\,d\theta \tag{8.56}$$

where the factor $\tfrac{1}{2}$ results from the linear relationship between the moment, M, and the rotation, $d\theta$, as explained in Fig. 2.12. Substituting the expression for $d\theta$ from Eq. 8.11 into Eq. 8.56, we obtain

$$dU = \frac{1}{2EI}M^2\,dx \tag{8.57}$$

where M is the bending moment on the element, I is the moment of inertia of the element cross section, and E is the modulus of elasticity. Thus, the total strain energy for a member of length l is

$$U = \frac{1}{2E}\int_{x=0}^{x=l}\frac{M^2}{I}dx \tag{8.58}$$

and the total strain energy stored in a structure of m members is given by

$$U = \sum_{i=1}^{m}\frac{1}{2E}\int_{x=0}^{x=l_i}\frac{M_i^2}{I_i}dx \tag{8.59}$$

where the i subscript is associated with the ith member. It is noted that M_i must be expressed in terms of the externally applied loads in order that the differentiation indicated in Eq. 8.55 can be carried out.

Equation 8.55 represents a general formulation of Castigliano's second theorem. This theorem was applied to truss problems in Section 7.7 where the strain energy resulted from the axial deformations of the truss members, as expressed in Eq. 7.24. In the application discussed this far in this section, the strain energy is associated with planar bending deformations, and Eq. 8.59 gives the appropriate expression for the strain energy. For nonplanar structures, such as those discussed in Section 8.8, Eq. 8.55 remains valid; however, the strain energy expression must include the individual contributions from axial and torsional effects, and bending and shear effects in two separate planes.

EXAMPLE 8.17

Use Castigliano's second theorem to develop the displacement–force equations for the structure shown. From these equations, extract the structure flexibility matrix. The structure coordinate system and the member notation are given on the figure. EI = constant.

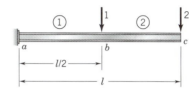

Moment Expressions

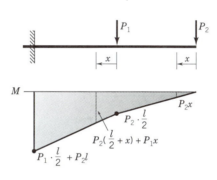

Region Function	Member ①; $a-b$ $0 \leq x \leq l/2$	Member ②; $b-c$ $0 \leq x \leq l/2$
M	$P_2\left(\dfrac{l}{2}+x\right)+P_1x$	P_2x

Strain Energy

$$U = \sum_{i=1}^{m} \frac{1}{2E} \int_{x=0}^{x=l_i} \frac{M_i^2}{I_i} dx \qquad (8.59)$$

$$i = 1: \quad U_1 = \frac{1}{2EI}\int_{x=0}^{x=l/2}\left[P_1x + P_2\left(\frac{l}{2}+x\right)\right]^2 dx = \frac{1}{EI}\left[\frac{P_1^2l^3}{48} + \frac{5P_1P_2l^3}{48} + \frac{7P_2^2l^3}{48}\right]$$

$$i = 2: \quad U_2 = \frac{1}{2EI}\int_{x=0}^{x=l/2}[P_2x]^2 dx = \frac{1}{EI}\left[\frac{P_2^2l^3}{48}\right]$$

$$U = U_1 + U_2 = \frac{1}{48EI}[P_1^2l^3 + 5P_1P_2l^3 + 8P_2^2l^3]$$

Displacement–Force Equations and Flexibility Matrix

$$\frac{\partial U}{\partial P_i} = \Delta_i \quad (i = 1, 2) \tag{8.55}$$

$$\frac{\partial U}{\partial P_1} = \Delta_1 \Rightarrow \frac{1}{EI}\left[\frac{P_1 l^3}{24} + \frac{5 P_2 l^3}{48}\right] = \Delta_1$$

$$\frac{\partial U}{\partial P_2} = \Delta_2 \Rightarrow \frac{1}{EI}\left[\frac{5 P_1 l^3}{48} + \frac{P_2 l^3}{3}\right] = \Delta_2$$

Or, in matrix form,

$$\frac{l^3}{EI}\begin{bmatrix} \frac{1}{24} & \frac{5}{48} \\ \frac{5}{48} & \frac{1}{3} \end{bmatrix}\begin{Bmatrix} P_1 \\ P_2 \end{Bmatrix} = \begin{Bmatrix} \Delta_1 \\ \Delta_2 \end{Bmatrix}$$

From this equation, it is clear that

$$[D] = \frac{l^3}{EI}\begin{bmatrix} \frac{1}{24} & \frac{5}{48} \\ \frac{5}{48} & \frac{1}{3} \end{bmatrix}$$

8.10 REFERENCE

Au, T., and Christiano, P., *Structural Analysis,* Chapter 6, Prentice-Hall, Englewood Cliffs, N.J., 1987.

8.11 SUGGESTED PROBLEMS

8.1 through 8.4 Use the integration method to determine the expressions for slope and deflection throughout each of the following structures for the indicated loadings. EI = constant for each structure.

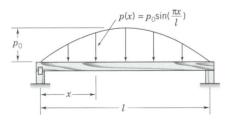

8.1

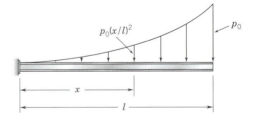

8.2

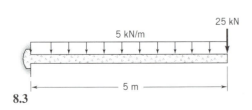

8.3

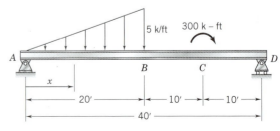

8.4

8.5 through 8.26   Use the moment–area method to determine the indicated displacement quantities for each of the indicated structures and loadings.

8.5   Vertical deflection and rotation at point A. $I = 2,000$ in.^{4}; $E = 10 \times 10^3$ ksi.

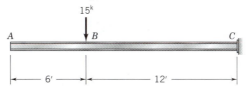

8.6   Vertical deflection and rotation at point B. $E = 200$ GPa; $I = 1000 \times 10^{-6}$ m^4 = 1000×10^6 mm^4.

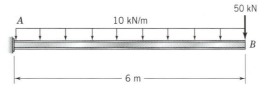

8.7   Vertical deflection and rotation at points C and E. $I = 3,000$ in.^{4}; $E = 3 \times 10^3$ ksi.

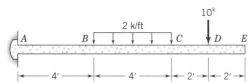

8.8   Vertical deflection at points C and E; rotation at point B. $I = 1,500$ in.^{4}; $E = 29 \times 10^6$ psi.

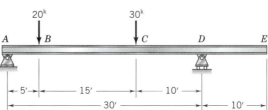

8.9   Location and magnitude of the maximum vertical deflection. $E = 200$ GPa; $I = 200 \times 10^6$ mm^4.

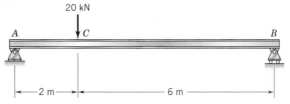

8.10   Location and magnitude of the maximum vertical deflection. What is the maximum value of P so that this deflection does not exceed 1 inch? $I = 1,200$ in.^{4}; $E = 29 \times 10^3$ ksi.

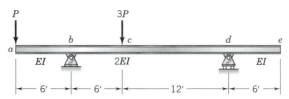

8.11 Vertical deflection and rotation at points b and d and the location and magnitude of maximum vertical deflection. $I = 2,500$ in.^{4}; $E = 29 \times 10^3$ ksi.

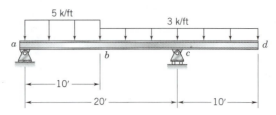

8.12 Vertical deflection and rotation at point B. $I = 200 \times 10^6$ mm^4; $E = 70$ GPa.

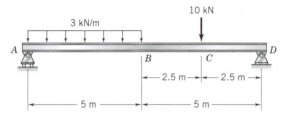

8.13 Vertical deflection and angle change at point B. $E = 29 \times 10^6$ psi; $I = 1,000$ in.^{4}

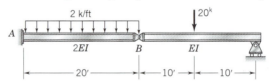

8.14 Vertical deflection at point D. $E = 200$ GPa; $I = 50 \times 10^6$ mm^4.

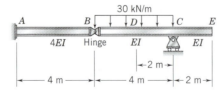

8.15 Vertical deflection and rotation at point C. $E = 29 \times 10^3$ ksi; $I = 1,500$ in.^{4}

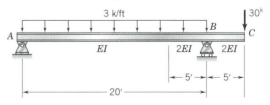

8.16 Vertical deflection and rotation at center of span bc. $I = 1,500$ in.^{4}; $E = 29 \times 10^3$ ksi.

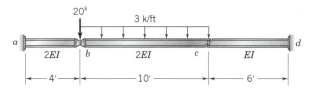

8.17 Vertical deflection at point C and the slope at point E. $I = 1,000$ in.^{4}; $E = 29 \times 10^3$ ksi.

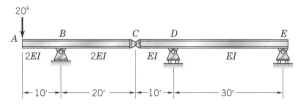

8.18 Midspan vertical deflection for spans AB and CD. $I = 300 \times 10^6$ mm^4; $E = 200$ GPa.

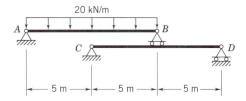

8.19 Midspan vertical deflection for each stringer and the girder. *Note:* Assume stringers are simply supported between floorbeam support points. $I = 1,250$ in.^{4}; $E = 29 \times 10^3$ ksi.

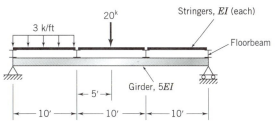

8.20 Vertical deflection and rotation at point f: $I = 1,000$ in.^{4}; $E = 29 \times 10^3$ ksi.

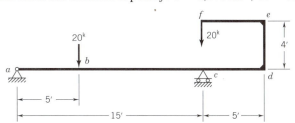

8.21 Vertical and horizontal deflections at point C. $E = 200$ GPa; $I = 300 \times 10^6$ mm^4.

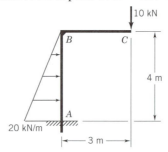

8.22 Rotation at point A and horizontal deflection at point D. $E = 200$ GPa; $I = 500 \times 10^6$ mm^4.

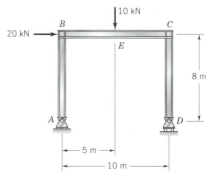

8.23 Horizontal and vertical deflections at, and the rotation on each side of, point b. $I = 800$ in.^{4}; $E = 29 \times 10^3$ ksi.

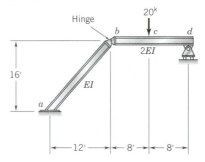

8.24 Horizontal deflection and rotation at point b and vertical deflection at point c. $I = 1,800$ in.^{4}; $E = 29 \times 10^3$ ksi.

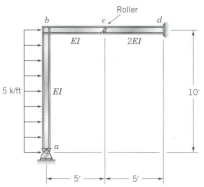

8.25 Horizontal deflection at point *b* and vertical deflection at point *c*. $I = 1{,}000$ in.^{4}; $E = 29 \times 10^3$ ksi.

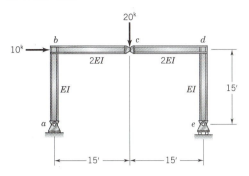

8.26 Horizontal deflection at points *B* and *C* and the rotation at point *D*. $E = 29 \times 10^3$ ksi; $I = 2{,}000$ in.^{4}

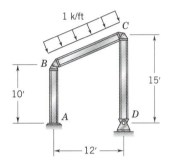

8.27 through 8.34   Use the conjugate beam method to determine the indicated displacement quantities for each of the structures and loadings that are designated.

8.27 Vertical deflection and rotation at point *A* for the structure and loading of Problem 5.

8.28 Vertical deflection and rotation at point *B* for the structure and loading of Problem 6.

8.29 Vertical deflections at points *C* and *E*, and the rotation at point *B*, for the structure and loading of Problem 8.

8.30 Location and magnitude of the maximum vertical deflection for the structure and loading of Problem 9.

8.31 Vertical deflection and angle change at point *B* for the structure and loading of Problem 13.

8.32 Vertical deflection at point *D* for the structure and loading of Problem 14.

8.33 Vertical deflection and rotation at point *C* for the structure and loading of Problem 15.

8.34 Vertical deflection and rotation at center of span *bc* for the structure and loading of Problem 16.

8.35 through 8.40   Use the conjugate beam method to construct the complete slope and deflection diagrams for the prescribed structure and loading.

8.35 Structure and loading of Problem 5.

8.36 Structure and loading of Problem 7.

8.37 Structure and loading of Problem 8.

8.38 Structure and loading of Problem 9.

8.39 Structure and loading of Problem 14.

8.40 Structure and loading of Problem 17.

8.41 through 8.60 Use the complementary virtual work method to determine the indicated displacement quantities for each of the structures and loadings that are designated.

8.41 Vertical deflection at point B for the structure and loading of Problem 5.

8.42 Vertical deflection at point C for the structure and loading of Problem 9.

8.43 Vertical deflection at point C for the structure and loading shown. $E = 29 \times 10^3$ ksi; $I = 200$ in.^{4}

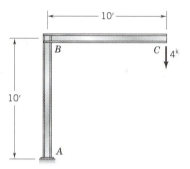

8.44 Horizontal deflection at point B for the structure and loading shown. $E = 200$ GPa; $I = 500 \times 10^6$ mm^4.

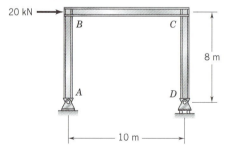

8.45 Vertical deflection and rotation at point B for the structure and loading of Problem 6.

8.46 Vertical deflection and rotation at point D for the structure and loading of Problem 7.

8.47 Vertical deflection at point C and rotation at point E for the structure and loading of Problem 8.

8.48 Vertical deflection at point c for the structure and loading of Problem 10.

8.49 Rotation at points A and D for the structure and loading of Problem 12.

8.50 Vertical deflection and angle change at point B for the structure and loading of Problem 13.

8.51 Vertical deflection at point B and rotation at point E for the structure and loading of Problem 14.

8.52 Vertical deflection and rotation at the center of span bc for the structure and loading of Problem 16.

8.53 Vertical deflection and rotation at point f for the structure and loading of Problem 20.

8.54 Horizontal deflection at point e for the structure and loading of Problem 20.

8.55 Horizontal and vertical deflection at point C for the structure and loading of Problem 21.

8.56 Horizontal deflection at point C for the structure and loading of Problem 22.

8.57 Horizontal and vertical deflection at point B for the structure and loading of Problem 22.

8.58 Horizontal deflection and rotation at point b for the structure and loading of Problem 24.

8.59 Horizontal deflection at point b and the vertical deflection at point c for the structure and loading of Problem 25.

8.60 Horizontal deflection at points B and C for the structure and loading of Problem 26.

8.61 and 8.62 Use the complementary virtual work method to determine the indicated displacement quantity for the structures and loadings shown. Include axial and flexural deformations.

8.61 Vertical deflection at point b. $A_{abc} = 10.0$ in.^{2}, $A_{cd} = 0.5$in.^{2}, $I_{abc} = 800$ in.^{4}, $E = 29 \times 10^3$ ksi.

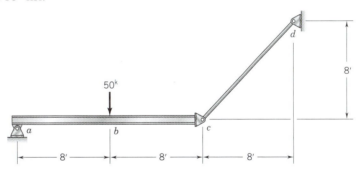

8.62 Vertical deflection at point c. $A_{abc} = 14.0$ in.^{2}, $A_{bd} = 2.0$ in.^{2}, $I_{abc} = 400$ in.^{4}, $E = 29 \times 10^3$ ksi.

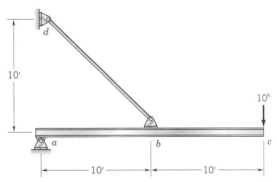

8.63 and 8.64 Use the complementary virtual work method to determine the indicated displacement quantity for the structures and loadings shown. Include shearing and flexural deformations.

8.63 Vertical deflection at point b. The beam is a wide-flange member. $A_{web} = 5.25$ in.^{2}, $I = 600$ in.^{4}, $E = 29 \times 10^3$ ksi, $G = 12 \times 10^3$ ksi.

8.64 Vertical deflection at the beam mid-span. The beam is a wide-flange member. $A_{web} = 7.5$ in.^{2}, $I = 625$ in.^{4}, $E = 29 \times 10^3$ ksi, $G = 12 \times 10^3$ ksi.

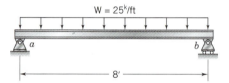

8.65 Use the complementary virtual work method to determine the displacement along the y-axis at point d for the structure and loading of Problem 3.41. Assume that each member has the same area and that E is a constant.

8.66 Use the complementary virtual work method to determine the displacement at the point and in the direction of the load for the structure and loading of Problem 3.42. Assume that each member has the same area and that E is a constant.

8.67 Determine the rotation about the x axis at point b and the vertical deflection at point g for the structure and loading of Problem 3.37. The following section and material properties apply to all members: $A = 4$ in.^{2}; $I_x = 200$ in.^{4}; $I_y = I_z = 500$ in.^{4} (in terms of member coordinates); $E = 29 \times 10^3$ ksi; $v = 0.3$.

8.68 Determine the displacements in the $x, y,$ and z directions at point e for the structure and loading of Problem 3.40. The following section and material properties apply to all members: $A = 6$ in.^{2}; $I_x = 400$ in.^{4}; $I_y = I_z = 1,000$ in.^{4} (in terms of member coordinates: member y axis normal to structure xz plane); $E = 29 \times 10^3$ ksi; $v = 0.3$.

8.69 Use Castigliano's second theorem to determine the vertical deflection at point C and the slope of point E for the structure and loading of Problem 17.

8.70 Use Castigliano's second theorem to determine the horizontal deflection at point B and vertical deflection at point E for the structure and loading of Problem 22.

8.71 Use Castigliano's second theorem to develop the displacement–force equations for the structure of Problem 8 employing the structure coordinate system shown below. From these equations, determine the vertical deflections at points B and C for the loading of Problem 8.

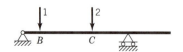

8.72 Use Castigliano's second theorem to develop the displacement–force equations for the structure of Problem 22 using the coordinate system shown below. From these equations, determine the horizontal deflection at B and the vertical deflection at E for the loading of Problem 22.

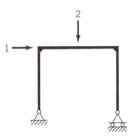

Part Four

Analysis of Statically Indeterminate Structures

Chapter 9

More Basic Concepts of Structural Analysis

The Petronas Towers, Kuala Lumpur, Malaysia (courtesy The Thornton-Tomasetti Group Inc., photo by Michael Goodman).

9.1   REQUIREMENTS AND LIMITATIONS OF EQUILIBRIUM

As was stated in Section 2.4, equilibrium refers to the condition of a structure when it is at rest and remains at rest during the application of loading. For equilibrium to be realized, there must be a balance of the force tendencies that would act to disrupt the structure in any way. For planar structures, equilibrium is ensured by satisfying the three equations of static equilibrium as they were expressed in Eq. 2.5.

It is to be emphasized that equilibrium is a requirement that must be satisfied. Thus, the analysis must lead to a set of reactions and internal forces that satisfy the equations of static equilibrium. If these equations are sufficient for the complete structural analysis, the structure is said to be *statically determinate*. However, if there are more independent reaction components, and/or internal forces than can be determined from the application of the equations of equilibrium, then the structure is said to be *statically indeterminate*. This does not imply that a solution does not exist. As long as the structure is stable, a solution will exist; however, the conditions of static equilibrium are insufficient for completing the solution.

Consider the beam structure shown in Fig. 9.1. According to the criteria developed in Section 3.3, this structure is statically indeterminate because there is one more reaction component than there are equations of equilibrium available for the solution. Application of these equations leads to the obvious result that $R_{ax} = 0$ and that

$$R_{ay} + R_{by} = 500$$
$$M_{az} - 10R_{by} = -2\,500 \tag{9.1}$$

A unique solution for the reactions is not possible in this case because there are three unknown reactions and only two equations. However, equilibrium solutions can be determined by rewriting the equations in the form

$$R_{ay} = 500 - R_{by}$$
$$M_{az} = -2\,500 + 10R_{by} \tag{9.2}$$

Here, for any assigned value of R_{by}, the two equations can be solved for R_{ay} and M_{az}. Figure 9.2 shows the results of such solutions as R_{by} is allowed to take on values ranging from 0 kN to +500 kN. For instance, if $R_{by} = +100$ kN then R_{ay} and M_{az} are +400 kN and −1 500 kN · m , respectively. This plot shows that there is an infinite number of solutions—each resulting in values for R_{ay} and M_{az} that, along with the assigned value of R_{by}, satisfy the equilibrium requirements.

Thus, it can be concluded that equilibrium is a requirement that the solution must satisfy. However, there is a limitation in that equilibrium considerations alone give no clue regarding which one of the infinite array of possible equilibrium solutions is the correct one.

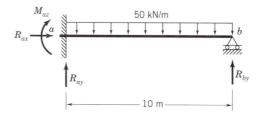

Figure 9.1 Statically indeterminate beam structure.

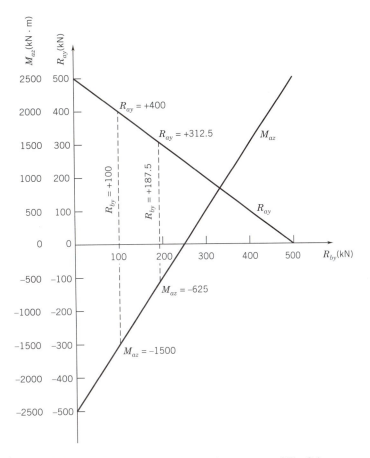

Figure 9.2 Equilibrium solutions for structure of Fig. 9.1.

9.2 STATIC INDETERMINACIES; REDUNDANCIES

In studying the detailed methods of analysis for statically indeterminate structures, it is essential that the criteria for the statical classification of structures be clearly understood. This topic has already been given careful consideration, and the reader is urged to review Sections 3.3, 3.5, 3.7, 4.8, 4.14, and 5.4. These sections develop the statical classification criteria for several different types of structures with regard to external, internal, and overall static determinancy and stability.

In essence, all of the criteria that are developed in these sections involve a comparison between the number of independent unknown force quantities and the number of independent equations of equilibrium that are available for the solution of these unknowns. The criteria always take the following forms:

1. If there are more equations than there are unknowns, the structure is statically unstable.
2. If there is the same number of equations as unknowns, the structure is statically determinate.
3. If there are fewer equations than unknowns, the structure is statically indeterminate.

The first of these criteria is absolute. The second and third criteria give conditions that are necessary, but not sufficient, for the statical classification of the structure.

The unknown force quantities must be arranged so as to ensure the stability of the structure.

Thus, the criteria that have been developed indicate that a structure is statically indeterminate when there are more reaction components available and/or member forces present than are necessary for the stability of the structure. The degree of external indeterminacy is equal to the number of reaction components that are available in excess of the number that is required for external stability. These excess reaction components are called *redundants* because they are unnecessary for the stability of the structure. The degree of internal indeterminacy is given by the number of internal force components that are present in excess of those that are needed for internal stability. These are also called *redundants* since they are not required for a stable structure.

In certain methods of statically indeterminate analysis, it is necessary for analysts to identify explicitly the reaction components or internal force components that they wish to select as the redundants. These are then conceptually removed from the structure, and the statically determinate structure that remains is called the *primary structure*. However, it is essential that the redundants be selected so that the primary structure is stable.

As an example, consider the beam structure of Fig. 9.3*a,* which is the same structure that was considered in the previous section. Since $r_a = 4 > 3$, the structure is statically indeterminate externally to the first degree. Thus, there is one redundant, or unnecessary, reaction component. If R_{ax} is selected as the redundant, the primary structure is as shown in Fig. 9.3*b*. This structure is both unstable and indeterminate. However, if R_{by} is taken as the redundant, then the primary structure is statically determinate and stable, as shown in Fig. 9.3*c*.

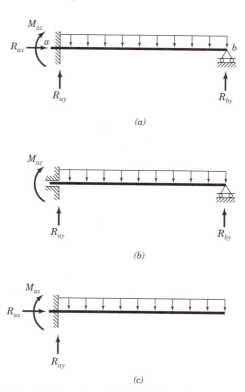

Figure 9.3 Selection of redundant reaction. *(a)* Statically indeterminate structure. *(b)* Unstable and statically indeterminate primary structure. *(c)* Stable and statically determinate primary structure.

9.3   REQUIREMENTS AND LIMITATIONS OF COMPATIBILITY

As was described in Chapter 2, compatibility places constraints on the displacements of a structure to ensure that its individual elements fit together properly and that the structure conforms to the displacement boundary conditions prescribed at the support points. Certain internal compatibility conditions are implicitly satisfied through the formulation of the method of analysis. Other internal conditions must be explicitly expressed whenever the structure is separated into free-body diagrams. Here, care must be taken to make certain that the equations that are formulated for the analysis retain the required compatibility at the points where the structure has been severed. External compatibility conditions require compliance between the displacements that the structure undergoes at its support points and the displacements that the support mechanisms permit to occur.

It is important to note that compatibility is a requirement that must be satisfied. Thus, the analysis must lead to a deformed configuration that will adhere to all of the deformation constraints imposed on the structure. However, in general, there is an infinite number of deformed patterns that will satisfy the compatibility conditions. Therefore, satisfaction of compatibility is not enough to ensure the correct solution.

As an example, consider again the beam of Section 9.1. At that point, the analysis could not be completed by statics alone because of the statically indeterminate nature of the structure. However, if the reaction component R_{by} is removed, as shown in Fig. 9.3c, the resulting primary structure is statically determinate. The reactions can now be determined from statics, and the deflections, which include a deflection of point b, Δ_{b1}, of $-62\,500/EI$ (downward),[1] can be determined by one of the methods

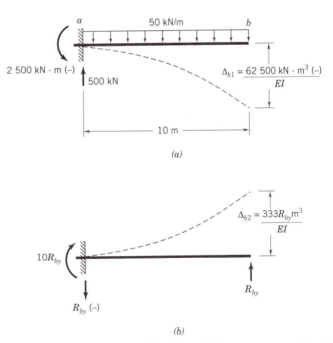

Figure 9.4 Primary structure for structure of Fig. 9.1. *(a)* Primary structure subjected to given loading. *(b)* Primary structure subjected to redundant reaction.

---

[1] Deflection in meters if EI has units of $kN \cdot m^2$.

presented in Chapter 8. Comparing this solution, which is shown in Fig. 9.4a, with the given structure of Fig. 9.1, we see that the designated boundary condition at point b is violated. That is, the vertical restraint required by the support point b is not maintained.

To remedy this problem, we allow the primary structure to be acted upon by the redundant reaction R_{by}, as shown in Fig. 9.4b. The solution for this case includes a deflection of point b, Δ_{b2}, equal to $+333 \, R_{by}/EI$ (upward),[2] which can also be determined by one of the methods of Chapter 8.

For a solution that includes the proper loading and also satisfies the designated boundary conditions at point b, the solutions shown in Figs. 9.4a and 9.4b must be superimposed so that the final vertical displacement at point b, Δ_b, is equal to zero. Thus, we have the compatibility equation

$$\Delta_b = \Delta_{b1} + \Delta_{b2} = 0 \tag{9.3}$$

or, upon substitution and rearrangement,

$$\frac{333 R_{by}}{EI} = \frac{62\ 500}{EI} \tag{9.4}$$

from which

$$R_{by} = +187.5 \text{ kN} \tag{9.5}$$

Returning to Eqs. 9.2, and using $R_{by} = +187.5$ kN, we obtain $R_{ay} = 312.5$ kN and $M_{az} = -625$ kN $\cdot$ m, as shown in Fig. 9.2.

The final solution clearly satisfies the requirements of equilibrium given in Eqs. 9.2 and the requirements of compatibility stated as Eq. 9.5. These equations collectively form a set of three equations that must be solved simultaneously for the unknown reaction components.

9.4   KINEMATIC INDETERMINACIES; REDUNDANCIES

It is convenient at this point to introduce the notion of *kinematic indeterminacy* in a fashion that is analogous to the concept of static indeterminacy. As static indeterminacy deals with the number of force quantities that must be determined in order to render the equilibrium solution complete, kinematic indeterminacy refers to the number of displacement quantities (kinematic degrees of freedom) that are necessary to define the deformational response of the structure.

To illustrate the concept, consider the structure previously studied in Section 9.1. This structure is shown again in Fig. 9.5a along with a qualitative deflected shape. Since the structure is fixed at point a and vertically restrained at point b and the axial deformation is zero in this case, there is only one kinematic degree of freedom—θ_b, the rotation at point b. Therefore, the structure is kinematically indeterminate to the first degree.

Extending the analogy with static determinacy, we identify the rotation, θ_b, as a *kinematic redundancy*, and if it is removed from the structure ($\theta_b = 0$), the resulting *primary structure* is said to be *kinematically determinate*. Figure 9.5b shows the kinematically determinate primary structure for our example case. It should be noted that a structure is rendered kinematically determinate by setting the displacements equal to zero or some predetermined values. The latter case would be appropriate, for example, if there were prescribed settlement displacements.

---

[2]Deflection in meters if EI has units of kN $\cdot$ m^2.

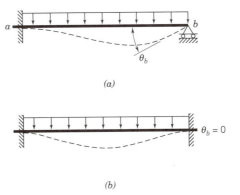

Figure 9.5 Selection of redundant displacement. *(a)* Kinematically indeterminate structure with redundant displacement, θ_b. *(b)* Kinematically determinate primary structure.

9.5   ALTERNATIVE FORM OF ANALYSIS

In Eqs. 9.1, the conditions of equilibrium were applied to the structure of Fig. 9.1. These conditions were found to be inadequate in determining a unique solution, and the compatibility condition of Eq. 9.3 was used to complete the solution.

As an alternative approach, let's begin with compatibility considerations. The structure of Fig. 9.1 is shown again in Fig. 9.6a. In this case, an end moment of M_{bz} has been added which must eventually be set equal to zero. It is clear that this structure is statically indeterminate. Next, the statically determinate primary structure of Fig. 9.3c is separately loaded with w, R_{by}, and M_{bz} as shown in the successive figures of Fig. 9.6b. The displacement quantities in these figures were determined by the methods of Chapter 8.

The separate solutions of Fig. 9.6b must be superimposed to obtain the correct boundary conditions for the given structure of Fig. 9.6a. Thus, we have

$$\Delta_{b1} + \Delta_{b2} + \Delta_{b3} = \Delta_b = 0$$
$$\theta_{b1} + \theta_{b2} + \theta_{b3} = \theta_b \tag{9.6}$$

Substitution of the displacement quantities from Fig. 9.6b into Eqs. 9.6 yields

$$\frac{R_{by}l^3}{3EI} + \frac{M_{bz}l^2}{2EI} = \frac{wl^4}{8EI}$$
$$\frac{R_{by}l^2}{2EI} + \frac{M_{bz}l}{EI} = \frac{wl^3}{6EI} + \theta_b \tag{9.7}$$

Solving Eqs. 9.7 for M_{bz} and R_{by} and then applying statics to structure of Fig. 9.6a, we obtain

$$R_{ay} = \frac{wl}{2} + \frac{3k_b}{2l}\theta_b$$

$$M_{az} = -\frac{wl^2}{12} - \frac{k_b}{2}\theta_b$$

$$R_{by} = \frac{wl}{2} - \frac{3k_b}{2l}\theta_b \tag{9.8}$$

$$M_{bz} = -\frac{wl^2}{12} + k_b\theta_b$$

where $k_b = 4EI/l$, as defined in Section 8.2.

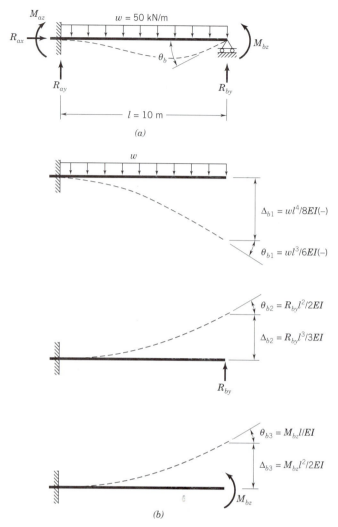

Figure 9.6 Statically indeterminate beam structure. (a) Loading and deformation. (b) Statically determinate primary structure with individual loadings.

Equations 9.8 express all of the response quantities in terms of the single kinematic degree of freedom θ_b. The solution represented by these equations satisfies the required compatibility conditions and the overall equilibrium requirements for the structure. Figure 9.7 gives the member-end forces associated with a range of $EI\theta_b$ values from zero to $1\,200\ \text{kN}\cdot\text{m}^2$ for the structure shown in Fig. 9.6a. For example, for $EI\theta_b = 400\ \text{kN}\cdot\text{m}^2$, $R_{ay} = +274\ \text{kN}$, $M_{az} = -496.7\ \text{kN}\cdot\text{m}$, $R_{by} = +226\ \text{kN}$, and $M_{bz} = -256.7\ \text{kN}\cdot\text{m}$.

These equations also provide the member-end forces associated with the kinematically determinate primary structure shown in Fig. 9.5b. With $\theta_b = 0$, these member-end forces have the values shown in Fig. 9.8a, and these are consistent with the values shown in Fig. 9.7 for $EI\theta_b = 0$. In lieu of the negative signs for M_{az} and M_{bz}, the moments are shown in the opposite directions from the positive directions of Fig. 9.6a. Although this solution is a valid equilibrium solution, it violates the required force boundary condition on the moment at point b.

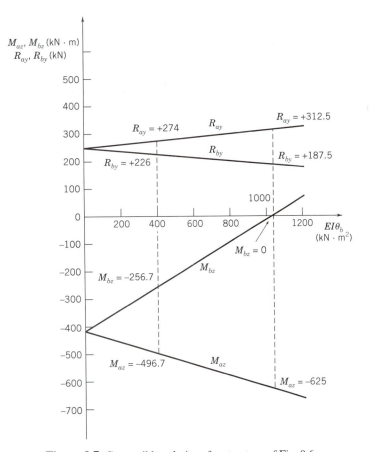

Figure 9.7 Compatible solutions for structure of Fig. 9.6a.

To remedy this problem, the primary structure is subjected to the redundant displacement θ_b. Based on Eq. 8.19, and noting that $k_b = 4EI/l$, the associated end moments and reactions take on the values shown in Fig. 9.8b.

For a solution that reflects the proper loading and satisfies the designated force boundary conditions at point b, the solution shown in Figs. 9.8a and 9.8b must be superimposed so that the final moment at point b is zero. Thus, we have the equilibrium equation

$$M_{bz1} + M_{bz2} = 0 \tag{9.9}$$

or, upon substitution,

$$-416.7 + 0.4EI\theta_b = 0 \tag{9.10}$$

from which

$$EI\theta_b = 1\,041.7 \text{ kN} \cdot \text{m}^2 \tag{9.11}$$

Returning to Eqs. 9.8 and taking $EI\theta_b = 1\,041.7$ kN $\cdot$ m^2, we obtain $R_{ay} = +312.5$ kN, $M_{az} = -625$ kN $\cdot$ m, $R_{by} = 187.5$ kN, and $M_{bz} = 0$ kN $\cdot$ m, and these values are shown in Fig. 9.7.

In this formulation of the problem, compatibility considerations gave rise to an infinite array of possible solutions, as demonstrated in Fig. 9.7. However, compatibility alone could not sort out the correct solution. Instead, the full satisfaction of equilibrium was necessary to extract the appropriate solution.

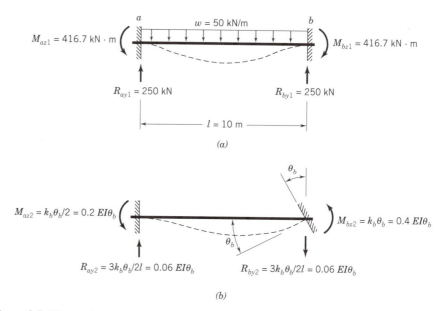

Figure 9.8 Kinematic primary structure of Fig. 9.5b. *(a)* Primary structure subjected to given loading. *(b)* Primary structure subjected to redundant rotation.

The final solution clearly satisfies the partial requirements of equilibrium and the compatibility conditions expressed through Eqs. 9.8 and the augmenting equilibrium condition of Eq. 9.9.

9.6 STATIC VERSUS KINEMATIC INDETERMINACY

The example problem treated in the previous sections of this chapter had one static redundancy, R_{by}, and one kinematic redundancy, θ_b. This agreement in the order of static and kinematic indeterminancy is not usually the case. Table 9.1 shows several examples of the determination of indeterminacy. Within the scope of these simple examples, no clear trends are evident; however, for larger frame structures, the degree of static indeterminacy is normally greater than the degree of kinematic indeterminacy.

One additional observation is crucial. In establishing the statically determinate primary structure, there are a number of alternative selection patterns for the redundant force quantities. However, in developing the kinematically determinate primary structure, no selection process is necessary. Instead, all of the displacement quantities that are necessary to describe the structure's response automatically become redundant quantities.

9.7 COMPATIBILITY METHOD OF ANALYSIS

Two different methods of analysis are available for the solution of statically indeterminate structures. This section will deal with the first—the so-called *compatibility method*. This method is straightforward and easy to understand, and it serves as an effective method for certain classes of problems. As the name suggests, it is based on the solution of a set of equations that express compatibility relationships throughout the structure. This method is also referred to as the *force method* or the *flexibility method*, since the unknowns in the governing equations are forces and the coefficients of these unknown forces are flexibility (displacement) quantities.

Table 9.1 Comparison of Static and Kinematic Indeterminacies

Structure	Static Classification (Redundant Forces Shown)	Kinematic Classification (Redundant Displs. Shown)	Degree of Indeterminacy Static	Degree of Indeterminacy Kinematic
(a)			1	5
(b)			4	2
(c)			1	8
(d)			3	6

Note: In evaluating kinematic redundants, axial deformations are frequently omitted for flexural structures. This would reduce the kinematic indeterminacy to 3, 1, and 3 for structures *a*, *b*, and *d*, respectively.

It is important to note that the final solution must satisfy both the conditions of compatibility and the requirements of equilibrium. However, when the method is properly formulated, the compatibility equations represent a superposition of a set of partial solutions, each of which satisfies the requirements of equilibrium.

The illustrative problem discussed earlier in Sections 9.1 and 9.3 serves as an example of the compatibility method. Equation 9.3 states a condition of compatibility concerning the vertical displacement at point b. The individual displacement quantities that are required are determined from the individual analyses shown in Fig. 9.4, each of which satisfies equilibrium. In the final form of the compatibility equation, Eq. 9.4, the coefficient of R_{by} is a flexibility quantity, and this equation allows for the solution of the redundant reaction R_{by}. This equation, along with the overall equilibrium equations, Eqs. 9.1, provides the three equations needed for the determination of the reactions R_{ay}, M_{az}, and R_{by}.

For large structures, there are many redundant force quantities and, therefore, a number of compatibility equations must be simultaneously solved for the corresponding number of redundant forces. Since there are several acceptable patterns for the selection of the redundant forces, there is no unique formulation of the problem. This variability is a detriment to computer formulation because it is not possible to automate the procedure for generating the compatibility equations.

Some classical compatibility methods of analysis are presented in Chapter 10. The formal matrix compatibility approach, the flexibility method, is presented in Chapter 15.

9.8   EQUILIBRIUM METHOD OF ANALYSIS

The second method of analysis of statically indeterminate structures, which is equally applicable for statically determinate structures, is the *equilibrium method*. This method is also straightforward conceptually and has advantages for certain types of structures. As the name indicates, this method is based on the solution of a set of equations that express the equilibrium requirements for the structure. This approach is also called the *displacement method* or the *stiffness method*, because the unknown quantities in the governing equations are displacements, and the coefficients of these unknowns are stiffness (force) quantities.

Again, it is stressed that the final solution must satisfy both the requirements of equilibrium and the conditions of compatibility. The solution of the governing equations of equilibrium will ensure the former, but care must be exercised in developing these equations to make certain that none of the compatibility conditions are violated.

The problem treated in Section 9.5 provides an example of the equilibrium method. Equation 9.9 gives the equilibrium requirement regarding the moment at point b. The individual moment quantities that are needed in this equation are determined from the individual analyses shown in Fig. 9.8, each of which fully satisfies the conditions of compatibility. In the final form of the equilibrium equation, Eq. 9.10, the coefficient of θ_b is a stiffness quantity, and this equation allows for the solution of the redundant displacement, θ_b. This equation, along with Eqs. 9.8, which represent overall equilibrium and compatibility, provides the relationships needed to evaluate all the reaction components.

For larger structures, there are, of course, many redundant displacement quantities—one corresponding to each kinematic degree of freedom. Therefore, a number of equilibrium equations must be solved simultaneously for the displacement quantities. However, unlike the compatibility method, there is no selectivity in choosing the redundants—all the kinematic degrees of freedom are taken as redundant displacements. This is a major advantage for the automation that is needed in computer programming.

Some classical equilibrium methods are developed in Chapters 11 and 12. The formal matrix equilibrium procedure, which is referred to as the stiffness method, is developed in Chapter 14.

9.9   BEHAVIORAL CHARACTERISTICS OF STATICALLY INDETERMINATE STRUCTURES

Courses of instruction in structural analysis generally treat first the area of statically determinate structures, as was done in Part Two of this book. This treatment typically includes stress analysis, which leads to the determination of the forces that act on any element of the structure, and deformation analysis, which leads to the determination of the displacements of any point on the structure. Students who master these tools are ready to consider statically indeterminate structures where the stress analysis and the deformation analysis are integral parts of a solution that satisfies the required equilibrium and compatibility conditions.

One might legitimately ask, however, why indeterminate structures are used. They are obviously more difficult to analyze than determinate ones, so why aren't statically determinate structures used exclusively? And, if there are specific advantages in statically indeterminate structures, why are determinate structures used in certain cases? As it turns out, specific advantages and disadvantages are associated with each type of structure.

The major advantages of statically indeterminate structures are manifested in three ways. First, a statically indeterminate structure displays greater stiffness in resisting load than does a comparable statically determinate structure. For example, consider the two structures shown in Figs. 9.9a and 9.9b, in which the individual elements of each structure have the same cross-sectional dimensions and lengths. The post and lintel

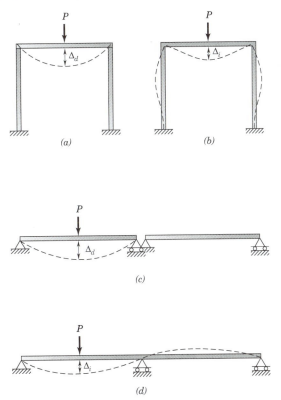

Figure 9.9 Comparative responses for statically determinate and indeterminate structures. (a) Post and lintel. (b) Portal frame. (c) Two simply supported beams. (d) Two-span continuous beam.

arrangement shows a beam member (lintel) that is supported atop two vertical struts (posts). When the load P is applied at midspan of the beam, the beam deflects with a vertical displacement of Δ_d at the load point. If the beam is integrally connected to the column, a portal frame is formed. In this case, the same load P will cause a vertical displacement of $\Delta_i < \Delta_d$ at the load point. Computing the stiffness for each case, the load per unit displacement, we see $P/\Delta_i > P/\Delta_d$. That is, the stiffness of the indeterminate portal frame is greater than that of the determinate post and lintel system. The increased stiffness of the portal frame stems from the fact that the ends of the beam are restrained by the columns. Thus, as the beam deflects, the columns assist in resisting the load. This behavior differs from the post and lintel construction, where the beam simply deflects and the struts are merely passive supports for the beam.

A similar situation exists when the two simply supported beams of Fig. 9.9c are compared with the two-span continuous beam of Fig. 9.9d. Again, if the individual elements in each system have the same lengths and cross-sectional dimensions, then $\Delta_i < \Delta_d$, and thus $P/\Delta_i > P/\Delta_d$. Again the statically determinate case reduces to a simply supported beam, with the loaded span carrying the entire load. However, the statically indeterminate beam provides an end restraint for the loaded span, and the unloaded span participates in resisting the load.

The second advantage of a statically indeterminate structure is that it will have lower stress intensities than would a comparable statically determinate structure. Consider Fig. 9.10, which shows the moment diagrams for the structures and loading conditions, respectively, given in Fig. 9.9. For both statically determinate cases, the full static moment M_s acts on the midspan section of the loaded span. However, for the statically indeterminate cases, the midspan moment on the loaded span is less than the static moment because of the effects of the negative restraining moments at

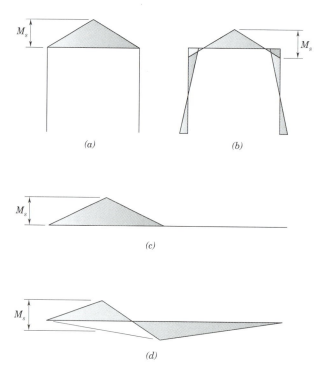

Figure 9.10 Comparative moment diagrams for statically determinate and indeterminate structures. (a) Post and lintel. (b) Portal frame. (c) Two simply supported beams. (d) Two-span continuous beam.

the ends of the loaded span. The reduced moments for the indeterminate structures mean that the stress levels are lower, and thus smaller beam sections can be selected than would be required for the determinate structures. This, of course, leads to economies in the design of the structure.

The third advantageous feature of a statically indeterminate structure is related to safety. Recall from the nature of the criteria for statical classification that statically indeterminate structures have redundant force quantities. That is, there are either internal forces or external reactions that are not needed for stability, and if these force components are removed, the structure will not become unstable. This redundancy represents safety in that upon failure of a joint, member, or support to carry an assigned force, the structure will not necessarily collapse. For example, consider the frame of Fig. 9.9*b*. The frame and the associated moment diagram are redrawn in Fig. 9.11*a*. Again, M_s is the full static moment. Now, suppose that because of repeated cyclical-type loading, a crack develops in the beam under the load such that the moment capacity is severely decreased. The resulting deflected structure and the redistributed moment diagram are shown in Fig. 9.11*b*. The deflected structure reflects the fact that the structure is under distress; however, collapse does not occur. Therefore, if the difficulty is observed and the structure is taken out of service, injuries can be prevented and repairs can be made. If a similar crack developed in the beam of Fig. 9.9*a*, failure would be immediate.

For complex structures, the redistribution process that accompanies partial failure can be complicated to trace; however, again, it is the presence of redundant quantities that precludes failure. In fact, modern design specifications recognize that redistribution of moment can occur when a structure is overloaded and, in fact, depend on this reserve capacity as part of the factor of safety with respect to failure.

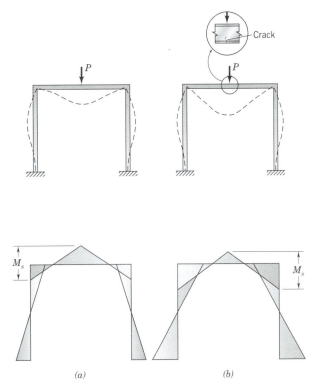

Figure 9.11 Redistribution of moment in a redundant structure. *(a)* Statically indeterminate frame. *(b)* Partially collapsed frame.

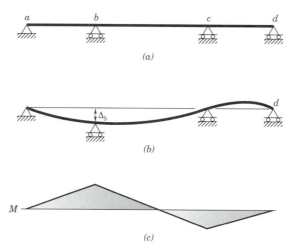

Figure 9.12 Relative settlement in three-span continuous beam. *(a)* Three-span continuous beam. *(b)* Deformations caused by relative settlement. *(c)* Moments induced by settlement.

Actually, all of these advantages are a result of the inherent differences between determinate and indeterminate structures. That is, the integral attachment of the individual structural members acts to reduce the deflections and stress intensities and to preclude failure. This property is referred to as *continuity* because the separate members are continuously attached where they connect.

The primary disadvantage of statically indeterminate structures results from the same feature that underlies its advantages—continuity. As an example, consider the statically indeterminate three-span continuous beam shown in Fig. 9.12a. Should this structure undergo a support settlement of Δ_b at point b relative to the other supports, it would assume the deformed pattern in Fig. 9.12b, which, in turn, would induce the bending moments shown in Fig. 9.12c. Thus, without any loading on the structure, sizable moments would be introduced at the interior support points. With these moments superimposed upon those that result from the design loads, it is possible that the structure will be overstressed at one of the interior supports. For this reason, a statically indeterminate beam bridge cannot be used at a site where the foundation conditions are such that relative support settlements cannot be precluded.

In situations where such settlements can occur, a statically determinate structure should be employed, such as the cantilever structure shown in Fig. 9.13a. When the support at point b settles in this case, the structure merely realigns itself as shown in Fig. 9.13b. Since the members remain straight, no moments are induced.

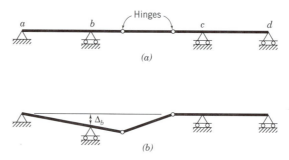

Figure 9.13 Relative settlement in cantilever structure. *(a)* Cantilever structure. *(b)* Relative settlement pattern.

The potential disadvantage of a statically indeterminate structure has been illustrated here for a specific structure and a distinct deformation. However, the same phenomenon occurs for any statically indeterminate structure. That is, any induced deformations, such as relative settlements, fabrication errors, or member length changes caused by temperature variations, will induce internal forces throughout the structure. Deformations that can be predicted, such as those that result from temperature change, can be included as a loading condition in the analysis; however, those that cannot be predicted may cause serious structural distress.

Chapter 10

Method of Consistent Deformations (and Other Compatibility Methods)

The Eads Bridge, St. Louis, Mo. (reprinted from Modern Steel Construction, Third Quarter, 1974, courtesy of American Institute of Steel Construction, Chicago, Ill.).

10.1 NATURE OF COMPATIBILITY METHODS

Any method of structural analysis must lead to a solution that satisfies all of the requisite conditions of compatibility and equations of equilibrium. The methods that are broadly classified as *compatibility methods* are those in which the key relationships used in the solution are compatibility equations that are formulated through the superposition of a set of partial solutions, each of which satisfies the requirements of equilibrium.

Many methods can be classified as compatibility methods. In this chapter, we focus on the *method of consistent deformations*. This technique is a system flexibility approach that requires that certain displacements, each of which are determined from considering the structural system as a whole, be forced to be consistent with the requirements of compatibility. The illustrative problem discussed in Sections 9.1 and 9.3 serves as an example of the method of consistent deformations. There, one compatibility condition was needed to solve for one redundant reaction quantity. For larger structures, there are many redundant force quantities and, therefore, a number of compatibility equations must be solved simultaneously for the corresponding number of redundant forces. Since there are several acceptable patterns for the selection of redundant forces, there is no unique formulation of the problem. This variability is a deterrent to computer formulation because it is not possible to automate the procedure for selecting the redundants and generating the corresponding compatibility equations.

In addition to the system approach, there are element flexibility approaches, which consider the compatibility of the individual structural elements. This procedure is developed in Chapter 15.

10.2 REDUNDANCIES: EXTERNAL VERSUS INTERNAL

As described in Chapter 9, redundant forces are those that can be removed from the structure without impairing the stable integrity of the structure. These redundant forces may be either external or internal. In the former case, the redundants are reaction forces, whereas in the latter, the redundant forces are member forces. In each case, these forces are unnecessary for the stability of the structure.

This chapter considers only planar structures, and thus the nature and number of redundancies can be determined by examining the provisions of Sections 3.3, 3.5, 4.8, and 5.4.

10.3 DETERMINATION OF REDUNDANT REACTIONS

10.3.1 Single Redundant Reaction

The simple propped cantilever beam solution that was discussed in Sections 9.1 through 9.3 was an application of the method of consistent deformations. The structure of that example is portrayed in a more general sense in Fig. 10.1a. The objective of the analysis is to determine the four independent reaction components, R_1 through R_4, and the internal member forces for member ab. The considerations of Section 3.3 reveal that the structure is statically indeterminate to the first degree. That is, there is one redundant reaction component. In this case, R_1 is taken as the redundant reaction, with the remaining reactions being sufficient in number and arrangement to ensure the external stability of the structure.

Upon removal of R_1, the statically determinate *primary structure* of Fig. 10.1b remains. Since this structure is statically determinate, the reactions R_{20} through R_{40} can be determined, where the 0 subscript indicates that these quantities are associated with the actual loading on the primary structure. Corresponding to this arrangement, there is a displacement Δ_{10} at the point and in the direction of the released redundant. This displacement is, of course, in violation of the prescribed boundary condition for point b of the original structure, which requires that $\Delta_1 = 0$, as shown in Fig. 10.1a. Thus, the solution of the primary structure must be altered to meet the boundary conditions.

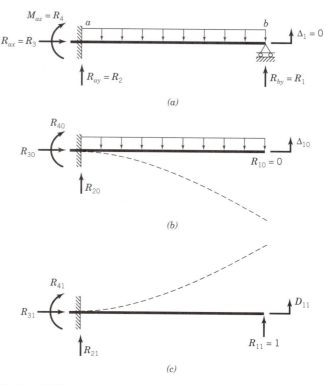

Figure 10.1 Consistent deformation analysis of a statically indeterminate beam structure. *(a)* Statically indeterminate structure. *(b)* Statically determinate primary structure. *(c)* Unit value of R_1.

This alteration is accomplished by introducing a unit value of the redundant reaction ($R_{11} = 1$) on the primary structure, as is shown in Fig. 10.1c. Here, the reactions R_{21} through R_{41} result from a static analysis of the primary structure, with the 1 subscript indicating that these values are associated with a unit value of R_1. The displacement corresponding to the released redundant in this case is identified as D_{11}, which is the *flexibility coefficient* (see Section 2.10) that expresses the deflection at the point and in the direction of R_1 that is caused by a unit value of R_1.

Since D_{11} is the displacement that results from a unit value of R_1, superposition can be used to obtain the displacement for any other value of R_1 by simply multiplying D_{11} by the magnitude of R_1. Thus, the deflection at the point and in the direction of the released redundant that is caused by the redundant reaction R_1 is identified as Δ_{1R} and is given by

$$\Delta_{1R} = D_{11}R_1 \tag{10.1}$$

For the desired solution of Fig. 10.1a, the solutions of Figs. 10.1b and 10.1c must be superimposed. Specifically, the displacements Δ_{10} and Δ_{1R} are combined to give the final displacement, Δ_1. Thus, we have

$$\Delta_{10} + D_{11}R_1 = \Delta_1 \tag{10.2}$$

Solving for R_1, we have

$$R_1 = \frac{(\Delta_1 - \Delta_{10})}{D_{11}} \tag{10.3}$$

In this case, all displacements are considered to be positive when upward. Thus, Δ_{10} is actually negative, as shown in Fig. 10.1b. Also, Δ_1 is zero, as shown in Fig. 10.1a, because of the nonyielding constraint associated with the redundant reaction.

It should be noted that Eq. 10.2 is a compatibility equation that has units of displacement. Since D_{11} also has units of displacement, the quantity R_1 is unitless. That is, Eq. 10.3 merely gives the magnitude of the redundant reaction, or the factor by which the unit load solution of Fig. 10.1c must be multiplied so that when it is combined with the primary solution of Fig. 10.1b, the correct solution of Fig. 10.1a is obtained.

Once R_1 has been determined, statics could be applied to determine the nonredundant reactions. There is, however, a more general approach. The superposition pattern expressed in Eq. 10.2 for displacements holds for all other aspects of the solution. Thus, to determine one of the nonredundant reactions, such as R_q, we have

$$R_q = R_{q0} + R_{q1}R_1 \qquad (10.4)$$

Or, in a more general form, if S is taken as any response quantity of interest, such as a reaction force or any internal force component in member ab, then

$$S = S_0 + S_1R_1 \qquad (10.5)$$

where S_0 is the value of S on the primary structure when the actual loading of the given structure is applied, and S_1 is the value of S on the primary structure when a unit value of R_1 is applied.

10.3.2 Two Redundant Reactions

Consider next the continuous beam structure of Fig. 10.2a. Application of the criteria of Section 3.3 reveals that the structure is twice statically indeterminate. One way to reduce the given structure to a statically determinate primary structure is to remove the two interior reactions, as shown in Fig. 10.2b. These redundant reaction components are identified as R_1 and R_2. The primary structure can now be analyzed by the methods of statics, and the displacements Δ_{10} and Δ_{20}, which correspond to the lines of action of the redundant reactions, can be determined by the methods of Chapter 8. Because Δ_{10} and Δ_{20} are in violation of the boundary conditions of the original structure, it is necessary to modify the solution of the primary structure until the displacements at these points are compatible with the prescribed boundary conditions. The required modification is accomplished by introducing, in turn, unit values of the redundant reactions on the primary structure and determining the effects that these individual loading cases have on the displacements where compatibility is to be restored. These unit load cases are shown in Fig. 10.2c; they can be analyzed in accordance with static considerations and the desired displacements determined by the methods of Chapter 8. Each of these displacements is shown in Fig. 10.2c as D_{ij}, which is the flexibility coefficient that expresses the displacement at the point and in the direction of the redundant reaction R_i that is caused by a unit value of the redundant reaction R_j.

The total displacements at the points and in the directions of the redundant reactions that are caused by the combined effects of the redundant reactions are identified as Δ_{1R} and Δ_{2R} and are determined from superposition to be

$$\Delta_{1R} = D_{11}R_1 + D_{12}R_2$$
$$\Delta_{2R} = D_{21}R_1 + D_{22}R_2 \qquad (10.6)$$

These displacements must be combined with Δ_{10} and Δ_{20} of Fig. 10.2b to yield the desired displacements of the original structure, as defined in Fig. 10.2a. This combination is expressed by compatibility equations in the form

$$\Delta_{10} + D_{11}R_1 + D_{12}R_2 = \Delta_1$$
$$\Delta_{20} + D_{21}R_1 + D_{22}R_2 = \Delta_2 \qquad (10.7)$$

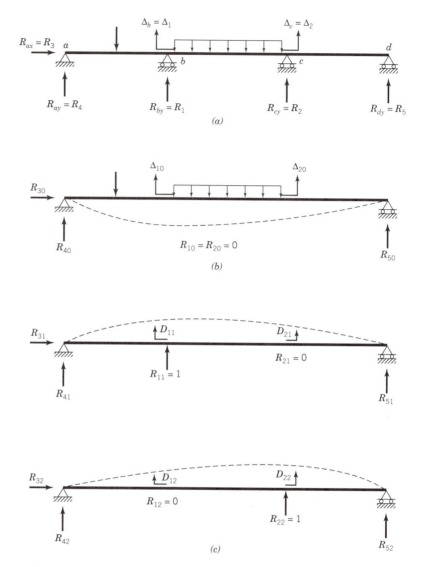

Figure 10.2 Statically indeterminate continuous beam. *(a)* Statically indeterminate beam. *(b)* Statically determinate primary structure. *(c)* Unit values of the redundant reactions.

or in matrix form

$$\begin{bmatrix} D_{11} & D_{12} \\ D_{21} & D_{22} \end{bmatrix} \begin{Bmatrix} R_1 \\ R_2 \end{Bmatrix} = \begin{Bmatrix} \Delta_1 - \Delta_{10} \\ \Delta_2 - \Delta_{20} \end{Bmatrix}; \tag{10.8}$$

In this form, the square matrix of D terms is the *structure flexibility matrix* (see Section 2.10) involving the displacements on the primary structure that are associated with the directions of R_1 and R_2. Again, since upward displacements are considered positive, Δ_{10} and Δ_{20} are negative, as indicated in Fig. 10.2*b*.

The solution of Eq. 10.8 gives the magnitudes of the redundant reactions. These reactions can be placed on the original structure, and the remaining reactions can be determined from statics. Or, as a more general procedure, the same superposition pattern that is expressed in Eq. 10.7 is used for determining any other response quantity of interest, such as reaction, moment, or shear. For instance, if S is taken as such a response quantity, then

$$S = S_0 + S_1 R_1 + S_2 R_2 \qquad (10.9)$$

where S_0 is the value of S on the primary structure when the actual loading of the given structure is applied, and S_i is the value of S on the primary structure when a unit value of R_i is applied.

It is noted that superposition plays a vital role in the development of the method of consistent deformations. Because, as was pointed out in Section 2.9, superposition is valid only for structures that display a linear response to loading, it is concluded that the method of consistent deformations can be applied only to this class of structures.

For more redundant reactions, the same procedure is followed. Each redundant requires a matching compatibility equation, and these collectively lead to a set of simultaneous equations in the form of Eq. 10.8. The solution of this equation gives the magnitudes of the redundant reactions, and the final solution is then formed by superposition similar to that expressed in Eq. 10.9.

It is stressed that the final solution satisfies all of the requirements of compatibility and equilibrium. The solution for the redundant forces stems directly from a satisfaction of the compatibility equations, and each individual solution of the primary structure results from a statically determinate analysis that satisfies the requirements of equilibrium. Thus, the superposition indicated in Eq. 10.9 is consistent with the requirements of compatibility and equilibrium.

10.4 APPLICATION OF THE METHOD OF CONSISTENT DEFORMATIONS

The application of the method of consistent deformations requires the formulation of compatibility equations of the type given by Eq. 10.7. This requires the determination of the Δ_{i0} displacements and the D_{ij} flexibility coefficients. These are all displacement quantities for the statically determinate primary structure, which are determined by any of the methods presented in Chapter 8.

The example problems given in this section illustrate the application of the method. The first two examples involve beam- or frame-type structures. The method can be applied to other types of structures, however, and the third example illustrates its application to a truss structure.

EXAMPLE 10.1   Determine the reactions, and construct the moment diagram for the frame structure given. The quantity EI is the same for each member.

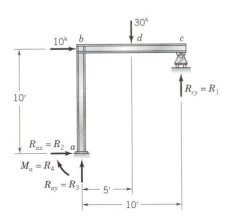

Structure Classification

Application of the criteria of Sections 3.3 or 5.4 shows that the structure is statically indeterminate to the first degree.

Primary Structure and Loadings

Select R_{cy} as the redundant reaction R_1, which produces a simple cantilever-type system as the primary structure.

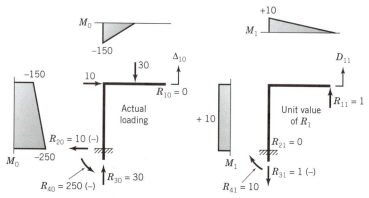

Reactions in kips ; moments in kip-ft

Displacement Calculations

The moment–area method is used because it is especially useful for a cantilever-type structure.

The Displacement Δ_{10}

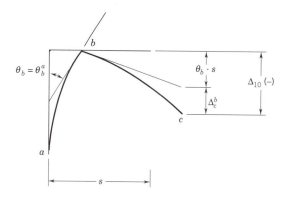

$$\theta_b = \theta_b^a = \left[\frac{250 + 150}{2EI}\right]10 = \frac{2,000 \text{ ft}^2\text{-k}}{EI}$$

$$\Delta_c^b = \left(\frac{150}{2EI}\right)5(8.33) = \frac{3,124 \text{ ft}^3\text{-k}}{EI}$$

$$\Delta_{10} = \left(\frac{2,000}{EI} \times 10\right) + \left(\frac{3,124}{EI}\right) = \frac{23,124 \text{ ft}^3\text{-k}}{EI}(-)$$

The Flexibility Coefficient D_{11}

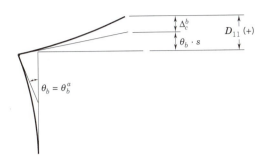

$$\theta_b = \theta_b^a = \left(\frac{10}{EI}\right)10 = \frac{100 \text{ ft}^2\text{-k}}{EI}$$

$$\Delta_c^b - \left(\frac{10}{2EI}\right)10(6.67) = \frac{333.5 \text{ ft}^3\text{-k}}{EI}$$

$$D_{11} = \left(\frac{100}{EI} \times 10\right) + \left(\frac{333.5}{EI}\right) = \frac{1{,}333.5 \text{ ft}^3\text{-k}}{EI}(+)$$

Determination of Reactions

The redundant reaction, R_1, is determined by imposing displacement compatibility at point c through the principle of superposition.

$$\Delta_{10} + D_{11}R_1 = \Delta_1 \tag{10.2}$$

Since $\Delta_1 = 0$,

$$R_1 = -\frac{\Delta_{10}}{D_{11}} = \frac{23{,}124}{1{,}333.5} = 17.34$$

Note: In this case, the quantity EI cancels. Thus, the final results are not dependent on the actual values of E and I.

The same superposition pattern can be used to determine the remaining reactions.

$$R_q = R_{q0} + R_{q1}R_1 \tag{10.4}$$

That is, for $q = 2, 3, 4$

$$R_2 = -10 + (0 \times 17.34) = -10 \text{ kips}$$

$$R_3 = 30 + (-1 \times 17.34) = +12.66 \text{ kips}$$

$$R_4 = -250 + (10 \times 17.34) = -76.6 \text{ kip-ft}$$

Final Moments

Again, superposition provides the final moments at any point on the structure.

$$M = M_0 + M_1 \cdot R_1 \tag{10.5}$$

Moment value for $R_1 = 1$ on primary structure

Moment value for actual loading on primary structure

In matrix form, for key points on the frame,

$$
\begin{Bmatrix} M_a \\ M_b \\ M_c \\ M_d \end{Bmatrix} = \begin{Bmatrix} -250 \\ -150 \\ 0 \\ 0 \end{Bmatrix} + \begin{Bmatrix} +10 \\ +10 \\ 0 \\ +5 \end{Bmatrix} \times 17.34 = \begin{Bmatrix} -76.6 \\ +23.4 \\ 0 \\ +86.7 \end{Bmatrix} \text{ (kip-ft)}
$$

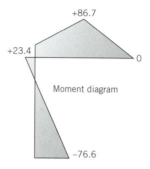

Moment diagram

EXAMPLE 10.2

Determine the reactions, and construct the final shear and moment diagrams for the structure and loading given. The quantity EI is the same for each span.

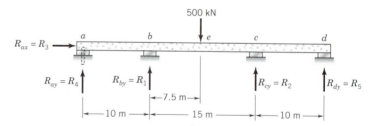

Structure Classification

Application of the criteria of Section 3.3 or 5.4 reveals that the structure is statically indeterminate to the second degree.

Primary Structure and Loadings

Select R_{by} and R_{cy} as redundant reactions, which produce a simply supported beam as the primary structure.

Actual Loading

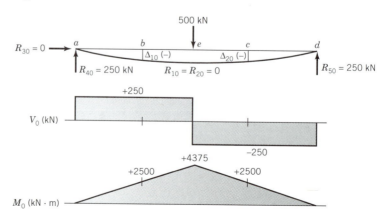

Unit Value of $R_1 = R_{11}$

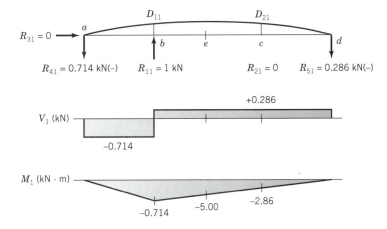

Unit Value of $R_2 = R_{22}$

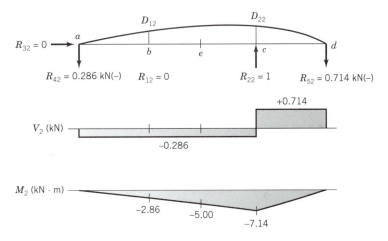

Displacement Calculations

Conjugate beam method is used because of its efficiency in determining two displacements along the length of the structure.

The Displacements Δ_{10} *and* Δ_{20}

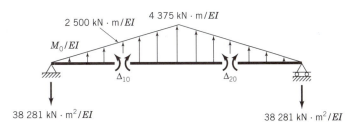

$$\Delta_{10} = -\left(\frac{38\,281}{EI}\right)10 + \left(\frac{2\,500}{2EI}\right)(10)(3.33) = \frac{-341\,185 \text{ kN} \cdot \text{m}^3}{EI}$$

$$\Delta_{20} = \Delta_{10} \text{ by symmetry}$$

Both are negative, since upward displacements are positive.

The Flexibility Coefficients D_{11} and D_{21}

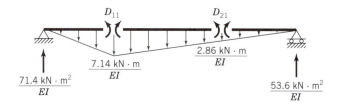

$$D_{11} = \left(\frac{71.4}{EI}\right)10 - \left(\frac{7.14}{2EI}\right)(10)(3.33) = \frac{+595.1 \text{ kN} \cdot \text{m}^3}{EI}$$

$$D_{21} = \left(\frac{53.6}{EI}\right)10 - \left(\frac{2.86}{2EI}\right)(10)(3.33) = \frac{+488.4 \text{ kN} \cdot \text{m}^3}{EI}$$

The Flexibility Coefficients D_{12} and D_{22}

$$D_{12} = D_{21} \text{ by Maxwell's law (Eq. 2.72)}$$

$$D_{22} = D_{11} \text{ by symmetry}$$

Redundant Reactions

The requirements of vertical support compatibility at points b and c forms the basis of the following superposition equations:

$$\Delta_{10} + D_{11}R_1 + D_{12}R_2 = \Delta_1$$
$$\Delta_{20} + D_{21}R_1 + D_{22}R_2 = \Delta_2$$

(10.7)

Since $\Delta_1 = \Delta_2 = 0$, we have

$$\begin{bmatrix} D_{11} & D_{12} \\ D_{21} & D_{22} \end{bmatrix} \begin{Bmatrix} R_1 \\ R_2 \end{Bmatrix} = \begin{Bmatrix} -\Delta_{10} \\ -\Delta_{20} \end{Bmatrix};$$

Substitution for the displacement quantities and, noting that EI cancels, leads to

$$\begin{bmatrix} 595.1 & 488.4 \\ 488.4 & 595.1 \end{bmatrix} \begin{Bmatrix} R_1 \\ R_2 \end{Bmatrix} = \begin{Bmatrix} 341\,185 \\ 341\,185 \end{Bmatrix}$$

Solution for the redundant reactions gives

$$\begin{Bmatrix} R_1 \\ R_2 \end{Bmatrix} = \begin{Bmatrix} 314.9 \\ 314.9 \end{Bmatrix}$$

The remaining reactions are readily determined from statics or from superposition (Eq. 10.9). The final values for the reactions are

$$R_{ax} = R_3 = 0, \quad R_{ay} = R_4 = -64.9 \text{ kN}, \quad R_{by} = R_1 = 314.9 \text{ kN}$$
$$R_{cy} = R_2 = 314.9 \text{ kN}, \quad R_{dy} = R_5 = -64.9 \text{ kN}$$

Final Shears and Moments

The same superposition format used to determine the redundant reactions can be used to find any other response quantity of interest. In this case,

$$S = S_0 + S_1 R_1 + S_2 R_2 \tag{10.9}$$

where S can be interpreted as the shears or moments. In matrix form, for key points on the structure,

$$
\begin{Bmatrix} V_{a-b} \\ V_{b-e} \\ V_{e-c} \\ V_{c-d} \end{Bmatrix} = \begin{Bmatrix} +250 \\ +250 \\ -250 \\ -250 \end{Bmatrix} + \begin{Bmatrix} -0.714 \\ +0.286 \\ +0.286 \\ +0.286 \end{Bmatrix} 314.9 + \begin{Bmatrix} -0.286 \\ -0.286 \\ -0.286 \\ +0.714 \end{Bmatrix} 314.9 = \begin{Bmatrix} -64.9 \\ +250.0 \\ -250.0 \\ +64.9 \end{Bmatrix} \text{(kN)}
$$

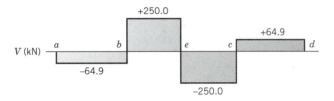

$$
\begin{Bmatrix} M_a \\ M_b \\ M_e \\ M_c \\ M_d \end{Bmatrix} = \begin{Bmatrix} 0 \\ +2\,500 \\ +4\,375 \\ +2\,500 \\ 0 \end{Bmatrix} + \begin{Bmatrix} 0 \\ -7.14 \\ -5.00 \\ -2.86 \\ 0 \end{Bmatrix} 314.9 + \begin{Bmatrix} 0 \\ -2.86 \\ -5.00 \\ -7.14 \\ 0 \end{Bmatrix} 314.9 = \begin{Bmatrix} 0 \\ -649 \\ +1\,226 \\ -649 \\ 0 \end{Bmatrix} \text{(kN} \cdot \text{m)}
$$

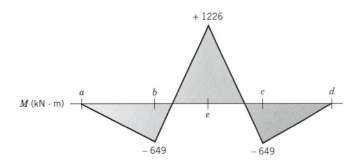

EXAMPLE 10.3   Determine the reactions and bar forces for the statically indeterminate truss shown. The quantity EA is the same for each member.

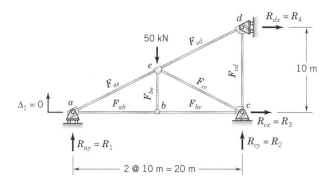

Structure Classification

Application of the criteria of Sections 3.3 and 4.8 shows that the structure is statically determinate internally and statically indeterminate externally to the first degree.

Primary Structure and Loadings

Select R_1 as the redundant reaction.

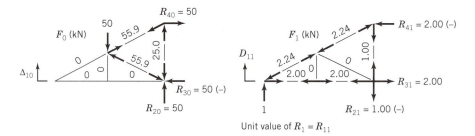

Unit value of $R_1 = R_{11}$

Displacement Calculations

The complementary virtual work method is used to obtain the required displacement quantities. The fundamental equation is

$$\sum_{i=1}^{n} (\delta P)_i \Delta_i = \sum_{j=1}^{m} \left(\delta F_p \cdot \frac{F_\Delta l}{EA} \right)_j \qquad (7.18)$$

The following table describes how this equation is applied.

Displ. Quantity	δF_P	F_Δ	Equation 7.18
Δ_{10}	F_1	F_0	$1 \cdot \Delta_{10} = \dfrac{1}{EA} \sum F_1 F_0 l$
D_{11}	F_1	F_1	$1 \cdot D_{11} = \dfrac{1}{EA} \sum F_1^2 l$

The desired summations are carried out in the following table:

Member	l m	F_0 kN	F_1 kN	$F_1 F_0 l$ $(\text{kN})^2 \cdot \text{m}$	$F_1^2 l$ $(\text{kN})^2 \cdot \text{m}$
ab	10.0	0	+2.00	0	40.0
bc	10.0	0	+2.00	0	40.0
ae	11.2	0	−2.24	0	56.2
ed	11.2	+55.9	−2.24	−1 402.4	56.2
eb	5.0	0	0	0	0
ec	11.2	−55.9	0	0	0
cd	10.0	−25.0	+1.00	−250.0	10.0
Σ				−1 652.4	+202.4

$$\therefore \Delta_{10} = \frac{-1\,652.4}{EA} \left(\frac{\text{kN} \cdot \text{m}}{EA} \right); \qquad D_{11} = \frac{202.4}{EA} \left(\frac{\text{kN} \cdot \text{m}}{EA} \right)$$

Calculation of Redundant Reaction

The imposition of vertical displacement compatibility at point a leads to the following super-position equation:

$$\Delta_{10} + D_{11} R_1 = \Delta_1 \tag{10.2}$$

Since $\Delta_1 = 0$,

$$R_1 = -\frac{\Delta_{10}}{D_{11}}$$

$$R_1 = -\frac{(-1\,652.4)}{202.4} = +8.16$$

Note: In this case, the quantity EA cancels. Thus, the final results are not dependent on the actual values of E and A.

Member Forces and Reactions

The member forces can be determined by extending the table used above and by applying the superposition pattern given by Eq. 10.5.

Member rs		$(F_{rs})_1 \cdot R_1$ kN	$F_{rs} = (F_{rs})_0 + (F_{rs})_1 \cdot R_1$ kN
ab		+16.32	+16.32
bc		+16.32	+16.32
ae		−18.28	−18.28
ed		−18.28	+37.62
eb		0	0
ec		0	−55.90
cd		+8.16	−16.84

where $(F_{rs})_1 = F_1$ force in member rs, and $(F_{rs})_0 = F_0$ force in member rs.

The reactions are determined from superposition according to Eq. 10.4.

Reaction q	R_{q0} kN	R_{q1} kN	$R_{q1} \cdot R_1$ kN	$R_q = R_{q0} + R_{q1} \cdot R_1$ kN
1	—	1.00	+8.16	+8.16
2	50	−1.00	−8.16	+41.84
3	−50	2.00	+16.32	−33.68
4	50	−2.00	−16.32	+33.68

where $R_{q1} = R_q$ for F_1 system, and $R_{q0} = R_q$ for F_0 system.

Final Results

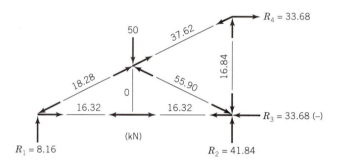

10.5 SUPPORT SETTLEMENTS AND ELASTIC SUPPORTS

It was illustrated in Chapter 9 that support settlements in statically indeterminate structures induce forces throughout the structure. Examples 10.1, 10.2, and 10.3 represented situations in which the supports were nonyielding; however, the development of Section 10.3 included provisions for treating the problem of support settlements.

For instance, consider the structure shown in Fig. 10.1. The compatibility equation given as Eq. 10.2 includes the displacement Δ_1, which is the upward vertical displacement at point b for the actual structure. This displacement appears on the right-hand side of Eq. 10.3, which is the expression for the magnitude of the redundant reaction, R_1. In Fig. 10.1, the vertical support at point b is a nonyielding support for which $\Delta_1 = 0$. If there is a downward settlement of Δ_{1S}, then $\Delta_1 = -\Delta_{1S}$, and Eq. 10.3 would have the form

$$R_1 = \frac{(-\Delta_{1S} - \Delta_{10})}{D_{11}} \tag{10.10}$$

Thus, it is clear that the magnitude of the redundant reaction is affected by the support settlement. Once R_1 has been established, the final values of other response quantities are determined in accordance with Eq. 10.5.

In a similar manner, if downward settlements of Δ_{1S} and Δ_{2S} are prescribed for the structure of Fig. 10.2, then $\Delta_1 = -\Delta_{1S}$ and $\Delta_2 = -\Delta_{2S}$, and Eq. 10.8 becomes

$$\begin{bmatrix} D_{11} & D_{12} \\ D_{21} & D_{22} \end{bmatrix} \begin{Bmatrix} R_1 \\ R_2 \end{Bmatrix} = \begin{Bmatrix} -\Delta_{1S} & -\Delta_{10} \\ -\Delta_{2S} & -\Delta_{20} \end{Bmatrix} \tag{10.11}$$

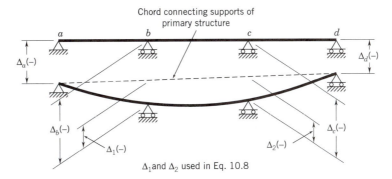

Chord connecting supports of
primary structure

Δ_1 and Δ_2 used in Eq. 10.8

Figure 10.3 Settlement of continuous beam. Note that upward displacements are positive.

The solution of Eq. 10.11 gives the magnitudes of the redundant reactions R_1 and R_2, and other response quantities of interest can be determined from Eq. 10.9. If there are prescribed settlements for the end support points, then Δ_1 and Δ_2 must be interpreted as the displacements relative to the chord connecting the ends of the beam. This situation is illustrated in Fig. 10.3.

Frequently, structures are mounted on supports that are not rigid; however, there is not a prescribed settlement. Instead, the support movement depends on the magnitude of the reaction force. The simplest representation of this kind of support is called an *elastic support,* in which settlement is a linear function of the reaction force. This arrangement is shown in Fig. 10.4*a* for the beam that was previously considered in Fig. 10.1. Here, it is clear that

$$R_1 = k_s(-\Delta_1) \tag{10.12}$$

where k_s is the elastic spring constant, as shown in Fig. 10.4*b*. Solving Eq. 10.12 for Δ_1 and substituting the result in Eq. 10.2, we obtain

$$\Delta_{10} + D_{11}R_1 = -\frac{R_1}{k_s} \tag{10.13}$$

from which

$$R_1 = \frac{-\Delta_{10}}{(D_{11} + 1/k_s)} \tag{10.14}$$

For a structure with multiredundant reactions, each reaction provided by an elastic support must be represented by an expression of the type given in Eq. 10.12. Therefore, if the supports at points *b* and *c* on Fig. 10.2*a* are provided by elastic supports, as shown in Fig. 10.5, then Eq. 10.8 takes the form

$$\begin{bmatrix} \left(D_{11} + \dfrac{1}{k_{s1}}\right) & D_{12} \\ D_{21} & \left(D_{22} + \dfrac{1}{k_{s2}}\right) \end{bmatrix} \begin{Bmatrix} R_1 \\ R_2 \end{Bmatrix} = \begin{Bmatrix} -\Delta_{10} \\ -\Delta_{20} \end{Bmatrix} \tag{10.15}$$

This arrangement would correspond to the case in which R_1 and R_2 are provided by beams that span in a direction perpendicular to continuous beam *abcd*. Here, the elastic constants, k_{s1} and k_{s2}, would be a function of the properties of the supporting beams.

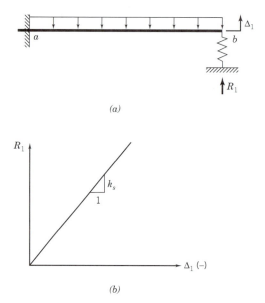

(a)

(b)

Figure 10.4 Elastic support. *(a)* Indeterminate beam with elastic support at point *b*. *(b)* Elastic support stiffness.

If points a and d also have elastic supports, Eq. 10.15 is still valid. However, care must be exercised to make certain that the flexibility coefficients D_{ij} and the displacements Δ_{i0} include the effects of the displacements at points a and d of the primary structure.

In either case, the redundant reactions are provided through the solution of Eq. 10.15, and any other response quantities are determined from Eq. 10.9.

In the examples of Section 10.4, where there were no support displacements, numerical substitutions were not necessary for common values of EI or EA. These quantities were left in symbolic form in the expressions for the displacements Δ_{i0} and D_{ij}. Because $\Delta_i = 0$ in these cases, the common EI or EA terms canceled uniquely when Eqs. 10.3 and 10.8 were solved for the redundant reactions. However, when support displacements are admitted, and $\Delta_i \neq 0$, then these common EI or EA terms no longer cancel, and they must be included in the determination of the redundant reactions.

Thus, in the absence of support movement, the member forces and reactions of a statically indeterminate beam or truss are dependent on the relative EI or EA values. However, when support movement exists, the member forces and reactions depend on the absolute EI or EA values. This fact is illustrated in the following problems.

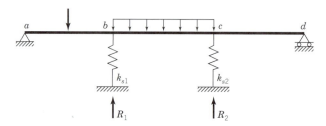

Figure 10.5 Continuous beam on elastic supports.

EXAMPLE 10.4

Consider the continuous beam of Example 10.2, and determine the moment diagram for the following set of support settlements:

$$\Delta_a = -27.5 \text{ mm}; \quad \Delta_b = -47.5 \text{ mm}; \quad \Delta_c = -22 \text{ mm}; \quad \Delta_d = -10 \text{ mm}$$

Note: Positive displacements are upward; therefore, these settlements are downward.

Settlement Pattern

The displacements of points b and c relative to the chord connecting points a and d (Δ_1 and Δ_2, respectively) are needed.

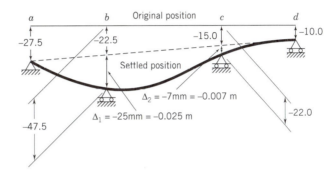

Structure Classification, Primary Structure, and Determination of Flexibility Coefficients

Same as Example 10.2.

Compatibility Equations

The compatibility conditions at points b and c provide the following superposition equations:

$$\Delta_{10} + D_{11}R_1 + D_{12}R_2 = \Delta_1$$
$$\Delta_{20} + D_{21}R_1 + D_{22}R_2 = \Delta_2 \tag{10.7}$$

Redundant Reactions

In this case, $\Delta_{10} = \Delta_{20} = 0$, and thus

$$\begin{bmatrix} D_{11} & D_{12} \\ D_{21} & D_{22} \end{bmatrix} \begin{Bmatrix} R_1 \\ R_2 \end{Bmatrix} = \begin{Bmatrix} \Delta_1 \\ \Delta_2 \end{Bmatrix}$$

Substitution for the appropriate displacement quantities gives

$$\frac{1}{EI}\begin{bmatrix} 595.1 & 488.4 \\ 488.4 & 595.1 \end{bmatrix} \begin{Bmatrix} R_1 \\ R_2 \end{Bmatrix} = \begin{Bmatrix} -0.025 \\ -0.007 \end{Bmatrix}$$

Solving for the redundant reactions, we obtain

$$\begin{Bmatrix} R_1 \\ R_2 \end{Bmatrix} = EI \begin{Bmatrix} -99.1 \times 10^{-6} \\ 69.5 \times 10^{-6} \end{Bmatrix} \left(\frac{EI}{kN \cdot m^2} \right)$$

Note: R_1 and R_2 are unitless; they merely give the magnitudes of the redundant reactions.

Final Moments

$$M = M_0 + M_1 R_1 + M_2 R_2 \tag{10.9}$$

where, in this case, $M_0 = 0$ and M_1 and M_2 are the moment diagram ordinates given in Example 10.2. In matrix form, for key points on the structure,

$$\begin{Bmatrix} M_a \\ M_b \\ M_e \\ M_c \\ M_d \end{Bmatrix} = \begin{Bmatrix} 0 \\ -7.14 \\ -5.00 \\ -2.86 \\ 0 \end{Bmatrix} (-99.1 \; EI \times 10^{-6}) + \begin{Bmatrix} 0 \\ -2.86 \\ -5.00 \\ -7.14 \\ 0 \end{Bmatrix} (+69.5 \; EI \times 10^{-6})$$

$$= \begin{Bmatrix} 0 \\ 508.8 \\ 148.0 \\ -212.8 \\ 0 \end{Bmatrix} EI \times 10^{-6} \left(\frac{EI}{m} \right)$$

Note: The reactions and the moments are dependent on the flexural stiffness, EI. For the specific case of $I = 3000 \times 10^{-6}\,\text{m}^4$ and $E = 200 \times 10^9\,\text{Pa} = 200 \times 10^6\,\text{kN}/\text{m}^2$, we obtain

$$\begin{Bmatrix} M_a \\ M_b \\ M_e \\ M_c \\ M_d \end{Bmatrix} = \begin{Bmatrix} 0 \\ 508.8 \\ 148.0 \\ -212.8 \\ 0 \end{Bmatrix} (200 \times 10^6)(3\,000 \times 10^{-6}) 10^{-6} = \begin{Bmatrix} 0 \\ 305.3 \\ 88.8 \\ -127.7 \\ 0 \end{Bmatrix} (\text{kN} \cdot \text{m})$$

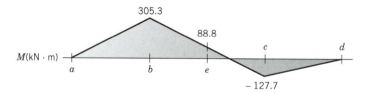

EXAMPLE 10.5

Consider again the beam and loading of Example 10.2, but, in this case, assume that points b and c are supported by elastic supports that have stiffnesses of $k_{s1} = k_{s2} = 0.006EI$ (units EI/m^3). Determine the reactions carried by the elastic supports.

Structure Classification, Primary Structure, and Determination of Displacement Quantities and Flexibility Coefficients

Same as Example 10.2.

Compatibility Equations

The compatibility conditions at points b and c lead to the following superposition equations:

$$\Delta_{10} + D_{11}R_1 + D_{12}R_2 = \Delta_1 = -\frac{R_1}{k_{s1}}$$

$$\Delta_{20} + D_{21}R_1 + D_{22}R_2 = \Delta_2 = -\frac{R_2}{k_{s2}}$$

Redundant Reactions

$$\begin{bmatrix} \left(D_{11} + \dfrac{1}{k_{s1}}\right) & D_{12} \\ D_{21} & \left(D_{22} + \dfrac{1}{k_{s2}}\right) \end{bmatrix} \begin{Bmatrix} R_1 \\ R_2 \end{Bmatrix} = \begin{Bmatrix} -\Delta_{10} \\ -\Delta_{20} \end{Bmatrix} \tag{10.15}$$

Substitution leads to

$$\frac{1}{EI}\begin{bmatrix} (595.1 + 166.7) & 488.4 \\ 488.4 & (595.1 + 166.7) \end{bmatrix} \begin{Bmatrix} R_1 \\ R_2 \end{Bmatrix} = \frac{1}{EI}\begin{Bmatrix} 341\,185 \\ 341\,185 \end{Bmatrix}$$

from which

$$\begin{Bmatrix} R_1 \\ R_2 \end{Bmatrix} = \begin{Bmatrix} 272.9 \\ 272.9 \end{Bmatrix}$$

Comparing the results with those of Example 10.2, we see that the magnitudes of the interior redundant reactions are reduced by the softening effect of the elastic supports.

10.6 SELECTION OF INTERNAL MOMENTS AS REDUNDANTS

10.6.1 Single Redundant Moment

Consider the propped cantilever beam shown in Fig. 10.6a. As already noted, this structure is statically indeterminate to the first degree. However, instead of identifying an external reaction component as the redundant, we now select the internal moment at point a, M_1, as the redundant. The primary structure is then formed by removing the redundant moment (inserting a hinge as a moment release at point a), which essentially reduces member ab to a simply supported beam.

When the primary structure is subjected to the actual loading, as depicted in Fig. 10.6b, a slope discontinuity of θ_{10} occurs at point a that violates the internal compatibility condition for beam continuity. Here, the subscript 0 is again used to indicate that a response is associated with the actual loading on the primary structure. To restore slope compatibility at point a, a unit value of the redundant moment is introduced ($M_{11} = 1$) on the primary structure, as shown in Fig. 10.6c. The second 1 subscript denotes that this is the value of M_1 associated with a unit value of the redundant moment. This loading produces a slope discontinuity at point a of D_{11}, which is the flexibility coefficient expressing the slope discontinuity corresponding to the moment release at M_1 caused by a unit value of M_1.

The final slope discontinuity, θ_1, is determined by superimposing the effect of the actual loading on the effect of the unit moment amplified by the magnitude of the redundant moment. That is,

$$\theta_{10} + D_{11}M_1 = \theta_1 \tag{10.16}$$

Thus, the redundant moment is

$$M_1 = \frac{\theta_1 - \theta_{10}}{D_{11}} \tag{10.17}$$

In this case, a clockwise rotation is taken as positive; therefore, both θ_{10} and D_{11} are positive, as shown. When there is no slope discontinuity at point a, such as the case presented in Fig. 10.6a, θ_1 is zero. Having determined the magnitude of the redundant moment, M_1, we can readily determine the final solution from applying statics to the primary structure or through superposition in the form

$$S = S_0 + S_1 M_1 \tag{10.18}$$

where S is any response quantity of interest, S_0 is the value of S when the actual loading is applied to the primary structure, and S_1 is the value of S when a unit value of M_1 is applied to the primary structure.

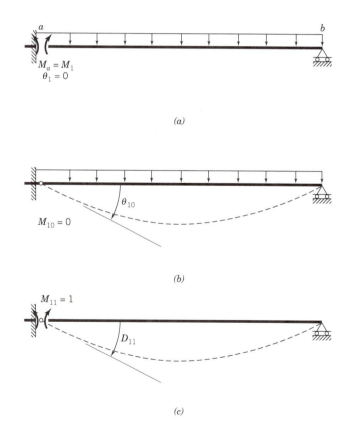

(a)

(b)

(c)

Figure 10.6 Consistent deformation analysis using an internal support moment as the redundant. (a) Statically indeterminate structure. (b) Statically determinate primary structure. (c) Unit value of M_1.

EXAMPLE 10.6

Determine the internal bending moments at the supports for the structure shown. Relative I values are given.

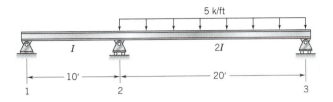

Structure Classification

Application of the criteria of Section 3.3 or 5.4 shows that the structure is statically indeterminate to the first degree.

Primary Structure and Loadings

The internal support moments are shown as follows. The force boundary conditions at points 1 and 3 require that $M_1 = M_3 = 0$, and M_2 is taken as the redundant moment.

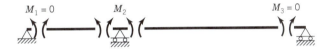

The primary structure is, therefore, two simply supported beams.

Actual Loading

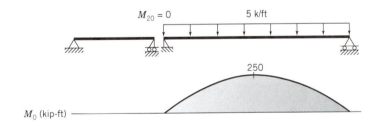

Unit value of M_2

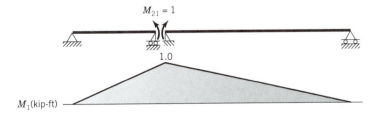

Displacement Calculations

The moment–area method is used.

θ_{20}:

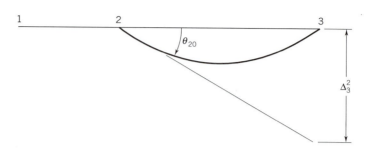

$$\theta_{20} = \frac{\Delta_3^2}{20} = \frac{\frac{2}{3}\left(\frac{250}{2EI}\right)20 \times 10}{20} = 833.3 \text{ ft}^2\text{-k}/EI$$

D_{22}:

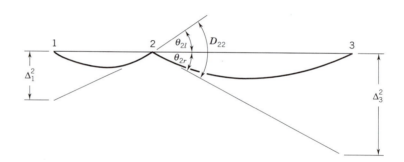

$$\theta_{2l} = \frac{\Delta_1^2}{10} = \frac{\frac{1}{2}\left(\frac{1}{EI}\right)10 \times 6.67}{10} = 3.33 \text{ ft}^2\text{-k}/EI$$

$$\theta_{2r} = \frac{\Delta_3^2}{20} = \frac{\frac{1}{2}\left(\frac{1}{2EI}\right)20 \times 13.33}{20} = 3.33 \text{ ft}^2\text{-k}/EI$$

$$D_{22} = \theta_{2l} + \theta_{2r} = 6.67 \text{ ft}^2\text{-k}/EI$$

Compatibility Equation and Solution for Redundant Moment

Rotational compatibility is expressed through the following superposition equation:

$$\theta_{20} + D_{22}M_2 = \theta_2$$

$$M_2 = -\frac{\theta_{20}}{D_{22}} = \frac{-833.3}{6.67} = -125$$

Therefore, the final support moments are $M_1 = 0$; $M_2 = -125$ kip-ft; and $M_3 = 0$.

10.6.2 Multiple Redundant Moments for Continuous Beams: The Three-Moment Equation

Consider the m-span continuous beam shown in Fig. 10.7a. Based on the criteria of Section 5.4, this structure is externally indeterminate to the $(m-1)$th degree. Likewise, the overall classification leads to the conclusion that the structure is indeterminate to the $(m-1)$th degree. Both of these structural classifications are shown in the figure. Actually, many different combinations can be used in selecting redundants that will render this structure statically determinate. The most obvious approach, which follows the procedure of Section 10.3, is to select the $(m-1)$ interior reaction components as the redundants, as shown in Fig. 10.7b. The primary structure is then a simply supported beam whose span is equal to the total length of the structure. In this case, the required compatibility for the given structure is violated at each of the interior support points. An alternative approach is to select the internal moments at the $(m-1)$ interior support points as the redundants, as depicted in Fig. 10.7c. For this selection, the primary structure is a series of m simply supported beams, and in this case compatibility for the given structure is relaxed at each of the interior support points where continuity of slope is violated. The advantage of this selection of redundants is that the required displacement quantities are easily determined for the m simply supported beams.

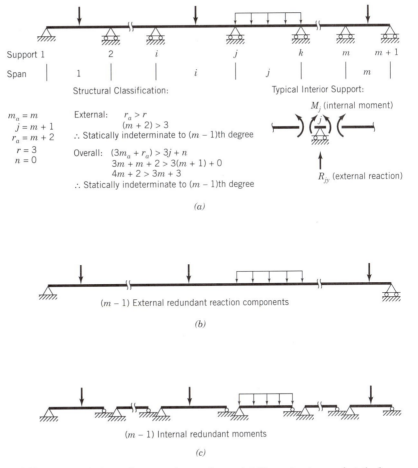

(a)

(b)

(c)

Figure 10.7 Statically indeterminate continuous beam. (a) Given structure and statical classification. (b) Primary structure based on external redundant reactions. (c) Primary structure based on internal redundant moments.

Attention is now focused on the two-span section that reaches over the supports i, j, and k, which is shown in Fig. 10.8a. When the internal moments are removed and the continuous beam is transformed into a series of simply supported beams, there is a slope discontinuity over each of the interior supports. For the actual loading on the primary structure, this discontinuity at support j is θ_{j0}, as indicated on Fig. 10.8b. Since this discontinuity is at variance with the compatibility requirements for the original structure, it must be corrected in order to obtain the correct solution. This correction is accomplished by introducing, through three separate loading conditions, unit values of M_i, M_j, and M_k. The discontinuities in the slope at point j corresponding to each of these unit moment cases are shown as D_{ji}, D_{jj}, and D_{jk}, respectively, which are flexibility coefficients for the primary structure. Applying superposition, we multiply each of these discontinuities by the actual value of the respective redundant moment and combine them with the discontinuity θ_{j0} to obtain the total discontinuity, θ_j. Since θ_j is zero in the given structure, the final compatibility equation at point j becomes

$$D_{ji}M_i + D_{jj}M_j + D_{jk}M_k + \theta_{j0} = \theta_j = 0 \tag{10.19}$$

In this equation, positive discontinuities are taken to be clockwise, as shown in Fig. 10.8.

The required rotation quantities can be determined by the moment–area principle, or any of the other methods presented in Chapter 8. The quantity θ_{j0} depends on the loads that act on spans i and j, and the expressions for the flexibility coefficients are

$$D_{ji} = \frac{l_i}{6EI_i}$$

$$D_{jj} = \frac{l_i}{3EI_i} + \frac{l_j}{3EI_j} \tag{10.20}$$

$$D_{jk} = \frac{l_j}{6EI_j}$$

where l_i and l_j are the lengths for spans i and j, and I_i and I_j are the corresponding moments of inertia. Substitution of Eq. 10.20 into 10.19 and a rearrangement of terms leads to

$$\left(\frac{l_i}{I_i}\right)M_i + \left(\frac{2l_i}{I_i} + \frac{2l_j}{I_j}\right)M_j + \left(\frac{l_j}{I_j}\right)M_k = -6E\theta_{j0} \tag{10.21}$$

It is possible to develop equations for the quantity θ_{j0} for various loading arrangements on spans i and j. This has been done in some textbooks, but it will not be done here.

For the complete structure, the slope discontinuities that occur at all of the interior supports must be corrected. Thus, for the structure of Fig. 10.7, Eq. 10.21 must be applied with $j = 2, 3, \ldots, m$. However, each of the $(m - 1)$ compatibility equations involves only three moments—the moment at the point where compatibility is being considered and the moments at the far ends of the spans to the left and to the right. For this reason, Eq. 10.21 is commonly referred to as the *three-moment equation*. The original presentation was given in 1857 by the French engineer Clapeyron, although he did not develop it as a special case of the method of consistent deformations.

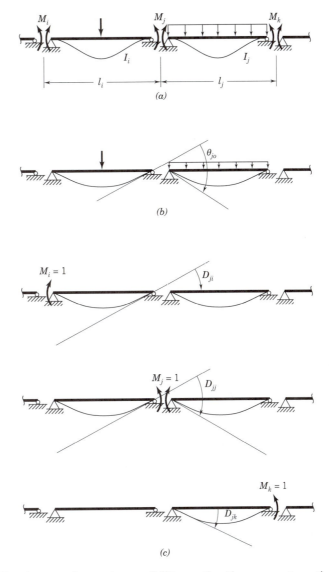

Figure 10.8 Development of general compatibility equation (three-moment equation). (a) Statically indeterminate continuous beam. (b) Statically determinate primary structure. (c) Unit values of redundant moments.

10.6.3 Application of Three-Moment Equation

The three-moment equation is particularly useful in determining the internal support moments of a continuous beam. Thus, for the arrangement given in Fig. 10.7a, the equation would be applied at each of the $(m - 1)$ interior support points. This would provide the $(m - 1)$ compatibility equations that are required for the determination of the $(m - 1)$ redundant moments.

For fixed-ended beams, such as the one shown in Fig. 10.9, there is an additional redundant moment at each fixed end. Although a special form of the three-moment equation could be formulated for this case, a convenient artifice is to replace the fixed end by an imaginary end span of zero length. The three-moment equation is then applied at the end support points as well as at the interior points. Thus, an additional compatibility equation is gained for each additional redundant moment.

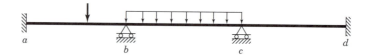

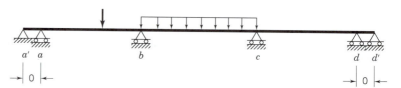

Figure 10.9 Treatment of fixed-end beam.

One of the major advantages of selecting the internal moments as the redundants in the method of consistent deformations, which leads to the three-moment equation, is that the flexibility matrix used in the solution of the redundant moments is a banded matrix. This greatly simplifies the solution of the simultaneous equations.

Note that strict adherence to the sign convention that was introduced through the derivations of Section 10.6.2 is necessary. That is, positive moments are those that cause compression on the top fibers of the beam, and interior support discontinuities are positive when clockwise.

Since the solution of the resulting compatibility equations yields member-end moments, the shear and moment diagrams are easily constructed in accordance with the method presented in Section 5.11.

EXAMPLE 10.7

Determine the support moments for the structure given in Example 10.6 by applying the three-moment equation.

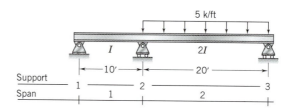

Structure Classification

Application of the criteria of Section 3.3 or 5.4 shows that the structure is statically indeterminate to the first degree.

Three-Moment Equation for Support j

$$\left(\frac{l_i}{I_i}\right)M_i + 2\left(\frac{l_i}{I_i} + \frac{l_j}{I_j}\right)M_j + \left(\frac{l_j}{I_j}\right)M_k = -6E\theta_{j0} \tag{10.21}$$

This equation is applied at point 2 ($j = 2$).

Point 2

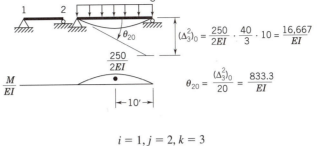

$$(\Delta_3^2)_0 = \frac{250}{2EI} \cdot \frac{40}{3} \cdot 10 = \frac{16{,}667}{EI}$$

$$\theta_{20} = \frac{(\Delta_3^2)_0}{20} = \frac{833.3}{EI}$$

$$i = 1, j = 2, k = 3$$
$$l_1 = 10', I_1 = I$$
$$l_2 = 20', I_2 = 2I$$

$$\left(\frac{10}{I}\right)\overset{0}{\underset{\uparrow}{M_1}} + 2\left(\frac{10}{I} + \frac{20}{2I}\right)M_2 + \left(\frac{20}{2I}\right)\overset{0}{\underset{\uparrow}{M_3}} = -6E\left(\frac{833.3}{EI}\right) \tag{10.21}$$

$$\frac{40}{I}M_2 = -\frac{5{,}000}{I}$$

Solution for Moment

$$\frac{40}{I}M_2 = -\frac{5{,}000}{I}$$
$$M_2 = -125 \text{ kip-ft}$$

EXAMPLE 10.8   Determine the support moments for the structure shown by application of the three-moment equation.

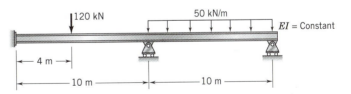

Structure Classification

Structure is statically indeterminate to the second degree.

Modified Structure

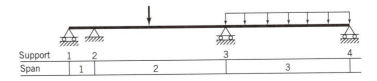

Three-Moment Equation for Support j

$$l_i M_i + 2(l_i + l_j) M_j + l_j M_k = -6EI\theta_{j0} \qquad (10.21)$$

This equation must be applied at points 2 and 3.

Point 2

$$i = 1, \quad j = 2, \quad k = 3$$
$$l_1 = 0$$
$$l_2 = 10 \text{ m}$$

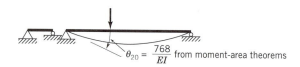

$\theta_{20} = \dfrac{768}{EI}$ from moment-area theorems

$$0 \cdot M_1 + 2(0 + 10)M_2 + 10 \cdot M_3 = -6EI\left(\frac{768}{EI}\right) \qquad (10.21)$$
$$20M_2 + 10M_3 = -4\,608$$

Point 3

$$i = 2, \quad j = 3, \quad k = 4$$
$$l_2 = 10 \text{ m}$$
$$l_3 = 10 \text{ m}$$

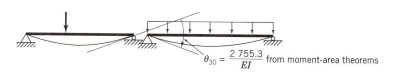

$\theta_{30} = \dfrac{2\,755.3}{EI}$ from moment-area theorems

$$10 \cdot M_2 + 2(10 + 10)M_3 + 10 \cdot M_4 = -6EI\left(\frac{2\,755.3}{EI}\right) \qquad (10.21)$$
$$10M_2 + 40M_3 = 16\,532$$

Solution for Moments

$$20M_2 + 10M_3 = -4\,608$$
$$10M_2 + 40M_3 = -16\,532$$

Solving simultaneously,

$$M_2 = -27.2 \text{ kN} \cdot \text{m}; \quad M_3 = -406.5 \text{ kN} \cdot \text{m}$$

10.7 SELF-STRAINING PROBLEMS

Support settlements, temperature change, or fabrication errors induce forces within the system, without the application of external forces. For this reason, problems of this type are referred to as *self-straining problems.*

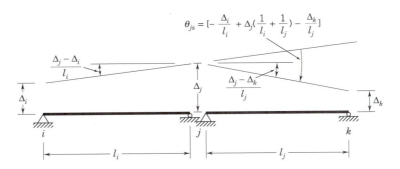

Figure 10.10 Chord discontinuity from support displacements.

10.7.1 Settlement

For the case of support settlements, consider again the two-span section shown in Fig. 10.8. If the supports displace as shown in Fig. 10.10, where upward displacements are taken as positive, then there is a chord discontinuity of θ_{js} at support j that is given by the expression

$$\theta_{js} = \left[-\frac{\Delta_i}{l_i} + \Delta_j \left(\frac{1}{l_i} + \frac{1}{l_j} \right) - \frac{\Delta_k}{l_j} \right] \tag{10.22}$$

where Δ_i, Δ_j, and Δ_k are the displacements at the support points. This rotation must be included in the compatibility equation at point j, and thus Eq. 10.19 takes the form

$$D_{ji}M_i + D_{jj}M_j + D_{jk}M_k + \theta_{js} + \theta_{j0} = \theta_j = 0 \tag{10.23}$$

Substitution of Eqs. 10.20 and 10.22 into Eq. 10.23 gives

$$\left(\frac{l_i}{I_i} \right)M_i + \left(\frac{2l_i}{I_i} + \frac{2l_j}{I_j} \right)M_j + \left(\frac{l_j}{I_j} \right)M_k = 6E\left[\frac{\Delta_i}{l_i} - \Delta_j \left(\frac{1}{l_i} + \frac{1}{l_j} \right) + \frac{\Delta_k}{l_j} \right] - 6E\theta_{j0} \tag{10.24}$$

For the special case of $I_i = I_j = I$, Eq. 10.24 reduces to

$$l_iM_i + 2(l_i + l_j)M_j + l_jM_k = 6EI\left[\frac{\Delta_i}{l_i} - \Delta_j \left(\frac{1}{l_i} + \frac{1}{l_j} \right) + \frac{\Delta_k}{l_j} \right] - 6EI\theta_{j0} \tag{10.25}$$

Moreover, Eq. 10.24 or 10.25 replaces Eq. 10.21 as the basic three-moment equation when support settlements are present.

10.7.2 Temperature Change

For the effects of temperature change, consider the single-span structure shown in Fig. 10.11a. The temperature gradient ΔT is given by

$$\Delta T = T_b - T_t \tag{10.26}$$

where T_b and T_t are the temperatures at the bottom and top fibers of the beam, respectively. For the element shown in Fig. 10.11b, it is clear that

$$d\theta = \frac{\alpha(T_b - T_t)dx}{h} = \frac{\alpha\Delta T \, dx}{h} \tag{10.27}$$

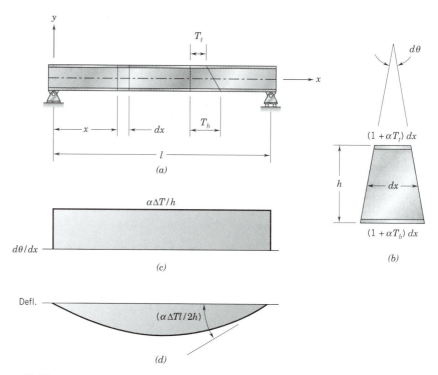

Figure 10.11 Temperature gradient on beam. *(a)* Gradient. *(b)* Strained element. *(c)* Induced curvatures. *(d)* Deflected beam with end rotation.

where α is the coefficient of thermal expansion and h is the member depth. Therefore, the curvature is

$$\frac{d\theta}{dx} = \frac{\alpha\Delta T}{h} \tag{10.28}$$

This curvature, of course, plays the same role as M/EI for a beam under loading and, therefore, for constant temperature gradient along the member, the curvature diagram of Fig. 10.11c results. Application of the moment–area method leads to end slopes of

$$\theta = \frac{\alpha\Delta T l}{2h} \tag{10.29}$$

which are shown in Fig. 10.11d. In Eq. 10.29, positive ΔT produces positive curvature.

Returning to Fig. 10.8, and replacing the loading of Fig. 10.8a with separate temperature gradients for each span, we see that

$$\theta_{j0} = \left(\frac{\alpha\Delta T l}{2h}\right)_i + \left(\frac{\alpha\Delta T l}{2h}\right)_j \tag{10.30}$$

where the subscripts refer to the two separate spans. For common α, ΔT, and h, Eq. 10.30 becomes

$$\theta_{j0} = \left(\frac{\alpha\Delta T}{2h}\right)(l_i + l_j) \tag{10.31}$$

Equation 10.21 remains valid for the temperature problem, with θ_{j0} of Eq. 10.31 being used. Of course, in the case of combined loading and temperature change, θ_{j0} must include the superposition of the two effects, and if settlement is involved, Eq. 10.24 must be used.

10.7.3 Fabrication Errors

For beam structures, fabrication errors include initial crookedness of the members and the associated end slopes. These end slopes must be included in the determination of θ_{j0}.

EXAMPLE 10.9

Determine the support moments for the structure shown for the given temperature gradient by applying the three-moment equation. Here, $E = 30 \times 10^3$ ksi; $I = 1{,}500$ in.^{4}; $EI = 45 \times 10^6$ k-in.$^2 = 312.5 \times 10^3$ k-ft^2; $\alpha = 0.000{,}006{,}5 / °F$; and $\Delta T = T_b - T_t = 10 - 60 = -50°F$.

$T_t = +60° \text{ F}$

$h - 18''$

$T_b = +10° \text{ F}$

Temperature gradient

Modified Structure

Support	1		2		3	4
Span		1		2		3

Three-Moment Equation for Support j

$$l_i M_i + 2(l_i + l_j)M_j + l_j M_k = 6EI\theta_{j0} \qquad (10.21)$$

This equation must be applied at points 2 and 3, where

$$\theta_{j0} = \left(\frac{\alpha \Delta T}{2h}\right)(l_i + l_j) \qquad (10.31)$$

Point 2

$$i = 1, \quad j = 2, \quad k = 3$$
$$l_1 = 20'$$
$$l_2 = 30'$$

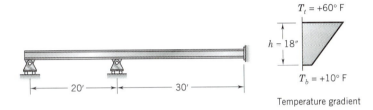

$$\theta_{20} = \frac{(0.000{,}006{,}5)(-50)}{2(18/12)}(20 + 30)$$
$$= -0.005{,}42$$

$$\overset{0}{\uparrow}$$
$$20 \cdot M_1 + 2(20 + 30)M_2 + 30 \cdot M_3 = -6EI(-0.005,42)$$
$$100M_2 + 30M_3 = 10,163$$

Point 3

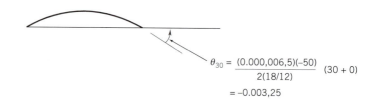

$$i = 2, \quad j = 3, \quad k = 4$$
$$l_2 = 30'$$
$$l_3 = 0'$$

$$\theta_{30} = \frac{(0.000,006,5)(-50)}{2(18/12)}(30 + 0)$$
$$= -0.003,25$$

$$30 \cdot M_2 + 2(30 + 0)M_3 + 0 \cdot M_4 = -6EI(-0.003,25)$$
$$30M_2 + 60M_3 = 6,094$$

Solution for Moments

$$100M_2 + 30M_3 = 10,163$$
$$30M_2 + 60M_3 = 6,094$$

Solving simultaneously,

$$M_2 = 83.7 \text{ ft-kips}; \qquad M_3 = 59.8 \text{ ft-kips}$$

EXAMPLE 10.10   Determine the support moments for the structure of Example 10.9 for the lack-of-fit situation described. Use the three-moment equation, and disregard the temperature gradient previously given.

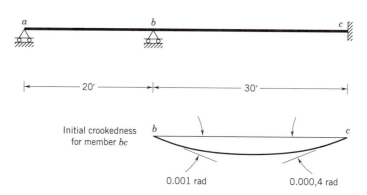

Initial crookedness for member bc

0.001 rad   0.000,4 rad

Three-Moment Equation for Support j

As in the previous example, Eq. 10.21 is used, with θ_{j0} given directly from the initial crookedness for member bc.

Point 2

$$100M_2 + 30M_3 = -6EI(0.001) = -1,875$$

Point 3

$$30M_2 + 60M_3 = -6EI(0.000,4) = -750$$

Solution for Moments

$$100M_2 + 30M_3 = -1,875$$
$$30M_2 + 60M_3 = -750$$

Solving simultaneously,

$$M_2 = -17.65 \text{ ft-kips}; \qquad M_3 = -3.67 \text{ ft-kips}$$

10.8 DETERMINATION OF REDUNDANT MEMBER FORCES FOR TRUSS STRUCTURES

Section 10.3 dealt with the problem of determining the redundant reactions for a structure that is statically indeterminate externally, whereas Section 10.6 examined the continuous beam problem by selecting internal moments as the redundants. The method of consistent deformations is equally useful for the determination of redundant member forces for truss structures that are statically indeterminate internally.

Consider the truss shown in Fig. 10.12a. The objective is to determine the independent reaction components and the F set of bar forces. Application of the provisions of Section 4.8 reveals that this structure is internally indeterminate to the first degree. Thus, there is one redundant member. In the current case, the member bf is taken as the redundant member. When the load-carrying capacity of this member is released, the remaining members are sufficient in number and arrangement to ensure the internal stability of the structure. This release is accomplished by envisioning a telescoping sleevelike fixture along the member length that permits the member to retain its position and alignment within the structure without its being able to transmit any force.

Upon release of member bf, the statically determinate primary structure of Fig. 10.12b is formed. Since this structure is determinate, the reactions and the F_0 set of bar forces can be determined. Again, the 0 subscript indicates that these quantities are associated with the actual loading on the primary structure. For this loading, there is an overlap, Δ_{10}, of the cut member within the sleeve, which is in

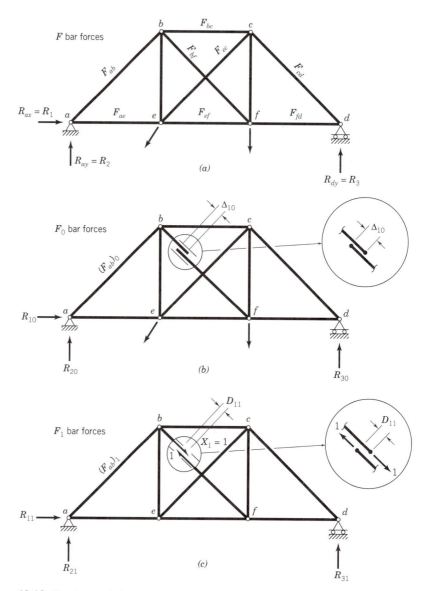

Figure 10.12 Consistent deformation analysis of internally indeterminate truss. *(a)* Internally statically indeterminate truss. *(b)* Statically determinate primary structure. *(c)* Unit value of X_1.

violation of the prescribed continuity for this member in the original structure. Thus, this solution of the primary structure must be modified in order to restore the required continuity to member bf.

The required modification is brought about by introducing a unit value of the redundant member force on the primary structure, as is shown in Fig. 10.12c. Here, the F_1 set of bar forces results from a static analysis of the primary structure, with the subscript 1 indicating that these values are associated with a unit value of the redundant $X_1 = (F_{bf})_1$. Here, too, there is an overlap within the sleeve of the cut member that is denoted by D_{11}.

Since D_{11} is the overlapping displacement associated with a unit value of X_1, superposition can be used to determine the corresponding displacement that occurs when the cut member is subjected to a force equal to the actual value of the redundant member force X_1. This is represented by Δ_{1R} and is given by

$$\Delta_{1R} = D_{11}X_1 \tag{10.32}$$

For the final solution of Fig. 10.12a, the solutions of Figs. 10.12b and 10.12c must be superimposed. Specifically, the displacements Δ_{10} and Δ_{1R} must be combined to restore the continuity of the severed member. Thus,

$$\Delta_{10} + D_{11}X_1 = 0 \tag{10.33}$$

Solving for X_1, we obtain

$$X_1 = \frac{-\Delta_{10}}{D_{11}} \tag{10.34}$$

The final values of other response quantities can now be determined from statics, or they can be determined by superposition. In the latter case, if S is taken as the response quantity of interest, such as a reaction or internal bar force, then

$$S = S_0 + S_1X_1 \tag{10.35}$$

If there are two redundant member forces, the continuity equations take the form

$$\begin{bmatrix} D_{11} & D_{12} \\ D_{21} & D_{22} \end{bmatrix} \begin{Bmatrix} X_1 \\ X_2 \end{Bmatrix} = \begin{Bmatrix} -\Delta_{10} \\ -\Delta_{20} \end{Bmatrix} \tag{10.36}$$

where D_{ij} is a flexibility coefficient, which gives the displacement along the release of the ith redundant member that is associated with a unit force in the jth redundant member of the primary structure, and Δ_{i0} is the displacement along the ith redundant member that is caused by the actual loading on the primary structure.

Solution of Eq. 10.36 gives the magnitudes of the redundant forces X_1 and X_2. The final value of any other response quantity S is then determined by superposition to be

$$S = S_0 + S_1X_1 + S_2X_2 \tag{10.37}$$

If there are more than two redundant members, the same general procedure is followed. Each redundant member requires a continuity equation. These equations collectively provide a set of simultaneous equations in the form of Eq. 10.36. The solution of these equations gives the magnitudes of the redundant member forces, and the final solution is then completed by superposition similar to that expressed in Eq. 10.37. As explained earlier, the final solution will inherently satisfy all of the requirements of compatibility and equilibrium.

In the example problem that follows, the required displacement quantities are determined from the complementary virtual work method. It is noted that the summations must include the terms corresponding to the cut member, since internal virtual work is done on this member in the determination of some displacement quantities.

EXAMPLE 10.11 Determine the member forces for the structure given. The quantity EA is the same for each member of the structure.

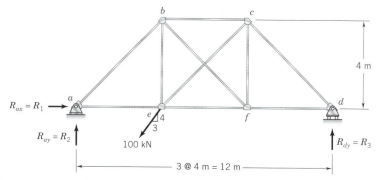

Structure Classification

Application of the criteria of Section 4.8 shows that the structure is statically indeterminate internally to the first degree.

Primary Structure and Loadings

Select F_{bf} as a redundant member force.

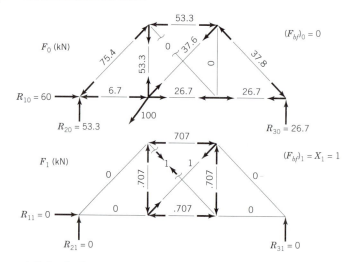

Displacement Calculations

The fundamental complementary virtual work expression is used to obtain the required displacement quantities.

$$\sum_{i=1}^{n} (\delta P)_i \Delta_i = \sum_{j=1}^{m} \left(\delta F_P \cdot \frac{F_\Delta l}{EA} \right)_j \tag{7.18}$$

This equation is applied, as described in the following table:

Displacement Quantity	δF_P	F_Δ	Equation 7.18
Δ_{10}	F_1	F_0	$1 \cdot \Delta_{10} = \dfrac{1}{EA} \sum F_1 F_0 l$
D_{11}	F_1	F_1	$1 \cdot D_{11} = \dfrac{1}{EA} \sum F_1^2 l$

The required summations are formed in the following table:

Bar rs	l m	F_0 kN	F_1 kN	$F_1 F_0 l$ kN² · m	$F_1^2 l$ kN² · m	$(F_{rs})_1 \cdot X_1$ kN	$F_{rs} = (F_{rs})_0$ $+ (F_{rs})_1 \cdot X_1$ kN
ab	5.66	−75.4	0	0	0	0	−75.4
bc	4	−53.3	−0.707	150.7	2.0	5.0	−48.3
cd	5.66	−37.8	0	0	0	0	−37.8
ae	4	−6.7	0	0	0	0	−6.7
ef	4	26.7	−0.707	−75.5	2.0	5.0	31.7
fd	4	26.7	0	0	0	0	26.7
be	4	53.3	−0.707	−150.7	2.0	5.0	58.3
bf	5.66	0	1	0	5.66	−7.1	−7.1
ce	5.66	37.6	1	212.8	5.66	−7.1	30.5
cf	4	0	−0.707	0	2.0	5.0	5.0
Σ				137.3	19.32		

In this table, the first six columns are used to form the summations used in Eq. 7.18 to determine Δ_{10} and D_{11}. The last two columns are used to determine the final member forces after the magnitude of the redundant X_1 is established. Here, $(F_{rs})_0$ and $(F_{rs})_1$ are the values of F_0 and F_1 for member rs, respectively.

Note: The redundant member bf must be included in the tabulation.

$$1 \cdot \Delta_{10} = \frac{1}{EA} \sum_{j=1}^{m} (F_1 F_0 l)_j = \frac{137.3}{EA}; \quad \Delta_{10} = \frac{137.3}{EA} \left(\frac{kN \cdot m}{EA} \right)$$

$$1 \cdot D_{11} = \frac{1}{EA} \sum_{j=1}^{m} (F_1^2 l)_j = \frac{19.32}{EA}; \quad D_{11} = \frac{19.32}{EA} \left(\frac{kN \cdot m}{EA} \right)$$

Calculation of Redundant and Member Forces

The requirements of compatibility for member bf form the basis for the following superposition equation:

$$\Delta_{10} + D_{11} X_1 = 0 \tag{10.33}$$

Solution for the redundant X_1 gives

$$X_1 = \frac{-\Delta_{10}}{D_{11}} \tag{10.34}$$

from which

$$X_1 = \frac{-137.3}{19.32} = -7.1$$

The final bar forces are determined from

$$S = S_0 + S_1 X_1 \tag{10.35}$$

or, specifically for the bar force in member *rs*,

$$F_{rs} = (F_{rs})_0 + (F_{rs})_1 \cdot X_1$$

These forces are calculated in the last two columns of the table.

10.8.1 Temperature Change or Fabrication Error for a Truss

Consider again the structure of Fig. 10.12. If member *bf* has a specified length change Δ_1 that is a result of *temperature change* or *fabrication error,* then the continuity requirement that was previously expressed as Eq. 10.33 becomes

$$\Delta_{10} + D_{11}X_1 = \Delta_1 \tag{10.38}$$

Solving for X_1, we obtain

$$X_1 = \frac{\Delta_1 - \Delta_{10}}{D_{11}} \tag{10.39}$$

If there are two redundant member forces, the continuity equations take the form

$$\begin{bmatrix} D_{11} & D_{12} \\ D_{21} & D_{22} \end{bmatrix} \begin{Bmatrix} X_1 \\ X_2 \end{Bmatrix} = \begin{Bmatrix} \Delta_1 - \Delta_{10} \\ \Delta_2 - \Delta_{20} \end{Bmatrix} \tag{10.40}$$

where, in addition to the terms defined after Eq. 10.36, Δ_i is the specified change in member length for the *i*th redundant member. Equations 10.39 and 10.40 give the magnitudes of the redundant member forces. The final member forces and reactions are given by superposition using Eq. 10.35 for a single redundant or Eq. 10.37 for the case with two redundants.

A specified length change that occurs as a result of temperature variation or fabrication error plays the same role in internally indeterminate structures as does support movement in externally indeterminate structures. In the absence of such length changes, it is sufficient to use relative *EA* values in the computations. When length changes are included, however, the absolute *EA* values must be used.

It should be noted that length changes due to temperature effects or fabrication errors are not necessarily limited to the redundant members. The effects of length changes for the nonredundant members are included in the calculation of the Δ_{i0} quantities for the primary structure.

10.9 COMPATIBILITY EQUATIONS BY ENERGY METHODS

Castigliano's second theorem was developed in Section 2.18.2, and it was stated there that this theorem is useful in determining individual displacement quantities or in writing compatibility equations for statically indeterminate analysis. The application concerning the determination of individual displacements was illustrated in Sections 7.7 and 8.10. The use of the theorem in generating the governing compatibility equations for the analysis of statically indeterminate structures is illustrated in this section.

Recall from Section 2.18.2 that the mathematical expression for Castigliano's second theorem has the form

$$\frac{\partial U}{\partial P_i} = \Delta_i \tag{10.41}$$

where U is the total strain energy for the structure, and P_i and Δ_i represent the ith externally applied load and the corresponding displacement, respectively.

10.9.1 Beam- and Frame-Type Structures in Flexure

In the case of a beam- or frame-type structure, which is composed of m members and for which the strain energy results from flexure, it was shown in Section 8.9 that the strain energy is

$$U = \sum_{j=1}^{m} \frac{1}{2E} \int_{x=0}^{x=l_j} \frac{M_j^2}{I_j} \, dx \qquad (10.42)$$

where M_j is the bending moment as a function of x, l_j is the member length, I_j is the moment of inertia, and E is the modulus of elasticity. The moment M_j must be expressed in terms of the applied P loads so that the differentiation of Eq. 10.41 can be carried out. Equation 10.42 can become very complicated when it is expressed in terms of the P loads, and the differentiation called for in Eq. 10.41 can be tedious.

EXAMPLE 10.12   Determine the reactions and the vertical displacements at points b and d for the given structure.

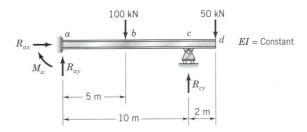

Structure Classification

Reference to the criteria of Sections 3.3 and 5.4 reveals that the structure is statically indeterminate to the first degree.

Primary Structure

Select $M_a = P_1$ as a redundant reaction component. For this loading, $R_{ax} = 0$.

The load vector $\{P\}$ contains the applied loads on the primary structure, which includes the redundant moment, $M_a = P_1$. The displacement vector $\{\Delta\}$ includes the displacements corresponding to each term in the load vector. The reactions are calculated from the applied loads $\{P\}$.

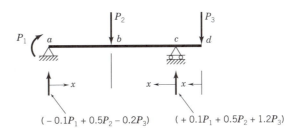

$$\{P\} = \begin{Bmatrix} P_1 \\ P_2 \\ P_3 \end{Bmatrix} = \begin{Bmatrix} M_a \\ 100 \\ 50 \end{Bmatrix}; \quad \{\Delta\} = \begin{Bmatrix} \Delta_1 \\ \Delta_2 \\ \Delta_3 \end{Bmatrix} = \begin{Bmatrix} 0 \\ v_b \\ v_d \end{Bmatrix}$$

The bending moment for each region of the beam is determined in the following table:

Region / Function	a – b $0 \le x \le 5$	d – c $0 \le x \le 2$	c – b $0 \le x \le 5$
M	$P_1 + (-0.1P_1 + 0.5P_2 - 0.2P_3)x$	$-P_3x$	$-P_3(2 + x)$ $+(0.1P_1 + 0.5P_2 + 1.2P_3)x$

Strain Energy

$$U = \sum_{j=1}^{m} \frac{1}{2EI_j} \int_0^{l_j} M_j^2 \, dx \tag{10.42}$$

$$U = \frac{1}{2EI} \int_0^5 [P_1 + (-0.1P_1 + 0.5P_2 - 0.2P_3)x]^2 dx + \frac{1}{2EI} \int_0^2 [-P_3x]^2 dx$$

$$+ \frac{1}{2EI} \int_0^5 [-P_3(2 + x) + (0.1P_1 + 0.5P_2 + 1.2P_3)x]^2 dx$$

Compatibility Equations from Castigliano's Second Theorem

$$\frac{\partial U}{\partial P_i} = \Delta_i \quad (i = 1, 2, 3) \tag{10.41}$$

$$\frac{\partial U}{\partial P_1} = \Delta_1 \Rightarrow \frac{1}{EI}(3.333P_1 + 6.250P_2 - 3.333P_3) = \Delta_1$$

$$\frac{\partial U}{\partial P_2} = \Delta_2 \Rightarrow \frac{1}{EI}(6.250P_1 + 20.833P_2 - 12.500P_3) = \Delta_2$$

$$\frac{\partial U}{\partial P_3} = \Delta_3 \Rightarrow \frac{1}{EI}(-3.333P_1 - 12.500P_2 + 11.000P_3) = \Delta_3$$

Determination of Reactions and Displacements

In matrix form, the compatibility equations are as follows:

$$\frac{1}{EI} \begin{bmatrix} 3.333 & 6.250 & -3.333 \\ 6.250 & 20.833 & -12.500 \\ -3.333 & -12.500 & 11.000 \end{bmatrix} \begin{Bmatrix} M_a \\ 100 \\ 50 \end{Bmatrix} = \begin{Bmatrix} 0 \\ v_b \\ v_c \end{Bmatrix}$$

Solution of the first equation for M_a yields

$$M_a = -137.5 \text{ kN} \cdot \text{m}$$

Then, from statics, $R_{ax} = 0$, $R_{ay} = +53.75$ kN, and $R_{cy} = +96.25$ kN. Then, using the computed value for M_a and the second and third of the preceding equations, we obtain

$$v_b = 598.9/EI$$

$$v_c = -241.7/EI$$

10.9.2 Truss-Type Structures

For a truss structure, which is composed of m pin-connected members that each carry pure tension, the strain energy was found in Section 7.6 to be given by

$$U = \sum_{j=1}^{m} \frac{F_j^2 l_j}{2E_j A_j} \qquad (10.43)$$

In this expression, F_j is the member force, which must be expressed in terms of the P loads that are applied to the primary structure, l_j is the member length, A_j is the member cross-sectional area, and E_j is the modulus of elasticity. It is noted that the total strain energy must include the strain energy stored in any cut members. Again, Eq. 10.43 can be complicated when it is expressed in terms of the P loads, and its differentiation can be tedious.

EXAMPLE 10.13   Consider the truss structure shown below. Use Castigliano's second theorem to generate the compatibility equations relating the displacements associated with the redundant actions and the kinematic degrees of freedom to their respective forces. Assume EA is the same for each member.

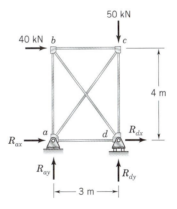

Structure Classification

Examination of the structure according to the criteria of Section 4.8 shows that it is statically indeterminate externally to the first degree and internally to the first degree.

Primary Structure

Select R_{dx} as the redundant reaction component and F_{bd} as the redundant member force. The applied forces, including the redundants, are assembled in $\{P\}$ and the corresponding displacements are arranged in $\{\Delta\}$.

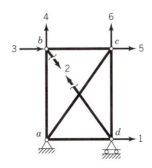

$$\{P\} = \begin{Bmatrix} P_1 \\ P_2 \\ P_3 \\ P_4 \\ P_5 \\ P_6 \end{Bmatrix} = \begin{Bmatrix} R_{dx} \\ F_{bd} \\ 40 \\ 0 \\ 0 \\ -50 \end{Bmatrix}; \quad \{\Delta\} = \begin{Bmatrix} \Delta_1 \\ \Delta_2 \\ \Delta_3 \\ \Delta_4 \\ \Delta_5 \\ \Delta_6 \end{Bmatrix} = \begin{Bmatrix} 0 \\ 0 \\ \Delta_3 \\ \Delta_4 \\ \Delta_5 \\ \Delta_6 \end{Bmatrix}$$

Member Forces for Primary Structure

The bar forces and reactions are established from the $\{P\}$ forces, where $R_{dx} = P_1 =$ the redundant reaction, and $F_{bd} = P_2 =$ the redundant member force (this is an internal force for the original structure, but it is an external force for the primary structure).

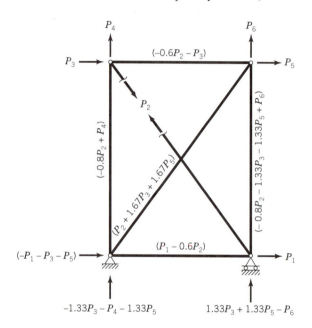

Strain Energy

$$U = \sum_{j=1}^{m} \frac{F_j^2 l_j}{2 E_j A_j} = \frac{1}{2EA} \sum_{j=1}^{m} F_j^2 l_j \tag{10.43}$$

$$U = \frac{1}{2EA}[(-0.8P_2 + P_4)^2(4) + (-0.6P_2 - P_3)^2(3)$$

$$+ (P_2 + 1.67P_3 + 1.67P_5)^2(5) + (P_2)^2(5)$$

$$+ (-0.8P_2 - 1.33P_3 - 1.33P_5 + P_6)^2(4)$$

$$+ (P_1 - 0.6P_2)^2(3)]$$

Note: Cut bar must be included in strain energy expression.

Compatibility Equations from Castigliano's Second Theorem

$$\frac{\partial U}{\partial P_1} = \Delta_1 \Rightarrow \frac{1}{EA}[(P_1 - 0.6P_2)(1)(3)] = \Delta_1$$

$$\frac{1}{EA}(3.00P_1 - 1.80P_2) = \Delta_1$$

$$\frac{\partial U}{\partial P_2} = \Delta_2 \Rightarrow \frac{1}{EA}[(-0.8P_2 + P_4)(-0.8)(4) + (-0.6P_2 - P_3)(-0.6)(3)$$

$$+ (P_2 + 1.67P_3 + 1.67P_5)(1)(5) + (P_2)(1)(5)$$

$$+ (-0.8P_2 - 1.33P_3 - 1.33P_5 + P_6)(-0.8)(4)$$

$$+ (P_1 - 0.6P_2)(-0.6)(3)]$$

$$\frac{1}{EA}(-1.80P_1 + 17.28P_2 + 14.41P_3 - 3.20P_4 + 12.61P_5 - 3.20P_6) = \Delta_2$$

and so on, for

$$\frac{\partial U}{\partial P_i} = \Delta_i \quad \text{for } i = 3, 4, 5, 6$$

The resulting equations are

$$\frac{1}{EA}\begin{bmatrix} 3.00 & -1.80 & 0 & 0 & 0 & 0 \\ -1.80 & 17.28 & 14.41 & -3.20 & 12.61 & -3.20 \\ 0 & 14.41 & 24.02 & 0 & 21.02 & -5.32 \\ 0 & -3.20 & 0 & 4.00 & 0 & 0 \\ 0 & 12.61 & 21.02 & 0 & 21.02 & -5.32 \\ 0 & -3.20 & -5.32 & 0 & -5.32 & 4.00 \end{bmatrix}\begin{Bmatrix} P_1 \\ P_2 \\ P_3 \\ P_4 \\ P_5 \\ P_6 \end{Bmatrix} = \begin{Bmatrix} \Delta_1 \\ \Delta_2 \\ \Delta_3 \\ \Delta_4 \\ \Delta_5 \\ \Delta_6 \end{Bmatrix}$$

Because $\Delta_1 = \Delta_2 = 0$, the first two equations can be used to solve for P_1 and P_2 (the redundant forces) in terms of P_3 through P_6. Then, using these expressions for P_1 and P_2 along with the given values of P_3 through P_6, we are able to determine Δ_3 through Δ_5 from the last four equations.

Of course, once all of the displacements are known, the individual member forces can be easily determined from the member force–displacement relationships.

10.9.3 Structures Subjected to Combined Axial Forces and Flexure

For structures in which both flexure and axial forces are present, the strain energy for each member results from summing the individual strain energies of Eqs. 10.42 and 10.43. Equation 10.41 must be applied to each displacement degree of freedom for which a compatibility relation is desired on the primary structure. This leads to a set of displacement–force equations, and the solution of these equations leads to the redundant forces and the displacements associated with the applied loads.

10.10 SUGGESTED PROBLEMS

10.1 through 10.6 Analyze each of the statically indeterminate structures given for the specified loading. Use the method of consistent deformations, selecting external reaction components as the redundant forces. Construct the shear and moment diagrams for each member.

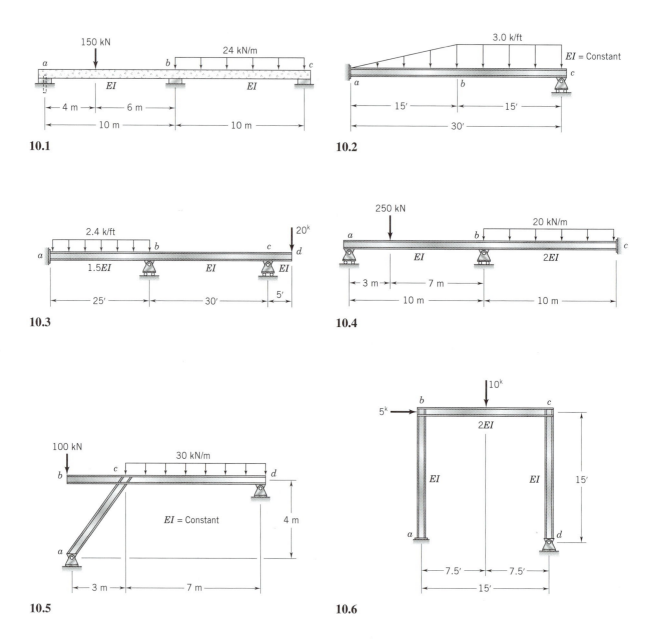

10.1

10.2

10.3

10.4

10.5

10.6

10.7 through 10.13 Use the method of consistent deformations to determine the reactions and member forces for each of the statically indeterminate trusses given, subject to the prescribed loading.

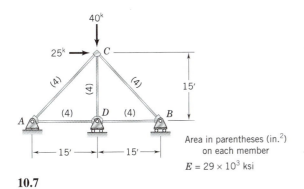

40^k

25^k →

C

(4) (4) (4)

A (4) D (4) B

15'

← 15' → ← 15' →

Area in parentheses (in.²)
on each member
$E = 29 \times 10^3$ ksi

10.7

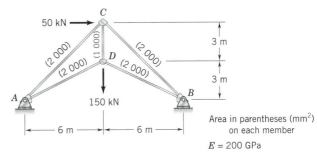

C

50 kN →

(2 000)   (1 000)   (2 000)

(2 000) D (2 000)

A B

150 kN

3 m

3 m

← 6 m → ← 6 m →

Area in parentheses (mm²)
on each member
$E = 200$ GPa

10.8

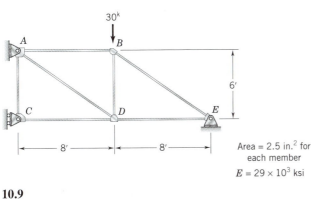

30^k

A B

C D E

6'

← 8' → ← 8' →

Area = 2.5 in.² for
each member
$E = 29 \times 10^3$ ksi

10.9

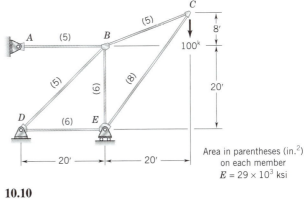

C

(5)

A (5) B 100^k

8'

(5)   (6)   (8)

D (6) E

20'

← 20' → ← 20' →

Area in parentheses (in.²)
on each member
$E = 29 \times 10^3$ ksi

10.10

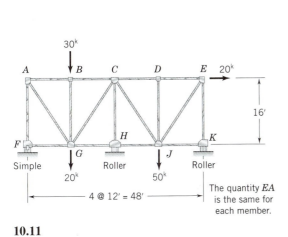

30^k

A B C D E 20^k

16'

F G H J K

Simple   20^k Roller   50^k Roller

← 4 @ 12' = 48' →

The quantity EA
is the same for
each member.

10.11

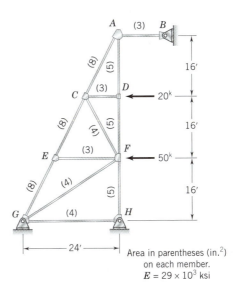

A (3) B

(8)   (5)

16'

C (3) D 20^k

(8)   (4)   (5)

16'

E (3) F

50^k

(8)   (4)   (5)

16'

G (4) H

← 24' →

Area in parentheses (in.²)
on each member.
$E = 29 \times 10^3$ ksi

10.12

10.13 The quantity *EA* is the same for each member of the truss, and EA_{cable} is one-tenth of the *EA* values for the truss members.

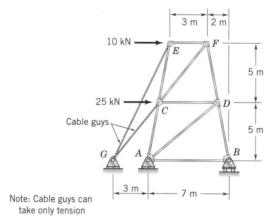

Note: Cable guys can
take only tension

10.14 Analyze the structure and loading of Problem 1 if the rigid support at point *c* is replaced by the elastic support shown below.

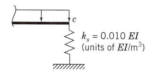

10.15 Analyze the structure and loading of Problem 4 if the rigid supports of points *a* and *b* are replaced by elastic supports with stiffnesses of $k_s = 0.005EI$ (units EI/m^3).

10.16 Determine the reactions and construct the shear and moment diagrams for the structure of Problem 1 if, in addition to the given loading, point *c* settles 0.012 m downward. Take $EI = 700\,000$ kN $\cdot$ m^2.

10.17 Determine the reactions, shears, and moments for the structure of Problem 2 if point *a* experiences a rotation of 0.001 radian clockwise. Do not include the loads of Problem 2, and take the value of *EI* to be 30×10^6 k-in^2.

10.18 Consider the structure of Problem 4 in the absence of any applied loads. Determine the shears and moments throughout the structure for the following pattern of support settlements: Point *a*, 15 mm downward; point *b*, 25 mm downward; point *c*, 0.000 5 radian counterclockwise. Express your results in terms of *EI*.

10.19 Determine the reactions and member forces for the truss of Problem 8 if the support at point *B* settles downward by 25 mm and to the right by 15 mm. Disregard the loads given in Problem 8.

10.20 Determine the reactions and member forces for the truss of Problem 9 if, in addition to the given load, the support at point *E* settles downward by 0.75 inch.

10.21 Determine the reactions and member forces for the truss of Problem 7 if the following vertical support settlements occur: point *A*, 0.50 inch upward; point *B*, 0.50 inch downward; point *D*, 0.75 inch downward. Disregard the loads given in Problem 7.

10.22 Determine the reactions and member forces for the structure of Problem 12 if, in addition to the given loads, point G settles 0.75 inch downward and point H is uplifted 0.25 inch.

10.23 Determine the reactions and the member forces for the truss and loading of Problem 7 if the rigid support of point D is replaced by the elastic support shown below. Compare your solution with that obtained in Problem 7.

10.24 Determine the reactions and the member forces for the truss and loading of Problem 8 if the rigid support of point B is replaced by the elastic support shown. Compare your solution with that obtained in Problem 8.

10.25 Determine the reactions and member forces for the structure of Problem 7 if members AC and CB are each subjected to a temperature increase of $50\,°F$. Disregard the loads given in Problem 7, and take $\alpha_t = 0.000,006,5/°F$.

10.26 Determine the reactions and member forces for the structure of Problem 9 if, in addition to the specified load, member AD is 0.75 in. too long but is forced to fit.

10.27 through 10.36 Use the three-moment equation formulation to determine the support moments for each of the given structures for the indicated loading.

10.27 The structure and loading of Problem 1.

10.28 The structure and loading of Problem 3.

10.29 The structure and loading of Problem 4.

10.30 The structure and loading of Problem 1 with the settlement prescribed in Problem 16.

10.31 The structure and loading of Problem 2 with the support movement described in Problem 17.

10.32 The structure of Problem 4 with the support settlements designated in Problem 18. Disregard the loads of Problem 4, and express the results in terms of EI.

10.33 The structure of Problem 1 with a temperature gradient of $\Delta T = T_b - T_t = 30\,°C$ applied to both spans. Disregard the given loads of Problem 1, and take $\alpha_t = 0.000\,012/°C$, $h = 40$ mm, and $EI = 500\,000$ kN $\cdot$ m^2.

10.34 The structure of Problem 3 with a temperature gradient of $\Delta T = T_b - T_t = -50\,°F$ acting on span ab. Assume that the loads of Problem 3 remain on the structure, and take $\alpha_t = 0.000,006,5/°F$, $EI = 25 \times 10^6$ k-in.^{2}, and $h = 15$ in.

10.35 The structure of Problem 4 in which member ab has an initial crookedness as noted below, but is forced into place. Disregard the prescribed loads of Problem 4, and take $EI = 400\,000$ kN $\cdot$ m^2.

10.36 The structure of Problem 3 in which member bc has the following initial crookedness but is forced to fit. Do not include the loads prescribed in Problem 3, and take $EI = 30 \times 10^6$ k-in^2. *Hint:* Convert initial crookedness to end slopes only.

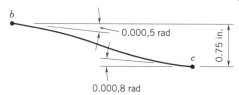

10.37 through 10.40 Use the conventional method of consistent deformations to determine the member forces for each of the statically indeterminate trusses shown for the given loadings.

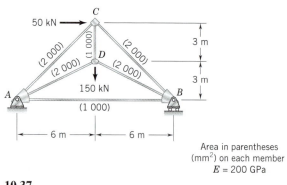

Area in parentheses
(mm^2) on each member
$E = 200$ GPa

10.37

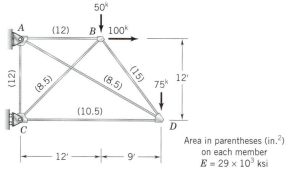

Area in parentheses (in.^{2})
on each member
$E = 29 \times 10^3$ ksi

10.38

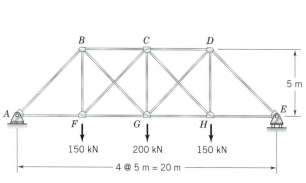

The quantity EA is the same for each member.

10.39

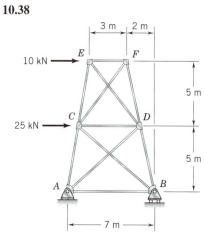

The quantity EA is the same
for each member.

10.40

10.41 Determine the reactions and member forces for the structure of Problem 37 if members *AC* and *CB* are subjected to a temperature increase of 30°C. Disregard the loading of Problem 37, and take $\alpha_t = 0.000\ 012/°C$.

10.42 Determine the reactions and member forces for the structure of Problem 38 if, in addition to the given loads, member *BD* is 0.50 in. too short but is forced to fit.

10.43 Determine the reactions and member forces for the structure of Problem 39 if members *AB, BC, CD,* and *DE* are subjected to a temperature increase of 25°C. Disregard the loading of Problem 39, and take $\alpha_t = 0.000\ 012/°C$. Express your results in terms of *EA*.

10.44 and 10.45 Use Castigliano's second theorem to determine the complete set of reactions and the indicated displacements for each of the structures and loadings indicated.

10.44 The structure and loading of Problem 2, with the desired displacements being the vertical displacement at point *b* and the rotation at point *c*.

10.45 The structure and loading of Problem 5, with the desired displacements being the vertical displacements at point *b* and at the middle of span *cd*.

10.46 and 10.47 Use Castigliano's second theorem to generate the required compatibility equations for each of the designated structures for the specified loadings. The selection of the redundant force components and the free-joint displacements to be included are indicated by the suggested primary structure and coordinate system.

10.46 The structure and loading of Problem 7 with the primary structure and coordinate system indicated.

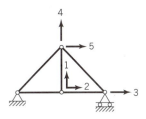

10.47 The structure and loading shown, along with the primary structure and coordinate system specified.

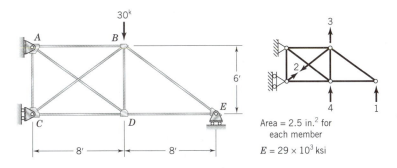

Chapter 11

Slope Deflection Method (and Other Equilibrium Methods)

The Bank of China, Hong Kong (courtesy Leslie E. Robertson Associates, R.L.L.P. Structural Engineer, photo by John Nye).

11.1 NATURE OF EQUILIBRIUM METHODS

For pedagogic reasons, the analysis of statically determinate structures and statically indeterminate structures is presented separately in this book. This separation follows the classical presentations for teaching the fundamentals of structural analysis. This approach is indeed logical for the method of consistent deformations presented in Chapter 10, where redundant forces are identified and determined. As will be seen, however, *equilibrium methods* are not uniquely identified with statically indeterminate analysis, but are equally applicable to statically determinate systems. When the conventional methods of static analysis are applied to statically determinate structures, only the member forces are determined. The desired displacements must then be calculated separately by the methods that are appropriate for the structure type. However, equilibrium methods are more comprehensive in that they readily produce the complete set of structure displacements and member forces.

To reiterate, all methods of structural analysis must ultimately produce a solution that satisfies both equilibrium and compatibility. Equilibrium methods derive their name from the fact that the fundamental equations used in the formation are equations of equilibrium. The conditions of compatibility, which include the appropriate displacement boundary conditions, are taken into account as the equations of equilibrium are developed. Since the resulting equilibrium equations are composed of stiffness quantities, and since the solution of these equations yields displacements, the equilibrium method is alternatively referred to as the *stiffness method* or the *displacement method*.

Many methods fall into the general category of equilibrium methods. In this chapter, the emphasis is on the *slope deflection method*. This is one of the classical methods of analysis for statically indeterminate beam- and frame-type structures. Chapter 14 will present a completely general matrix formulation of the equilibrium method, applicable to all classes of structures.

The concept of kinematic indeterminacy, which was introduced in Section 9.4, is of particular importance for equilibrium methods. As was explained previously, a kinematically determinate primary structure is formed by restraining each structure displacement component, and the number of restraints gives the degree of kinematic indeterminacy. In the case of beam- and frame-type structures, since forces are applied between the member ends, the individual members are under load in the restrained position. Since the primary structure does not conform to the structure originally given, the artificially imposed restraints must be relaxed. As this relaxation takes place, equilibrium must be satisfied for each displacement degree of freedom. This requires the inclusion of the loads applied directly at the joints as well as the effects of the loads applied along the member lengths.

The methods of this chapter require direct solutions to the equilibrium equations. This requirement makes it difficult to use these methods for hand solutions, except for small problems. Hand calculations are more easily handled by using iterative solutions for the equilibrium equations. One such iterative approach is considered in Chapter 12.

11.2 THE SLOPE DEFLECTION EQUATION

The slope deflection method was presented by G. A. Maney in 1915 as a method of analysis for rigid-jointed beam and frame structures. The method is an equilibrium method that accounts for flexural deformations but ignores axial and shear deformations.

The member force–deformation equations that are needed for the slope deflection method will be developed by considering member *AB* of Fig. 11.1*a*. This member, of length *l* and with constant *EI*, experiences the moments shown in Fig. 11.1*b* and undergoes the deformed configuration displayed in Fig. 11.1c. The member-end moments (M_{AB} and M_{BA}), the member-end rotations (θ_A, θ_B), and the chord rotation (ψ_{AB}) are all taken as positive for the clockwise orientations shown. This convention for moments is at variance with that used for constructing moment diagrams, but this difference will be reconciled as needed in the derivation that follows and in the problem applications.

Application of the second moment–area theorem leads to

$$\Delta_A^B = -\frac{M_{AB}l^2}{6EI} + \frac{M_{BA}l^2}{3EI} - \frac{Pabx_A}{2EI} \tag{11.1}$$

and

$$\Delta_B^A = \frac{M_{AB}l^2}{3EI} - \frac{M_{BA}l^2}{6EI} + \frac{Pabx_B}{2EI} \tag{11.2}$$

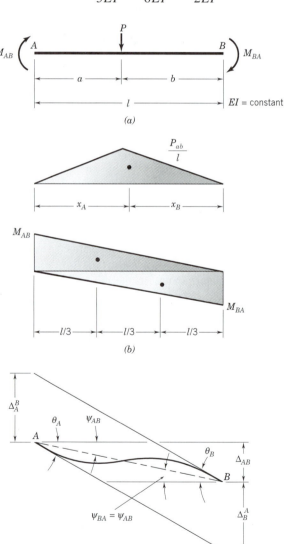

Figure 11.1 Derivation of slope deflection equation. *(a)* Member *AB* under load with end moments. *(b)* Moment diagrams. *(c)* Deformed configuration.

where Δ_A^B and Δ_B^A are as shown in Fig. 11.1c, and x_A and x_B position the centroid of the static moment diagram for the P load as noted in Fig. 11.1b. It should be noted that the moment diagrams of Fig. 11.1b are plotted according to the sign convention introduced in Section 5.3, and that the signs of the individual terms of Eqs. 11.1 and 11.2 depend on how the individual moment diagram components contribute to the positive quantities of Δ_A^B and Δ_B^A as shown in Fig. 11.1c.

From the boundary conditions at ends A and B of member AB, it is clear that

$$\theta_A - \psi_{AB} = \frac{\Delta_B^A}{l} \tag{11.3}$$

and

$$\theta_B - \psi_{AB} = \frac{\Delta_A^B}{l} \tag{11.4}$$

Substitution of Eqs. 11.1 and 11.2 into Eqs. 11.3 and 11.4 leads to

$$\theta_A - \psi_{AB} = \frac{M_{AB}l}{3EI} - \frac{M_{BA}l}{6EI} + \frac{Pabx_B}{2EIl} \tag{11.5}$$

and

$$\theta_B - \psi_{AB} = -\frac{M_{AB}l}{6EI} + \frac{M_{BA}l}{3EI} - \frac{Pabx_A}{2EIl} \tag{11.6}$$

Noting from the area properties given inside the back cover that $x_A = (l + a)/3$ and $x_B = (l + b)/3$, and simultaneously solving Eqs. 11.5 and 11.6 for M_{AB} and M_{BA}, we obtain

$$M_{AB} = \frac{2EI}{l}(2\theta_A + \theta_B - 3\psi_{AB}) - \frac{Pab^2}{l^2} \tag{11.7}$$

$$M_{BA} = \frac{2EI}{l}(2\theta_B + \theta_A - 3\psi_{AB}) + \frac{Pa^2b}{l^2} \tag{11.8}$$

If member AB is fixed at its ends, then $\theta_A = \theta_B = \psi_{AB} = 0$, and Eqs. 11.7 and 11.8 reduce to

$$M_{AB} = FEM_{AB} = -\frac{Pab^2}{l^2} \tag{11.9}$$

$$M_{BA} = FEM_{BA} = +\frac{Pa^2b}{l^2} \tag{11.10}$$

where FEM_{AB} and FEM_{BA} are the *fixed-end moments* at ends A and B, respectively, of span AB. These are the member-end moments for the beam as a kinematically determinate primary structure, and they reflect the character of the static moment diagram associated with the loading on the member. A table inside the back cover gives the fixed-end moments for several loading conditions. Superposition can be employed to establish the fixed-end moments for combined loadings.

Equations 11.7 and 11.8 can be written in a general form for any loading as

$$M_{AB} = \frac{2EI}{l}(2\theta_A + \theta_B - 3\psi_{AB}) + FEM_{AB} \tag{11.11}$$

$$M_{BA} = \frac{2EI}{l}(2\theta_B + \theta_A - 3\psi_{AB}) + FEM_{BA} \tag{11.12}$$

In some instances, it is convenient to replace ψ_{AB} with Δ_{AB}/l, where Δ_{AB} is the transverse displacement between the member ends as shown in Fig. 11.1c.

It should be noted that for $P = 0$ and $\psi_{AB} = 0$, Eqs. 11.11 and 11.12 reduce to the member-end moment–rotation relationships given by Eq. 8.19.

Equations 11.11 and 11.12 are called the *slope deflection equations*, and they can be further generalized as a single equation in the form

$$M_{nf} = 2EK_{nf}(2\theta_n + \theta_f - 3\psi_{nf}) + FEM_{nf} \tag{11.13}$$

where the subscripts n and f refer to the near end and far end, respectively, of member nf. The quantity K_{nf} is referred to as the stiffness factor for member nf and is equal to I/l for the member.

11.3 INTERPRETATION OF THE SLOPE DEFLECTION EQUATION

The slope deflection equation shows that the moment at the end of a member is dependent on a number of quantities. It is easier to see these dependencies if Eq. 11.13 is written in the form

$$M_{nf} = \frac{4EI}{l}\theta_n + \frac{2EI}{l}\theta_f - \frac{6EI}{l^2}\Delta_{nf} + FEM_{nf} \tag{11.14}$$

In this form, it is clear that each of the first three terms involves a coupling of a stiffness term, which reflects the member characteristics E, I, and l, and a displacement quantity. The fourth term reflects the transverse loading on the member.

Figure 11.2 illustrates the physical meaning of each of the four terms in the expanded slope deflection equation. For example, Fig. 11.2b shows the moment at the near end associated with the rotation θ_n, while $\theta_f = \Delta_{nf} = FEM_{nf} = 0$. Similarly, Figs. 11.2c, d, and e show the near-end moment associated with separate inducements of θ_f, Δ_{nf}, and FEM_{nf}, respectively. The total moment on the member end is thus seen to be the superposition of these individual effects.

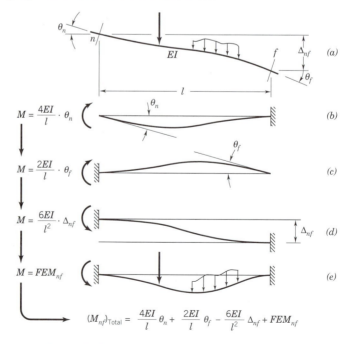

Figure 11.2 Interpretation of slope deflection equation.

11.4 SLOPE DEFLECTION METHOD FOR BEAM PROBLEMS

The slope deflection method is used to determine the end moments for statically indeterminate beams and frames. The general procedure used in solving a problem by this
method will be explained by using several example problems.

The first example is the two-span beam shown in Fig. 11.3a, which is statically indeterminate to the first degree. A qualitative sketch of the deflected structure is shown in Fig.
11.3b. This sketch aids in establishing the controlling displacement boundary conditions
and compatibility conditions. From this sketch, it is clear that the boundary conditions require that points a, b, and c are nondeflecting supports, and thus $\Delta_{ab} = \Delta_{bc} = 0$.
Also, since none of the supports provides rational restraint, there will be rotations θ_a, θ_b,
and θ_c at the three supports, respectively. Thus, the structure is kinematically indeterminate to the third degree. At point b, there is a compatibility condition that requires that θ_b
be the common rotation for the b end of both members framing into joint b.

Based on the compatibility conditions and the boundary conditions expressed earlier, the slope deflection equation (Eq. 11.13) can be written for each member end
throughout the structure. Noting that $K_{ab} = K_{ba}$ and $K_{bc} = K_{cb}$, we have

$$
\begin{aligned}
M_{ab} &= 2EK_{ab}(2\theta_a + \theta_b) + FEM_{ab} \\
M_{ba} &= 2EK_{ab}(2\theta_b + \theta_a) + FEM_{ba} \\
M_{bc} &= 2EK_{bc}(2\theta_b + \theta_c) \\
M_{cb} &= 2EK_{bc}(2\theta_c + \theta_b)
\end{aligned}
\tag{11.15}
$$

The fixed-end moments, FEM_{ab} and FEM_{ba}, depend on the loading on span ab. Since
there is no loading on span bc, $FEM_{bc} = FEM_{cb} = 0$.

A moment equilibrium equation can be written for each joint using the free-body
diagrams of the joints that are shown in Fig. 11.3c, where all member-end moments
are shown to act in the positive sense. These equations must reflect the force boundary
conditions, such as the moment-free supports at points a and c, as well as the internal
equilibrium condition at point b. Thus, these equations take the form

$$
\sum M_a = M_{ab} = 0
$$

$$
\sum M_b = M_{ba} + M_{bc} = 0
\tag{11.16}
$$

$$
\sum M_c = M_{cb} = 0
$$

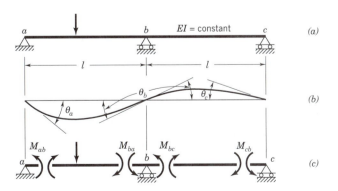

Figure 11.3 Beam problem by slope deflection method.

The equations of equilibrium can be written in terms of displacement quantities by substituting Eqs. 11.15 into Eqs. 11.16. This leads to

$$4EK_{ab}\theta_a + 2EK_{ab}\theta_b = -FEM_{ab}$$
$$2EK_{ab}\theta_a + (4EK_{ab} + 4EK_{bc})\theta_b + 2EK_{bc}\theta_c = -FEM_{ba} \qquad (11.17)$$
$$2EK_{bc}\theta_b + 4EK_{bc}\theta_c = 0$$

In matrix form, these equations take the form

$$\begin{bmatrix} 4EK_{ab} & 2EK_{ab} & 0 \\ 2EK_{ab} & (4EK_{ab} + 4EK_{bc}) & 2EK_{bc} \\ 0 & 2EK_{bc} & 4EK_{bc} \end{bmatrix} \begin{Bmatrix} \theta_a \\ \theta_b \\ \theta_c \end{Bmatrix} = \begin{Bmatrix} -FEM_{ab} \\ -FEM_{ba} \\ 0 \end{Bmatrix} \qquad (11.18)$$

The elements of the square coefficient matrix are *stiffness coefficients* (see Section 2.10) that express joint moment per unit of joint rotation. These elements depend entirely on quantities that are related to the structure, and they collectively form the so-called *structure stiffness matrix*. The vector on the right-hand side of Eq. 11.18 contains the fixed-end moment terms, and these are easily determined once the loads are specified. A few typical cases are given in the table presented inside the rear cover. Thus, Eq. 11.18 can be solved for the displacements enumerated in the displacement vector. These displacement quantities can now be substituted into Eq. 11.15 to determine the member-end moments.

If there are any support settlements, the Δ_{nf} quantity for each member must reflect these conditions. The slope deflection equations are altered accordingly, and the equations of equilibrium have additional terms in the vector on the right-hand side of Eq. 11.18. As before, Eq. 11.18 is solved for the displacements, and the member-end moments are determined from Eq. 11.13.

The problem discussed here has only three unknown displacement quantities. For large beam problems, there would be additional displacements to consider, but the general approach would be unaltered.

11.4.1 Application of the Method to Beam Problems

In reviewing the conceptual presentation of the slope deflection method in Section 11.4, it is crucial to understand the underlying logic of the step-by-step procedure. The qualitative sketch of the deflected structure revealed the governing displacement boundary conditions, the appropriate compatibility conditions, and the unknown displacement quantities. The slope deflection equation was then written for each member end in terms of these unknown displacements. Also, corresponding to each structure displacement, or kinematic degree of freedom, an equation of equilibrium was written. That is, each joint where rotational freedom existed required an equilibrium equation involving the moments acting on that joint. Thus, there were as many equations of equilibrium as there were unknown displacements. This ensured the determination of the unknown displacements through the simultaneous solution of the equilibrium equations after they had been expressed in terms of these displacements. With the displacements now known, the moment at each member end could be determined from the slope deflection equation. Two examples of beam problems follow to illustrate how the method is applied.

EXAMPLE 11.1   Determine the end moments and construct the shear and moment diagrams for the following structure. This structure was previously considered as Example 10.8. EI = constant.

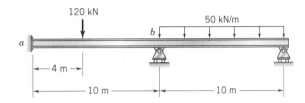

Compatibility and Boundary Conditions

Here, θ_b and θ_c represent the two kinematic degrees of freedom.

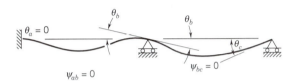

Moment Equations

$$M_{nf} = 2EK_{nf}(2\theta_n + \theta_f - 3\psi_{nf}) + FEM_{nf} \tag{11.13}$$

$$K_{ab} = K_{bc} = I/10 \text{ m} = K$$

$$FEM_{ab} = \frac{-120 \times 4 \times 6^2}{10^2} = -172.8 \text{ kN} \cdot \text{m}$$

$$FEM_{ba} = \frac{120 \times 4^2 \times 6}{10^2} = +115.2 \text{ kN} \cdot \text{m}$$

$$FEM_{bc} = \frac{-50 \times 10^2}{12} = -416.7 \text{ kN} \cdot \text{m}; \quad FEM_{cb} = +416.7 \text{ kN} \cdot \text{m}$$

$$M_{ab} = 2EK(\theta_b) - 172.8$$

$$M_{ba} = 2EK(2\theta_b) + 115.2$$

$$M_{bc} = 2EK(2\theta_b + \theta_c) - 416.7$$

$$M_{cb} = 2EK(2\theta_c + \theta_b) + 416.7$$

Equilibrium Equations

Unknown moments are assumed to act positively (clockwise) on member ends.

At joint b: $M_{ba} + M_{bc} = 0$

At joint c: $M_{cb} = 0$

Substitution of moment equations into equilibrium equations yields

$$8EK\theta_b + 2EK\theta_c = 301.5$$

$$2EK\theta_b + 4EK\theta_c = -416.7$$

In matrix form,

$$\begin{bmatrix} 8 & 2 \\ 2 & 4 \end{bmatrix} \begin{Bmatrix} EK\theta_b \\ EK\theta_c \end{Bmatrix} = \begin{Bmatrix} 301.5 \\ -416.7 \end{Bmatrix}$$

Solution for Displacements

Solution of the above equation yields

$$\begin{Bmatrix} EK\theta_b \\ EK\theta_c \end{Bmatrix} = \begin{Bmatrix} 72.8 \\ -140.6 \end{Bmatrix} \ kN \cdot m$$

Final Moments

Substitution of $EK\theta$ values into the moment equations yields the final moments

$$M_{ab} = 2(72.8) - 172.8 = -27.2 \ kN \cdot m$$

$$M_{ba} = 4(72.8) + 115.2 = +406.5 \ kN \cdot m$$

$$M_{bc} = 4(72.8) + 2(-140.6) - 416.7 = -406.5 \ kN \cdot m$$

$$M_{cb} = 4(-140.6) + 2(72.8) + 416.7 = \ 0 \ kN \cdot m$$

Shear and Moment Diagrams

The moments are shown to act on the member ends according to the slope deflection sign convention introduced at the beginning of Section 11.2. The shear and moment diagrams are then constructed according to the method of Section 5.11 using the sign convention of Section 5.3.

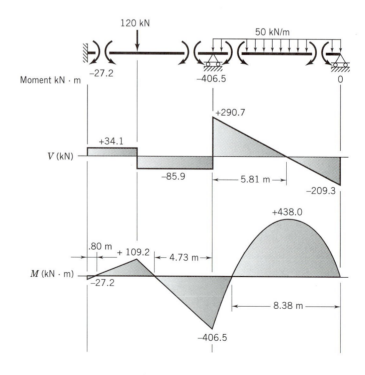

EXAMPLE 11.2

Determine the end moments for the structure of Example 11.1 if point b settles 0.03 m. Disregard the member loads given in Example 11.1.

$$E = 200 \times 10^9 \, \text{Pa} = 200 \, \text{GN/m}^2; \quad I = 2\,000 \times 10^{-6} \, \text{m}^4$$

Compatibility and Boundary Conditions

Here, θ_b and θ_c are the two kinematic degrees of freedom.

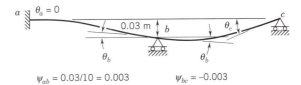

$$\psi_{ab} = 0.03/10 = 0.003 \qquad \psi_{bc} = -0.003$$

Moment Equations

All FEMs are zero (no loads are present); therefore, Eq. 11.13 reduces to

$$M_{nf} = 2EK_{nf}(2\theta_n + \theta_f) - 6EK_{nf}\psi_{nf}$$

$$K_{ab} = K_{bc} = K = \frac{2\,000 \times 10^{-6} \, \text{m}^4}{10 \, \text{m}} = 2\,000 \times 10^{-7} \, \text{m}^3$$

$$EK = 200\frac{\text{GN}}{\text{m}^2} \times 2\,000 \times 10^{-7} \, \text{m}^3 = 40 \, \text{MN} \cdot \text{m}$$

$$6EK\psi_{ab} = 6 \times 40 \, \text{MN} \cdot \text{m} \times 0.003 = 720 \, \text{kN} \cdot \text{m}; \quad 6EK\psi_{bc} = -720 \, \text{kN} \cdot \text{m}$$

$$M_{ab} = 2EK(\theta_b) - 720$$
$$M_{ba} = 2EK(2\theta_b) - 720$$
$$M_{bc} = 2EK(2\theta_b + \theta_c) + 720$$
$$M_{cb} = 2EK(2\theta_c + \theta_b) + 720$$

Equilibrium Equations

Same equilibrium equations as for Example 11.1.

$$M_{ba} + M_{bc} = 0$$
$$M_{cb} = 0$$

Substitution of moment equations into equilibrium equations yields

$$8EK\theta_b + 2EK\theta_c = 0$$
$$2EK\theta_b + 4EK\theta_c = -720$$

Solution for Displacements

$$EK\theta_b = +51.4; \quad EK\theta_c = -205.7 \quad (\text{kN} \cdot \text{m})$$

Final Moments

$$M_{ab} = 2(51.4) - 720 = -617.2 \, \text{kN} \cdot \text{m}$$
$$M_{ba} = 4(51.4) - 720 = -514.2 \, \text{kN} \cdot \text{m}$$
$$M_{bc} = 4(51.4) + 2(-205.7) + 720 = +514.2 \, \text{kN} \cdot \text{m}$$
$$M_{cb} = 4(-205.7) + 2(51.4) + 720 = 0 \, \text{kN} \cdot \text{m}$$

Moment Diagram

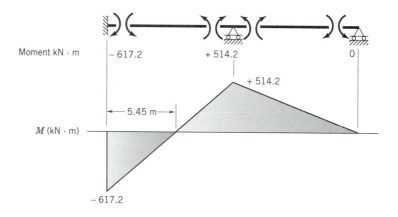

11.5 SLOPE DEFLECTION METHOD FOR FRAME PROBLEMS

The general procedure used to solve frame problems by the slope deflection method is essentially the same as with beam problems. For example, consider the frame shown in Fig. 11.4a, which is statically indeterminate to the first degree and kinematically indeterminate to the third degree.

The qualitative deflected shape of Fig. 11.4b shows that consideration of the boundary conditions and the assumption that the members are inextensible leads to $\Delta_{ab} = \Delta_{bc} = 0$. The rotations θ_a and θ_c develop at the pinned supports, and compatibility requires that the rotation θ_b be common to the top of member ab and the left end of member bc.

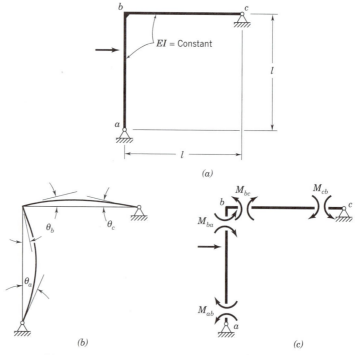

Figure 11.4 Frame problem by slope deflection method.

The remainder of this frame problem is the same as that which was presented for the beam problem in the previous section. The slope deflection equations and the equations of equilibrium, Eqs. 11.15 and 11.18, respectively, are the same as in the beam problem. Equation 11.18 is solved for the displacements, and Eqs. 11.15 yield the final end moments.

If the boundary conditions of the previous frame are changed by fixing point a against rotation and placing a roller at point c, as shown in Fig. 11.5a, the response of the structure is quite different. Again, the first step is to construct a qualitative sketch of the deflected structure that will reflect the governing boundary conditions and compatibility requirements. As seen in Fig. 11.5b, the boundary condition at point a

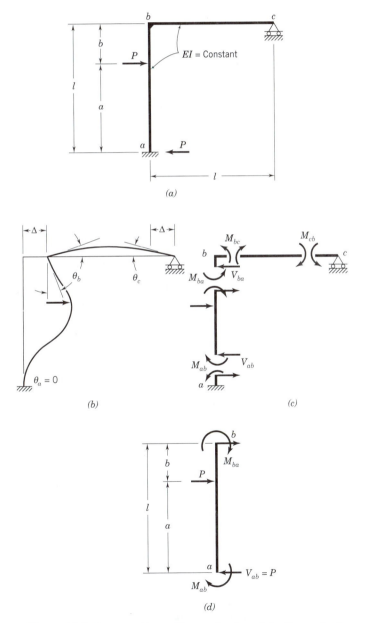

Figure 11.5 Frame problem with sway by slope deflection method.

requires that $\theta_a = 0$. In addition to the rotations θ_b and θ_c, the frame sways through the displacement Δ. Therefore, the structure is again kinematically indeterminate to the third degree. Since member extensions are ignored, Δ is the common horizontal displacement at points b and c, while these points do not displace vertically. Thus, the individual member distortions required in the slope deflection equations are $\Delta_{ab} = \Delta$ and $\Delta_{bc} = 0$. At point b, compatibility requires that θ_b be common to both members framing into joint b.

The slope deflection equation can now be written for each member end. These are

$$M_{ab} = 2EK_{ab}\left(\theta_b - 3\frac{\Delta}{l}\right) + FEM_{ab}$$

$$M_{ba} = 2EK_{ab}\left(2\theta_b - 3\frac{\Delta}{l}\right) + FEM_{ba} \qquad (11.19)$$

$$M_{bc} = 2EK_{bc}(2\theta_b + \theta_c)$$

$$M_{cb} = 2EK_{bc}(2\theta_c + \theta_b)$$

As in the previous frame problem, equilibrium equations can be written for moment equilibrium at joints b and c. These are

$$\sum M_b = M_{ba} + M_{bc} = 0$$
$$\qquad (11.20)$$
$$\sum M_c = M_{cb} = 0$$

Consideration of the free-body diagram of member ab shown in Fig. 11.5d shows that moment equilibrium about point b requires that

$$M_{ab} + M_{ba} + V_{ab}l - Pb = 0 \qquad (11.21)$$

where V_{ab} is the column shear at the base of member ab. However, equilibrium for the entire structure requires that $V_{ab} = P$. Thus, Eq. 11.21 can be written as

$$P = V_{ab} = \frac{Pb - M_{ab} - M_{ba}}{l}$$

or

$$M_{ab} + M_{ba} = Pb - Pl \qquad (11.22)$$

Equations 11.20 and 11.22 are the governing equations of equilibrium.

Substitution of Eqs. 11.19 into Eqs. 11.20 and 11.22 yields the equilibrium equations in terms of displacements.

$$(4EK_{ab} + 4EK_{bc})\theta_b + 2EK_{bc}\theta_c - 6EK_{ab}\frac{\Delta}{l} = -FEM_{ba}$$

$$2EK_{bc}\theta_b + 4EK_{bc}\theta_c = 0 \qquad (11.23)$$

$$6EK_{ab}\theta_b \qquad -12EK_{ab}\frac{\Delta}{l} = -FEM_{ab} - FEM_{ba}$$
$$+ Pb - Pl$$

Multiplication of the third of the above equations by $(-1/l)$ and substitution of the expressions for the fixed-end moments from the table inside the back cover lead to the following matrix form of the equations of equilibrium:

$$\begin{bmatrix} (4EK_{ab} + 4EK_{bc}) & 2EK_{bc} & \dfrac{-6EK_{ab}}{l} \\[3mm] 2EK_{bc} & 4EK_{bc} & 0 \\[3mm] \dfrac{-6EK_{ab}}{l} & 0 & \dfrac{12EK_{ab}}{l^2} \end{bmatrix} \begin{Bmatrix} \theta_b \\[3mm] \theta_c \\[3mm] \Delta \end{Bmatrix} = \begin{Bmatrix} \dfrac{-Pa^2b}{l^2} \\[3mm] 0 \\[3mm] \dfrac{Pa^2(l+2b)}{l^3} \end{Bmatrix} \quad (11.24)$$

The solution of Eq. 11.24 provides the displacement quantities, which, when substituted into Eqs. 11.19, yield the member-end moments.

11.5.1 Application of the Method to Frame Problems

The discussion in Section 11.4.1 regarding beam problems is relevant for frame problems as well, but other considerations are required for frames that are free to sway. In reviewing the conceptual presentation of Section 11.5, we see that once again a qualitative sketch of the deflected structure is used to identify the unknown displacements (kinematic degrees of freedom). The slope deflection equations then provide the member-end moments in terms of these displacements. There is once again an equation of equilibrium for each rotational kinematic degree of freedom and, in the case of a frame that is free to sway, there is a sway equation of equilibrium associated with each translational degree of freedom. These sway equations involve column shears and moments. The solution of the equations of equilibrium would then provide the displacements from which the member-end moments can then be determined from the slope deflection equation.

The solution for a nonsway frame problem is similar to the beam examples already presented. The following example is for a frame that is free to sway.

EXAMPLE 11.3

Determine the end moments for all members of the frame shown below and construct the moment diagrams.

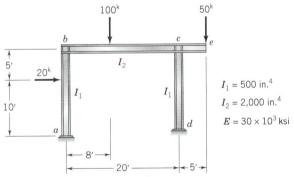

$I_1 = 500 \text{ in.}^4$

$I_2 = 2{,}000 \text{ in.}^4$

$E = 30 \times 10^3 \text{ ksi}$

Compatibility and Boundary Conditions

Here, θ_b, θ_c, and Δ identify the three kinematic degrees of freedom.

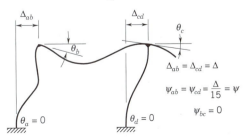

$\Delta_{ab} = \Delta_{cd} = \Delta$

$\psi_{ab} = \psi_{cd} = \dfrac{\Delta}{15} = \psi$

$\psi_{bc} = 0$

Moment Equations

$$M_{nf} = 2EK_{nf}(2\theta_n + \theta_f - 3\psi_{nf}) + FEM_{nf} \tag{11.13}$$

$$K_{ab} = K_{cd} = I_1/15 = K; \quad K_{bc} = I_2/20 = 4I_1/20 = 3K$$

$$FEM_{ab} = \frac{-20 \times 10 \times 5^2}{15^2} = -22.2'^{-k}$$

$$FEM_{ba} = \frac{+20 \times 10^2 \times 5}{15^2} = +44.4'^{-k}$$

$$FEM_{bc} = \frac{-100 \times 8 \times 12^2}{20^2} = -288'^{-k}$$

$$FEM_{cb} = \frac{+100 \times 8^2 \times 12}{20^2} = +192'^{-k}$$

$$FEM_{cd} = FEM_{dc} = 0$$

$$M_{ce} = -(50 \times 5) = -250'^{-k} \quad \text{(from statics)}$$

$$M_{ab} = 2EK(\theta_b - 3\psi) - 22.2$$

$$M_{ba} = 2EK(2\theta_b - 3\psi) + 44.4$$

$$M_{bc} = 2E \cdot 3K(2\theta_b + \theta_c) - 288$$

$$M_{cb} = 2E \cdot 3K(2\theta_c + \theta_b) + 192$$

$$M_{cd} = 2EK(2\theta_c - 3\psi)$$

$$M_{dc} = 2EK(\theta_c - 3\psi)$$

Equilibrium Equations

At joint b:

$$M_{ba} + M_{bc} = 0$$

At joint c:

$$M_{cb} + M_{cd} - 250 = 0$$

Columns ab and cd:

$$H_a = -\left(\frac{M_{ba} + M_{ab} - 100}{15}\right)$$

$$H_d = -\left(\frac{M_{cd} + M_{dc}}{15}\right)$$

but from consideration of the entire structure,

$$H_a + H_d = 20$$

$$\therefore M_{ab} + M_{ba} + M_{cd} + M_{dc} = -200$$

Substitution of moment equations into the equilibrium equations yields

$$16EK\theta_b + 6EK\theta_c - 6EK\psi = 243.6$$

$$6EK\theta_b + 16EK\theta_c - 6EK\psi = 58.0$$

$$6EK\theta_b + 6EK\theta_c - 24EK\psi = -222.2$$

In matrix form,

$$\begin{bmatrix} 16 & 6 & -6 \\ 6 & 16 & -6 \\ -6 & -6 & 24 \end{bmatrix} \begin{Bmatrix} EK\theta_b \\ EK\theta_c \\ EK\psi \end{Bmatrix} = \begin{Bmatrix} 243.6 \\ 58.0 \\ 222.2 \end{Bmatrix}$$

Solution for Displacements

Solution of the above equations yields

$$\begin{Bmatrix} EK\theta_b \\ EK\theta_c \\ EK\psi \end{Bmatrix} = \begin{Bmatrix} +20.14 \\ +1.58 \\ +14.69 \end{Bmatrix} \text{ ft-k}$$

Final Moments

Substitution of $EK\theta_b$, $EK\theta_c$, and $EK\psi$ values into the moment equations gives the final moments.

$$M_{ab} = 2(20.14) - 6(14.69) - 22.2 = -70.06'^{-k}$$

$$M_{ba} = 4(20.14) - 6(14.69) + 44.4 = +36.82'^{-k}$$

$$M_{bc} = 12(20.14) + 6(1.58) - 288 = -36.84'^{-k}$$

$$M_{cb} = 12(1.58) + 6(20.14) + 192 = +331.80'^{-k}$$

$$M_{cd} = 4(1.58) - 6(14.69) = -81.82'^{-k}$$

$$M_{dc} = 2(1.58) - 6(14.69) = -84.98'^{-k}$$

$$M_{ce} = -250'^{-k} \quad \text{(from statics)}$$

Moment Diagrams

(Columns viewed from right side.)

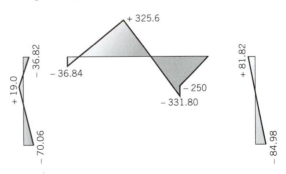

Moments in ft-k—detailed construction of M diagrams is not shown here. Follow the method of Example 5.11.

11.6 TEMPERATURE CHANGE

In defining the fixed-end moments in Eqs. 11.9 and 11.10, it became clear that these are member-end moments that are induced through the application of transverse forces on a member that is fixed at its ends. However, the application of transverse forces is not the only action that can introduce fixed-end moments. For instance, consider the temperature gradient shown in Fig. 11.6a. As was shown in Eq. 10.28, this gradient introduces a constant curvature of $\alpha \, \Delta T / h$, as depicted in Fig. 11.6b, where α is the coefficient of thermal expansion, h is the member depth, and ΔT is the difference in temperature between the bottom and top beam fibers.

Returning to Eqs. 11.1 and 11.2 and replacing the load terms with those associated with the temperature-induced curvature ($d\theta / dx$ for temperature problem plays the same role as M / EI for loading case), we obtain

$$\Delta_A^B = -\frac{M_{AB} l^2}{6EI} + \frac{M_{BA} l^2}{3EI} - \frac{\alpha \Delta T l^2}{2h} \tag{11.25}$$

$$\Delta_B^A = \frac{M_{AB} l^2}{3EI} - \frac{M_{BA} l^2}{6EI} + \frac{\alpha \Delta T l^2}{2h} \tag{11.26}$$

Proceeding through the detailed derivation as before, we obtain, according to the notation introduced in Eq. 11.13,

$$FEM_{nf} = -EI\alpha \, \Delta T / h \tag{11.27}$$

$$FEM_{fn} = EI\alpha \, \Delta T / h$$

Axial deformations associated with temperature variations induce length changes if the element is not restrained axially. However, if the element is restrained, axial forces are induced. In beam-type problems this is readily included; however, for frame-type problems the basic force–deformation relation should be broadened to include axial effects. These effects are touched on in Example 11.4 and are treated more fully in Chapter 14.

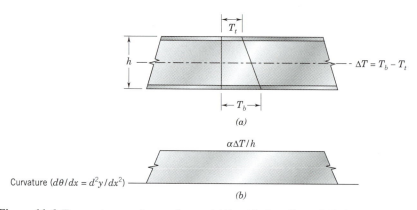

Figure 11.6 Temperature gradient on beam. *(a)* Specified gradient. *(b)* Induced curvature.

EXAMPLE 11.4

Determine the end moments that result from the indicated temperature gradient on the beam structure shown.

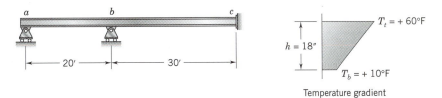

$$E = 30 \times 10^3 \text{ ksi}; \quad I = 1{,}500 \text{ in.}^4; \quad EI = 45 \times 10^6 \text{ k-in.}^2$$
$$\alpha = 0.000{,}006{,}5/°F; \quad \Delta T = T_b - T_t = 10 - 60 = -50°F$$

Compatibility and Boundary Conditions

Here, θ_a and θ_b represent the two kinematic degrees of freedom.

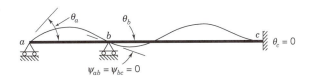

Moment Equations

$$M_{nf} = 2EK_{nf}(2\theta_n + \theta_f - 3\psi_{nf}) + FEM_{nf} \tag{11.13}$$
$$K_{ab} = I/20 = K; \quad K_{bc} = I/30 = 0.67K$$

$$FEM_{ab} = FEM_{bc} = -\frac{EI\alpha\Delta T}{h} \tag{11.27}$$

$$= -\frac{(45 \times 10^6)(6.5 \times 10^{-6})(-50)}{18}$$

$$= +812.5 \text{ in.-kips} = +66.7 \text{ ft-kips}$$

$$FEM_{ba} = FEM_{cb} = -66.7 \text{ ft-kips}$$
$$M_{ab} = 2EK(2\theta_a + \theta_b) + 66.7$$
$$M_{ba} = 2EK(2\theta_b + \theta_a) - 66.7$$
$$M_{bc} = 2E(0.67K)(2\theta_b) + 66.7$$
$$M_{cb} = 2E(0.67K)(\theta_b) - 66.7$$

Equilibrium Equations

$$M_{ab} = 0$$
$$M_{ba} + M_{bc} = 0$$

Substitution of the moment equations into the equilibrium equations yields

$$4EK\theta_a + 2EK\theta_b = -66.7$$
$$2EK\theta_a + 6.67EK\theta_b = 0$$

Solution for Displacements

Solution of the above equations yields

$$EK\theta_a = -19.61; \quad EK\theta_b = +5.89 \quad \text{(ft-kips)}$$

Final Moments

Substitution of the $EK\theta$ values into the moment equations gives the final moments

$$M_{ab} = 4(-19.61) + 2(5.89) + 66.7 = 0'^{-k}$$

$$M_{ba} = 4(5.89) + 2(-19.61) - 66.7 = -82.4'^{-k}$$

$$M_{bc} = 2.67(5.89) + 66.7 = +82.4'^{-k}$$

$$M_{cb} = 1.33(5.89) - 66.7 = -58.9'^{-k}$$

Moment Diagram and Deflected Structure

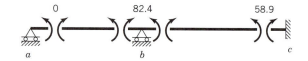

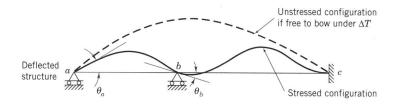

Notes:

1. The average temperature variation over the depth of the cross section is $+35°F$. Since this structure is not restrained against axial deformations, the beam will elongate by an amount given by

$$\Delta L = \alpha L T_{avg} = (0.000,006,5)(50 \times 12)(35) = 0.137 \text{ inch}$$

2. If the structure were restrained against horizontal movement at point a, then an axial compression force would be induced of magnitude

$$F = EA\alpha T_{avg} = (30 \times 10^3)A(0.000,006,5)(35) = 6.83A \text{ kips}$$

where A is the cross-sectional area in square inches.

3. It is evident that this axial force can be significant. This force, coupled with the beam deflections, will affect the bending moments along the beam. These are considered to be higher-order effects.

11.7 PROBLEM VARIATIONS

This section covers a few variations in the problems already discussed. As a first example, the two-story frame shown in Fig. 11.7a is considered. For this structure, there are six unknown displacements, as shown in Fig. 11.7b.

The slope deflection method requires that the slope deflection equation be written for each member end in the structure. These equations give the end moments in terms of θ_b, θ_c, θ_d, θ_e, $\psi_{ab} = \psi_{fe} = \Delta_1/h_1$, and $\psi_{bc} = \psi_{ed} = (\Delta_2 - \Delta_1)/h_2$. The solution requires six equilibrium equations—four of these involve moment equilibrium at the joints where rotation occurs and two are related to shear equilibrium for each story that is free to sway. The first sway equation requires that the base shears in columns ab and fe equal to 30^k, and the second sway equation requires that the base shears in columns bc and de are equal to 10^k.

As a second example, consider the bent with sloping legs shown in Fig. 11.8a. The deflected structure shown in Fig. 11.8b shows that there are three unknown displacement quantities.

A solution by the slope deflection method requires that a ψ term be evaluated for each member. As the frame sways laterally through the displacement Δ, each member experiences relative end displacements. Because axial distortions are ignored in the slope deflection method, the top of each column must displace normal to its original direction through a horizontal displacement of Δ, as shown in Fig. 11.8b. Thus, the shaded triangles at points b and c are similar to the triangles formed by the sloping columns and their projections. These shaded triangles are shown in an enlarged view in Fig. 11.8c, from which the following ψ terms are determined:

$$\psi_{ab} = \frac{\Delta_{ab}}{l_1} = \frac{(l_1/h)\cdot\Delta}{l_1} = \frac{\Delta}{h}$$

$$\psi_{cd} = \frac{\Delta_{cd}}{l_2} = \frac{(l_2/h)\cdot\Delta}{l_2} = \frac{\Delta}{h} \tag{11.28}$$

$$\psi_{bc} = \frac{\Delta_{bc}}{l} = \frac{[(r_1+r_2)/h]\Delta}{l} = \frac{r_1+r_2}{lh}\Delta \quad \text{(negative)}$$

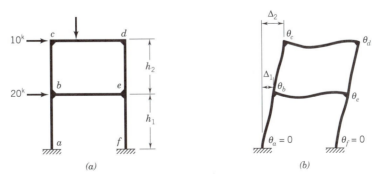

(a)   (b)

Figure 11.7 Two-story frame.

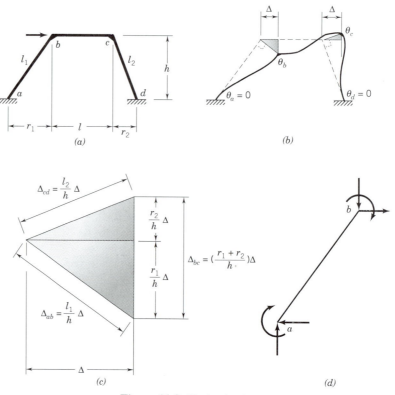

Figure 11.8 Sloping leg bent.

It is thus seen that all three ψ terms are a function of the sway displacement Δ.

Three equations of equilibrium are required for the solution of the three displacements. Two of these equations stem from moment equilibrium at joints b and c, and the third equation requires that the horizontal reactions at points a and d satisfy the requirements of statics for the entire structure.

This latter equation results from taking a free-body diagram of each column and expressing the horizontal base reaction in terms of the end moments on the column and the vertical load carried by the column. Figure 11.8d shows such a free-body diagram for column ab. The vertical load on the column is determined from the end shear on member bc.

11.8 EQUILIBRIUM EQUATIONS BY ENERGY METHODS

Castigliano's first theorem provides an alternative procedure for generating the equations of equilibrium. This theorem was presented in mathematical form as Eq. 2.57, which is repeated here as

$$\frac{\partial U}{\partial \Delta_i} = P_i \tag{11.29}$$

where U is the total strain energy stored in the system, P_i is the ith externally applied load, and Δ_i is the corresponding displacement.

For a beam- or frame-type structure that is composed of m members, and for which the strain energy results from flexure, it is shown in Section 8.9 that

$$U = \sum_{j=1}^{m} \frac{1}{2E} \int_{x=0}^{x=l_j} \frac{M_j^2}{I_j} dx \tag{11.30}$$

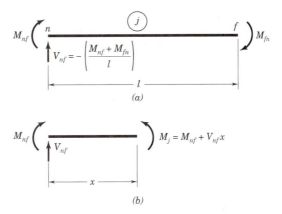

Figure 11.9 Typical *j*th member. *(a)* End moments and associated left-end shear. *(b)* Bending moment at distance *x* along *j*th member.

where M_j is the bending moment at distance x along the jth member whose length is l_j, I_j is the moment of inertia, and E is the modulus of elasticity.

In the current context, only joint loaded structures are considered. For this case, a typical jth member connecting points n and f is shown in Fig. 11.9a together with its end moments and associated left-end shear. Figure 11.9b shows that the moment at distance x from end n is given by

$$M_j = M_{nf} + V_{nf}x \tag{11.31}$$

Using Eq. 11.13 and noting that $FEM_{nf} = 0$ (no member loads), we have

$$M_{nf} = 2EK_{nf}(2\theta_n + \theta_f - 3\psi_{nf}) \tag{11.32}$$

where θ_n and θ_f are the joint rotations at ends n and f, respectively, and ψ_{nf} is the rotation of the chord connecting the member ends. Further, applying Eq. 11.13 to both member ends in conjunction with the equilibrium requirement for the left-end shear shown in Fig. 11.9a, we find that

$$
\begin{aligned}
V_{nf} &= -\left[\frac{M_{nf} + M_{fn}}{l}\right] \\
&= -\frac{6EK_{nf}\theta_n}{l} - \frac{6EK_{nf}\theta_f}{l} + \frac{12EK_{nf}\psi_{nf}}{l}
\end{aligned}
\tag{11.33}
$$

Substitution of Eqs. 11.32 and 11.33 into Eq. 11.31, and the resulting expression into Eq. 11.30, and completion of the indicated integration yields

$$U = \sum_{j=1}^{m} [2EK_{nf}(\theta_n^2 + \theta_f^2 + \theta_n\theta_f - 3\theta_n\psi_{nf} - 3\theta_f\psi_{nf} + 3\psi_{nf}^2)]_j \tag{11.34}$$

When member loads are included, the preceding derivation is complicated by the need to include the effects of these loads in Eq. 11.31. This procedure is straightforward, but it will not be treated here.

For the jth member, compatibility equations are required to express the member displacements (θ_n, θ_f, and ψ_{nf}) in terms of the structure displacements, Δ_i, to accommodate the differentiation required in Eq. 11.29.

EXAMPLE 11.5

Use Castigliano's first theorem to generate the equations of joint equilibrium for the structure shown below. Consider flexural effects only.

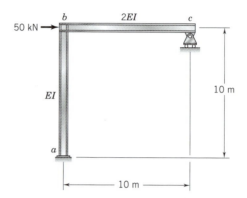

Coordinate Systems

Structure

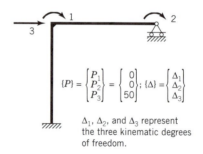

$$\{P\} = \begin{Bmatrix} P_1 \\ P_2 \\ P_3 \end{Bmatrix} = \begin{Bmatrix} 0 \\ 0 \\ 50 \end{Bmatrix}; \{\Delta\} = \begin{Bmatrix} \Delta_1 \\ \Delta_2 \\ \Delta_3 \end{Bmatrix}$$

Δ_1, Δ_2, and Δ_3 represent the three kinematic degrees of freedom.

Member

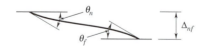

Member	n	f
ab	a	b
bc	b	c

Note: Rotations are positive when clockwise for both structure and member systems in accordance with assumptions in derivation of Eqs. 11.13 and 11.34.

Compatibility and Boundary Conditions

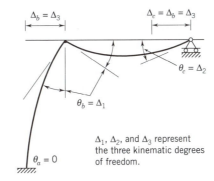

Δ_1, Δ_2, and Δ_3 represent the three kinematic degrees of freedom.

Strain Energy

$$U = \sum_{j=1}^{m} [2EK_{nf}\,(\theta_n^2 + \theta_f^2 + \theta_n\theta_f - 3\theta_n\psi_{nf} - 3\theta_f\psi_{nf} + 3\psi_{nf}^2)]_j \qquad (11.34)$$

$$K_{ab} = \frac{I}{10} = K; \quad K_{bc} = \frac{2I}{10} = 2K$$

Member ab

$$\theta_n = \theta_a = 0; \quad \theta_f = \theta_b = \Delta_1; \quad \psi_{nf} = \frac{\Delta_b}{10} = \frac{\Delta_3}{10}$$

$$U_{ab} = 2EK\left[\Delta_1^2 - 3\Delta_1\left(\frac{\Delta_3}{10}\right) + 3\left(\frac{\Delta_3}{10}\right)^2\right]$$

Member bc

$$\theta_n = \theta_b = \Delta_1; \quad \theta_f = \theta_c = \Delta_2; \quad \psi_{nf} = 0$$

$$U_{bc} = 2E(2K)[\Delta_1^2 + \Delta_2^2 + \Delta_1\Delta_2]$$

The total strain energy is, therefore,

$$U = U_{ab} + U_{bc} = 2EK[3\Delta_1^2 + 2\Delta_2^2 + 2\Delta_1\Delta_2 - 0.3\Delta_1\Delta_3 + 0.03\Delta_3^2]$$

Equations of Equilibrium

$$\frac{\partial U}{\partial \Delta_i} = P_i \qquad (11.29)$$

$$\frac{\partial U}{\partial \Delta_1} = 2EK[6\Delta_1 + 2\Delta_2 - 0.3\Delta_3] = P_1 = 0$$

$$\frac{\partial U}{\partial \Delta_2} = 2EK[2\Delta_1 + 4\Delta_2] = P_2 = 0$$

$$\frac{\partial U}{\partial \Delta_3} = 2EK[-0.3\Delta_1 + 0.06\Delta_3] = P_3 = 50$$

Or, in matrix form

$$EK\begin{bmatrix} 12 & 4 & -0.6 \\ 4 & 8 & 0 \\ -0.6 & 0 & 0.12 \end{bmatrix} \begin{Bmatrix} \Delta_1 \\ \Delta_2 \\ \Delta_3 \end{Bmatrix} = \begin{Bmatrix} 0 \\ 0 \\ 50 \end{Bmatrix}$$

11.9 SUGGESTED PROBLEMS

11.1 through 11.34 Analyze the given statically indeterminate structures by the slope deflection method and construct the shear and moment diagrams for all members. Take *EI* as a constant for all problems.

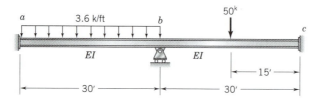

11.1

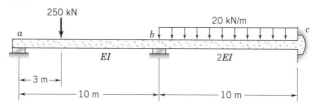

11.2

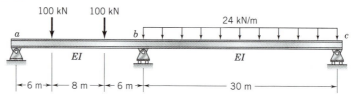

11.3

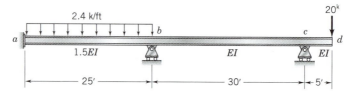

11.4

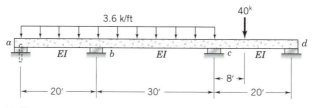

11.5

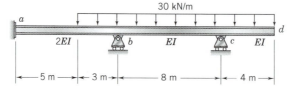

11.6

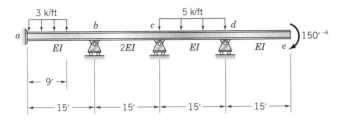

11.7

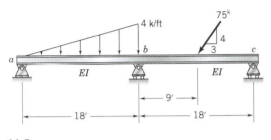

11.8

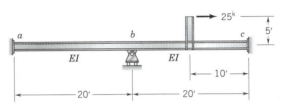

11.9

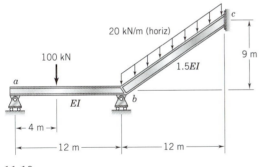

11.10

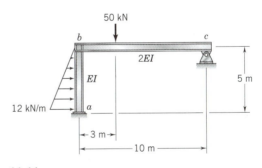

11.11

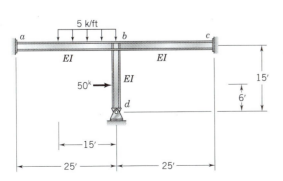

11.12

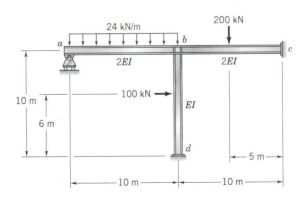

11.13

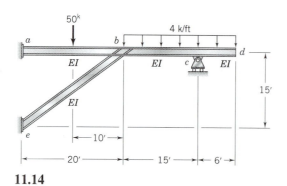

11.14

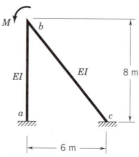

11.15

11.16 The structure of Problem 4 with a settlement of 1.5 inches at support b if $E = 30 \times 10^3$ ksi and $I = 1,000$ in.^{4} Disregard the loads of Problem 4.

11.17 The structure and loading of Problem 8 with settlements at support a and support c of 0.75 and 0.35 inch, respectively, if $E = 30 \times 10^3$ ksi and $I = 1,200$ in.^{4}

11.18 The structure of Problem 10 if the support at point c settles 15 mm downward and rotates 0.0007 radian counterclockwise. Take $EI = 500\ 000$ kN · m^2.

11.19 The structure of Problem 13 with a clockwise rotation of 0.001 radian at support c and a settlement of 0.03 m at support d if $E = 200 \times 10^9$ Pa and $I = 1$ 000×10^{-6} m^4. Disregard the loads of Problem 13.

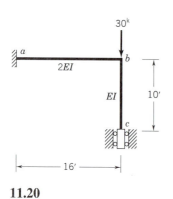

11.20

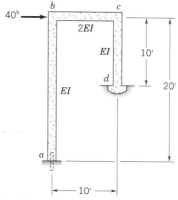

11.21

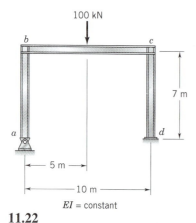

11.22

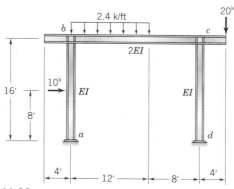

11.23

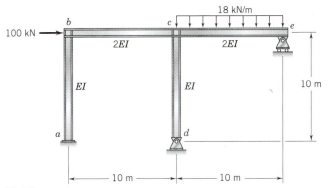

11.24

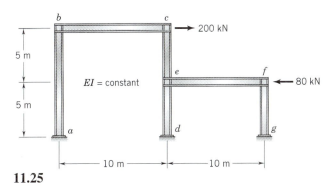

11.25

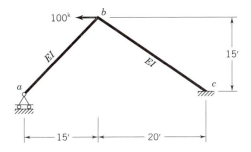

11.26

11.27

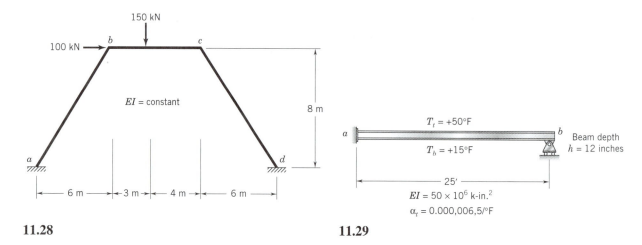

11.28

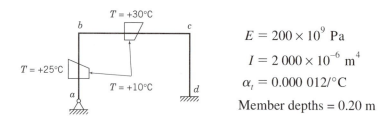

11.29

11.30 The structure of Problem 21 if, in addition to the loads, point a shifts 0.5 inch downward and 0.35 inch to the right. Take $EI = 35 \times 10^6$ k-in.^{2}.

11.31 The structure of Problem 1 with the following temperature gradients:
(a) $T_t = -40$ °F, $T_b = -10$ °F for span ab only.
(b) $T_t = -40$ °F, $T_b = -10$ °F for both spans.
For both cases, $E = 30 \times 10^6$ psi, $I = 1,500$ in.^{4}, $\alpha_t = 0.000,006,5/$°F, and $h = 15$ in.

11.32 The structure of Problem 10 with temperature changes for span bc of $T_t = +30$ °C and $T_b = +5$ °C. Take $I = 2\,500 \times 10^{-6}$ m^4, $E = 200$ GPa, $\alpha_t = 0.000\,012/$°C, and $h = 0.25$ m.

11.33 The structure of Problem 22 with the following temperature changes for members ab and bc:

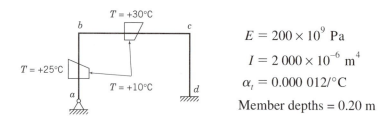

$E = 200 \times 10^9$ Pa

$I = 2\,000 \times 10^{-6}$ m^4

$\alpha_t = 0.000\,012/$°C

Member depths = 0.20 m

11.34 The structure of Problem 12 with the following temperature changes for span ab:

$$T_t = -50°\text{F}, \quad T_b = -15°\text{F}$$

Disregard the loads given in Problem 12, and take $EI = 50 \times 10^6$ k-in.^{2}, $\alpha_t = 0.000,006,5/$°F, and $h = 18$ in.

11.35 through 11.38 Use Castigliano's first theorem to establish the equations of joint equilibrium for the structures and loadings specified. Use the structure coordinate system shown in each case, and consider flexural effects only.

11.35 The structure and loading of Problem 20.

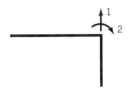

11.36 The structure and loading of Problem 15.

11.37 The structure and loading of Problem 25.

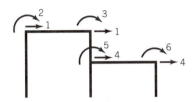

11.38 The structure and loading of Problem 27.

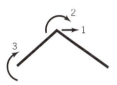

Chapter 12

Moment Distribution Method

The Akashi Kaikyo Bridge, Japan (courtesy Japan National Tourist Organization).

12.1 SOLUTION TECHNIQUES FOR EQUILIBRIUM METHODS

In Chapter 11, equilibrium methods were considered. For both the slope deflection and energy methods for beam and frame structures, a set of equilibrium equations resulted that was expressed in terms of displacement quantities. In solving these equations, the analyst is seeking the set of displacements that simultaneously satisfies all of the equilibrium conditions.

The methods for solving the equilibrium equations for the desired deformations were not discussed in Chapter 11; it was merely implied that a solution existed. In essence, the solution process requires inverting the stiffness matrix, or solving the associated simultaneous equations. For small problems, where there are only a few unknowns, simple methods of elimination or Cramer's rule can be applied. For large problems, numerous techniques are available for reaching a solution. Some of these methods are direct in that they produce the correct results after applying the appropriate algorithm. Other methods are iterative; that is, they are based on a step-by-step procedure. The initial step gives an approximation to the solution, and each subsequent step acts to refine the solution. The analyst can stop whenever the desired degree of accuracy is achieved.

In the preceding discussion, it was implied that the complete set of equilibrium equations is generated, and the solution centers on the problem of matrix inversion. However, there are iterative methods that lead to a solution without actually developing the complete set of equilibrium equations. It is this type of iterative solution for beam and frame structures that is considered in this chapter.

12.2 ITERATIVE METHODS

As stated in the previous section, an iterative solution results from the application of a step-by-step procedure in which each step acts to refine the results from the previous step. In many cases, it can be shown that the iterative process converges to the correct solution if enough steps are taken. However, the analyst has the option of terminating the procedure when the desired accuracy has been achieved.

To illustrate an iterative solution, consider the continuous beam problem of Example 11.1. This beam is shown again in Fig. 12.1 along with its deflected structure, and the governing equations of equilibrium, as formulated in Example 11.1, are

$$8EK\theta_b + 2EK\theta_c = 301.5 \tag{12.1}$$

$$2EK\theta_b + 4EK\theta_c = -416.7 \tag{12.2}$$

The solution requires a set of joint rotations, θ_b and θ_c, that will simultaneously satisfy the equilibrium equations.

The iterative technique that will be employed requires that each of the equilibrium equations be used to solve for one deformation quantity based on an assumed value for the other deformation quantity. For convenience, Eqs. 12.1 and 12.2 are rewritten as follows:

$$EK\theta_b = 37.7 - 0.25(EK\theta_c) \tag{12.3}$$

$$EK\theta_c = -104.2 - 0.5(EK\theta_b) \tag{12.4}$$

As an initial step, it is assumed that $EK\theta_c = 0$ and Eq. 12.3 is used to solve for $EK\theta_b = 37.7$. Using $EK\theta_b = 37.7$, we then use Eq. 12.4 to solve for $EK\theta_c = -123.1$. This new value for $EK\theta_c$ is now used in Eq. 12.3 to obtain a new value for $EK\theta_b$. The

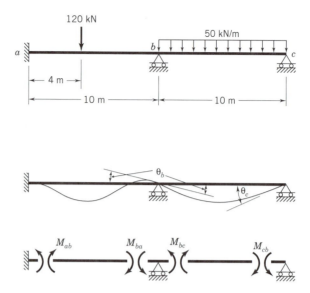

Figure 12.1 Example problem for iterative solution.

procedure continues, as summarized in Table 12.1, until the same $EK\theta$ values satisfy both equilibrium equations. The final results of $EK\theta_b = 72.8$ and $EK\theta_c = -140.6$ agree with those determined in Example 11.1.

The same procedure can be used when there are more than two unknowns. For instance, if there are n unknown deformations, each of the n equations should be solved for one of the unknowns in terms of the remaining $(n-1)$ deformation quantities. Then each equation is used, in turn, to determine the value of one deformation quantity from the latest values of the other $(n-1)$ deformations. The process continues until a consistent set of deformation quantities is obtained that satisfies all of the governing equilibrium equations.

The procedure outlined here is the so-called *Gauss–Seidel iteration method.* There are, of course, other techniques that can be employed for an iterative solution, and there are procedures that will hasten convergence. It will be seen in subsequent sections that the technique illustrated here has a convenient physical interpretation that attaches a special value to it.

Table 12.1 Example Iterative Solution

		Assumed Value		Solve for	
Step	Eq. No.	$EK\theta_b$	$EK\theta_c$	$EK\theta_b$	$EK\theta_c$
1	12.3	—	0	37.7	—
2	12.4	37.7	—	—	−123.1
3	12.3	—	−123.1	68.5	—
4	12.4	68.5	—	—	−138.5
5	12.3	—	−138.5	72.3	—
6	12.4	72.3	—	—	−140.4
7	12.3	—	−140.4	72.8	—
8	12.4	72.8	—	—	−140.6
9	12.3	—	−140.6	72.8	—

12.3 PHYSICAL INTERPRETATION OF ITERATIVE SOLUTIONS

Each step of the iterative process developed in Section 12.2 has a physical significance. To illustrate this point, we will recall some of the equations used in Example 11.1. The moment equations are as follow:

$$M_{ab} = 2(EK\theta_b) - 172.8$$

$$M_{ba} = 4(EK\theta_b) + 115.2$$

$$M_{bc} = 4(EK\theta_b) + 2(EK\theta_c) - 416.7 \tag{12.5}$$

$$M_{cb} = 4(EK\theta_c) + 2(EK\theta_b) + 416.7$$

where these are member-end moments, as shown in Fig. 12.1, and the equilibrium equations are

$$M_{ba} + M_{bc} = 0 \tag{12.6}$$

$$M_{cb} = 0 \tag{12.7}$$

Table 12.2 gives the values for $EK\theta_b$ and $EK\theta_c$ for each step of the iteration process as determined in Table 12.1. Also shown in Table 12.2 are the moments corresponding to each step as determined by Eqs. 12.5 with the appropriate $EK\theta$ values. An examination of the joint rotations and moments in Table 12.2 reveals the following:

- Initially, neither Eq. 12.6 nor Eq. 12.7 is satisfied. That is, $M_{ba} \neq -M_{bc}$ and $M_{cb} \neq 0$.
- From the initial step to Step 1, joint b is allowed to rotate while joint c remains clamped. In the process, Eq. 12.6 ($M_{ba} = -M_{bc}$) is satisfied[1] while Eq. 12.7 remains unsatisfied.
- From Step 1 to Step 2, joint b is clamped in a rotated position while joint c is permitted to rotate. As a result, Eq. 12.7 ($M_{bc} = 0$) is satisfied,[1] and Eq. 12.6 is not satisfied.

Table 12.2   Moments for Iterative Solution

Step	$EK\theta_b$	$EK\theta_c$	M_{ab}	M_{ba}	M_{bc}	M_{cb}
Initial	0	0	− 172.8	+ 115.2	− 416.7	+ 416.7
1	37.7	0	− 97.4	+ 266.0	− 265.9	+ 492.1
2	37.7	− 123.1	− 97.4	+ 266.0	− 512.1	− 0.3
3	68.5	− 123.1	− 35.8	+ 389.2	− 388.9	+ 61.3
4	68.5	− 138.5	− 35.8	+ 389.2	− 419.7	− 0.3
5	72.3	− 138.5	− 28.2	+ 404.4	− 404.5	+ 7.3
6	72.3	−140.4	− 28.2	+ 404.4	− 408.3	− 0.3
7	72.8	−140.4	− 27.2	+ 406.4	− 406.3	+ 0.7
8, 9	72.8	− 140.6	− 27.2	+ 406.4	− 406.7	− 0.1

[1]Values of appropriate moments are enclosed within broken lines of Table 12.2. Satisfaction is within a few tenths of a kN · m. Exact satisfaction is not achieved because of round-off error.

- From Step 2 to Step 3, joint c is clamped in its rotated position while joint b is again permitted to rotate. This action satisfies[2] Eq. 12.6 ($M_{ba} = -M_{bc}$), but Eq. 12.7 is, again, not satisfied.

This pattern continues: At each step, one of the equilibrium conditions is satisfied, and one is not. In going to the next step, the joint where equilibrium is not satisfied is allowed to rotate; this produces equilibrium at this joint, but disturbs equilibrium at the other. The process continues until both joints are in equilibrium, at which time the correct moments are obtained.

This process of alternatively clamping and releasing joints until equilibrium is achieved at each joint is the physical process that is followed according to the iterative solution. This suggests a physical analog to the iterative procedure. Such a technique is called the *moment distribution method*, and it is developed in the next section.

12.4   MOMENT DISTRIBUTION METHOD FOR BEAM PROBLEMS

When the iterative-type solution outlined in the previous sections is carried out with emphasis on the physical process of alternatively clamping and releasing joints until equilibrium is achieved, it is known as the *moment distribution method*. This method was developed by Professor Hardy Cross, who taught it to his students at the University of Illinois in the late 1920s. The method was formally presented by Cross in published form in 1930. In the 1930s, the moment distribution method emerged as the dominant method for frame analysis, and it enjoyed this position through the 1950s. With the advent of the digital computer, direct solutions became feasible. Moment distribution remains an important method for hand solutions for small problems.

To illustrate the essential features of the moment distribution method, consider the two-span continuous beam shown in Fig. 12.2a. This beam is clamped at points a and c, but is free to rotate at point b. However, it is initially assumed that joint b is also clamped, and thus the loading on span ab causes fixed-end moments at ends a and b as shown in Fig. 12.2b. If joint b is released, it will rotate under the unbalanced moment M_b. As rotation occurs, end moments develop at the b end of members ab and bc. Rotation continues as shown in Fig. 12.2c until the summation of the end moments is equal and opposite to the unbalanced moment M_b. That is,

$$M'_{ba} + M'_{bc} = -M_b \tag{12.8}$$

where M'_{ba} and M'_{bc} are the end moments that develop as a result of the rotation θ_b at joint b.

The fundamental slope deflection equation, Eq. 11.13, can be used to express the end moments that develop from member-end rotations in the form

$$M'_{nf} = 2EK_{nf}(2\theta_n + \theta_f) \tag{12.9}$$

Since $\theta_a = \theta_c = 0$, the moments that develop at the b end of members ab and bc as joint b rotates through the angle θ_b are given by

$$M'_{ba} = 2EK_{ba}(2\theta_b) = 4EK_{ba}\theta_b$$

$$M'_{bc} = 2EK_{bc}(2\theta_b) = 4EK_{bc}\theta_b \tag{12.10}$$

---

[2]Values of appropriate moments are enclosed within broken lines of Table 12.2. Satisfaction is within a few tenths of a kN · m. Exact satisfaction is not achieved because of round-off error.

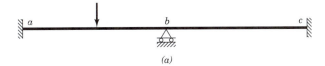

(a)

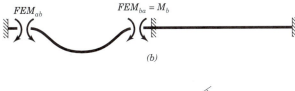

(b)

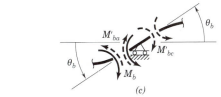

(c)

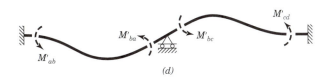

(d)

Figure 12.2 Fundamentals of moment distribution.

Substitution of Eqs. 12.10 into Eq. 12.8 yields

$$4E\theta_b(K_{ba} + K_{bc}) = -M_b \tag{12.11}$$

from which

$$\theta_b = \frac{-M_b}{4E(K_{ba} + K_{bc})}$$

Substituting this expression for θ_b into Eqs. 12.10, we obtain

$$M'_{ba} = -M_b\left(\frac{K_{ba}}{K_{ba} + K_{bc}}\right) \tag{12.12}$$

and

$$M'_{bc} = -M_b\left(\frac{K_{bc}}{K_{ba} + K_{bc}}\right) \tag{12.13}$$

or, in general,

$$M'_{bi} = -M_b\left(\frac{K_{bi}}{\displaystyle\sum_j K_{bj}}\right) \tag{12.14}$$

where i is the far end of a specific member framing into joint b, and j sums over all members connected to joint b. Equation 12.14 can be rewritten as

$$M'_{bi} = -M_b \cdot DF_{bi} \qquad (12.15)$$

where

$$DF_{bi} = \frac{K_{bi}}{\sum_j K_{bj}} \qquad (12.16)$$

The quantity DF_{bi} relates to the stiffness (a measure of flexural stiffness as defined in Section 8.2—referred to as stiffness factor when used in Eq. 11.13) of member bi to the summation of the stiffnesses of all members framing into joint b. This quantity is called the *distribution factor* for member bi, and it indicates the portion of the unbalanced moment M_b that is distributed to member bi as joint b is allowed to rotate. The negative sign on the right-hand side of Eq. 12.15 simply indicates that the moment M'_{bi} acts in opposition to the unbalanced moment M_b.

As joint b rotates and moments develop at the b end of all members framing into joint b, moments also develop at points a and c as shown in Fig. 12.2d. Using Eq. 12.9 and recalling that $\theta_a = \theta_c = 0$, we get

$$M'_{ab} = 2EK_{ab}(\theta_b)$$
$$M'_{cb} = 2EK_{cb}(\theta_b) \qquad (12.17)$$

Since $K_{bi} = K_{ib}$, a comparison of Eqs. 12.10 and 12.17 shows that

$$M'_{ab} = \tfrac{1}{2}M'_{ba}$$
$$M'_{cb} = \tfrac{1}{2}M'_{bc} \qquad (12.18)$$

or, in general,

$$M'_{ib} = \tfrac{1}{2}M'_{bi} \qquad (12.19)$$

Equation 12.19 can be expressed as

$$M'_{ib} = C_{bi} \cdot M'_{bi} \qquad (12.20)$$

where C_{bi} is called the *carry-over factor*. This expresses the factor that must be applied to M'_{bi} to determine the moment M'_{ib} that is "carried over" to the i end of member bi as joint b rotates. In this case,

$$C_{bi} = \tfrac{1}{2} \qquad (12.21)$$

For nonprismatic members and other special cases, the carry-over factor may have other values.

12.5 APPLICATION OF MOMENT DISTRIBUTION METHOD TO BEAM PROBLEMS

The analysis of a statically indeterminate beam by the moment distribution method follows a very systematic procedure. The initial step is to determine the stiffness,

$K = I/l$, for each member throughout the structure. These stiffnesses are then used to determine the distribution factors in accordance with Eq. 12.16 at each joint that will be released during the moment distribution process. It is clear from Eq. 12.16 that only relative stiffness quantities are needed to determine the distribution factors. These relative stiffnesses simply show the relative magnitudes of the member stiffnesses for the members that frame into a joint. The fixed-end moments are then determined. These moments correspond to the situation in which all support joints are fixed against rotation. If there are joints that should not be fixed, the moment distribution process of sequential release, balance, and carry-over is followed until each released joint is in equilibrium.

The moment distribution accounting process registers member-end moments in accordance with the slope deflection sign convention. There are, of course, companion moments that act on the joints, but the method focuses on the member-end moments.

It should be noted that moment distribution is not an approximate method of analysis. If enough cycles are employed, the procedure will converge to the exact solution. In most cases, however, the analyst will stop when the carry-over moments are small when compared with the original fixed-end moments. In all cases, the procedure should not be terminated until all joints are balanced.

EXAMPLE 12.1   Determine the end moments for the structure of Example 11.1. This problem was also considered earlier in this chapter when iteration methods were discussed. The structure and loading are shown. EI = constant.

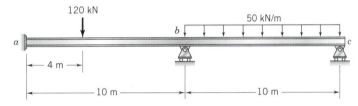

Stiffnesses and Relative Stiffnesses

Stiffness $K = I/l$; relative stiffness is circled.

$$K_{ab} = K_{ba} = \left(\frac{I}{l}\right)_{ab} = \left(\frac{I}{10}\right) = Ⓚ$$

$$K_{bc} = K_{cb} = \left(\frac{I}{l}\right)_{bc} = \left(\frac{I}{10}\right) = Ⓚ$$

Distribution Factors

$$DF_{bi} = \frac{K_{bi}}{\sum\limits_{j} K_{bj}} \qquad\qquad (12.16)$$

At joint b

$$DF_{ba} = \frac{K_{ba}}{K_{ba} + K_{bc}} = \frac{K}{K + K} = 0.500$$

$$DF_{bc} = \frac{K_{bc}}{K_{ba} + K_{bc}} = \frac{K}{K + K} = 0.500$$

At joint c

$$DF_{cb} = \frac{K_{cb}}{K_{cb}} = \frac{K}{K} = 1.000$$

Note: Joint a is fixed; therefore, it will not be released, and no distribution factor is needed.

Fixed-End Moments

See Example 11.1. $FEM_{ab} = -172.8$ kN · m ; $FEM_{ba} = +115.2$ kN · m; $FEM_{bc} = -416.7$ kN · m ; $FEM_{cb} = +416.7$ kN · m

Moment Distribution Process

The information computed above is displayed in a tabular arrangement as follows:

Joint	a	b		c
Member end	ab	ba	bc	cb
Rel K	K	K	K	K
DF	Not Released	0.500	0.500	0.500
FEM	−172.8	+115.2	−416.7	+416.7

At this stage, all joints are fixed, and thus the support conditions at points *b* and *c* are not satisfied. If joint *b* is released, there is an unbalanced moment of $M_b = (115.2 - 416.7) = -301.5$. This unbalanced moment is equilibrated by balancing moments that develop in members *ba* and *bc* according to the member distribution factors (Eq. 12.15), and these are shown in our tabulation in the following manner:

Member end	ab	ba	bc	cb
DF	—	0.500	0.500	0.500
FEM	−172.8	+115.2	−416.7	+416.7
		+150.7	+150.8	

Since the unbalanced moment is negative, the balancing moments are positive. A line is drawn beneath the balancing moments to indicate that the joint is balanced and that the sum of all the moments above the line at joint *b* is zero. As the balancing moments develop at the *b* end of the members framing into joint *b*, carry-over moments are introduced at the member ends at point *a* and *c* in accordance with Eq. 12.20. These carry-over moments are shown in the tabulation at the terminal of the arrows:

Member end	ab	ba	bc	cb
DF	—	0.500	0.500	0.500
FEM	−172.8 ←	+115.2	−416.7 →	+416.7
	+75.4	+150.7	+150.8	+75.4

The equilibrium condition at point *b* is now satisfied, but joint *c* remains erroneously fixed. Joint *b* is now fixed in its rotated position, and joint *c* is released. The accumulated unbalanced moment at joint *c* is $M_b = (+416.7 + 75.4) = +492.1$; it is distributed according to Eq. 12.15, and a carry-over moment is induced at the *b* end of member *bc* as required by Eq. 12.20, as shown next.

Member end	ab	ba	bc	cb
DF	—	0.500	0.500	0.500
FEM	−172.8	+115.2	−416.7	+416.7
	+75.4 ←—	+150.7	+150.8 —→	+75.4
			−246.1 ←—	−492.1

A line is drawn under the balancing moment to indicate that joint c is balanced. However, joint b is again unbalanced—the new unbalance of −246.1 will be distributed to members ba and bc when joint b is released. In this case, for the distribution factors at joint b to be correct, joint c must be reclamped before joint b is released.

The process continues until joints b and c are both in a released state and there is no unbalance at either joint. At this stage, the governing support conditions and equilibrium conditions are satisfied, and the final member-end moments are determined by summing the individual contributions.

The complete moment distribution is shown below:

Joint	a	b		c
Member end	ab	ba	bc	cb
Rel K	K	K	K	K
DF	Not Released	0.500	0.500	0.500
FEM	−172.8	+115.2	−416.7	+416.7
	+75.4	+150.7	+150.8	+75.4
			−246.1	−492.1
	+61.5	+123.0	+123.1	+61.5
			−30.8	−61.5
	+7.7	+15.4	+15.4	+7.7
			−3.8	−7.7
	+1.0	+1.9	+1.9	+1.0
			−0.5	−1.0
	+0.1	+0.2	+0.3	
Final moments (kN · m)	−27.1	+406.4	−406.4	0

These final moments agree with those determined earlier in Example 11.1. It is also of interest to compare the moment distribution procedure with the iterative solution given in Table 12.2. At any interim point in the moment distribution, the summation of the accumulated moments will correspond to the moments at some interim step in the iterative solution. For instance, if the moments are summed up to the dashed lines at the margins of the moment distribution table, the moments are $M_{ab} = -35.9$, $M_{ba} = +388.9$, $M_{bc} = -388.9$, $M_{cb} = +61.5$. These agree very closely with those given at Step 3 in Table 12.2. The final moments given in Table 12.2 are essentially the same as those determined in the moment distribution.

In a typical solution, the shear and moment diagrams would follow the moment distribution as was done in Example 11.1.

12.5.1 Support Settlements

The moment distribution process directly accounts only for the moments associated with member-end rotations. This is evident if Eq. 11.13 is written in the form

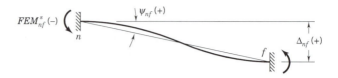

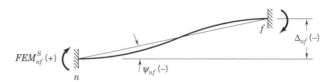

Figure 12.3 Fixed-end moments induced by settlement.

$$M_{nf} = 2EK_{nf}(2\theta_n + \theta_f) - \frac{6EK_{nf}\Delta_{nf}}{l} + FEM_{nf} \qquad (12.22)$$

where Δ_{nf} is the relative transverse member-end displacement for the member nf, which has the length l. In the problem just considered, Δ_{nf} was zero for all members. The FEM_{nf} terms were determined initially, and the subsequent moment distribution process yielded the additional moments, which resulted from the member-end rotation θ_n and θ_f for all members.

If settlement is present, then the members where Δ_{nf} is not zero should be given initial fixed-end moments that reflect the effects of both loading and settlement. That is, the total fixed-end moment, FEM_{nf}^T, is

$$FEM_{nf}^T = FEM_{nf}^S + FEM_{nf} \qquad (12.23)$$

where FEM_{nf}^S is the fixed-end moment induced by the relative settlement Δ_{nf} and is given by

$$FEM_{nf}^S = \frac{-6EK_{nf}\Delta_{nf}}{l} \qquad (12.24)$$

and FEM_{nf} is the usual fixed-end moment that results from the loading on the member. The total fixed-end moments correspond to the condition in which the structure is fixed against rotation at all points but has the prescribed support settlements induced. The moment distribution process then accounts for the moments that result from the member-end rotations.

It should be noted that Eqs. 12.22 and 12.24 indicate that negative end moments are induced by a positive Δ_{nf}, where a positive Δ_{nf} is associated with a clockwise chord rotation of $\psi_{nf} = \Delta_{nf}/l$. Conversely, a negative Δ_{nf}, which corresponds to a counterclockwise chord rotation ψ_{nf}, induces positive end moments. These relationships are illustrated in Fig. 12.3.

EXAMPLE 12.2

Determine the end moments for the structure of Example 12.1 if, in addition to the loading, point b settles 0.03 m.

$$E = 200 \times 10^9 \text{ Pa} = 200 \text{ GN/m}^2; \quad I = 2\,000 \times 10^{-6}\,\text{m}^4$$

Stiffnesses, Relative Stiffnesses, and Distribution Factors

Same as for Example 12.1.

Fixed-End Moments

From Loading

Same as for Example 12.1. $FEM_{ab} = -172.8$ kN $\cdot$ m; $FEM_{ba} = +115.2$ kN $\cdot$ m; $FEM_{bc} = -416.7$ kN $\cdot$ m; $FEM_{cb} = +416.7$ kN $\cdot$ m

From Settlement

Member ab

$$K_{ab} = \left(\frac{I}{l}\right)_{ab} = \frac{2\,000 \times 10^{-6} \text{ m}^4}{10 \text{ m}} = 2\,000 \times 10^{-7} \text{ m}^3 ; \Delta_{ab} = 0.03 \text{ m } (+)$$

$$FEM_{ab}^S = FEM_{ba}^S = \frac{-6EK_{ab}\Delta_{ab}}{l} \tag{12.24}$$

$$= \frac{-6 \times 200 \times 10^9 \times 2\,000 \times 10^{-7} \times 0.03}{10}$$

$$= -720\,000 \text{ N} \cdot \text{m} = -720 \text{ kN} \cdot \text{m}$$

Member bc

$$K_{bc} = \left(\frac{I}{l}\right)_{bc} = 2\,000 \times 10^{-7} \text{m}^3 ; \Delta_{bc} = 0.03 \text{ m } (-)$$

$$FEM_{bc}^S = FEM_{cb}^S = \frac{-6EK_{bc}\Delta_{bc}}{l} = +720 \text{ kN} \cdot \text{m} \tag{12.24}$$

Moment Distribution Process

Joint	a	b		c
Member end	ab	ba	bc	cb
Rel K	K	K	K	K
DF	Not Released	0.500	0.500	0.500
FEM	−172.8	+115.2	−416.7	+416.7
FEM^S	−720.0	−720.0	+720.0	+720.0
FEM^T	−892.8	−604.8	+303.3	+1136.7
			−568.4	−1136.7
	+217.5	+435.0	+434.9	+217.5
			−108.8	−217.5
	+27.2	+54.4	+54.4	+27.2
			−13.6	−27.2
	+3.4	+6.8	+6.8	+3.4
			−1.7	−3.4
		+0.9	+0.8	
Final moments (kN $\cdot$ m)	−644.7	−107.7	−107.7	0

These final moments compare favorably with those obtained from superimposing the results of Examples 11.1 and 11.2.

12.5.2 Temperature Effects

As was shown in Section 11.7, the flexural effects of temperature change are manifested through the introduction of fixed-end moments. These moments are expressed in the form of Eq. 11.27 and are included with those from other effects, such as member loads and/or settlement, to determine the total fixed-end moments. The moment distribution procedure then advances according to the normal process to determine the final member-end moments.

It should be noted that the axial effects of temperature change are not accounted for by this approach. For some structures, such as statically indeterminate frames, the axial distortions of some members induce flexure in other members, and the approach outlined above is not suitable.

12.6 MODIFICATIONS IN MOMENT DISTRIBUTION METHOD

Certain situations lend themselves to modifications in the moment distribution procedure. One situation in which a modified approach is possible is where an exterior span is simply supported at its exterior support point.

Consider again the structure given in Fig. 12.2a, but this time let point c be simply supported as shown in Fig. 12.4a. If point b is initially clamped against rotation, fixed-end moments develop in span ab as shown in Fig. 12.4b. Once again, if joint b is released, it will rotate under the unbalanced moment M_b, and equilibrating moments M'_{ba} and M'_{bc} will be induced along with the corresponding carry-over moments, as shown in Fig. 12.4c. However, in this case, $M'_{cb} = 0$, and thus Eq. 12.9 gives $M'_{cb} = 2EK_{bc}(2\theta_c + \theta_b) = 0$, from which $\theta_c = -\theta_b/2$. This value for θ_c, along with $\theta_a = 0$, when substituted into Eq. 12.9, gives

$$M'_{ba} = 2EK_{ba}(2\theta_b) = 4EK_{ba}\theta_b \tag{12.25}$$

and

$$M'_{bc} = 2EK_{bc}\left(2\theta_b - \frac{\theta_b}{2}\right) = 3EK_{bc}\theta_b \tag{12.26}$$

The expression for M'_{bc} can be written in the form

$$M'_{bc} = 4E\left(\frac{3}{4}K_{bc}\right)\theta_b = 4EK^m_{bc}\theta_b \tag{12.27}$$

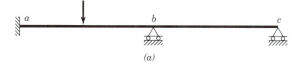

(a)

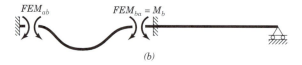

(b)

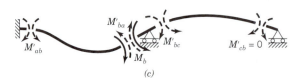

(c)

Figure 12.4 Modified stiffness.

where K_{bc}^m is the modified stiffness of member bc and is given by

$$K_{bc}^m = \tfrac{3}{4} K_{bc} \qquad (12.28)$$

Since the unbalanced moment M_b is again equilibrated by M'_{ba} and M'_{bc}, Eq. 12.8 remains valid. Substitution of Eqs. 12.25 and 12.27 into Eq. 12.8 gives

$$4E\theta_b(K_{ba} + K_{bc}^m) = -M_b \qquad (12.29)$$

Solving for θ_b and substituting the resulting expression into Eqs. 12.25 and 12.27, we obtain

$$M'_{ba} = -M_b\left(\frac{K_{ba}}{K_{ba} + K_{bc}^m}\right)$$

and $\qquad (12.30)$

$$M'_{bc} = -M_b\left(\frac{K_{bc}^m}{K_{ba} + K_{bc}^m}\right)$$

or, in general,

$$M'_{bi} = -M_b\left(\frac{K'_{bi}}{\displaystyle\sum_j K'_{bj}}\right) = -M_b \cdot DF'_{bi} \qquad (12.31)$$

In this expression, i represents the far end of a member that frames into joint b, and j ranges over all members framing into joint b. The quantity K'_{bi} can be either K_{bi} or K_{bi}^m, depending on the support condition at joint i; if i is a fixed end, then $K'_{bi} = K_{bi}$, whereas if i is simply supported, then $K'_{bi} = K_{bi}^m$. The summation on K'_{bj} includes K_{bj} if j is fixed and K_{bj}^m if j is simply supported. The quantity DF'_{bi} is again the distribution factor, which in this case includes the modified stiffnesses when appropriate.

There are other cases where modified stiffness can be employed. One such case is for a member that develops a symmetric deflection pattern about its centerline as a result of equal and opposite moments and rotations at its ends, as shown in Fig. 12.5a. In this case, it turns out that the modified stiffness is

$$K_{nf}^m = \tfrac{1}{2} K_{nf} \qquad (12.32)$$

Another modified stiffness applies for a member that develops an antisymmetric deflection pattern about its centerline in response to equal moments and rotations at its ends, as shown in Fig. 12.5b. In this case, the modified stiffness is

$$K_{nf}^m = \tfrac{3}{2} K_{nf} \qquad (12.33)$$

The modified stiffnesses reflected in Eqs. 12.32 and 12.33 are both developed by the method that was used to arrive at Eq. 12.28.

It should be emphasized that the distribution factors are determined at a joint by using stiffnesses of the members framing into that joint. From Eq. 12.31, we extract

$$DF'_{bi} = \frac{K'_{bi}}{\displaystyle\sum_j K'_{bj}} \qquad (12.34)$$

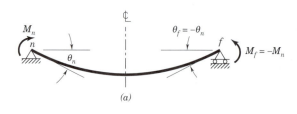

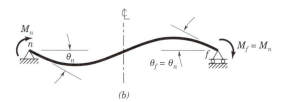

(a)

(b)

Figure 12.5 Other modified stiffness cases. *(a)* Symmetric case. *(b)* Antisymmetric case.

For each member, it is the support condition at the *far end* of the member that determines whether the normal stiffness or a modified stiffness is employed. The primes in Eq. 12.34 remind the analyst to select the appropriate stiffness for each member framing into the joint. If all members are fixed at the far ends, then the normal stiffness would be used for all members, and Eq. 12.34 would reduce to the basic definition of distribution factor given by Eq. 12.16.

It is helpful to consider the physical significance of the stiffness quantities used in getting distribution factors. In general, stiffness is the action required to cause a unit deformation—in this case, it is the moment that must be applied at the member end to induce a unit rotation at that end. Using Eq. 12.9 and the prescribed boundary conditions for each of the cases considered, we can determine

Table 12.3 Physical Meaning of Flexural Stiffness

Case	$\dfrac{M'_{nf}}{\theta_n}$	K^m_{nf}	Member
Normal	$4EK_{nf}$	—	
Simple support	$3EK_{nf} = 4E(K^m_{nf})$	$\frac{3}{4}K_{nf}$	
Symmetrical	$2EK_{nf} = 4E(K^m_{nf})$	$\frac{1}{2}K_{nf}$	
Antisymmetrical	$6EK_{nf} = 4E(K^m_{nf})$	$\frac{3}{2}K_{nf}$	

the moment per unit rotation, M'_{nf}/θ_n. Table 12.3 shows that M'_{nf}/θ_n is equal to $4EK_{nf}$ when point f is fixed, whereas M'_{nf}/θ_n is equal to $4EK^m_{nf}$ for the three special cases where point f is modified.

12.7 APPLICATION TO BEAM PROBLEMS USING MODIFICATIONS

The moment distribution process can be shortened considerably when modified stiffnesses are employed. There are, however, a few important changes in the solution process that must be followed.

First, the distribution factors must be determined from modified stiffnesses when appropriate. Second, the sequence in which the joints are released is important. Before a joint is released in the distribution process, care must be taken to make certain that the far end of each member is in a state that corresponds to that for which the distribution factors were computed.

The example problems that follow will illustrate the technique that must be used.

EXAMPLE 12.3

Determine the end moments for the structure of Example 12.1 using modified stiffnesses. EI = constant.

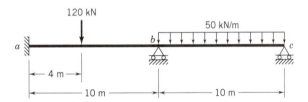

Stiffnesses and Relative Stiffnesses

Stiffness $= K = I/l$; relative stiffness is circled.

$$K_{ab} = K_{ba} = \left(\frac{I}{l}\right)_{ab} = \left(\frac{I}{10}\right) = \boxed{K}$$

$$K_{bc} = K_{cb} = \left(\frac{I}{l}\right)_{bc} = \left(\frac{I}{10}\right) = \boxed{K}$$

Modified Stiffnesses

Consider joint b. The far end of member bc is simply supported and, therefore, modified stiffness for the span bc may be used.

$$K^m_{bc} = \tfrac{3}{4}K_{bc} = \boxed{0.75K}$$

Distribution Factors

$$DF'_{bi} = \frac{K'_{bi}}{\sum_j K'_{bj}} \tag{12.34}$$

At joint b

$$DF'_{ba} = \frac{K_{ba}}{K_{ba} + K^m_{bc}} = \frac{K}{K + 0.75K} = 0.571$$

$$DF'_{bc} = \frac{K^m_{bc}}{K_{ba} + K^m_{bc}} = \frac{0.75K}{K + 0.75K} = 0.429$$

At joint c

$$DF'_{cb} = \frac{K_{cb}}{K_{cb}} = \frac{K}{K} = 1.000$$

Fixed-End Moments

Same as Example 12.1.

Moment Distribution Process

The moment distribution process begins with all joints fixed against rotation. Since a modified stiffness was used for span *bc* to establish the distribution factors at joint *b*, joint *b* cannot be released until joint *c* is simply supported. Thus, the distribution process *must* begin by releasing joint *c* and registering the carry-over moment to joint *b*. Once released, joint *c* is not reclamped. Joint *b* can now be released because the far-end conditions for all members are consistent with the distribution factors at *b*. There is a carry-over moment from *b* to *a*, but since point *c* is simply supported, there is no carry-over moment from *b* to *c*.

Joint	a	b		c
Member end	ab	ba	bc	cb
Rel K	K	K	K	K
Rel K^m			0.75K	
DF'	Not Released	0.571	0.429	1.000
FEM	−172.8	+115.2	−416.7	+416.7
			−208.4	−416.7
	+145.6	+291.2	+218.7	
Final moments (kN · m)	−27.2	+406.4	−406.4	0

EXAMPLE 12.4   Determine the end moments for the structure shown below. *EI* = constant.

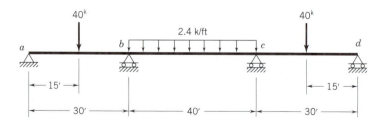

Preliminary Data

The stiffnesses, modified stiffnesses, distribution factors, and fixed-end moments are determined in the usual way and are included in the tabulation shown below. Since the structure and loading are symmetrical, $K^m_{bc} = K^m_{cb} = \frac{1}{2}K_{bc}$.

Moment Distribution Process

Joint	a	b		c		d
Member end	ab	ba	bc	cb	cd	dc
Rel K	$\dfrac{I}{30} = K$	K	$\dfrac{I}{40} = 0.75K$	$0.75K$	K	K
Rel K^m		$0.75K$	$0.375K$	$0.375K$	$0.75K$	
DF'	1.000	0.667	0.333	0.333	0.667	1.000
FEM	−150	+150	−320	+320	−150	+150
	+150	+75			−75	−150
		+63	+32	−32	−63	
Final moments (k-ft)	0	+288	−288	+288	−288	0

Joints a and d are initially released and balanced, and the carry-over moments are recorded at points b and c. Joints b and c are then simultaneously released and balanced. In the process, joints b and c experience equal and opposite rotations. Since there are no carry-over moments, the distribution process is complete.

12.8   MOMENT DISTRIBUTION METHOD FOR FRAME STRUCTURES

The moment distribution method can be applied to frame problems with ease. Thus far, only beam problems have been considered, and the distribution process was arranged in a table that was physically aligned with the horizontal orientation of the structure itself. In the case of frames, there are nonhorizontal members. Here, the table that contains the moment distribution process will not have a strong physical correspondence to the structure. This does not present a problem, as long as one has a complete understanding of the step-by-step nature of the moment distribution procedure and systematically catalogs each step in the table. The example problems that follow serve to illustrate the application of the method to frames.

Frame problems must be considered in two categories—one in which there is no chord rotation for the columns and another in which chord rotation of the columns occurs. In the first case, the frame is to be restrained against sway, whereas in the latter case it is free to sway.

12.8.1   Frames without Sway

Frames that are restrained against sway are solved in precisely the same fashion as are beam problems. The distribution factors are determined from the relative stiffnesses at each joint; the fixed-end moments are computed for all members; and the moment distribution procedure leads to the final moments. The final moments reflect a superposition of the fixed-end moments and the moments that are induced through joint rotations. As with beam problems, modified stiffnesses can be employed when the appropriate conditions are satisfied.

12.8.2   Frames with Sway

When a frame sways, there are moments induced by the chord rotations of the columns. These moments are initially introduced as fixed-end moments in the columns in the same fashion that the fixed-end moments associated with beam settlements were introduced in Section 12.5.2. In the case of settlement, the chord rotation was specified, and thus the fixed-end moments could be determined. However, since the column chord rotations are not known in advance, the column fixed-end moments cannot be computed.

The procedure that must be followed when a frame is subjected to sway is illustrated by considering the simple frame shown in Fig. 12.6a. A set of arbitrary fixed-end moments is introduced in column ab is shown in Fig. 12.6b, and a subsequent moment distribution leads to the moments shown in Fig. 12.6c. The free-body diagram of column ab shown in Fig. 12.6c reveals that there must be a base reaction, which must, in turn, be equilibrated on the entire structure by an applied lateral load of H at point b. However, since the desired solution is for a lateral load of P at point b, the solution of Fig. 12.6c must be multiplied by the factor P/H to obtain the correct solution, as is noted in Fig. 12.6c.

A frame need not be subjected to a lateral force in order to sway. For instance, consider the frame shown in Fig. 12.7a. Starting with the fixed-end moments attributable to the loading on member bc only, we distribute moments and obtain the solution corresponding to no sway, as shown in Fig. 12.7b. A free-body diagram of column ab

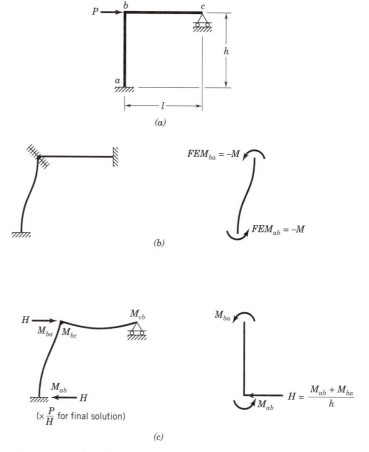

Figure 12.6 Frame sway induced by lateral force. (*a*) Given frame. (*b*) Fixed-end moments. (*c*) Final moments.

shows that this solution requires a base reaction of H at point a, and thus the solution is correct only if a restraining force of $R = H$ is applied at point c to equilibrate the entire structure and to prevent sway. However, since the boundary conditions at point c preclude such a restraining force, a sway solution corresponding to the loading shown in Fig. 12.7c must be superimposed on the solution of Fig. 12.7b to obtain the complete solution to the given problem. The sway solution required in Fig. 12.7c is determined in accordance with the procedure outlined earlier in this section and illustrated in Fig. 12.6.

The solution of Fig. 12.7b corresponds to the loading only and the associated joint rotations, but without sway. It turns out that this solution is correct only when sway is prevented by a restraining force. If sway is not prevented, a solution corresponding to sway and its associated joint rotations, as shown in Fig. 12.7c, must be superimposed on the no-sway solution to obtain the total solution. This latter solution indicates that the frame must sway until equilibrium is achieved by relieving the structure of the artificial restraining force at point c.

For multistory structures, a separate sway solution is necessary for each sway degree of freedom. Horizontal equilibrium conditions are then used to determine how much of each sway solution must be superimposed for the final solution.

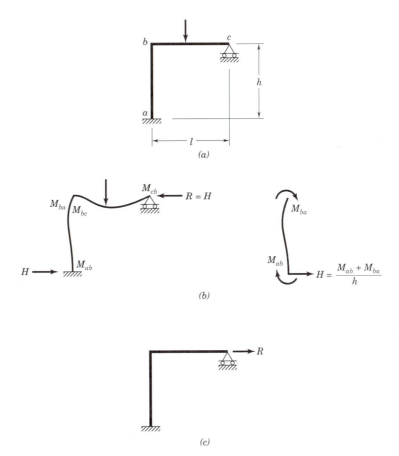

Figure 12.7 Frame sway induced by vertical load. (*a*) Given frame. (*b*) No-sway solution. (*c*) Required sway solution.

12.9 APPLICATION OF MOMENT DISTRIBUTION METHOD
TO FRAME PROBLEMS

This section presents several frame solutions by the moment distribution procedure. Example 12.5 illustrates the procedure for a frame that is restrained against sway. The solution to frame problems in which sway can occur is illustrated in Examples 12.6 and 12.7.

Care must be exercised in the carry-over phase of the distribution procedure for frames. The carry-over moment must be to the far end of the member, which is not necessarily directly adjacent within the tabulation, as was the case in beam problems.

EXAMPLE 12.5

Determine the end moments for all members of the frame shown below. The frame is restrained against sway at point a. Here, $I_{ab} = I_{bc} = I_{cd} = I$; and $I_{be} = I_{cf} = I$.

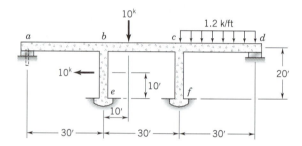

Stiffnesses and Relative Stiffnesses

Stiffness $= K = I/l$; relative stiffness circled.

$$K_{ab} = K_{ba} = K_{bc} = K_{cb} = K_{cd} = K_{dc} = I/30 = \boxed{K}$$

$$K_{be} = K_{eb} = K_{cf} = K_{fc} = I/20 = \boxed{1.5K}$$

Modified Stiffnesses

$$K_{ba}^{m} = \frac{3}{4}K_{ba} = \boxed{0.75K}$$

$$K_{cd}^{m} = \frac{3}{4}K_{cd} = \boxed{0.75K}$$

Distribution Factors

$$DF_{bi}' = \frac{K_{bi}'}{\displaystyle\sum_{j} K_{bj}'} \qquad (12.34)$$

At joint a

$$DF_{ab}' = \frac{K_{ab}}{K_{ab}} = 1.000$$

At joint b

$$DF_{ba}' = \frac{K_{ba}^{m}}{K_{ba}^{m} + K_{be} + K_{bc}} = \frac{0.75K}{0.75K + 1.5K + K} = 0.231$$

$$DF'_{bc} = \frac{K_{bc}}{K^m_{ba} + K_{be} + K_{bc}} = \frac{K}{3.25K} = 0.307$$

$$DF'_{be} = \frac{K_{be}}{K^m_{ba} + K_{be} + K_{bc}} = \frac{1.5K}{3.25K} = 0.462$$

Joints c and d

Determined from joints a and b and symmetry considerations.

Fixed-End Moments

$$FEM_{bc} = \frac{-10 \times 10 \times 20^2}{30^2} = -44.4'^{-k} \; ; \; FEM_{cb} = \frac{+10 \times 10^2 \times 20}{30^2} = +22.2'^{-k}$$

$$FEM_{cd} = \frac{-1.2 \times 30^2}{12} = -90.0'^{-k} \; ; \; FEM_{dc} = +90.0'^{-k}$$

$$FEM_{eb} = \frac{+10 \times 10 \times 10^2}{20^2} = +25.0'^{-k} \; ; \; FEM_{be} = -25.0'^{-k}$$

$$FEM_{ab} = FEM_{ba} = FEM_{cf} = FEM_{fc} = 0$$

Moment Distribution Process

Joint	a	b			e	f	c			d
Member end	ab	ba	bc	be	eb	fc	cf	cb	cd	dc
K	$\frac{I}{30} = K$	K	K	$\frac{I}{20} = 1.5K$	$1.5K$	$1.5K$	$1.5K$	K	K	K
K^m		0.75K							0.75K	
DF'	1.000	0.231	0.307	0.462	NR	NR	0.462	0.307	0.231	1.000
FEM	0	0	−44.4	−25.0	+25.0	0	0	+22.2	−90.0	+90.0
									−45.0	−90.0
			+17.3				+26.1	+52.1	+34.6	+26.1
		+12.0	+16.0	+24.1	+12.1				+8.0	
			−1.3				−1.9	−3.7	−2.5	−1.8
		+0.3	+0.4	+0.6						
Final moments (k-ft)	0	+12.3	−12.0	−0.3	+37.1	+24.2	+48.4	+62.3	−110.7	0

EXAMPLE 12.6   Determine the end moments for each member and the support reactions for the following frame.

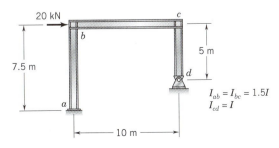

$I_{ab} = I_{bc} = 1.5I$
$I_{cd} = I$

Stiffnesses and Relative Stiffnesses

Stiffness $= K = I/l$; relative stiffness is circled.

$$K_{ab} = K_{ba} = \frac{1.5I}{7.5} = \boxed{K}$$

$$K_{bc} = K_{cb} = \frac{1.5I}{10} = \boxed{0.75K}$$

$$K_{cd} = K_{dc} = \frac{I}{5} = \boxed{K}$$

Modified Stiffnesses

$$K_{cd}^m = \frac{3}{4}K_{cd} = \boxed{0.75K}$$

Distribution Factors

$$DF_{bi}' = \frac{K_{bi}'}{\sum_j K_{bj}'} \tag{12.34}$$

At joint b

$$DF_{ba}' = \frac{K_{ba}}{K_{ba} + K_{bc}} = \frac{K}{K + 0.75K} = 0.571$$

$$DF_{bc}' = \frac{K_{bc}}{K_{ba} + K_{bc}} = \frac{0.75K}{K + 0.75K} = 0.429$$

At joint c

$$DF_{cb}' = \frac{K_{cb}}{K_{cb} + K_{cd}^m} = \frac{0.75K}{0.75K + 0.75K} = 0.500$$

$$DF_{cd}' = \frac{K_{cd}^m}{K_{cb} + K_{cd}^m} = \frac{0.75K}{0.75K + 0.75K} = 0.500$$

Fixed-End Moments

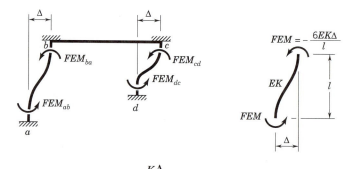

$$FEM \sim \frac{-K\Delta}{l}$$

Member ab

$$FEM_{ab} = FEM_{ba} \sim \frac{-K_{ab}\Delta}{l_{ab}} = \frac{(-1.5I/7.5)\Delta}{7.5} = -0.026\,7\Delta$$

Member cd

$$FEM_{cd} = FEM_{dc} \sim \frac{-K_{cd}\Delta}{l_{cd}} = \frac{-(I/5)\Delta}{5} = -0.040\,0\Delta$$

The sway Δ is not known, but the relationship between the fixed-end moments for the two columns may be established.

If we let $FEM_{ab} = FEM_{ba} = -100$ kN $\cdot$ m, then

$$FEM_{cd} = FEM_{dc} = -100\left(\frac{-0.040\,0\Delta}{-0.026\,7\Delta}\right) = -150 \text{ kN} \cdot \text{m}$$

Moment Distribution Process

Joint	a	b		c		d
Member end	ab	ba	bc	cb	cd	dc
K	K	K	0.75K	0.75K	K	K
K^m					0.75K	
DF'	NR	0.571	0.429	0.500	0.500	1.000
FEM	−100.0	−100.0	0	0	−150.00	−150.00
					+75.00	+150.00
			+18.8	+37.5	+37.50	
	+23.2	+46.4	+34.8	+17.4		
			−4.4	−8.7	−8.7	
	+1.3	+2.5	+1.9	+1.0		
			−0.3	−0.5	−0.5	
	+0.1	+0.2	+0.1			
Final moments (kN · m)	+75.4	−50.9	+50.9	+46.7	−46.7	0

Equilibrium Considerations

Consider a free-body diagram of each member.

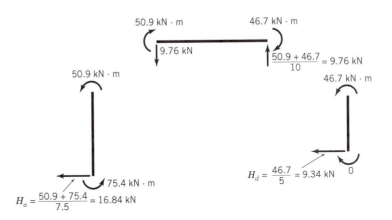

Moments and reactions for assumed *FEMs*:

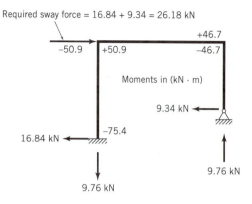

Final Solution

The above solution must be multiplied by (20/26.18) to obtain the final solution, which requires a sway force of 20 kN.

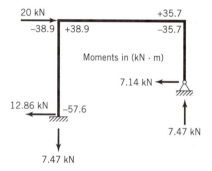

EXAMPLE 12.7

Determine the end moments for all members of the frame analyzed in Example 11.3.

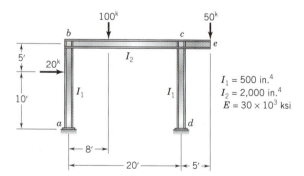

Stiffnesses and Relative Stiffnesses

Stiffness = $K = I/l$; relative stiffness is circled.

$$K_{ab} = K_{ba} = K_{cd} = K_{dc} = \frac{I_1}{15} = \boxed{K}$$

$$K_{bc} = K_{cb} = \frac{I_2}{20} = \frac{4I_1}{20} = \boxed{3K}$$

Part I Solution; No Sway

Distribution Factors

$$DF_{bi} = \frac{K_{bi}}{\displaystyle\sum_j K_{bj}} \qquad (12.16)$$

At joint b

$$DF_{ba} = \frac{K_{ba}}{K_{ba} + K_{bc}} = \frac{K}{4K} = 0.250$$

$$DF_{bc} = \frac{K_{bc}}{K_{ba} + K_{bc}} = \frac{3K}{4K} = 0.750$$

At joint c

$$DF_{cb} = \frac{K_{cb}}{K_{cb} + K_{cd}} = \frac{3K}{4K} = 0.750$$

$$DF_{cd} = \frac{K_{cd}}{K_{cb} + K_{cd}} = \frac{K}{4K} = 0.250$$

Fixed-end Moments

See Example 11.3. $FEM_{ab} = -22.2'^{-k}$; $FEM_{ba} = +44.4'^{-k}$; $FEM_{bc} = -288.0'^{-k}$;

$FEM_{cb} = +192.0'^{-k}$; $FEM_{cd} = FEM_{dc} = 0$; and $FEM_{ce} = -250'^{-k}$.

Moment Distribution Process

Joint	a	b		c			d
Member end	ab	ba	bc	cb	cd	ce	dc
K	K	K	3K	3K	K	—	K
DF	NR	0.250	0.750	0.750	0.250	0.000	NR
FEM	−22.2	+44.4	−288.0	+192.0	0	−250.0	0
	+30.5	+60.9	+182.7	+91.4			
			−12.5	−25.0	−8.4		−4.2
	+1.6	+3.1	+9.4	+4.7			
			−1.8	−3.5	−1.2		−0.6
	+0.2	+0.4	+1.4	+0.7			
			−0.3	−0.5	−0.2		−0.1
		+0.1	+0.2				
Final moments (ft-k)	+10.1	+108.9	−108.9	+259.8	−9.8	−250.0	−4.9

Part I Results

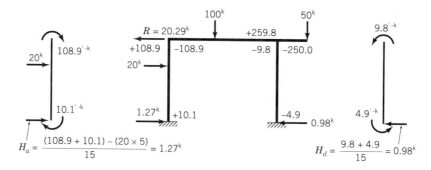

$$H_a = \frac{(108.9 + 10.1) - (20 \times 5)}{15} = 1.27^k \qquad H_d = \frac{9.8 + 4.9}{15} = 0.98^k$$

$R = 20.29^k$ is necessary to prevent sway.

Part II Solution; Sway

Modified Stiffnesses

$$K_{bc}^m = K_{cb}^m = \frac{3}{2}K_{bc} = \boxed{4.5K}$$

Distribution Factors

$$DF'_{bi} = \frac{K'_{bi}}{\displaystyle\sum_j K'_{bj}} \qquad (12.34)$$

At joint b

$$DF'_{ba} = \frac{K_{ba}}{K_{ba} + K_{bc}^m} = \frac{K}{5.5K} = 0.182$$

$$DF'_{bc} = \frac{K_{bc}^m}{K_{ba} + K_{bc}^m} = \frac{4.5K}{5.5K} = 0.818$$

At joint c

$$DF'_{cb} = 0.818 \; ; \; DF'_{cd} = 0.182$$

Fixed-End Moments

Assume $FEM_{ab} = FEM_{ba} = FEM_{cd} = FEM_{dc} = -100'^{-k}$.

Moment Distribution Process

Joint	a	b		c			d
Member end	ab	ba	bc	cb	cd	ce	dc
K	K	K	3K	3K	K	—	K
K^m			4.5K	4.5K			
DF'	NR	0.182	0.818	0.818	0.182	0.000	NR
FEM	−100.0	−100.0	0	0	−100.0	0	−100.0
	+9.1	+18.2	+81.8	+81.8	+18.2		+9.1
Final moments (ft-k)	−90.9	−81.8	+81.8	+81.8	−81.8	0	−90.9

Part II Results

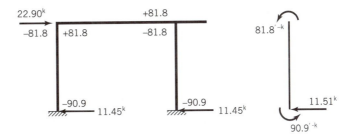

Overall Results

The final solution is determined as follows:

$$
\begin{pmatrix} \text{Part I} \\ \text{solution} \end{pmatrix} + \begin{pmatrix} \text{Part II} \\ \text{solution} \end{pmatrix} \times \frac{20.29}{22.90} = \begin{pmatrix} \text{Final} \\ \text{solution} \end{pmatrix}
$$

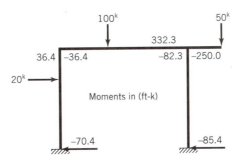

These results agree reasonably well with those determined in Example 11.3.

12.10 SUGGESTED PROBLEMS

12.1 through 12.10 Use the moment distribution method to determine the end moments for each of the beam structures and loading conditions indicated in problems 11.1 through 11.10 Construct the shear and moment diagrams for each member.

12.11 through 12.15 Use the moment distribution method to determine the end moments for each of the frame structures and loading conditions indicated in problems 11.11 through 11.15 Construct the shear and moment diagrams for each member.

12.16 through 12.19 Use the moment distribution method to determine the end moments for each of the structures and loadings in problems 11.16 through 11.19 with the settlements indicated.

12.20 through 12.22 Use the moment distribution method to determine the end moments for each of the frame structures with uninhibited sway shown, with the loading conditions indicated.

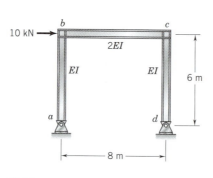

12.20

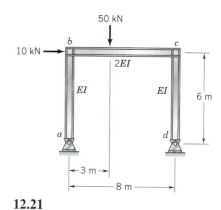

12.21

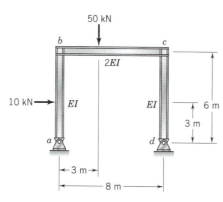

12.22

12.23 through 12.31 Use the moment distribution method to determine the end moments for each of the frame structures with uninhibited sway and loading conditions indicated in problems 11.20 through 11.28.

12.32 and 12.33 Use the moment distribution method to determine the end moments for each of the frame structures with uninhibited sway and temperature changes indicated in problems 11.33 and 11.34.

12.34 through 12.36 Use the moment distribution method along with the modified stiffnesses, when appropriate, to determine the end moments for each of the frame and loading conditions given.

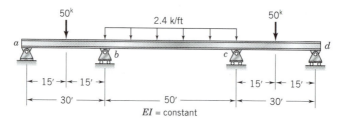

12.34

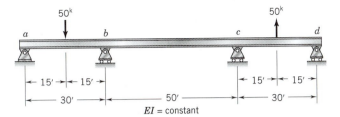

12.35

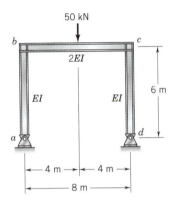

12.36

Part Five

Matrix Methods of Analysis

Chapter 13

Member Force–Deformation Relations

USX Tower (formerly the U. S. Steel Building), Pittsburgh, Pa. (courtesy U.S. Steel Group, a unit of USX Corporation).

13.1 SIGNIFICANCE OF MEMBER INTERACTIVE RELATIONSHIPS

In Chapters 3, 4, and 5, focus was given to the external and internal forces that were required to equilibrate the whole or part of a structure. Subsequently, in Chapters 7 and 8, consideration was given to how structures deform from the forces that they carry.

For structures as a whole, there is an important distinction between statically determinate and indeterminate systems. For statically determinate cases, the forces can be determined by considering static equilibrium alone, and the deformations are then determined as a function of these forces. Each member deforms according to its own force–deformation relationships, and the overall deformation of the structure will result from the accumulation of the individual member effects. However, for statically indeterminate situations, the determination of forces and deformations must be coupled since they are mutually dependent through the combined requirements of static equilibrium and compatibility. In this latter case, as was evident in Chapters 9 through 12, the analysis requires systematic synthesis of the member force–deformation relationships.

Therefore, in either statically determinate or indeterminate structures, the member force–deformation relationships must be known. These relationships have previously been established in a very limited and somewhat conceptualized fashion. In Section 7.2 the axial force–deformation relationships were established. Here a differential element was considered, and it served as a building block for the study of deflections for pin-connected frameworks in Chapter 7. Similarly, flexural force–deformation relationships were derived in Section 8.2. Again, these were based on a differential element, and they provided the basis for the study of deflections for beam- and frame-type structures in Chapter 8.

Currently, a more general approach to structural analysis is to be developed, and the fundamental building blocks are to be the individual members. Therefore, it is necessary to establish interactive relationships between member-end forces and the associated member-end deformations for the complete three-dimensional member.

As has been stipulated in previous sections of the text, we will limit our treatment to members that are not stressed beyond the elastic range.

13.2 STRUCTURE AND MEMBER COORDINATE SYSTEMS

In this section, we develop the force–deformation relationships for the full three-dimensional frame member that will serve as foundational material for the general matrix methods developed in Chapters 14 and 15. As a first step, coordinate systems must be established for the orderly identification of all structure quantities and member-end quantities.

A portion of a typical three-dimensional frame structure is shown in Fig. 13.1a along with the right-hand *XYZ structure coordinate system*. At each joint, the loading condition is specified by six force components, and the response is described by six displacement components. These force and displacement components are shown in their positive directions at joint *i* of Fig. 13.1a. The first three force and displacement components correspond to the direct forces and translational displacements, with the positive directions being along the positive coordinate axes. The remaining force and displacement components represent the moment forces and rotational displacements.

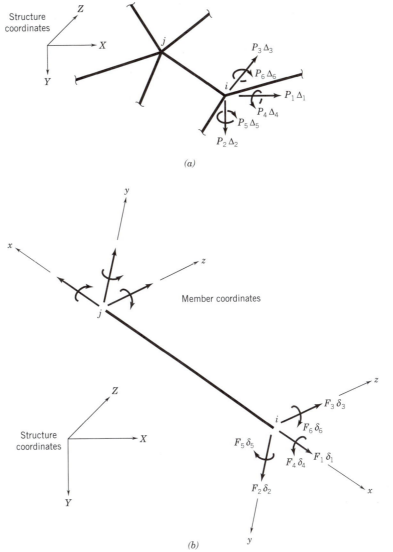

Figure 13.1 Structure and member coordinate systems. *(a)* Structure forces and displacements. *(b)* Member forces and displacements.

Here, the right-hand rule is employed to establish the positive directions. In matrix form, the joint forces and displacements can be expressed, respectively, as

$$\{P\}_i = \begin{Bmatrix} P_1 \\ P_2 \\ P_3 \\ P_4 \\ P_5 \\ P_6 \end{Bmatrix}_i ; \qquad \{\Delta\}_i = \begin{Bmatrix} \Delta_1 \\ \Delta_2 \\ \Delta_3 \\ \Delta_4 \\ \Delta_5 \\ \Delta_6 \end{Bmatrix}_i \qquad (13.1)$$

where the i subscript on a vector indicates that each element of the array is associated with joint i.

Figure 13.1*b* shows a typical member *ij* along with the positive *member coordinate system* at each end of the member. Each of these *xyz* sets of axes is a right-hand coordinate system. The figure also shows the right-hand *XYZ* structure coordinate system, which becomes important when the individual member characteristics are combined to establish the overall structure characteristics.

At the *i* end of the member, the local *xyz* axes are shown along with the positive member-end forces and displacements. The first three forces and displacements correspond to direct forces and translational displacements along the three coordinate axes, whereas the remaining three forces and displacements are moments and rotational displacements about the three coordinate axes. The positive directions for the moments and rotations are based on the right-hand rule. These forces and displacements can be written in matrix form as

$$\{F\}_{ij} = \begin{Bmatrix} F_1 \\ F_2 \\ F_3 \\ F_4 \\ F_5 \\ F_6 \end{Bmatrix}_{ij} ; \qquad \{\delta\}_{ij} = \begin{Bmatrix} \delta_1 \\ \delta_2 \\ \delta_3 \\ \delta_4 \\ \delta_5 \\ \delta_6 \end{Bmatrix}_{ij} \tag{13.2}$$

where the double subscript indicates that these arrays refer to the *i* end of member *ij*.

Similar sets of forces and displacements are present at the *j* end of the member. These are defined in precisely the same manner as those at the *i* end, and the vector representation would be the same as those given by Eqs. 13.2 with the *ij* subscript replaced by *ji*.

In subsequent sections of this chapter, the individual elements may carry a single member subscript, not the double subscript—that is, the axial component at the *i* end of member *ij* is represented by F_{1i} and not $(F_1)_{ij}$. This is done only when it is clear from the context that this force is at the *i* end of member *ij*.

13.3 MEMBER FLEXIBILITY MATRIX

We are now in a position to develop the flexibility matrix for a member with the full six-component member-end force and displacement vectors at the *i* and *j* ends, as shown in Fig. 13.1*b*.

As noted in Section 2.10, a flexibility coefficient is expressed in terms of a designated displacement that is caused by a specific force. There, the focus was on a structure, whereas, in this context, a member is under consideration. However, in either case, the entity for which the flexibility coefficients are to be determined must be restrained in a stable fashion in order for the required displacements to be determined. In other words, rigid-body motions must be precluded. For member *ij*, the required restraint could be accomplished in a variety of ways. In this instance, the member is restrained by imposing fixity at the *j* end, and the desired flexibility coefficients will singularly relate the individual displacement quantities at the *i* end to the corresponding force components at the *i* end. This will lead to the displacement-force relations for the *i* end of the member.

For our current development, consider a slightly simplified version of the member shown in Fig. 13.1*b*. Here, we take a member in the *xy* plane and consider only the forces and displacements shown in Fig. 13.2*a*. Also, as suggested at the end of the previous section, a single member subscript is used here. For example, for the axial force component, F_{1i} represents $(F_1)_{ij}$.

The total strain energy for this member can be expressed as

$$U = \int_{x=0}^{x=l} \left(\frac{F^2}{2EA} + \frac{M^2}{2EI_z} + \frac{\lambda V^2}{2GA} + \frac{T^2}{2GJ} \right) dx \qquad (13.3)$$

The first two terms within the parentheses correspond to the strain energy associated with axial and flexural deformations as previously expressed in Eqs. 7.23 and 8.64, respectively. Here, the quantities E, A, and I_z represent the modulus of elasticity, cross-sectional area, and moment of inertia about the z axis, respectively. The last two terms within the parentheses give the strain energies of shearing and torsional deformations, where λ is a constant that depends on the shape of the cross section, G is the shear modulus, and J is the torsional constant. The strain energy term for shear reflects the rationale introduced in Section 8.8.1, and the strain energy term corresponding to torsion ignores the effects of warping resistance, which can be significant in some cases. This is consistent with the treatment in Section 8.8.2.

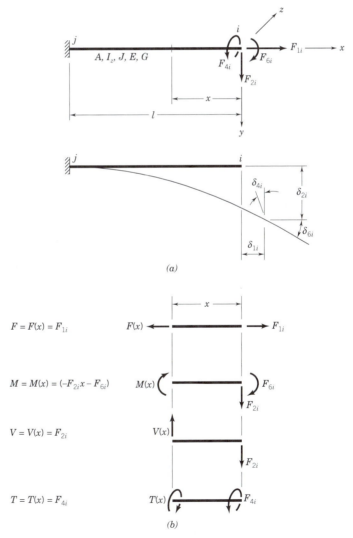

Figure 13.2 Forces and displacements for planar member. *(a)* Forces and displacements. *(b)* Internal forces at *x*.

The quantities F, M, V, and T in Eq. 13.3 represent the axial force, bending moment, shear force, and torsional moment, respectively, and each must be expressed in terms of the variable x as shown in Fig. 13.2b.

Castigliano's second theorem is now used to determine the desired displacement–force relationships. For example, using Eqs. 2.65 and 13.3, we obtain

$$\delta_{1i} = \frac{\partial U}{\partial F_{1i}} = \frac{\partial}{\partial F_{1i}}\left[\int_0^l \left(\frac{F^2}{2EA} + \cdots\right) dx\right] \tag{13.4}$$

Substitution into Eq. 13.4 of the expression for F given in Fig. 13.2b gives

$$\delta_{1i} = \frac{\partial}{\partial F_{1i}}\left[\int_0^l \left(\frac{F_{1i}^2}{2EA}\right) dx\right] = \frac{F_{1i}l}{EA} \tag{13.5}$$

Similarly,

$$\delta_{2i} = \frac{\partial}{\partial F_{2i}}\left[\int_0^l \left(\cdots + \frac{M^2}{2EI_z} + \frac{\lambda V^2}{2GA} + \cdots\right) dx\right] \tag{13.6}$$

Using the expressions for M and V from Fig. 13.2b, we can express Eq. 13.6 as

$$\delta_{2i} = \frac{\partial}{\partial F_{2i}}\left[\int_0^l \left(\frac{[-F_{2i}x - F_{6i}]^2}{2EI_z} + \frac{\lambda F_{2i}^2}{2GA}\right) dx\right] \tag{13.7}$$

$$= \frac{F_{2i}l^3}{3EI_z} + \frac{F_{6i}l^2}{2EI_z} + \frac{\lambda F_{2i}l}{GA} \tag{13.8}$$

In like fashion

$$\delta_{4i} = \frac{F_{4i}l}{GJ} \tag{13.9}$$

and

$$\delta_{6i} = \frac{F_{2i}l^2}{2EI_z} + \frac{F_{6i}l}{EI_z} \tag{13.10}$$

Equations 13.5, 13.8, 13.9, and 13.10 can be capsulized in a single matrix equation in the form

$$\begin{Bmatrix} \delta_1 \\ \delta_2 \\ \delta_4 \\ \delta_6 \end{Bmatrix}_{ij} = \begin{bmatrix} \dfrac{l}{EA} & 0 & 0 & 0 \\ 0 & \dfrac{l^3}{3EI_z} + \dfrac{\lambda l}{GA} & 0 & \dfrac{l^2}{2EI_z} \\ 0 & 0 & \dfrac{l}{GJ} & 0 \\ 0 & \dfrac{l^2}{2EI_z} & 0 & \dfrac{l}{EI_z} \end{bmatrix} \begin{Bmatrix} F_1 \\ F_2 \\ F_4 \\ F_6 \end{Bmatrix}_{ij} \tag{13.11}$$

For most members, the quantity $\lambda l / GA$ is very much smaller than $l^3/3EI$ and is thus ignored. Henceforth, shearing deformations will be neglected by setting $\lambda l/GA$ equal to zero.

For the complete member shown in Fig. 13.1b, which includes bending about the y axis and shearing forces in the xz plane (ignoring shear deformations), Eq. 13.11 can be expanded to the form

$$
\begin{Bmatrix} \delta_1 \\ \delta_2 \\ \delta_3 \\ \delta_4 \\ \delta_5 \\ \delta_6 \end{Bmatrix}_{ij} = \begin{bmatrix} \dfrac{l}{EA} & 0 & 0 & 0 & 0 & 0 \\ 0 & \dfrac{l^3}{3EI_z} & 0 & 0 & 0 & \dfrac{l^2}{2EI_z} \\ 0 & 0 & \dfrac{l^3}{3EI_y} & 0 & \dfrac{-l^2}{2EI_y} & 0 \\ 0 & 0 & 0 & \dfrac{l}{GJ} & 0 & 0 \\ 0 & 0 & \dfrac{-l^2}{2EI_y} & 0 & \dfrac{l^3}{EI_y} & 0 \\ 0 & \dfrac{l^2}{2EI_z} & 0 & 0 & 0 & \dfrac{l}{EI_z} \end{bmatrix} \begin{Bmatrix} F_1 \\ F_2 \\ F_3 \\ F_4 \\ F_5 \\ F_6 \end{Bmatrix}_{ij}
\tag{13.12}
$$

This equation can be written in the abbreviated form

$$
\{\delta\}_{ij} = [d]^j_{ii}\{F\}_{ij}
\tag{13.13}
$$

where $[d]^j_{ii}$ is the flexibility matrix relating the displacements at the i end (first subscript) to the forces at the i end (second subscript) for a member that connects points i and j (superscript). This is referred to as the *direct flexibility matrix* for member ij.

13.4 MEMBER STIFFNESS MATRIX

As was done in the previous section for the derivation of the member flexibility matrix, a simplified case is initially considered in developing the member stiffness matrix. Figure 13.3a shows a member in the xy plane along with the member-end forces and displacements that are to be included. This is the same member that was shown in Fig.13.2a; however, in the present case, the member is not restrained at point j.

From Eq. 13.5

$$
F_{1i} = \frac{EA}{l}\delta_{1i}
\tag{13.14}
$$

where F_{1i} is the axial force at the i end of the member, and δ_{1i} is the axial displacement at the i end of the member *relative* to the j end as shown in Fig. 13.2a. In this case, Fig. 13.3b shows that the *relative* axial displacement is $(\delta_{1i} + \delta_{1j})$, and thus Eq. 13.14 becomes

$$
F_{1i} = \frac{EA}{l}\delta_{1i} + \frac{EA}{l}\delta_{1j}
\tag{13.15}
$$

From equilibrium considerations,

$$
F_{1j} = F_{1i} = \frac{EA}{l}\delta_{1i} + \frac{EA}{l}\delta_{1j}
\tag{13.16}
$$

Solving Eqs. 13.8 and 13.10 simultaneously for F_{2i} and F_{6i} and recalling that the shear deformation term, $\lambda\, l/GA$, can be ignored, we obtain

$$
F_{2i} = \frac{12EI_z}{l^3}\delta_{2i} - \frac{6EI_z}{l^2}\delta_{6i}
$$

$$
F_{6i} = -\frac{6EI_z}{l^2}\delta_{2i} + \frac{4EI_z}{l}\delta_{6i}
\tag{13.17}
$$

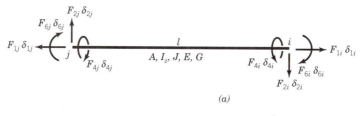

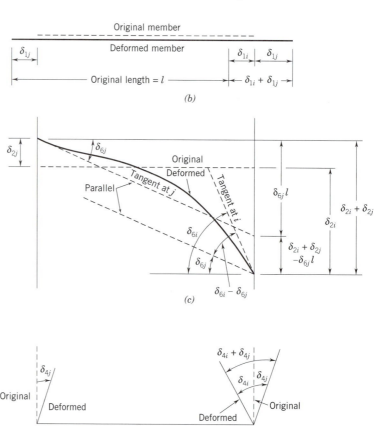

Figure 13.3 Forces and displacements for planar member. *(a)* Forces and displacements. *(b)* Axial action. *(c)* Flexural action. *(d)* Torsional action.

where F_{2i} and F_{6i} are the shearing force and the moment, respectively, at the i end of the member and δ_{2i} and δ_{6i} are the corresponding displacements at the i end of the member *relative* to the tangent at the j end as shown in Fig. 13.2a. Figure 13.3c shows that in this case the appropriate *relative* transverse and rotational displacements, respectively, are $(\delta_{2i} + \delta_{2j} - \delta_{6j}l)$ and $(\delta_{6i} - \delta_{6j})$, and thus Eq. 13.17 becomes

$$F_{2i} = \frac{12EI_z}{l^3}\delta_{2i} + \frac{12EI_z}{l^3}\delta_{2j} - \frac{6EI_z}{l^2}\delta_{6i} - \frac{6EI_z}{l^2}\delta_{6j} \tag{13.18}$$

$$F_{6i} = -\frac{6EI_z}{l^2}\delta_{2i} - \frac{6EI_z}{l^2}\delta_{2j} + \frac{4EI_z}{l}\delta_{6i} + \frac{2EI_z}{l}\delta_{6j} \tag{13.19}$$

Equilibrium then requires that

$$F_{2j} = F_{2i} = \frac{12EI_z}{l^3}\delta_{2i} + \frac{12EI_z}{l^3}\delta_{2j} - \frac{6EI_z}{l^2}\delta_{6i} - \frac{6EI_z}{l^2}\delta_{6j} \qquad (13.20)$$

and

$$F_{6j} = -(F_{6i} + F_{2i}l)$$
$$= -\frac{6EI_z}{l^2}\delta_{2i} - \frac{6EI_z}{l^2}\delta_{2j} + \frac{2EI_z}{l}\delta_{6i} + \frac{4EI_z}{l}\delta_{6j} \qquad (13.21)$$

From Eq. 13.9,

$$F_{4i} = \frac{GJ}{l}\delta_{4i} \qquad (13.22)$$

in which F_{4i} is the torsional force at the i end of the member and δ_{4i} is the rotational displacement at the i end of the member *relative* to the j end as illustrated in Fig. 13.2a. In this situation, Fig. 13.3d shows that the *relative* rotational displacement is $(\delta_{4i} + \delta_{4j})$. Accordingly, Eq. 13.22 takes the form

$$F_{4i} = \frac{GJ}{l}\delta_{4i} + \frac{GJ}{l}\delta_{4j} \qquad (13.23)$$

Equilibrium requires that

$$F_{4j} = F_{4i} = \frac{GJ}{l}\delta_{4i} + \frac{GJ}{l}\delta_{4j} \qquad (13.24)$$

Equations 13.15, 13.16, 13.18 through 13.21, 13.23, and 13.24 can be collected in a single matrix equation as

$$
\left\{
\begin{Bmatrix} F_1 \\ F_2 \\ F_4 \\ F_6 \end{Bmatrix}_{ij} \\
\begin{Bmatrix} F_1 \\ F_2 \\ F_4 \\ F_6 \end{Bmatrix}_{ji}
\right\}
=
\left[
\begin{array}{cccc:cccc}
\frac{EA}{l} & 0 & 0 & 0 & \frac{EA}{l} & 0 & 0 & 0 \\
0 & \frac{12EI_z}{l^3} & 0 & \frac{-6EI_z}{l^2} & 0 & \frac{12EI_z}{l^3} & 0 & \frac{-6EI_z}{l^2} \\
0 & 0 & \frac{GJ}{l} & 0 & 0 & 0 & \frac{GJ}{l} & 0 \\
0 & \frac{-6EI_z}{l^2} & 0 & \frac{4EI_z}{l} & 0 & \frac{-6EI_z}{l^2} & 0 & \frac{2EI_z}{l} \\
\hdashline
\frac{EA}{l} & 0 & 0 & 0 & \frac{EA}{l} & 0 & 0 & 0 \\
0 & \frac{12EI_z}{l^3} & 0 & \frac{-6EI_z}{l^2} & 0 & \frac{12EI_z}{l^3} & 0 & \frac{-6EI_z}{l^2} \\
0 & 0 & \frac{GJ}{l} & 0 & 0 & 0 & \frac{GJ}{l} & 0 \\
0 & \frac{-6EI_z}{l^2} & 0 & \frac{2EI_z}{l} & 0 & \frac{-6EI_z}{l^2} & 0 & \frac{4EI_z}{l}
\end{array}
\right]
\left\{
\begin{Bmatrix} \delta_1 \\ \delta_2 \\ \delta_4 \\ \delta_6 \end{Bmatrix}_{ij} \\
\begin{Bmatrix} \delta_1 \\ \delta_2 \\ \delta_4 \\ \delta_6 \end{Bmatrix}_{ji}
\right\}
\quad (13.25)
$$

For the complete three-dimensional member shown in Fig. 13.1b, which includes bending in the xz plane, Eq. 13.25 takes on the following expanded form:

$$
\left\{
\begin{array}{c}
\left\{
\begin{array}{c} F_1 \\ F_2 \\ F_3 \\ F_4 \\ F_5 \\ F_6 \end{array}
\right\}_{ij} \\[2pt]
\hline
\left\{
\begin{array}{c} F_1 \\ F_2 \\ F_3 \\ F_4 \\ F_5 \\ F_6 \end{array}
\right\}_{ji}
\end{array}
\right\}
=
\left[
\begin{array}{cccccc|cccccc}
\dfrac{EA}{L} & & & & & & \dfrac{EA}{l} & & & & & \\[6pt]
& \dfrac{12EI_z}{l^3} & & & & -\dfrac{6EI_z}{l^2} & & \dfrac{12EI_z}{l^3} & & & & -\dfrac{6EI_z}{l^2} \\[6pt]
& & \dfrac{12EI_y}{l^3} & & \dfrac{6EI_y}{l^2} & & & & -\dfrac{12EI_y}{l^3} & & -\dfrac{6EI_y}{l^2} & \\[6pt]
& & & \dfrac{GJ}{l} & & & & & & \dfrac{GJ}{l} & & \\[6pt]
& & \dfrac{6EI_y}{l^2} & & \dfrac{4EI_y}{l} & & & & -\dfrac{6EI_y}{l^2} & & -\dfrac{2EI_y}{l} & \\[6pt]
& -\dfrac{6EI_z}{l^2} & & & & \dfrac{4EI_z}{l} & & -\dfrac{6EI_z}{l^2} & & & & \dfrac{2EI_z}{l} \\[6pt]
\hline
\dfrac{EA}{l} & & & & & & \dfrac{EA}{l} & & & & & \\[6pt]
& \dfrac{12EI_y}{l^3} & & & & -\dfrac{6EI_z}{l^2} & & \dfrac{12EI_z}{l^3} & & & & -\dfrac{6EI_z}{l^2} \\[6pt]
& & -\dfrac{12EI_y}{l^3} & & -\dfrac{6EI_y}{l^2} & & & & \dfrac{12EI_y}{l^3} & & \dfrac{6EI_y}{l^2} & \\[6pt]
& & & \dfrac{GJ}{l} & & & & & & \dfrac{GJ}{l} & & \\[6pt]
& & -\dfrac{6EI_y}{l^2} & & -\dfrac{2EI_y}{l} & & & & \dfrac{6EI_y}{l^2} & & \dfrac{4EI_y}{l} & \\[6pt]
& -\dfrac{6EI_z}{l^2} & & & & \dfrac{2EI_z}{l} & & -\dfrac{6EI_z}{l^2} & & & & \dfrac{4EI_z}{l}
\end{array}
\right]
\left\{
\begin{array}{c}
\left\{
\begin{array}{c} \delta_1 \\ \delta_2 \\ \delta_3 \\ \delta_4 \\ \delta_5 \\ \delta_6 \end{array}
\right\}_{ij} \\[2pt]
\hline
\left\{
\begin{array}{c} \delta_1 \\ \delta_2 \\ \delta_3 \\ \delta_4 \\ \delta_5 \\ \delta_6 \end{array}
\right\}_{ji}
\end{array}
\right\}
\tag{13.26}
$$

In an abbreviated form, Eq. 13.26 becomes

$$
\left\{
\begin{array}{c} \{F\}_{ij} \\ \hline \{F\}_{ji} \end{array}
\right\}
=
\left[
\begin{array}{c|c} [k]_{ii}^{j} & [k]_{ij} \\ \hline [k]_{ji} & [k]_{jj}^{i} \end{array}
\right]
\left\{
\begin{array}{c} \{\delta\}_{ij} \\ \hline \{\delta\}_{ji} \end{array}
\right\}
\tag{13.27}
$$

The matrix $[k]_{ii}^{j}$ is a *direct stiffness matrix* for member ij. It relates the forces at the i end (first subscript) to the displacements at the i end (second subscript) for the member that connects points i and j (superscript). The matrix $\{k\}_{jj}^{i}$ is the direct stiffness matrix for member ij relating forces and displacements at the j end, and it is clear from Eq. 13.26 that $[k]_{ii}^{j} = \{k\}_{jj}^{i}$. The matrix $[k]_{ij}$ is a *cross stiffness matrix,* which relates the forces at the i end (first subscript) to the displacements at the j end (second subscript). Similarly, $[k]_{ji}$ is a cross stiffness matrix relating the forces at the j end to the displacements at the i end.

Equation 13.27 can be further simplified as

$$
\{F\} = [k]\{\delta\}
\tag{13.28}
$$

where $\{F\}$ and $\{\delta\}$ are the total member force and displacement vectors, respectively, that are shown in their expanded forms in Eq. 13.26, and $[k]$ is the *total member stiffness matrix.*

It should be noted that Eq. 13.28 cannot be solved for the displacements, $\{\delta\}$, because it does not constitute a set of independent equations. An examination of Eq. 13.26 clearly reveals dependencies. That is, $[k]^{-1}$ does not exist or, in other words, $[k]$ is a *singular matrix* that does not possess an inverse. Physically, this indicates that there is no unique set of displacements, $\{\delta\}$, associated with a given set of forces, $\{F\}$. For instance, the forces shown in Fig. 13.1a would not be changed by a rigid-body motion. However, if the member is restrained to preclude rigid-body movement, then there is a unique set of displacements that would accompany a prescribed set of forces. For instance, if the member shown in Fig. 13.1a is fixed at point j, then $\{\delta\}_{ji} = \{0\}$, and then from Eq. 13.27

$$\{F\}_{ij} = [k]^{j}_{ii}\{\delta\}_{ij} \tag{13.29}$$

Premultiplication by $[k]^{j-1}_{ii}$ gives

$$\{\delta\}_{ij} = [k]^{j-1}_{ii}\{F\}_{ij} \tag{13.30}$$

Comparing Eqs. 13.13 and 13.30, we see that

$$[k]^{j-1}_{ii} = [d]^{j}_{ii} \tag{13.31}$$

which can be verified by multiplication of the appropriate matrices from Eqs. 13.12 and Eq. 13.26 $([k]^{j}_{ii}[d]^{j}_{ii} = [I])$. This, of course, explains why member ij had to be fixed at end j in Section 13.3 when the flexibility coefficients were determined.

The matrix $[k]^{j}_{ii}$, or any other submatrix of $[k]$ in Eq. 13.26 for which an inverse exists—which would be the stiffness matrix corresponding to the displacement degrees of freedom for a properly restrained member—is referred to as a *reduced stiffness matrix*, $[k]_r$, in contrast to the *total stiffness matrix*, $[k]$.

An examination of Eq. 13.26 reveals some relevant comparisons with the less general treatments presented in earlier chapters of the book. If Eq. 13.26 is expanded into twelve individual force–displacement equations, the first and seventh of these equations confirm Eq. 7.11 for an axially loaded member; if only member-end rotations are considered, the sixth and twelfth equations sustain the form of Eq. 8.19; and, if both member-end rotations and transverse displacements are included, the sixth and twelfth equations show agreement with the slope deflection equation, Eq. 11.13, when the fixed-end moment terms are ignored.

One final note regarding the member stiffness matrix concerns the physical meaning of the individual stiffness coefficients. Each column of the stiffness matrix of Eq. 13.26 gives the end forces associated with a unit value of a single end displacement. For instance, the second column gives the array of member-end forces that are caused by a unit value of δ_{2i}, while other displacements are zero. This condition is shown in Fig. 13.4, and the results are verifiable by a number of methods.

13.5   SOME OBSERVATIONS ON THE FLEXIBILITY AND STIFFNESS MATRICES

Each of the flexibility coefficients that make up the flexibility matrix is a specified displacement quantity. That is, d_{rs} is the rth member-end displacement that results from the application of a unit value of the sth member-end force. Based on Maxwell's law, which is given by Eq. 2.72,

$$d_{sr} = d_{rs} \tag{13.32}$$

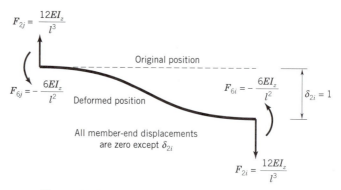

Figure 13.4 Physical meaning of stiffness coefficients.

Therefore, the flexibility matrix of Eq. 13.13, $[d]_{ii}^{j}$, is a symmetrical matrix as verified by Eq. 13.12. Taking the inverse of both sides of Eq. 13.31 yields

$$[k]_{ii}^{j} = [d]_{ii}^{j-1} \tag{13.33}$$

and since the inverse of a symmetrical matrix is itself a symmetrical matrix, we conclude that $[k]_{ii}^{j}$ is symmetrical and that

$$k_{sr} = k_{rs} \tag{13.34}$$

Of course, this holds true over the range of the elements for $[k]_{ii}^{j}$, the so-called reduced stiffness matrix, which includes the kinematic degrees of freedom of the restrained member. However, the member could be restrained by inhibiting different displacement components and, therefore, eventually all components could be included as part of a reduced stiffness matrix. Thus, Eq. 13.34 applies for all combinations of r and s, and this proves that the entire stiffness matrix is symmetrical.

Examination of Eqs. 13.26 and 13.27, therefore, clearly verifies that the direct stiffness matrices, $[k]_{ii}^{j}$ and $[k]_{jj}^{i}$, are symmetrical matrices and that the cross stiffness matrices have a transpose relationship. That is,

$$[k]_{ij} = [k]_{ji}^{T}$$

and

$$[k]_{ji} = [k]_{ij}^{T} \tag{13.35}$$

13.6 FLEXIBILITY–STIFFNESS TRANSFORMATIONS

Sections 13.4 and 13.5 traced the systematic development of the flexibility and stiffness matrices for member ij. The procedure used is as follows: The member was rendered stable by imposing displacement constraints at end j; for the resulting statically determinate structure, the displacements at end i were determined as a function of the forces at end i, thus establishing the flexibility matrix for the restrained structure; displacement compatibilities and equilibrium conditions were applied to the entire member to form the complete set of member-end force–deformation relationships from which the member stiffness matrix was extracted.

This procedure will now be accomplished through a formal matrix approach. The member introduced in Fig. 13.2a is used as a basis for the discussion. Figure 13.5a shows the member with the complete set of member coordinates, and Fig. 13.5b gives the restrained member, which is clamped at end j.

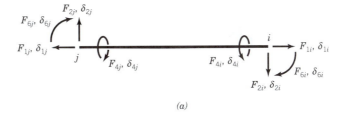

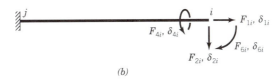

(a)

(b)

Figure 13.5 Member-end forces and displacements. *(a)* All member-end components. *(b)* Components corresponding to kinematic degrees of freedom for restrained member.

The member-end forces associated with the kinematic degrees of freedom for the restrained structure of Fig. 13.5*b*, $\{F\}_I$, are separated from the member-end forces that correspond to the support reactions, $\{F\}_{II}$. That is,

$$\{F\}_I = \begin{Bmatrix} F_{1i} \\ F_{2i} \\ F_{4i} \\ F_{6i} \end{Bmatrix}; \qquad \{F\}_{II} = \begin{Bmatrix} F_{1j} \\ F_{2j} \\ F_{4j} \\ F_{6j} \end{Bmatrix} \qquad (13.36)$$

The member-end displacement components are similarly separated into the vectors $\{\delta\}_I$ and $\{\delta\}_{II}$. According to this separation, Eq. 13.27 can be written in a partitioned form as

$$\begin{Bmatrix} \{F\}_I \\ \{F\}_{II} \end{Bmatrix} = \begin{bmatrix} [k]_{I,I} & [k]_{I,II} \\ [k]_{II,I} & [k]_{II,II} \end{bmatrix} \begin{Bmatrix} \{\delta\}_I \\ \{\delta\}_{II} \end{Bmatrix} \qquad (13.37)$$

where the subscripts on each $[k]$ submatrix designate, respectively, that it relates the $\{F\}$ vector (first subscript) to the $\{\delta\}$ vector (second subscript).

Since $\{\delta\}_{II} = \{0\}$, Eq. 13.37 can be expanded to

$$\{F\}_I = [k]_{I,I}\{\delta\}_I \qquad (13.38)$$

and

$$\{F\}_{II} = [k]_{II,I}\{\delta\}_I \qquad (13.39)$$

where $[k]_{I,I}$ is the reduced stiffness matrix, which was designated as $[k]_r$ in Section 13.4.

Since the restrained member of Fig. 13.5*b* is secure against rigid-body motions, $[k]_{I,I}$ is the inverse of the flexibility matrix $[d]$. That is,

$$[k]_{I,I} = [d]^{-1} \qquad (13.40)$$

As was stated earlier, the structure of Fig. 13.5b is a statically determinate structure. Therefore, the support forces (reactions) can be determined as a function of the applied forces through the application of the equations of equilibrium. These equations can be written in the form

$$\{F\}_{\mathrm{II}} = [b]\{F\}_{\mathrm{I}} \tag{13.41}$$

where the coefficients of the applied forces are gathered into the equilibrium matrix $[b]$. Substitution of Eq. 13.40 into Eq. 13.38 and of the resulting expression into Eq. 13.41 gives

$$\{F\}_{\mathrm{II}} = [b][d]^{-1}\{\delta\}_{\mathrm{I}} \tag{13.42}$$

Comparing Eqs. 13.39 and 13.42 we see that

$$[k]_{\mathrm{II,\,I}} = [b][d]^{-1} \tag{13.43}$$

Since the total stiffness matrix is symmetrical, as was verified in Section 13.5, we have

$$[k]_{\mathrm{I,\,II}} = [k]_{\mathrm{II,\,I}}^{T} = ([b][d]^{-1})^{T} = [d]^{-1}[b]^{T} \tag{13.44}$$

In the above matrix manipulation, it is noted that $([d]^{-1})^{T} = [d]^{-1}$ because of the symmetrical nature of $[d]$.

In forming $[k]_{\mathrm{II,II}}$, we return to Eq. 13.41 and note that $[b]$ is the equilibrium matrix that relates the $\{F\}_{\mathrm{II}}$ forces to the $\{F\}_{\mathrm{I}}$ forces. Thus, based on the interpretation of the subscripts on the $[k]$ submatrices, it is clear that

$$[k]_{\mathrm{II,\,II}} = [b][k]_{\mathrm{I,\,II}} \tag{13.45}$$

or, substituting Eq. 13.44 into Eq. 13.45, we have

$$[k]_{\mathrm{II,\,II}} = [b][d]^{-1}[b]^{T} \tag{13.46}$$

Collection of the individual submatrices from Eqs. 13.40, 13.43, 13.44, and 13.46 according to the requirements of Eq. 13.37 leads to

$$[k] = \left[\begin{array}{c|c} [d]^{-1} & [d]^{-1}[b]^{T} \\ \hline [b][d]^{-1} & [b][d]^{-1}[b]^{T} \end{array}\right] \tag{13.47}$$

This equation expresses the general transformation from the flexibility matrix for the restrained structure to the total stiffness matrix that includes all degrees of freedom, including those corresponding to rigid-body motion. The number of components in $\{F\}_{\mathrm{II}}$ is limited by the requirements that the restrained structure be statically determinate; however, there is no restriction on the number of components in $\{F\}_{\mathrm{I}}$.

In the presentation given here, the focus was on an individual member. However, this member is treated as a structure in itself, and therefore the procedure is easily extended to more complex structures in which the total structure stiffness matrix is determined from the flexibility matrix for the stable system.

EXAMPLE 13.1

Use the flexibility matrix for the restrained structure of Fig. 13.2a (Eq. 13.11) to develop the total stiffness matrix for the member shown below.

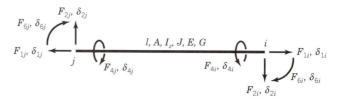

Flexibility Matrix

The member is fixed at point j.

The flexibility matrix relates $\{\delta\}_I$ to $\{F\}_I$ as follows:

$$\{\delta\}_I = [d]\{F\}_I$$

$$
\begin{Bmatrix} \delta_1 \\ \delta_2 \\ \delta_4 \\ \delta_6 \end{Bmatrix}_{ij}
=
\begin{bmatrix}
\dfrac{l}{EA} & 0 & 0 & 0 \\[8pt]
0 & \dfrac{l^3}{3EI_z} & 0 & \dfrac{l^2}{2EI_z} \\[8pt]
0 & 0 & \dfrac{l}{GJ} & 0 \\[8pt]
0 & \dfrac{l^2}{2EI_z} & 0 & \dfrac{l}{EI_z}
\end{bmatrix}
\begin{Bmatrix} F_1 \\ F_1 \\ F_1 \\ F_1 \end{Bmatrix}_{ij}
\tag{13.11}
$$

Note: $\{F\}_I$ and $\{\delta\}_I$ are defined by Eq. 13.36. The quantity $\lambda l / GA$ is set equal to zero.

Equilibrium Matrix

The equilibrium matrix, $[b]$, relates the support forces, $\{F\}_{II}$, to free-end forces, $\{F\}_I$.

$$\{F\}_{II} = [b]\{F\}_I$$

$$
\begin{aligned}
F_{1j} &= F_{1i} \\
F_{2j} &= F_{2i} \\
F_{4j} &= F_{4i} \\
F_{6j} &= -F_{2i}l - F_{6i}
\end{aligned}
\qquad
\begin{Bmatrix} F_1 \\ F_2 \\ F_4 \\ F_6 \end{Bmatrix}_j
=
\begin{bmatrix}
1 & 0 & 0 & 0 \\
0 & 1 & 0 & 0 \\
0 & 0 & 1 & 0 \\
0 & -l & 0 & -1
\end{bmatrix}
\begin{Bmatrix} F_1 \\ F_2 \\ F_4 \\ F_6 \end{Bmatrix}_i
\tag{13.41}
$$

Note: $\{F\}_I$ and $\{F\}_{II}$ are defined by Eq. 13.36.

Stiffness Matrices

$$
[k]_{I,I} = [d]^{-1} =
\begin{bmatrix}
\dfrac{EA}{l} & 0 & 0 & 0 \\[8pt]
0 & \dfrac{12EI_z}{l^3} & 0 & \dfrac{-6EI_z}{l^2} \\[8pt]
0 & 0 & \dfrac{GJ}{l} & 0 \\[8pt]
0 & \dfrac{-6EI_z}{l^2} & 0 & \dfrac{4EI_z}{l}
\end{bmatrix}
\tag{13.40}
$$

$$[k]_{\mathrm{II,I}} = [b][d]^{-1} = \begin{bmatrix} \dfrac{EA}{l} & 0 & 0 & 0 \\[2ex] 0 & \dfrac{12EI_z}{l^3} & 0 & \dfrac{-6EI_z}{l^2} \\[2ex] 0 & 0 & \dfrac{GJ}{l} & 0 \\[2ex] 0 & \dfrac{-6EI_z}{l^2} & 0 & \dfrac{2EI_z}{l} \end{bmatrix} \qquad (13.43)$$

$$[k]_{\mathrm{I,II}} = [d]^{-1}[b]^T = [k]_{\mathrm{II,I}}^T \qquad (13.44)$$

$$[k]_{\mathrm{II,II}} = [b][d]^{-1}[b]^T = \begin{bmatrix} \dfrac{EA}{l} & 0 & 0 & 0 \\[2ex] 0 & \dfrac{12EI_z}{l^3} & 0 & \dfrac{-6EI_z}{l^2} \\[2ex] 0 & 0 & \dfrac{GJ}{l} & 0 \\[2ex] 0 & \dfrac{-6EI_z}{l^2} & 0 & \dfrac{4EI_z}{l} \end{bmatrix} \qquad (13.46)$$

The total stiffness matrix is assembled according to Eq. 13.47, and it is seen to agree with Eq. 13.25.

13.7 ALTERNATIVE MEMBER COORDINATE SYSTEMS

The selection of a member coordinate system is somewhat arbitrary. The one selected in this text, which was first introduced in Fig. 13.1b, is shown again in Fig. 13.6a. This system has the advantage that the member-end forces for a planar structure (xy plane) are consistent with conventions established earlier in the text. Specifically, tension forces are positive at both ends of the member; end shear forces are positive when they act down at the right end and up at the left end; and end moments are positive when they act clockwise on the member ends. This coordinate system also has the advantage that if the member is rotated end-for-end about the z axis, the member axes at each end retain the same orientation with respect to the global axes. One disadvantage is that there is a different relationship between member and structure coordinate systems at each member end.

However, the adopted coordinate system is not the one that is most commonly used in computer applications. Instead, the one shown in Fig. 13.6b is employed. The major advantage of this system is that there is a common relationship between member and structure coordinate systems at each end of the member. When this system is used, care must be observed in interpreting the resulting member-end forces. For instance, a tension force would be given as a positive force at end i but a negative force at end j. This, of course, is not a serious disadvantage, but it does underscore the fact that computer results must be carefully examined and properly interpreted.

Both of the systems described in Fig. 13.6 are right-handed systems. Other variations are possible, but most systems employed are right-handed in nature.

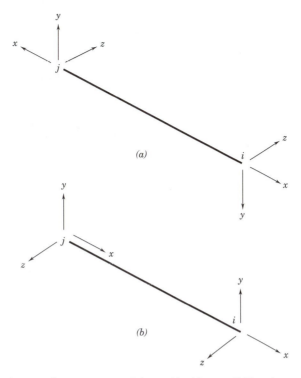

Figure 13.6 Member coordinate systems. *(a)* As used in this text. *(b)* Popular system for modern computer programs.

13.8 SUGGESTED PROBLEMS

13.1 Use the complementary virtual work method to determine the complete flexibility matrix for the structure shown with the prescribed boundary conditions. Use the coordinate system indicated.

13.2 Use Castigliano's second theorem to determine the complete stiffness matrix for the structure shown using the given coordinate system. Compare your results with appropriately extracted terms from Eq. 13.25. *Hint:* Generate force–displacement equations and solve for the forces (stiffness coefficients) for the appropriate unit displacements.

13.3 and 13.4 Consider the member shown below with the indicated member-end actions and deformations.

13.3 Use Castigliano's second theorem to determine the flexibility coefficients d_{11}, d_{12}, d_{21}, and d_{22} for the given member subject to the support conditions shown below. Assemble these flexibility coefficients into the flexibility matrix $[d]$.

13.4 Write the total stiffness matrix $[k]$ in accordance with the member-end actions and deformations noted for the given member by selecting the appropriate terms from Eq. 13.25. Extract the appropriate reduced stiffness matrix $[k]_r$ for the restrained member of Problem 3, and show that $[k]_r[d] = [I]$, where $[d]$ is the flexibility matrix of Problem 3.

13.5 through 13.7 Consider the member shown below with the indicated member-end actions and deformations.

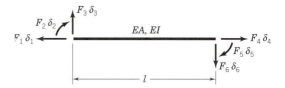

13.5 Use the complementary virtual work method to determine the flexibility coefficients d_{11}, d_{12}, d_{13}, d_{21}, d_{22}, d_{23}, d_{31}, d_{32}, and d_{33} for the given member subject to the support conditions shown below. Assemble these coefficients into the flexibility matrix $[d]$. Consider only the bending deformations associated with the shear forces F_3 and F_6.

13.6 Use Castigliano's second theorem to determine the full 6×6 stiffness matrix $[k]$ for the given member. Again, consider only flexural deformations associated with the shear forces.

13.7 Extract the reduced stiffness matrix $\{k\}_r$ from the results of Problem 6 for the restrained member of Problem 5. Show that $[k]_r[d] = [I]$, where $[d]$ is the flexibility matrix of Problem 5.

13.8 Use the flexibility matrix for the restrained structure of Problem 1 and the transformation procedures of Section 13.6 to establish the total stiffness matrix for the structure of Problem 2.

13.9 Use the flexibility matrix for the restrained structure of Problem 5 and the transformation procedures of Section 13.6 to generate the full 6×6 stiffness matrix called for in Problem 6.

Chapter 14

Stiffness Method

Verrazano Narrows Bridge, New York (courtesy NYC & Company—the Convention and Visitors Bureau).

14.1 FUNDAMENTAL CONCEPTS OF STIFFNESS METHOD

Throughout this textbook the importance of equilibrium has been stressed. It plays a central role in the analysis of any structure, whether statically determinate or indeterminate. In those methods that have been specifically identified as equilibrium methods, the solution results from simultaneously satisfying a number of equations that represent equilibrium throughout the structure. The solution of these equilibrium equations yields displacement quantities; thus, these techniques are sometimes identified as displacement methods.

When equilibrium methods are formulated in a general matrix format, the resulting equation contains the structure stiffness matrix, which is derived from a synthesis of the individual member stiffness matrices. For this reason, the matrix method, which results from a generalization of the equilibrium approach, is referred to as the *stiffness method.*

14.2 OVERVIEW OF STIFFNESS METHOD

Consider the planar framed structure given in Fig. 14.1a in which loads are admissible only at the points that are identified as nodes or joints on the structure. Figure 14.1b shows the structure forces and displacements that are operative at each of the four node points. The relationship between these structure forces and displacements is given by

$$\{P\} = [K]\{\Delta\} \tag{14.1}$$

where $\{P\}$ and $\{\Delta\}$ contain all of the structure forces and displacements, with the individual components as defined in Section 13.2. The matrix $[K]$ is the *total structure stiffness matrix.* In an expanded form, Eq. 14.1 can be written as

$$
\begin{Bmatrix}
(P_1)_1 \\
(P_2)_1 \\
(P_6)_1 \\
\cdot \\
\cdot \\
\cdot \\
\cdot \\
\cdot \\
\cdot \\
(P_1)_4 \\
(P_2)_4 \\
(P_6)_4
\end{Bmatrix}
= [K]
\begin{Bmatrix}
(\Delta_1)_1 \\
(\Delta_2)_1 \\
(\Delta_6)_1 \\
\cdot \\
\cdot \\
\cdot \\
\cdot \\
\cdot \\
\cdot \\
(\Delta_1)_4 \\
(\Delta_2)_4 \\
(\Delta_6)_4
\end{Bmatrix}
\tag{14.2}
$$

This matrix equation represents 12 simultaneous equations relating the 12 structure forces to the 12 structure displacements. Figure 14.1a shows that there are some structure displacements that are constrained by the boundary conditions. Separating these displacements into the vector $\{\Delta\}_{II}$ and collecting the remaining free displacements into the vector $\{\Delta\}_{I}$, we have

$$\{\Delta\}_{II}^{T} = \{(\Delta_1)_1 \ (\Delta_2)_1 \ (\Delta_6)_1 \ (\Delta_1)_3 \ (\Delta_2)_3 \ (\Delta_6)_3 \ (\Delta_2)_4\}$$
$$\{\Delta\}_{I}^{T} = \{(\Delta_1)_2 \ (\Delta_2)_2 \ (\Delta_6)_2 \ (\Delta_1)_4 \ (\Delta_6)_4\}$$

(14.3)

The load value can similarly be separated into the applied loads $\{P\}_I$, which correspond, component for component, with the free displacements, and the reaction forces $\{P\}_{II}$, which are associated with the constrained boundary displacements. These vectors take the form

$$\{P\}_{I}^{T} = \{(P_1)_2 \ (P_2)_2 \ (P_6)_2 \ (P_1)_4 \ (P_6)_4\}$$
$$\{P\}_{II}^{T} = \{(P_1)_1 \ (P_2)_1 \ (P_6)_1 \ (P_1)_3 \ (P_2)_3 \ (P_6)_3 \ (P_2)_4\}$$

(14.4)

Rearranging the rows and columns of Eq. 14.2 in accordance with the ordering of Eqs. 14.3 and 14.4, we have

$$\left\{ \begin{array}{c} \{P\}_I \\ \hline \{P\}_{II} \end{array} \right\} = \left[\begin{array}{c|c} [K]_{I,I} & [K]_{I,II} \\ \hline [K]_{II,I} & [K]_{II,II} \end{array} \right] \left\{ \begin{array}{c} \{\Delta\}_I \\ \hline \{\Delta\}_{II} \end{array} \right\}$$

(14.5)

It should be noted that Eq. 14.5 is not restricted to the illustrative problem of Fig. 14.1—it would apply for any structure in which $\{P\}_I$ and $\{\Delta\}_I$ represent the

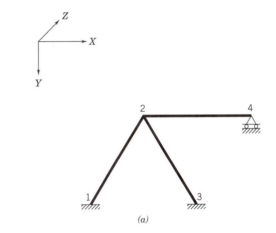

(a)

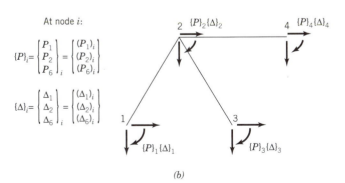

(b)

Figure 14.1 Typical planar frame structure. (a) Given structure and node points. (b) Structure loads and displacements.

applied forces and free displacements, respectively, and $\{P\}_{II}$ and $\{\Delta\}_{II}$ give the reaction forces and the constrained displacements, respectively.

Expansion of Eq. 14.5 gives

$$\{P\}_{I} = [K]_{I,I}\{\Delta\}_{I} + [K]_{I,II}\{\Delta\}_{II} \tag{14.6}$$

$$\{P\}_{II} = [K]_{II,I}\{\Delta\}_{I} + [K]_{II,II}\{\Delta\}_{II} \tag{14.7}$$

However, from the boundary conditions

$$\{\Delta\}_{II} = \{0\} \tag{14.8}$$

and thus Eq. 14.6 reduces to

$$\{P\}_{I} = [K]_{I,I}\{\Delta\}_{I} \tag{14.9}$$

This equation gives the relationship between the applied forces and the free displacements, and $[K]_{I,I}$ is the *reduced stiffness matrix.*

Equation 14.1 shows that a given set of structure displacements $\{\Delta\}$ produces a unique set of structure forces $\{P\}$. However, any attempt to solve this equation for $\{\Delta\}$ will fail because $[K]$ does not possess an inverse. That is, no unique set of displacements is associated with a given set of forces. This fact is illustrated by noting that a rigid-body motion of the entire structure will alter the displacements without changing the forces. This argument is similar to the one used in Section 13.4, in which the member stiffness matrix was found to be singular.

The situation represented by Eq. 14.9 is quite different. Here, we have the force–displacement relations for a structure that is kinematically stable and is restrained against rigid-body motion; thus, $[K]_{I,I}$ possesses an inverse, and

$$\{\Delta\}_{I} = [K]_{I,I}^{-1}\{P\}_{I} \tag{14.10}$$

If the reaction forces are desired, Eqs. 14.8 and 14.10 are substituted into Eq. 14.7 to obtain

$$\{P\}_{II} = [K]_{II,I}\{\Delta\}_{I} = [K]_{II,I}[K]_{I,I}^{-1}\{P\}_{I} \tag{14.11}$$

Before the analysis can develop in the direction presented in this section, it is necessary to generate the total structure stiffness matrix $[K]$ from which the reduced structure stiffness matrix $[K]_{I,I}$ is extracted. The following section of this chapter presents the technique for generating the total structure stiffness matrix from the individual member stiffness matrices.

Actually, it is not necessary that the boundary conditions prescribe zero displacements as is indicated by Eq. 14.8. Instead, a set of boundary conditions might be prescribed that includes some support settlements. In this case,

$$\{\Delta\}_{II} = \{\Delta\}_{S} \tag{14.12}$$

For this case, Eq. 14.10 is replaced by

$$\{\Delta\}_{I} = [K]_{I,I}^{-1}(\{P\}_{I} - [K]_{I,II}\{\Delta\}_{S}) \tag{14.13}$$

The reaction forces are then determined from

$$\{P\}_{II} = [K]_{II,I}[K]_{I,I}^{-1}(\{P\}_{I} - [K]_{I,II}\{\Delta\}_{S}) + [K]_{II,II}\{\Delta\}_{S} \tag{14.14}$$

14.3 GENERATION OF STRUCTURE STIFFNESS MATRIX

The total array of equilibrium equations for the entire structure is given by Eq. 14.1. This equation can be written in the form

$$
\begin{Bmatrix} \{P\}_1 \\ \cdot \\ \cdot \\ \cdot \\ \cdot \\ \cdot \\ \{P\}_i \\ \cdot \\ \cdot \\ \cdot \\ \cdot \\ \cdot \\ \{P\}_n \end{Bmatrix} = \begin{bmatrix} & & & \\ & & & \\ \cdots & [K]_{im} \cdots [K]_{ii} \cdots [K]_{ir} \cdots & \\ & & & \end{bmatrix} \begin{Bmatrix} \cdot \\ \cdot \\ \{\Delta\}_m \\ \cdot \\ \cdot \\ \{\Delta\}_i \\ \cdot \\ \{\Delta\}_r \\ \cdot \end{Bmatrix} \quad (14.15)
$$

In this form, $\{P\}_i$ and $\{\Delta\}_i$ are the structure force and displacement vectors at node i, respectively, for a structure containing n nodes. The ith node of the structure is shown in Fig. 14.2a with members $im \cdots ij \cdots ir$ framing into joint i. The matrix $[K]_{im}$ is a structure stiffness submatrix that relates the forces at node i to the displacements at node m. Figure 14.2b shows that the elements of this matrix will depend on the stiffness characteristics of the deformed member im. The submatrix $[K]_{ii}$ gives the forces at node i that are associated with the displacements at the same node. Figure 14.2c shows that these displacements induce deformations into all of the members that frame into node i, and thus the elements of this submatrix will contain a superposition of the stiffness characteristics of all members. This superposition can be expressed in the form

$$
[K]_{ii} = [K]_{ii}^m + \cdots + [K]_{ii}^j + \cdots + [K]_{ii}^r
$$

$$
= \sum_{j=m}^{r} [K]_{ii}^j \quad (14.16)
$$

where $[K]_{ii}^j$ relates the structure forces induced at node i, through the deformations of member ij, to the displacements of node i. The structure force vector at any given node can be expressed in terms of the structure displacements throughout the structure. For instance, extraction of the expression for $\{P\}_i$ from Eq. 14.15 and substitution of Eq. 14.16 gives

$$
\{P\}_i = \sum_{j=m}^{r} ([K]_{ii}^j \{\Delta\}_i + [K]_{ij} \{\Delta\}_j) \quad (14.17)
$$

where j ranges over the joints at the distant ends of all members framing into joint i.

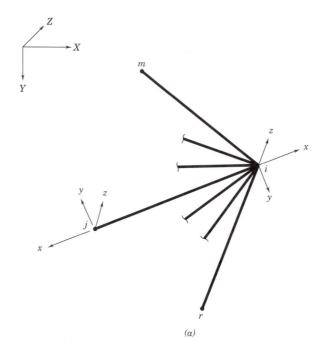

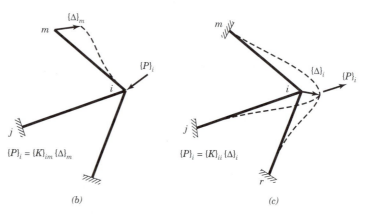

Figure 14.2 Stiffness relations for the ith node. (*a*) Members framing into node i. (*b*) $\{P\}_i = [K]_{im}\{\Delta\}_m$. (*c*) $\{P\}_i = [K]_{ii}\{\Delta\}_i$.

From Eq. 13.27, the member-end forces $\{F\}_{ij}$ can be expressed in terms of the member-end displacements $\{\delta\}_{ij}$ and $\{\delta\}_{ji}$ in the form

$$\{F\}_{ij} = [k]_{ii}^{j}\{\delta\}_{ij} + [k]_{ij}\{\delta\}_{ji} \tag{14.18}$$

where $[k]_{ii}^{j}$ and $[k]_{ij}$ are member stiffness submatrices that are defined in Section 13.4. If the individual members are to fit compatibly into the structure, there must be a relationship between the member-end displacements and the structure displacements. For instance, at node i

$$\{\delta\}_{ij} = [\beta]_{ij}\{\Delta\}_i \tag{14.19}$$

and at node j

$$\{\delta\}_{ji} = [\beta]_{ji}\{\Delta\}_j \tag{14.20}$$

where $[\beta]_{ij}$ and $[\beta]_{ji}$ are the *compatibility* or *connectivity matrices* at ends i and j, respectively. Substitution of Eqs. 14.19 and 14.20 into Eq. 14.18 gives

$$\{F\}_{ij} = [k]_{ii}^j[\beta]_{ij}\{\Delta\}_i + [k]_{ij}[\beta]_{ji}\{\Delta\}_j \tag{14.21}$$

From energy considerations, the work done at joint i must be the same whether expressed in structure quantities or member quantities. Thus, we can write

$$\tfrac{1}{2}\{P\}_i^T\{\Delta\}_i = \sum_{j=m}^{r} \tfrac{1}{2}\{F\}_{ij}^T\{\delta\}_{ij} \tag{14.22}$$

Substitution of Eq. 14.19 and cancellation of $\tfrac{1}{2}$ gives

$$\{P\}_i^T\{\Delta\}_i = \sum_{j=m}^{r} \{F\}_{ij}^T[\beta]_{ij}\{\Delta\}_i$$

Transposition of both sides and subsequent rearrangement give

$$\{\Delta\}_i^T\left(\{P\}_i - \sum_{j=m}^{r} [\beta]_{ij}^T\{F\}_{ij}\right) = 0 \tag{14.23}$$

Since $\{\Delta\}_i^T$ can be arbitrarily prescribed, the term in parentheses must vanish. This leads to

$$\{P\}_i = \sum_{j=m}^{r} [\beta]_{ij}^T\{F\}_{ij} \tag{14.24}$$

which states the condition of equilibrium at the ith node point. Substituting Eq. 14.21 into Eq. 14.24, we obtain

$$\{P\}_i = \sum_{j=m}^{r} ([\beta]_{ij}^T[k]_{ii}^j[\beta]_{ij}\{\Delta\}_i + [\beta]_{ij}^T[k]_{ij}[\beta]_{ji}\{\Delta\}_j) \tag{14.25}$$

A comparison of Eqs. 14.17 and 14.25 reveals that

$$[K]_{ii}^j = [\beta]_{ij}^T[k]_{ii}^j[\beta]_{ij} \tag{14.26}$$

$$[K]_{ij} = [\beta]_{ij}^T[k]_{ij}[\beta]_{ji} \tag{14.27}$$

Application of Eqs. 14.26 and 14.27 for all i and j along with the application of Eq. 14.16 leads to the submatrices that make up the total structure stiffness matrix of Eq. 14.15.

14.4 COMPATIBILITY MATRICES

Equations 14.19 and 14.20 express relationships between the member-end displacements at ends i and j of member ij and the structure displacements of nodes i and j, respectively. In each case, the $[\beta]$ matrix provides the connectivity between the member end and the structure node to which it is attached and ensures that the structure remains compatibly intact as it deforms. In the terminology of matrix algebra, $[\beta]$ is a transformation matrix, and its precise nature depends on the member type and the member orientation with regard to the structure coordinate system.

Consider the planar truss-type member shown in Fig. 14.3a. For this case, Eq. 14.19 becomes

$$(\delta_1)_{ij} = [\beta]_{ij} \begin{Bmatrix} \Delta_1 \\ \Delta_2 \end{Bmatrix}_i \tag{14.28}$$

This equation relates the axial member-end displacement to the two independent structure displacements at node i, and it can be further expanded to

$$(\delta_1)_{ij} = [l_{ix} \; m_{ix}] \begin{Bmatrix} \Delta_1 \\ \Delta_2 \end{Bmatrix}_i \tag{14.29}$$

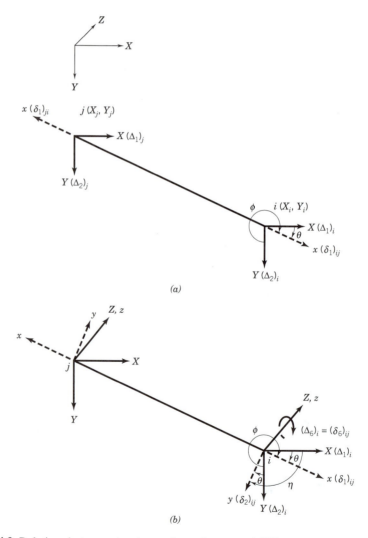

Figure 14.3 Relations between structure and member axes. (*a*) Planar truss-type member. (*b*) Planar beam-type member.

In this form, l and m are the direction cosines relating the member axis to the structure axes. Specifically, l_{ix} is the cosine of the angle between the X axis and the x axis, and m_{ix} is the cosine of the angle between the Y axis and the x axis—both angles being measured clockwise at the i end of the member. Reference to Fig. 14.3a shows that

$$l_{ix} = \cos \theta$$
$$m_{ix} = \cos \phi = \sin \theta \tag{14.30}$$

and thus Eq. 14.29 becomes

$$(\delta_1)_{ij} = [\cos \theta \quad \sin \theta] \left\{ \begin{array}{c} \Delta_1 \\ \Delta_2 \end{array} \right\}_i \tag{14.31}$$

For a full three-dimensional truss-type member, Eq. 14.29 takes the form

$$(\delta_1)_{ij} = [l_{ix} \quad m_{ix} \quad n_{ix}] \left\{ \begin{array}{c} \Delta_1 \\ \Delta_2 \\ \Delta_3 \end{array} \right\}_i \tag{14.32}$$

where n_{ix} is the cosine of the angle between the Z axis and the x axis, measured clockwise at the i end of the member. The direction cosines in Eqs. 14.29 and 14.32 are defined in general terms as

$$l_{ix} = \frac{X_i - X_j}{L_{ij}}; \quad m_{ix} = \frac{Y_i - Y_j}{L_{ij}}; \quad n_{ix} = \frac{Z_i - Z_j}{L_{ij}} \tag{14.33}$$

in which (X_i, Y_i, Z_i) and (X_j, Y_j, Z_j) are the structure coordinates of nodes i and j, respectively, and

$$L_{ij} = \sqrt{(X_i - X_j)^2 + (Y_i - Y_j)^2 + (Z_i - Z_j)^2} \tag{14.34}$$

At the j end of the truss member, the structure coordinate system remains the same, while the member x axis is reversed. Thus, $[\beta]_{ji} = -[\beta]_{ij}$.

For a planar beam-type member, the compatibility matrix takes on a more complicated format. Reference to Fig. 14.3b shows that in this case, Eq. 14.19 becomes

$$\left\{ \begin{array}{c} \delta_1 \\ \delta_2 \\ \delta_6 \end{array} \right\}_{ij} = [\beta]_{ij} \left\{ \begin{array}{c} \Delta_1 \\ \Delta_2 \\ \Delta_6 \end{array} \right\}_i \tag{14.35}$$

In terms of direction cosines, this relationship can be expressed as

$$\left\{ \begin{array}{c} \delta_1 \\ \delta_2 \\ \delta_6 \end{array} \right\}_{ij} = \left[\begin{array}{ccc} l_{ix} & m_{ix} & 0 \\ l_{iy} & m_{iy} & 0 \\ 0 & 0 & n_{iz} \end{array} \right] \left\{ \begin{array}{c} \Delta_1 \\ \Delta_2 \\ \Delta_6 \end{array} \right\}_i \tag{14.36}$$

where l_{iq}, m_{iq}, and n_{iq} are the direction cosines for the angles between the structure X, Y, and Z axes, respectively, and the member q axis. The zeros indicate that there are no compatibility relationships between member displacements and structure rotation or member rotation and structure displacements. For the planar member shown in Fig. 14.3b, we have

$$l_{ix} = \frac{X_i - X_j}{L_{ij}} = \cos\,\theta; \quad m_{ix} = \frac{Y_i - Y_j}{L_{ij}} = \cos\,\phi = \sin\,\theta;$$

$$l_{iy} = \cos\,\eta = -\sin\,\theta; \quad m_{iy} = \cos\,\theta; \quad n_{iz} = 1 \tag{14.37}$$

Using these direction cosines, we can write Eq. 14.36 as

$$
\begin{Bmatrix} \delta_1 \\ \delta_2 \\ \delta_6 \end{Bmatrix}_{ij} =
\begin{bmatrix} \cos\,\theta & \sin\,\theta & 0 \\ -\sin\,\theta & \cos\,\theta & 0 \\ 0 & 0 & 1 \end{bmatrix}
\begin{Bmatrix} \Delta_1 \\ \Delta_2 \\ \Delta_6 \end{Bmatrix}_i \tag{14.38}
$$

At the j end of a beam-type member, the structure coordinate system remains the same, as does the member z axis. However, the member x and y axes are reversed, and thus the direction cosines related to these axes have their signs reversed.

For a full three-dimensional arrangement of a beam-type member, Eq. 14.19 becomes

$$
\begin{Bmatrix} \delta_1 \\ \delta_2 \\ \delta_3 \\ \delta_4 \\ \delta_5 \\ \delta_6 \end{Bmatrix}_{ij} =
\begin{bmatrix}
l_{ix} & m_{ix} & n_{ix} & & & \\
l_{iy} & m_{iy} & n_{iy} & & 0 & \\
l_{iz} & m_{iz} & n_{iz} & & & \\
& & & l_{ix} & m_{ix} & n_{ix} \\
& 0 & & l_{iy} & m_{iy} & n_{iy} \\
& & & l_{iz} & m_{iz} & n_{iz}
\end{bmatrix}
\begin{Bmatrix} \Delta_1 \\ \Delta_2 \\ \Delta_3 \\ \Delta_4 \\ \Delta_5 \\ \Delta_6 \end{Bmatrix}_i \tag{14.39}
$$

The detailed procedures for determining the direction cosines for the three-dimensional situation are expanded in textbooks on matrix structural analysis.

14.5 APPLICATION OF STIFFNESS METHOD

The stiffness method is applied in accordance with a very orderly procedure. The step-by-step approach follows:

1. Number the nodal points of the structure. These nodes must include all load points, support points, and points where two or more members join together. This process implicitly introduces the structure forces and displacements in the form of Eq. 13.1; the member-end forces and displacements are defined by Eq. 13.2.

2. Form the individual compatibility matrices $[\beta]_{ij}$, which relate the member-end displacements of member ij to the structure displacements at node i for all combinations of i and j. The precise format of these matrices will depend on the arrays of member-end displacements and nodal displacements that must be connected, as explained in Section 14.4.

3. Establish the individual member stiffness matrices $[k]_{ii}^j$ and $[k]_{ij}$ in accordance with the designations of Eqs. 13.27 and 13.26 for all combinations of i and j throughout the structure. The makeup of these stiffness matrices will depend on the member-end forces and displacements that are assumed operable for the individual members.

4. Generate the structure stiffness submatrices $[K]_{ii}^j$ and $[K]_{ij}$ in accordance with Eqs. 14.26 and 14.27. These submatrices are combined to form the total structure stiffness matrix, as outlined by Eqs. 14.16 and 14.15.

5. Extract the reduced stiffness matrix $[K]_{LL}$ from the total stiffness matrix by deleting the rows and columns corresponding to the displacement components that are controlled through the boundary conditions.

6. Determine the free displacements $\{\Delta\}_L$ from Eq. 14.10 or 14.13. If desired, the member-end displacements may be determined from Eq. 14.19.

7. Calculate the final member-end forces from Eq. 14.21 or Eq. 14.18 and the reaction forces from Eq. 14.11 or 14.14.

The example problems that follow in this section illustrate the step-by-step procedure for analyzing a structure by the stiffness method.

It should be noted that the stiffness method makes no distinction between statically determinate and indeterminate structures. Thus, the method eliminates the need for the analyst to select redundant forces, and this precludes all of the problems that attend this decision-making process.

EXAMPLE 14.1

Determine the member-end forces and the displacements at the loaded point of the structure shown. Assume EA is the same for each member. The joint coordinates are given in parentheses.

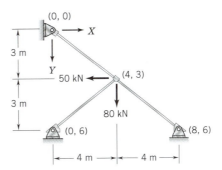

Selection of Node Points

Structure Forces and Displacements

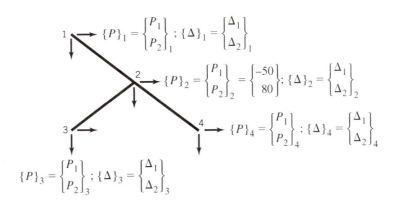

Member Forces and Displacements

$$(F_1)_{ij};\ (\delta_1)_{ij} \longleftarrow \overset{i}{\underline{\hspace{4cm}}} \overset{j}{\longrightarrow} (F_1)_{ji};\ (\delta_1)_{ji}$$

Compatibility Matrices

$$(\delta_1)_{ij} = [\beta]_{ij}\begin{Bmatrix}\Delta_1 \\ \Delta_2\end{Bmatrix}_i = \begin{bmatrix}l_{ix} & m_{ix}\end{bmatrix}\begin{Bmatrix}\Delta_1 \\ \Delta_2\end{Bmatrix}_i \tag{14.29}$$

where

$$l_{ix} = \frac{X_i - X_j}{L_{ij}}, \qquad m_{ix} = \frac{Y_i - Y_j}{L_{ij}} \tag{14.33}$$

Member 1-2: $l_{1x} = \dfrac{0-4}{5} = -0.8;\quad m_{1x} = \dfrac{0-3}{5} = -0.6$

$$[\beta]_{12} = [l_{1x}\ \ m_{1x}] = [0.8\ \ {-0.6}]$$
$$[\beta]_{21} = [0.8\ \ 0.6]$$

Member 2-3: $l_{2x} = \dfrac{4-0}{5} = 0.8;\quad m_{2x} = \dfrac{3-6}{5} = -0.6$

$$[\beta]_{23} = [0.8\ \ {-0.6}]$$
$$[\beta]_{32} = [{-0.8}\ \ 0.6]$$

Member 2-4: $l_{2x} = \dfrac{4-8}{5} = -0.8;\quad m_{2x} = \dfrac{3-6}{5} = -0.6$

$$[\beta]_{24} = [{-0.8}\ \ {-0.6}]$$
$$[\beta]_{42} = [\ \ 0.8\ \ \ 0.6]$$

Member Stiffnesses

$$\{F\}_{ij} = [k]_{ii}^{j}\{\delta\}_{ij} + [k]_{ij}\{\delta\}_{ji} \tag{13.27}$$

In this case,

$$[k]_{ii}^{j} = [k]_{ij} = \left(\frac{EA}{l}\right)_{ij} \tag{13.26}$$

Member 1-2: $[k]_{11}^{2} = [k]_{12} = \left(\dfrac{EA}{l}\right)_{12} = 0.2EA$

$$[k]_{22}^{1} = [k]_{21} = \left(\frac{EA}{l}\right)_{21} = 0.2EA$$

Member 2-3: $[k]_{22}^{3} = [k]_{23} = [k]_{33}^{2} = [k]_{32} = 0.2EA$

Member 2-4: $[k]_{22}^{4} = [k]_{24} = [k]_{44}^{2} = [k]_{42} = 0.2EA$

Structure Stiffness Matrix

Total Stiffness Equations

$$
\begin{Bmatrix} \{P\}_1 \\ \{P\}_2 \\ \{P\}_3 \\ \{P\}_4 \end{Bmatrix} = \begin{bmatrix} [K]_{11} & [K]_{12} & & \\ [K]_{21} & [K]_{22} & [K]_{23} & [K]_{24} \\ & [K]_{32} & [K]_{33} & \\ & [K]_{42} & & [K]_{44} \end{bmatrix} \begin{Bmatrix} \{\Delta\}_1 \\ \{\Delta\}_2 \\ \{\Delta\}_3 \\ \{\Delta\}_4 \end{Bmatrix}
\tag{14.15}
$$

Reduced Stiffness Equations

The rows and columns of the total stiffness matrix that correspond to the displacement constraints are deleted. That is,

$$
\{P\}_1 = [K]_{1,1}\{\Delta\}_1 \rightarrow \{P\}_2 = [K]_{22}\{\Delta\}_2
\tag{14.9}
$$

because

$$
\{\Delta\}_1 = \{\Delta\}_3 = \{\Delta\}_4 = \{0\}
$$

Required Submatrices

$$
[K]_{ii}^{j} = [\beta]_{ij}^{T}[k]_{ii}^{j}[\beta]_{ij}
\tag{14.26}
$$

$$
[K]_{ii} = \sum_{i} [K]_{ii}^{j}
\tag{14.16}
$$

$$
[K]_{22}^{1} = [\beta]_{21}^{T}[k]_{22}^{1}[\beta]_{21} = \begin{bmatrix} 0.8 \\ 0.6 \end{bmatrix}(0.2EA)[0.8 \quad 0.6] = EA\begin{bmatrix} 0.128 & 0.096 \\ 0.096 & 0.072 \end{bmatrix}
$$

$$
[K]_{22}^{3} = [\beta]_{23}^{T}[k]_{22}^{3}[\beta]_{23} = \begin{bmatrix} 0.8 \\ -0.6 \end{bmatrix}(0.2EA)[0.8 \quad -0.6] = EA\begin{bmatrix} 0.128 & -0.096 \\ -0.096 & 0.072 \end{bmatrix}
$$

$$
[K]_{22}^{4} = [\beta]_{24}^{T}[k]_{22}^{4}[\beta]_{24} = \begin{bmatrix} -0.8 \\ -0.6 \end{bmatrix}(0.2EA)[-0.8 \quad -0.6] = EA\begin{bmatrix} 0.128 & 0.096 \\ 0.096 & 0.072 \end{bmatrix}
$$

$$
[K]_{22} = [K]_{22}^{1} + [K]_{22}^{3} + [K]_{22}^{4} = EA\begin{bmatrix} 0.384 & 0.096 \\ 0.096 & 0.216 \end{bmatrix} = \frac{EA}{1000}\begin{bmatrix} 384 & 96 \\ 96 & 216 \end{bmatrix}
$$

Determination of Displacements

$$
\{\Delta\}_1 = [K]_{1,1}^{-1}\{P\}_1 \rightarrow \{\Delta\}_2 = [K]_{22}^{-1}\{P\}_2
\tag{14.10}
$$

$$
\{\Delta\}_2 = \frac{1000}{EA}\begin{bmatrix} 384 & 96 \\ 96 & 216 \end{bmatrix}^{-1}\begin{Bmatrix} -50 \\ 80 \end{Bmatrix} = \frac{1}{EA}\begin{Bmatrix} -250.65 \\ 481.77 \end{Bmatrix}
$$

Member Forces

$$
\{F\}_{ij} = [k]_{ii}^{j}[\beta]_{ij}\{\Delta\}_i + [k]_{ij}[\beta]_{ji}\{\Delta\}_j
\tag{14.21}
$$

Member 1-2: $\{F\}_{12} = [k]_{12}[\beta]_{21}\{\Delta\}_2 = (0.2EA)[0.8 \quad 0.6]\dfrac{1}{EA}\begin{Bmatrix} -250.65 \\ 481.77 \end{Bmatrix} = +17.71$ kN

Member 3-2: $\{F\}_{32} = [k]_{32}[\beta]_{23}\{\Delta\}_2 = (0.2EA)[0.8 \quad -0.6]\dfrac{1}{EA}\begin{Bmatrix} -250.65 \\ 481.77 \end{Bmatrix} = -97.92$ kN

Member 4-2: $\{F\}_{42} = [k]_{42}[\beta]_{24}\{\Delta\}_2 = (0.2EA)[-0.8 \quad -0.6]\dfrac{1}{EA}\begin{Bmatrix} -250.65 \\ 481.77 \end{Bmatrix} = -17.71$ kN

EXAMPLE 14.2

Determine the member-end forces for the structure shown, and construct the shear and moment diagrams. The joint coordinates are given in parentheses.

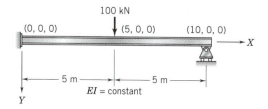

Selection of Node Points

Structure Forces and Displacements

Because of the nature of the applied structure forces, there will be no axial forces in the members and, therefore, no structure displacements along the X axis.

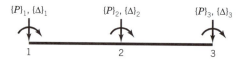

$$\{P\}_i = \begin{Bmatrix} P_2 \\ P_6 \end{Bmatrix}_i = \begin{Bmatrix} (P_2)_i \\ (P_6)_i \end{Bmatrix}$$

$$\{\Delta\}_i = \begin{Bmatrix} \Delta_2 \\ \Delta_6 \end{Bmatrix}_i = \begin{Bmatrix} (\Delta_2)_i \\ (\Delta_6)_i \end{Bmatrix}$$

Member Forces and Displacements

$$\{F\}_{ij} = \begin{Bmatrix} F_2 \\ F_6 \end{Bmatrix}_{ij} \; ; \quad \{F\}_{ji} = \begin{Bmatrix} F_2 \\ F_6 \end{Bmatrix}_{ji} \; ; \quad \{\delta\}_{ij} = \begin{Bmatrix} \delta_2 \\ \delta_6 \end{Bmatrix}_{ij} \; ; \quad \{\delta\}_{ji} = \begin{Bmatrix} \delta_2 \\ \delta_6 \end{Bmatrix}_{ji}$$

Member 1-2: $\{F\}_{12} \quad \{F\}_{21} \quad \{\delta\}_{12} \quad \{\delta\}_{21}$

Member 2-3: $\{F\}_{23} \quad \{F\}_{32} \quad \{\delta\}_{23} \quad \{\delta\}_{32}$

Compatibility Matrices

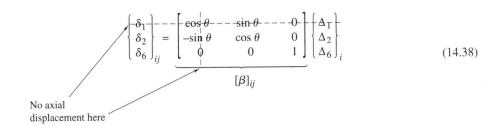

$$\left\{ \begin{array}{c} \delta_1 \\ \delta_2 \\ \delta_6 \end{array} \right\}_{ij} = \left[\begin{array}{ccc} \cos\theta & \sin\theta & 0 \\ -\sin\theta & \cos\theta & 0 \\ 0 & 0 & 1 \end{array} \right] \left\{ \begin{array}{c} \Delta_T \\ \Delta_2 \\ \Delta_6 \end{array} \right\}_i \tag{14.38}$$

$$\underbrace{\hphantom{\left[\begin{array}{ccc} \cos\theta & \sin\theta & 0 \end{array} \right]}}_{[\beta]_{ij}}$$

No axial displacement here

Member 1-2:

$$\cos\theta = l_{1x} = \frac{X_1 - X_2}{L_{12}} = \frac{0-5}{5} = -1; \qquad \sin\theta = 0 \tag{14.37}$$

$$[\beta]_{12} = \left[\begin{array}{ccc} -1 & 0 & 0 \\ 0 & -1 & 0 \\ 0 & 0 & 1 \end{array} \right]$$

$$[\beta]_{21} = \left[\begin{array}{ccc} -1 & 0 & 0 \\ 0 & 1 & 0 \\ 0 & 0 & 1 \end{array} \right]$$

Member 2-3:

$$\cos\theta = l_{2x} = \frac{X_2 - X_3}{L_{23}} = \frac{5-10}{5} = -1; \qquad \sin\theta = 0 \tag{14.37}$$

$$[\beta]_{23} = \left[\begin{array}{ccc} -1 & 0 & 0 \\ 0 & -1 & 0 \\ 0 & 0 & 1 \end{array} \right]$$

$$[\beta]_{32} = \left[\begin{array}{ccc} -1 & 0 & 0 \\ 0 & 1 & 0 \\ 0 & 0 & 1 \end{array} \right]$$

Member Stiffnesses

(From Eqs. 13.27 and 13.26)

$$\text{Member 1-2: } [k]^2_{11} = [k]^1_{22} = \left[\begin{array}{cc} \dfrac{12EI_z}{l^3} & -\dfrac{6EI_z}{l^2} \\[3mm] -\dfrac{6EI_z}{l^2} & \dfrac{4EI_z}{l} \end{array} \right]_{12} = EI \left[\begin{array}{cc} 0.096 & -0.24 \\ -0.24 & 0.80 \end{array} \right]$$

$$[k]_{12} = [k]_{21} = \begin{bmatrix} \dfrac{12EI_z}{l^3} & -\dfrac{6EI_z}{l^2} \\[3mm] -\dfrac{6EI_z}{l^2} & \dfrac{2EI_z}{l} \end{bmatrix}_{12} = EI \begin{bmatrix} 0.096 & -0.24 \\ -0.24 & 0.40 \end{bmatrix}$$

Member 2-3: $[k]_{22}^3 = [k]_{33}^2 = EI \begin{bmatrix} 0.096 & -0.24 \\ -0.24 & 0.80 \end{bmatrix}$

$$[k]_{23} = [k]_{32} = EI \begin{bmatrix} 0.096 & -0.24 \\ -0.24 & 0.40 \end{bmatrix}$$

Structure Stiffness Matrix

Total Stiffness Equations

$$\begin{Bmatrix} \begin{Bmatrix} P_2 \\ P_6 \end{Bmatrix}_1 \\ \begin{Bmatrix} P_2 \\ P_6 \end{Bmatrix}_2 \\ \begin{Bmatrix} P_2 \\ P_6 \end{Bmatrix}_3 \end{Bmatrix} = \begin{bmatrix} [K]_{11} & [K]_{12} & [0] \\ [K]_{21} & [K]_{22} & [K]_{23} \\ [0] & [K]_{32} & [K]_{33} \end{bmatrix} \begin{Bmatrix} \begin{Bmatrix} \Delta_2 \\ \Delta_6 \end{Bmatrix}_1 \\ \begin{Bmatrix} \Delta_2 \\ \Delta_6 \end{Bmatrix}_2 \\ \begin{Bmatrix} \Delta_2 \\ \Delta_6 \end{Bmatrix}_3 \end{Bmatrix} \tag{14.15}$$

Reduced Stiffness Equations

$$\{P\}_I = [K]_{I,I} \{\Delta\}_I \rightarrow \begin{Bmatrix} \begin{Bmatrix} P_2 \\ P_6 \end{Bmatrix}_2 \\ \{P_6\}_3 \end{Bmatrix} = [K]_{I,I} \begin{Bmatrix} \begin{Bmatrix} \Delta_2 \\ \Delta_6 \end{Bmatrix}_2 \\ \{\Delta_6\}_3 \end{Bmatrix} \tag{14.9}$$

because

$$(\Delta_2)_1 = (\Delta_6)_1 = (\Delta_2)_3 = 0$$

Required Submatrices

$$[K]_{ii}^j = [\beta]_{ij}^T [k]_{ii}^j [\beta]_{ij} \tag{14.26}$$

$$[K]_{ii} = \sum_j [K]_{ii}^j \tag{14.16}$$

$$[K]_{22}^1 = [\beta]_{21}^T [k]_{22}^1 [\beta]_{21} = \begin{bmatrix} 1 & 0 \\ 0 & 1 \end{bmatrix} EI \begin{bmatrix} 0.096 & -0.24 \\ -0.24 & 0.80 \end{bmatrix} \begin{bmatrix} 1 & 0 \\ 0 & 1 \end{bmatrix}$$

$$= EI \begin{bmatrix} 0.096 & -0.24 \\ -0.24 & 0.80 \end{bmatrix}$$

$$[K]_{22}^3 = [\beta]_{23}^T [k]_{22}^3 [\beta]_{23} = \begin{bmatrix} -1 & 0 \\ 0 & 1 \end{bmatrix} EI \begin{bmatrix} 0.096 & -0.24 \\ -0.24 & 0.80 \end{bmatrix} \begin{bmatrix} -1 & 0 \\ 0 & 1 \end{bmatrix}$$

$$= EI \begin{bmatrix} 0.096 & 0.24 \\ 0.24 & 0.80 \end{bmatrix}$$

$$[K]_{22} = [K]_{22}^1 + [K]_{22}^3 = EI \begin{bmatrix} 0.192 & 0 \\ 0 & 1.60 \end{bmatrix}$$

$$[K]_{33} = [K]_{33}^2 = [\beta]_{32}^T [k]_{33}^2 [\beta]_{32} = \begin{bmatrix} 1 & 0 \\ 0 & 1 \end{bmatrix} EI \begin{bmatrix} 0.096 & -0.24 \\ -0.24 & 0.80 \end{bmatrix} \begin{bmatrix} 1 & 0 \\ 0 & 1 \end{bmatrix}$$

$$= EI \begin{bmatrix} 0.096 & -0.24 \\ -0.24 & 0.80 \end{bmatrix}$$

$$[K]_{ij} = [\beta]_{ij}^T [k]_{ij} [\beta]_{ji} \tag{14.27}$$

$$[K]_{23} = [\beta]_{23}^T [k]_{23} [\beta]_{32} = \begin{bmatrix} -1 & 0 \\ 0 & 1 \end{bmatrix} EI \begin{bmatrix} 0.096 & -0.24 \\ -0.24 & 0.40 \end{bmatrix} \begin{bmatrix} 1 & 0 \\ 0 & 1 \end{bmatrix}$$

$$= EI \begin{bmatrix} -0.096 & 0.24 \\ -0.24 & 0.40 \end{bmatrix}$$

$$[K]_{32} = [\beta]_{32}^T [k]_{32} [\beta]_{23} = [K]_{23}^T = EI \begin{bmatrix} -0.096 & -0.24 \\ 0.24 & 0.40 \end{bmatrix}$$

Thus, the reduced stiffness matrix becomes

$$[K]_{I,I} = EI \begin{bmatrix} 0.192 & 0 & 0.24 \\ 0 & 1.60 & 0.40 \\ 0.24 & 0.40 & 0.80 \end{bmatrix}$$

Determination of Displacements

$$\{\Delta\}_I = [K]_{I,I}^{-1} \{P\}_I \tag{14.10}$$

$$\begin{Bmatrix} \begin{Bmatrix} \Delta_2 \\ \Delta_6 \end{Bmatrix}_2 \\ \{\Delta_6\}_3 \end{Bmatrix} = \frac{1}{EI} \begin{bmatrix} 0.192 & 0 & 0.24 \\ 0 & 1.60 & 0.40 \\ 0.24 & 0.40 & 0.80 \end{bmatrix}^{-1} \begin{Bmatrix} \begin{Bmatrix} P_2 \\ P_6 \end{Bmatrix}_2 \\ \{P_6\}_3 \end{Bmatrix}$$

$$= \frac{1}{EI} \begin{bmatrix} 9.1146 & 0.7813 & -3.125 \\ 0.7813 & 0.7813 & -0.6250 \\ -3.125 & -0.6250 & 2.500 \end{bmatrix} \begin{Bmatrix} 100 \\ 0 \\ 0 \end{Bmatrix}$$

$$= \frac{1}{EI} \begin{Bmatrix} 911.46 \\ 78.13 \\ -312.50 \end{Bmatrix}$$

Member Forces

$$\{F\}_{ij} = [k]_{ii}^j [\beta]_{ij} \{\Delta\}_i + [k]_{ij} [\beta]_{ji} \{\Delta\}_j \tag{14.21}$$

Member 1-2: $\{\Delta\}_1 = \{0\}$; $\{\Delta\}_2 = \frac{1}{EI} \begin{Bmatrix} 911.46 \\ 78.13 \end{Bmatrix}$

$$\{F\}_{12} = [k]_{12}[\beta]_{21}\{\Delta\}_2$$

$$= EI\begin{bmatrix} 0.096 & -0.24 \\ -0.24 & 0.40 \end{bmatrix}\begin{bmatrix} 1 & 0 \\ 0 & 1 \end{bmatrix}\frac{1}{EI}\begin{Bmatrix} 911.46 \\ 78.13 \end{Bmatrix}$$

$$= \begin{Bmatrix} 68.75 \text{ kN} \\ -187.50 \text{ kN} \cdot \text{m} \end{Bmatrix}$$

$$\{F\}_{21} = [k]_{22}^{1}[\beta]_{21}\{\Delta\}_2$$

$$= EI\begin{bmatrix} 0.096 & -0.24 \\ -0.24 & 0.80 \end{bmatrix}\begin{bmatrix} 1 & 0 \\ 0 & 1 \end{bmatrix}\frac{1}{EI}\begin{Bmatrix} 911.46 \\ 78.13 \end{Bmatrix}$$

$$= \begin{Bmatrix} 68.75 \text{ kN} \\ -156.25 \text{ kN} \cdot \text{m} \end{Bmatrix}$$

68.75 kN

187.50 kN · m   1   2   156.25 kN · m

68.75 kN

Member 2-3: $\{\Delta\}_2 = \dfrac{1}{EI}\begin{Bmatrix} 911.46 \\ 78.13 \end{Bmatrix}$; $\{\Delta\}_3 = \dfrac{1}{EI}\begin{Bmatrix} 0 \\ -312.50 \end{Bmatrix}$

$$\{F\}_{23} = [k]_{22}^{3}[\beta]_{23}\{\Delta\}_2 + [k]_{23}[\beta]_{32}\{\Delta\}_3$$

$$= EI\begin{bmatrix} 0.096 & -0.24 \\ -0.24 & 0.80 \end{bmatrix}\begin{bmatrix} -1 & 0 \\ 0 & 1 \end{bmatrix}\frac{1}{EI}\begin{Bmatrix} 911.46 \\ 78.13 \end{Bmatrix}$$

$$+ EI\begin{bmatrix} 0.096 & -0.24 \\ -0.24 & 0.40 \end{bmatrix}\begin{bmatrix} 1 & 0 \\ 0 & 1 \end{bmatrix}\frac{1}{EI}\begin{Bmatrix} 0 \\ -312.50 \end{Bmatrix}$$

$$= \begin{Bmatrix} -106.25 \\ 281.25 \end{Bmatrix} + \begin{Bmatrix} 75.00 \\ -125.00 \end{Bmatrix}$$

$$= \begin{Bmatrix} -31.25 \text{ kN} \\ 156.25 \text{ kN} \cdot \text{m} \end{Bmatrix}$$

$$\{F\}_{32} = [k]_{33}^{2}[\beta]_{32}\{\Delta\}_3 + [k]_{32}[\beta]_{23}\{\Delta\}_2$$

$$= EI\begin{bmatrix} 0.096 & -0.24 \\ -0.24 & 0.80 \end{bmatrix}\begin{bmatrix} 1 & 0 \\ 0 & 1 \end{bmatrix}\frac{1}{EI}\begin{Bmatrix} 0 \\ -312.50 \end{Bmatrix}$$

$$+ EI\begin{bmatrix} 0.096 & -0.24 \\ -0.24 & 0.40 \end{bmatrix}\begin{bmatrix} -1 & 0 \\ 0 & 1 \end{bmatrix}\frac{1}{EI}\begin{Bmatrix} 911.46 \\ 78.13 \end{Bmatrix}$$

$$= \begin{Bmatrix} 75.00 \\ -250.00 \end{Bmatrix} + \begin{Bmatrix} -106.25 \\ +250.00 \end{Bmatrix}$$

$$= \begin{Bmatrix} -31.25 \text{ kN} \\ 0 \quad \text{kN} \cdot \text{m} \end{Bmatrix}$$

Shear and Moment Diagrams

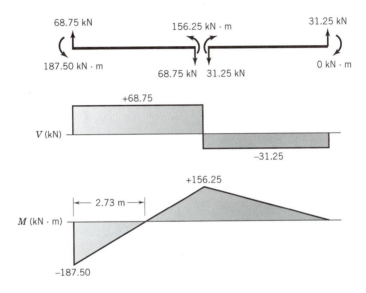

Note: We start with the forces at the left end of each member and use the static relations for shear and moment to obtain the forces at the right end of the member. These should be consistent with the computed forces at the right end.

The sign convention used for the shear and moment diagrams is that which was introduced in Chapter 5.

EXAMPLE 14.3

Determine the member-end forces for the given frame structure. The joint coordinates are noted in parentheses.

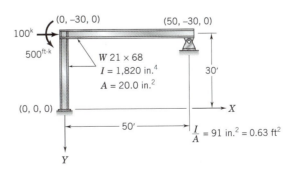

Selection of Node Points

Structure Forces and Displacements

$$\{P\}_i = \begin{Bmatrix} P_1 \\ P_2 \\ P_6 \end{Bmatrix}_i = \begin{Bmatrix} (P_1)_i \\ (P_2)_i \\ (P_6)_i \end{Bmatrix}$$

$$\{\Delta\}_i = \begin{Bmatrix} \Delta_1 \\ \Delta_2 \\ \Delta_6 \end{Bmatrix}_i = \begin{Bmatrix} (\Delta_1)_i \\ (\Delta_2)_i \\ (\Delta_6)_i \end{Bmatrix}$$

$$\{P\}_2 = \begin{Bmatrix} 100 \\ 0 \\ -500 \end{Bmatrix}$$

$$\{\Delta\}_3 = \begin{Bmatrix} 0 \\ 0 \\ (\Delta_6)_3 \end{Bmatrix}$$

$$\{\Delta\}_1 = \begin{Bmatrix} 0 \\ 0 \\ 0 \end{Bmatrix}$$

Member Forces and Displacements

$$\{F\}_{ij} = \begin{Bmatrix} F_1 \\ F_2 \\ F_6 \end{Bmatrix}_{ij} \;; \quad \{F\}_{ji} = \begin{Bmatrix} F_1 \\ F_2 \\ F_6 \end{Bmatrix}_{ji} \;; \quad \{\delta\}_{ij} = \begin{Bmatrix} \delta_1 \\ \delta_2 \\ \delta_6 \end{Bmatrix}_{ij} \;; \quad \{\delta\}_{ji} = \begin{Bmatrix} \delta_1 \\ \delta_2 \\ \delta_6 \end{Bmatrix}_{ji}$$

Member 1-2: $\{F\}_{12}$ $\{F\}_{21}$ $\{\delta\}_{12}$ $\{\delta\}_{21}$

Member 2-3: $\{F\}_{23}$ $\{F\}_{32}$ $\{\delta\}_{23}$ $\{\delta\}_{32}$

Compatibility Matrices

$$\begin{Bmatrix} \delta_1 \\ \delta_2 \\ \delta_6 \end{Bmatrix}_{ij} = \underbrace{\begin{bmatrix} \cos\theta & \sin\theta & 0 \\ -\sin\theta & \cos\theta & 0 \\ 0 & 0 & 1 \end{bmatrix}}_{[\beta]_{ij}} \begin{Bmatrix} \Delta_1 \\ \Delta_2 \\ \Delta_6 \end{Bmatrix}_i \qquad (14.38)$$

Member 1-2:

$$\cos\theta = l_{1x} = \frac{X_1 - X_2}{L_{12}} = \frac{0 - 0}{30} = 0$$

$$\sin\theta = m_{1x} = \frac{Y_1 - Y_2}{L_{12}} = \frac{0 - (-30)}{30} = 1 \qquad (14.37)$$

$$[\beta]_{12} = \begin{bmatrix} 0 & 1 & 0 \\ -1 & 0 & 0 \\ 0 & 0 & 1 \end{bmatrix}; \qquad [\beta]_{21} = \begin{bmatrix} 0 & -1 & 0 \\ 1 & 0 & 0 \\ 0 & 0 & 1 \end{bmatrix}$$

Member 2-3:

$$\cos\theta = l_{2x} = \frac{X_2 - X_3}{L_{23}} = \frac{0 - 50}{50} = -1$$

$$\sin\theta = m_{2x} = \frac{Y_2 - Y_3}{L_{23}} = \frac{-30 - (-30)}{50} = 0 \qquad (14.37)$$

$$[\beta]_{23} = \begin{bmatrix} -1 & 0 & 0 \\ 0 & -1 & 0 \\ 0 & 0 & 1 \end{bmatrix}; \qquad [\beta]_{32} = \begin{bmatrix} 1 & 0 & 0 \\ 0 & 1 & 0 \\ 0 & 0 & 1 \end{bmatrix}$$

Member Stiffnesses

(Extract from Eqs. 13.26 and 13.27.)

$$[k]_{ii}^j = \begin{bmatrix} \dfrac{EA}{l} & 0 & 0 \\ 0 & \dfrac{12EI_z}{l^3} & \dfrac{-6EI_z}{l^2} \\ 0 & \dfrac{-6EI_z}{l^2} & \dfrac{4EI_z}{l} \end{bmatrix}_{ij}$$

$$[k]_{ij} = \begin{bmatrix} \dfrac{EA}{l} & 0 & 0 \\ 0 & \dfrac{12EI_z}{l^3} & \dfrac{-6EI_z}{l^2} \\ 0 & \dfrac{-6EI_z}{l^2} & \dfrac{2EI_z}{l} \end{bmatrix}_{ij}$$

$$(13.26, 13.27)$$

Member 1-2: $l = 30'$

$$[k]_{11}^2 = [k]_{22}^1 = EI \begin{bmatrix} 0.052,9 & 0 & 0 \\ 0 & 0.000,444 & -0.006,67 \\ 0 & -0.006,67 & 0.133 \end{bmatrix}$$

$$[k]_{12} = [k]_{21} = EI \begin{bmatrix} 0.052,9 & 0 & 0 \\ 0 & 0.000,444 & -0.006,67 \\ 0 & -0.006,67 & 0.066,7 \end{bmatrix}$$

Member 2-3: $l = 50'$

$$[k]_{22}^3 = [k]_{33}^2 = EI \begin{bmatrix} 0.031,7 & 0 & 0 \\ 0 & 0.000,096,0 & -0.002,40 \\ 0 & -0.002,40 & 0.080,0 \end{bmatrix}$$

$$[k]_{23} = [k]_{32} = EI \begin{bmatrix} 0.031,7 & 0 & 0 \\ 0 & 0.000,096,0 & -0.002,40 \\ 0 & -0.002,40 & 0.040,0 \end{bmatrix}$$

Structure Stiffness Matrix

Total Stiffness Equations

$$
\left\{ \begin{array}{c} \left\{ \begin{array}{c} P_1 \\ P_2 \\ P_6 \end{array} \right\}_1 \\ \left\{ \begin{array}{c} P_1 \\ P_2 \\ P_6 \end{array} \right\}_2 \\ \left\{ \begin{array}{c} P_1 \\ P_2 \\ P_6 \end{array} \right\}_3 \end{array} \right\} = \left[\begin{array}{c:c:c} [K]_{11} & [K]_{12} & [0] \\ \hdashline [K]_{21} & [K]_{22} & [K]_{23} \\ \hdashline [0] & [K]_{32} & [K]_{33} \end{array} \right] \left\{ \begin{array}{c} \left\{ \begin{array}{c} \Delta_1 \\ \Delta_2 \\ \Delta_6 \end{array} \right\}_1 \\ \left\{ \begin{array}{c} \Delta_1 \\ \Delta_2 \\ \Delta_6 \end{array} \right\}_2 \\ \left\{ \begin{array}{c} \Delta_1 \\ \Delta_2 \\ \Delta_6 \end{array} \right\}_3 \end{array} \right\} \tag{14.15}
$$

Reduced Stiffness Equations

$$
\{P\}_{\mathrm{I}} = [K]_{\mathrm{I,I}}\{\Delta\}_{\mathrm{I}} \Rightarrow \left\{ \begin{array}{c} \left\{ \begin{array}{c} P_1 \\ P_2 \\ P_6 \end{array} \right\}_2 \\ \{P_6\}_3 \end{array} \right\} = [K]_{\mathrm{I,I}} \left\{ \begin{array}{c} \left\{ \begin{array}{c} \Delta_1 \\ \Delta_2 \\ \Delta_6 \end{array} \right\}_2 \\ \{\Delta_6\}_3 \end{array} \right\} \tag{14.9}
$$

because $\{\Delta\}_1 = \{0\}$ and $(\Delta_1)_3 = (\Delta_2)_3 = 0$

Required Submatrices

$$
[K]_{ii}^{j} = [\beta]_{ij}^{T}[k]_{ii}^{j}[\beta]_{ij} \tag{14.26}
$$

$$
[K]_{ii} = \sum_{i} [K]_{ii}^{j} \tag{14.16}
$$

$[K]_{22}^{1} = [\beta]_{21}^{T}[k]_{22}^{1}[\beta]_{21}$

$$
= \begin{bmatrix} 0 & 1 & 0 \\ -1 & 0 & 0 \\ 0 & 0 & 1 \end{bmatrix} \cdot EI \begin{bmatrix} 0.052{,}9 & 0 & 0 \\ 0 & 0.000{,}444 & -0.006{,}67 \\ 0 & -0.006{,}67 & 0.133 \end{bmatrix} \begin{bmatrix} 0 & -1 & 0 \\ 1 & 0 & 0 \\ 0 & 0 & 1 \end{bmatrix}
$$

$$
= EI \begin{bmatrix} 0.000{,}444 & 0 & -0.006{,}67 \\ 0 & 0.052{,}9 & 0 \\ -0.006{,}67 & 0 & 0.133 \end{bmatrix}
$$

$[K]_{22}^{3} = [\beta]_{23}^{T}[k]_{22}^{3}[\beta]_{23}$

$$
= \begin{bmatrix} -1 & 0 & 0 \\ 0 & -1 & 0 \\ 0 & 0 & 1 \end{bmatrix} \cdot EI \begin{bmatrix} 0.031{,}7 & 0 & 0 \\ 0 & 0.000{,}096{,}0 & -0.002{,}40 \\ 0 & -0.002{,}40 & 0.080{,}0 \end{bmatrix} \begin{bmatrix} -1 & 0 & 0 \\ 0 & -1 & 0 \\ 0 & 0 & 1 \end{bmatrix}
$$

$$
= EI \begin{bmatrix} 0.031{,}7 & 0 & 0 \\ 0 & 0.000{,}096{,}0 & 0.002{,}40 \\ 0 & 0.002{,}40 & 0.080{,}0 \end{bmatrix}
$$

$$[K]_{22} = [K]_{22}^1 + [K]_{22}^3$$

$$= EI \begin{bmatrix} 0.032,14 & 0 & -0.006,67 \\ 0 & 0.053,0 & 0.002,40 \\ -0.006,67 & 0.002,40 & 0.213 \end{bmatrix}$$

$$[K]_{33} = [K]_{33}^2 = [\beta]_{32}^T [k]_{33}^2 [\beta]_{32}$$

$$= \begin{bmatrix} 1 & 0 & 0 \\ 0 & 1 & 0 \\ 0 & 0 & 1 \end{bmatrix} \cdot EI \begin{bmatrix} 0.031,7 & 0 & 0 \\ 0 & 0.000,096,0 & -0.002,40 \\ 0 & -0.002,40 & 0.080,0 \end{bmatrix} \begin{bmatrix} 1 & 0 & 0 \\ 0 & 1 & 0 \\ 0 & 0 & 1 \end{bmatrix}$$

$$= EI \begin{bmatrix} 0.031,7 & 0 & 0 \\ 0 & 0.000,096,0 & -0.002,40 \\ 0 & -0.002,40 & 0.080,0 \end{bmatrix}$$

$$[K]_{ij} = [\beta]_{ij}^T [k]_{ij} [\beta]_{ji} \tag{14.27}$$

$$[K]_{23} = [\beta]_{23}^T [k]_{23} [\beta]_{32}$$

$$= \begin{bmatrix} -1 & 0 & 0 \\ 0 & -1 & 0 \\ 0 & 0 & 1 \end{bmatrix} \cdot EI \begin{bmatrix} 0.031,7 & 0 & 0 \\ 0 & 0.000,096,0 & -0.002,40 \\ 0 & -0.002,40 & 0.040,0 \end{bmatrix} \begin{bmatrix} 1 & 0 & 0 \\ 0 & 1 & 0 \\ 0 & 0 & 1 \end{bmatrix}$$

$$= EI \begin{bmatrix} -0.031,7 & 0 & 0 \\ 0 & -0.000,096,0 & 0.002,40 \\ 0 & -0.002,40 & 0.040,0 \end{bmatrix}$$

$$[K]_{32} = [\beta]_{32}^T [k]_{32} [\beta]_{23} = [K]_{23}^T$$

$$= EI \begin{bmatrix} -0.031,7 & 0 & 0 \\ 0 & -0.000,096,0 & -0.002,40 \\ 0 & -0.002,40 & 0.040,0 \end{bmatrix}$$

Thus, the reduced stiffness matrix becomes

$$[K]_{I,I} = EI \begin{bmatrix} 0.032,14 & 0 & -0.006,67 & 0 \\ 0 & 0.053,0 & 0.002,40 & 0.002,40 \\ -0.006,67 & 0.002,40 & 0.213 & 0.040,0 \\ 0 & 0.002,40 & 0.040,0 & 0.080,0 \end{bmatrix}$$

Determination of Displacements

$$\{\Delta\}_I = [K]_{I,I}^{-1} \{P\}_I \tag{14.10}$$

$$\left\{ \begin{Bmatrix} \Delta_1 \\ \Delta_2 \\ \Delta_6 \end{Bmatrix}_2 \\ \{\Delta_6\}_3 \right\} = [K]_{I,I}^{-1} \begin{Bmatrix} 100 \\ 0 \\ -500 \\ 0 \end{Bmatrix} = \frac{1}{EI} \begin{Bmatrix} 2,592 \\ 56.71 \\ -2,501 \\ 1,249 \end{Bmatrix}$$

Member Forces

$${F}_{ij} = [k]^{j}_{ii}[\beta]_{ij}{\Delta}_i + [k]_{ij}[\beta]_{ji}{\Delta}_j \qquad (14.21)$$

Member 1-2: ${\Delta}_1 = {0}$ $\qquad {\Delta}_2 = \dfrac{1}{EI}\begin{Bmatrix} 2{,}592 \\ 56.71 \\ -2{,}501 \end{Bmatrix}$

$[F]_{12} = [k]_{12}[\beta]_{21}{\Delta}_2$

$$= EI\begin{bmatrix} 0.052{,}9 & 0 & 0 \\ 0 & 0.000{,}444 & -0.006{,}67 \\ 0 & -0.006{,}67 & 0.066{,}7 \end{bmatrix}\begin{bmatrix} 0 & -1 & 0 \\ 1 & 0 & 1 \\ 0 & 0 & 1 \end{bmatrix}\frac{1}{EI}\begin{Bmatrix} 2{,}592 \\ 56.71 \\ -2{,}501 \end{Bmatrix}$$

$$= \begin{Bmatrix} -3.00^k \\ 17.83^k \\ -184.00'^{-k} \end{Bmatrix}$$

$[F]_{21} = [k]^{1}_{22}[\beta]_{21}{\Delta}_2$

$$= EI\begin{bmatrix} 0.052{,}9 & 0 & 0 \\ 0 & 0.000{,}444 & -0.006{,}67 \\ 0 & -0.006{,}67 & 0.133 \end{bmatrix}\begin{bmatrix} 0 & -1 & 0 \\ 1 & 0 & 0 \\ 0 & 0 & 1 \end{bmatrix}\frac{1}{EI}\begin{Bmatrix} 2{,}592 \\ 56.71 \\ -2{,}501 \end{Bmatrix}$$

$$= \begin{Bmatrix} -3.00^k \\ 17.83^k \\ -349.92'^{-k} \end{Bmatrix}$$

Member 2-3: ${\Delta}_2 = \dfrac{1}{EI}\begin{Bmatrix} 2{,}592 \\ 56.71 \\ -2{,}501 \end{Bmatrix}$ $\qquad {\Delta}_3 = \dfrac{1}{EI}\begin{Bmatrix} 0 \\ 0 \\ 1{,}249 \end{Bmatrix}$

${F}_{23} = [k]^{3}_{22}[\beta]_{23}{\Delta}_2 + [k]_{23}[\beta]_{32}{\Delta}_3$

$$= EI\begin{bmatrix} 0.031{,}7 & 0 & 0 \\ 0 & 0.000{,}096{,}0 & -0.002{,}40 \\ 0 & -0.002{,}40 & 0.080{,}0 \end{bmatrix}\begin{bmatrix} -1 & 0 & 0 \\ 0 & -1 & 0 \\ 0 & 0 & 1 \end{bmatrix}\frac{1}{EI}\begin{Bmatrix} 2{,}592 \\ 56.71 \\ -2{,}501 \end{Bmatrix}$$

$$+ EI\begin{bmatrix} 0.031{,}7 & 0 & 0 \\ 0 & 0.000{,}096{,}0 & -0.002{,}40 \\ 0 & -0.002{,}40 & 0.040{,}0 \end{bmatrix}\begin{bmatrix} 1 & 0 & 0 \\ 0 & 1 & 0 \\ 0 & 0 & 1 \end{bmatrix}\frac{1}{EI}\begin{Bmatrix} 0 \\ 0 \\ 1{,}249 \end{Bmatrix}$$

$$= \begin{Bmatrix} -82.17 + 0 \\ 6.00 - 3.00 \\ -199.95 + 49.96 \end{Bmatrix} = \begin{Bmatrix} -82.17^k \\ 3.00^k \\ -149.99'^{-k} \end{Bmatrix}$$

$$\{F\}_{32} = [k]^{2}_{33}[\beta]_{32}\{\Delta\}_{3} + [k]_{32}[\beta]_{23}\{\Delta\}_{2}$$

$$= EI \begin{bmatrix} 0.031,7 & 0 & 0 \\ 0 & 0.000,096,0 & -0.002,40 \\ 0 & -0.002,40 & 0.080,0 \end{bmatrix} \begin{bmatrix} 1 & 0 & 0 \\ 0 & 1 & 0 \\ 0 & 0 & 1 \end{bmatrix} \frac{1}{EI} \begin{Bmatrix} 0 \\ 0 \\ 1,249 \end{Bmatrix}$$

$$+ EI \begin{bmatrix} 0.031,7 & 0 & 0 \\ 0 & 0.000,096,0 & -0.002,40 \\ 0 & -0.002,40 & 0.040,0 \end{bmatrix} \begin{bmatrix} -1 & 0 & 0 \\ 0 & -1 & 0 \\ 0 & 0 & 1 \end{bmatrix} \frac{1}{EI} \begin{Bmatrix} 2,592 \\ 56.21 \\ -2,501 \end{Bmatrix}$$

$$= \begin{Bmatrix} 0 - 82.17 \\ -3.00 + 6.00 \\ 99.92 - 99.91 \end{Bmatrix} = \begin{Bmatrix} -82.17^{k} \\ +3.00^{k} \\ 0.01^{\prime-k} \end{Bmatrix}$$

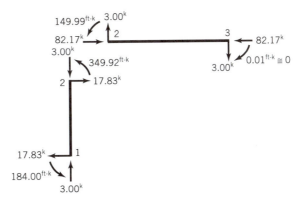

Notes:

- The shear and moment diagrams could be constructed in the normal fashion.
- If the normal frame analysis was used in which axial deformations are ignored, such as the slope deflection method, the member-end forces would be

$$\{F\}_{12} = \begin{Bmatrix} -3.10^{k} \\ 17.25^{k} \\ -172.5^{\prime-k} \end{Bmatrix}; \qquad \{F\}_{23} = \begin{Bmatrix} -82.75^{k} \\ +3.10^{k} \\ -155.0^{\prime-k} \end{Bmatrix}$$

14.6   LOADS APPLIED BETWEEN NODE POINTS—EQUIVALENT STRUCTURE FORCES

The stiffness method as it has been presented thus far admits the application of loads only at the structure nodes. This constitutes a serious constraint because it forces the analyst to assign a node to each point where a concentrated load is applied, and it precludes the application of distributed loads along the members. This limitation will now be relaxed.

Consider span *ij* of the planar beam in Fig. 14.4*a* with a transverse load acting along the member. Assume that the member is temporarily fixed at its ends, as shown in Fig. 14.4*b*. The fixed-end member forces corresponding to this condition

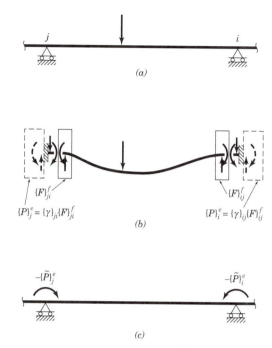

Figure 14.4 Loads applied along member lengths. *(a)* Span *ij* and loading. *(b)* Member fixed-end forces and equivalent structure forces. *(c)* Application of negative equivalent structure forces.

are determined in accordance with the formulae given in the table inside the back cover and are arranged in the fixed-end member force vectors $\{F\}^f_{ij}$ and $\{F\}^f_{ji}$. The cases given in the table are somewhat restrictive, and more complete treatments are available in standard matrix analysis textbooks. The cases given, however, are sufficient for our current considerations.

Equilibrium requires that the fixed-end forces at the *i* end of member *ij* be accompanied by a set of equivalent structure forces at mode *i*, $\{P\}^e_i$. These are shown in Fig. 14.4*b* and are expressed in the form

$$\{P\}^e_i = [\gamma]_{ij}\{F\}^f_{ij} \tag{14.40}$$

where $[\gamma]_{ij}$ evolves from considerations of static equilibrium. Comparing Eq. 14.40 with Eq. 14.24, we can see that

$$[\gamma]_{ij} = [\beta]^T_{ij} \tag{14.41}$$

where $[\beta]_{ij}$ is the compatibility matrix introduced in Section 14.4.

However, the given structure may not be capable of sustaining *all* of the required equivalent structure forces. For instance, in Fig. 14.4*b*, the vertical components of the equivalent forces at ends *i* and *j* can be sustained by the supports; however, the moment components cannot be sustained. Thus, those equivalent force components that cannot be sustained must be removed through the application of the negative of these components. These forces are denoted as $-\{\tilde{P}\}^e_i$ and are shown in Fig. 14.4*c*. The final solution is given by the superposition of Figs. 14.4*b* and 14.4*c*.

If there are applied loads at node i that correspond to the components of $\{\tilde{P}\}_i^e$, they must be included in the analysis. Thus, the stiffness analysis must consider the net load $\{P\}_i$ at each node as given by

$$\{P\}_i = \{P\}_i^a - \{\tilde{P}\}_i^e \qquad (14.42)$$

where $\{P\}_i^a$ are the applied loads.

In an actual structure, more than one loaded span may join at node $i,$ and thus

$$\{P\}_i^e = \sum_j [\beta]_{ij}^T\{F\}_{ij}^f \qquad (14.43)$$

where j ranges over all of the members framing into joint $i.$

The detailed application of the procedure is given in the example problem that follows. The general approach is the same as that outlined in Section 14.5 with the exceptions that the nodal numbering need not include all load points and the final member-end forces are given by

$$\{F\}_{ij} = [k]_{ii}^j[\beta]_{ij}\{\Delta\}_i + [k]_{ij}[\beta]_{ji}\{\Delta\}_j + \{F\}_{ij}^f \qquad (14.44)$$

EXAMPLE 14.4

Determine the member-end forces for the structure shown below. EI = constant.

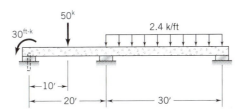

Selection of Node Points

Structure Forces and Displacements

$\{P\}_i = \begin{Bmatrix} P_2 \\ P_6 \end{Bmatrix}_i \qquad \{P\}_1 \qquad\qquad \{P\}_2 \qquad\qquad\qquad \{P\}_3$

$\{\Delta\}_i = \begin{Bmatrix} \Delta_2 \\ \Delta_6 \end{Bmatrix}_i \qquad \{\Delta\}_1 \qquad\qquad \{\Delta\}_2 \qquad\qquad\qquad \{\Delta\}_3$

Member Forces and Displacements

Member 1-2: $\{F\}_{12} = \begin{Bmatrix} F_2 \\ F_6 \end{Bmatrix}_{12}$; $\{\delta\}_{12} = \begin{Bmatrix} \delta_2 \\ \delta_6 \end{Bmatrix}_{12}$

Member 2-3: $\{F\}_{23} = \begin{Bmatrix} F_2 \\ F_6 \end{Bmatrix}_{23}$; $\{\delta\}_{23} = \begin{Bmatrix} \delta_2 \\ \delta_6 \end{Bmatrix}_{23}$

Compatibility Matrices

$$\begin{Bmatrix} \delta_2 \\ \delta_6 \end{Bmatrix}_{ij} = \begin{bmatrix} \cos\theta & 0 \\ 0 & 1 \end{bmatrix} \begin{Bmatrix} \Delta_2 \\ \Delta_6 \end{Bmatrix}_i \tag{14.38}$$

Member 1-2: $\cos\theta = l_{1x} = \dfrac{X_1 - X_2}{L_{12}} = \dfrac{0 - 20}{20} = -1$

$$[\beta]_{12} = \begin{bmatrix} -1 & 0 \\ 0 & 1 \end{bmatrix} \qquad [\beta]_{21} = \begin{bmatrix} 1 & 0 \\ 0 & 1 \end{bmatrix}$$

Member 2-3: $\cos\theta = l_{2x} = \dfrac{X_2 - X_3}{L_{23}} = \dfrac{20 - 50}{30} = -1$

$$[\beta]_{23} = \begin{bmatrix} -1 & 0 \\ 0 & 1 \end{bmatrix} \qquad [\beta]_{32} = \begin{bmatrix} 1 & 0 \\ 0 & 1 \end{bmatrix}$$

Fixed-end Forces and Equivalent Structure Forces

$\{F\}_{12}^f = \begin{Bmatrix} +25 \\ -125 \end{Bmatrix}$ $\qquad \{F\}_{21}^f = \begin{Bmatrix} -25 \\ +125 \end{Bmatrix}$ $\qquad \{F\}_{23}^f = \begin{Bmatrix} +36 \\ -180 \end{Bmatrix}$ $\qquad \{F\}_{32}^f = \begin{Bmatrix} -36 \\ +180 \end{Bmatrix}$

$$\{P\}_i^e = \sum_j [\beta]_{ij}^T \{F\}_{ij}^f \tag{14.43}$$

$$\{P\}_1^e = [\beta]_{12}^T \{F\}_{12}^f = \begin{bmatrix} -1 & 0 \\ 0 & 1 \end{bmatrix} \begin{Bmatrix} +25 \\ -125 \end{Bmatrix} = \begin{Bmatrix} -25 \\ -125 \end{Bmatrix}$$

$$\{P\}_2^e = [\beta]_{21}^T \{F\}_{21}^f + [\beta]_{23}^T \{F\}_{23}^f$$

$$= \begin{bmatrix} 1 & 0 \\ 0 & 1 \end{bmatrix} \begin{Bmatrix} -25 \\ +125 \end{Bmatrix} + \begin{bmatrix} -1 & 0 \\ 0 & 1 \end{bmatrix} \begin{Bmatrix} +36 \\ -180 \end{Bmatrix} = \begin{Bmatrix} -61 \\ -55 \end{Bmatrix}$$

$$\{P\}_3^e = [\beta]_{32}^T \{F\}_{32}^f = \begin{bmatrix} 1 & 0 \\ 0 & 1 \end{bmatrix} \begin{Bmatrix} -36 \\ +180 \end{Bmatrix} = \begin{Bmatrix} -36 \\ +180 \end{Bmatrix}$$

Components of $\{P\}_i^e$ sustained by supports:

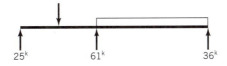

Loads for stiffness analysis:

$$\{P\}_i = \{P\}_i^a - \{\tilde{P}\}_i^e \tag{14.42}$$

Member Stiffnesses

(Extract from Eqs. 13.26 and 13.27)

Member 1-2:

$$[k]_{11}^2 = [k]_{22}^1 = \begin{bmatrix} \dfrac{12EI_z}{l^3} & \dfrac{-6EI_z}{l^2} \\[3mm] \dfrac{-6EI_z}{l^2} & \dfrac{4EI_z}{l} \end{bmatrix}_{l=20'} = EI \begin{bmatrix} 0.001,5 & -0.015 \\ -0.015 & 0.200 \end{bmatrix}$$

$$[k]_{12} = [k]_{21} = \begin{bmatrix} \dfrac{12EI_z}{l^3} & \dfrac{-6EI_z}{l^2} \\[3mm] \dfrac{-6EI_z}{l^2} & \dfrac{2EI_z}{l} \end{bmatrix}_{l=20'} = EI \begin{bmatrix} 0.001,5 & -0.015 \\ -0.015 & 0.100 \end{bmatrix}$$

Member 2-3:

$$[k]_{22}^3 = [k]_{33}^2 = \begin{bmatrix} \dfrac{12EI_z}{l^3} & \dfrac{-6EI_z}{l^2} \\[3mm] \dfrac{-6EI_z}{l^2} & \dfrac{4EI_z}{l} \end{bmatrix}_{l=30'} = EI \begin{bmatrix} 0.000,444 & -0.006,67 \\ -0.006,67 & 0.133 \end{bmatrix}$$

$$[k]_{23} = [k]_{32} = \begin{bmatrix} \dfrac{12EI_z}{l^3} & \dfrac{-6EI_z}{l^2} \\[3mm] \dfrac{-6EI_z}{l^2} & \dfrac{2EI_z}{l} \end{bmatrix}_{l=30'} = EI \begin{bmatrix} 0.000,444 & -0.006,67 \\ -0.006,67 & 0.066,7 \end{bmatrix}$$

Structure Stiffness Matrix

Total Stiffness Equations

$$
\begin{Bmatrix}
\begin{Bmatrix} P_2 \\ P_6 \end{Bmatrix}_1 \\[6pt]
\begin{Bmatrix} P_2 \\ P_6 \end{Bmatrix}_2 \\[6pt]
\begin{Bmatrix} P_2 \\ P_6 \end{Bmatrix}_3
\end{Bmatrix}
=
\begin{bmatrix}
[K]_{11} & [K]_{12} & [0] \\
[K]_{21} & [K]_{22} & [K]_{23} \\
[0] & [K]_{32} & [K]_{33}
\end{bmatrix}
\begin{Bmatrix}
\begin{Bmatrix} \Delta_2 \\ \Delta_6 \end{Bmatrix}_1 \\[6pt]
\begin{Bmatrix} \Delta_2 \\ \Delta_6 \end{Bmatrix}_2 \\[6pt]
\begin{Bmatrix} \Delta_2 \\ \Delta_6 \end{Bmatrix}_3
\end{Bmatrix}
\tag{14.15}
$$

Reduced Stiffness Equations

$$
[P]_{\mathrm{I}} = [K]_{\mathrm{I,I}}\{\Delta\}_1 \Rightarrow
\begin{Bmatrix} (P_6)_1 \\ (P_6)_2 \\ (P_6)_3 \end{Bmatrix}
= [K]_{\mathrm{I,I}}
\begin{Bmatrix} (\Delta_6)_1 \\ (\Delta_6)_2 \\ (\Delta_6)_3 \end{Bmatrix}
\tag{14.9}
$$

because $(\Delta_2)_1 = (\Delta_2)_2 = (\Delta_2)_3 = 0$

Required Submatrices

$$
[K]_{ii}^{j} = [\beta]_{ij}^{T}[k]_{ii}^{j}[\beta]_{ij}
\tag{14.26}
$$

$$
[K]_{ii} = \sum_i [K]_{ii}^{j}
\tag{14.16}
$$

$$
[K]_{11}^{2} = [\beta]_{12}^{T}[k]_{11}^{2}[\beta]_{12}
$$

$$
= \begin{bmatrix} -1 & 0 \\ 0 & 1 \end{bmatrix} \cdot EI
\begin{bmatrix} 0.001{,}5 & -0.015 \\ -0.015 & 0.200 \end{bmatrix}
\begin{bmatrix} -1 & 0 \\ 0 & 1 \end{bmatrix}
$$

$$
= EI \begin{bmatrix} 0.001{,}5 & 0.015 \\ 0.015 & 0.200 \end{bmatrix}
$$

$$
[K]_{22}^{1} = [\beta]_{21}^{T}[k]_{22}^{1}[\beta]_{21}
$$

$$
= \begin{bmatrix} 1 & 0 \\ 0 & 1 \end{bmatrix} \cdot EI
\begin{bmatrix} 0.001{,}5 & -0.015 \\ -0.015 & 0.200 \end{bmatrix}
\begin{bmatrix} 1 & 0 \\ 0 & 1 \end{bmatrix}
$$

$$
= EI \begin{bmatrix} 0.001{,}5 & -0.015 \\ -0.015 & 0.200 \end{bmatrix}
$$

$$
[K]_{22}^{3} = [\beta]_{23}^{T}[k]_{22}^{3}[\beta]_{23}
$$

$$
= \begin{bmatrix} -1 & 0 \\ 0 & 1 \end{bmatrix} \cdot EI
\begin{bmatrix} 0.000{,}444 & -0.006{,}67 \\ -0.006{,}67 & 0.133 \end{bmatrix}
\begin{bmatrix} -1 & 0 \\ 0 & 1 \end{bmatrix}
$$

$$
= EI \begin{bmatrix} 0.000{,}444 & 0.006{,}67 \\ 0.006{,}67 & 0.133 \end{bmatrix}
$$

$$[K]_{33}^2 = [\beta]_{32}^T [k]_{33}^2 [\beta]_{32}$$

$$= \begin{bmatrix} 1 & 0 \\ 0 & 1 \end{bmatrix} \cdot EI \begin{bmatrix} 0.000,444 & -0.006,67 \\ -0.006,67 & 0.133 \end{bmatrix} \begin{bmatrix} 1 & 0 \\ 0 & 1 \end{bmatrix}$$

$$= EI \begin{bmatrix} 0.000,444 & -0.006,67 \\ -0.006,67 & 0.133 \end{bmatrix}$$

$$[K]_{11} = [K]_{11}^2 = EI \begin{bmatrix} 0.001,5 & 0.015 \\ 0.015 & 0.200 \end{bmatrix}$$

$$[K]_{22} = [K]_{22}^1 + [K]_{22}^3 = EI \begin{bmatrix} 0.001,94 & -0.008,33 \\ -0.008,33 & 0.333 \end{bmatrix}$$

$$[K]_{33} = [K]_{33}^2 = EI \begin{bmatrix} 0.000,444 & -0.006,67 \\ -0.006,67 & 0.133 \end{bmatrix}$$

$$[K]_{ij} = [\beta]_{ij}^T [k]_{ij} [\beta]_{ji} \tag{14.27}$$

$$[K]_{12} = [\beta]_{12}^T [k]_{12} [\beta]_{21}$$

$$= \begin{bmatrix} -1 & 0 \\ 0 & 1 \end{bmatrix} \cdot EI \begin{bmatrix} 0.001,5 & -0.015 \\ -0.015 & 0.100 \end{bmatrix} \begin{bmatrix} 1 & 0 \\ 0 & 1 \end{bmatrix}$$

$$= EI \begin{bmatrix} -0.001,5 & 0.015 \\ -0.015 & 0.100 \end{bmatrix}$$

$$[K]_{21} = [\beta]_{21}^T [k]_{21} [\beta]_{12} = [K]_{12}^T = EI \begin{bmatrix} -0.001,5 & -0.015 \\ -0.015 & 0.100 \end{bmatrix}$$

$$[K]_{23} = [\beta]_{23}^T [k]_{23} [\beta]_{32}$$

$$= \begin{bmatrix} -1 & 0 \\ 0 & 1 \end{bmatrix} \cdot EI \begin{bmatrix} 0.000,444 & -0.006,67 \\ -0.006,67 & 0.066,7 \end{bmatrix} \begin{bmatrix} 1 & 0 \\ 0 & 1 \end{bmatrix}$$

$$= EI \begin{bmatrix} -0.000,444 & 0.006,67 \\ -0.006,67 & 0.066,7 \end{bmatrix}$$

$$[K]_{32} = [\beta]_{32}^T [k]_{32} [\beta]_{23} = [K]_{23}^T = EI \begin{bmatrix} -0.000,444 & -0.006,67 \\ 0.006,67 & 0.066,7 \end{bmatrix}$$

Thus, the reduced stiffness becomes

$$[K]_{I,I} = EI \begin{bmatrix} 0.200 & 0.100 & 0 \\ 0.100 & 0.333 & 0.066,7 \\ 0 & 0.066,7 & 0.133 \end{bmatrix}$$

Determination of Displacements

$$\{\Delta\}_I = [K]_{I,I}^{-1} \{P\}_I \tag{14.10}$$

$$
\begin{Bmatrix} (\Delta_6)_1 \\ (\Delta_6)_2 \\ (\Delta_6)_3 \end{Bmatrix} = \frac{1}{EI} \begin{bmatrix} 0.200 & 0.100 & 0 \\ 0.100 & 0.333 & 0.066,7 \\ 0 & 0.066,7 & 0.133 \end{bmatrix}^{-1} \begin{Bmatrix} (P_6)_1 \\ (P_6)_2 \\ (P_6)_3 \end{Bmatrix}
$$

$$
= \frac{1}{EI} \begin{bmatrix} 6.002 & -2.004 & 1.005 \\ -2.004 & 4.007 & -2.009 \\ 1.005 & -2.009 & 8.527 \end{bmatrix} \begin{Bmatrix} 95 \\ 55 \\ -180 \end{Bmatrix}
$$

$$
= \frac{1}{EI} \begin{Bmatrix} 279.07 \\ 391.63 \\ -1,549.88 \end{Bmatrix}
$$

Member Forces

$$
\{F\}_{ij} = [k]_{ii}^{j}[\beta]_{ij}\{\Delta\}_i + [k]_{ij}[\beta]_{ji}\{\Delta\}_j + \{F\}_{ij}^{f} \tag{14.44}
$$

Member 1-2: $\{\Delta\}_1 = \dfrac{1}{EI}\begin{Bmatrix} 0 \\ 279.07 \end{Bmatrix}$ $\{\Delta\}_2 = \dfrac{1}{EI}\begin{Bmatrix} 0 \\ 391.63 \end{Bmatrix}$

$$
\{F\}_{12} = [k]_{11}^{2}[\beta]_{12}\{\Delta\}_1 + [k]_{12}[\beta]_{21}\{\Delta\}_2 + \{F\}_{12}^{f}
$$

$$
= EI\begin{bmatrix} 0.001,5 & -0.015 \\ -0.015 & 0.200 \end{bmatrix}\begin{bmatrix} -1 & 0 \\ 0 & 1 \end{bmatrix}\frac{1}{EI}\begin{Bmatrix} 0 \\ 279.07 \end{Bmatrix}
$$

$$
+ EI\begin{bmatrix} 0.001,5 & -0.015 \\ -0.015 & 0.100 \end{bmatrix}\begin{bmatrix} 1 & 0 \\ 0 & 1 \end{bmatrix}\frac{1}{EI}\begin{Bmatrix} 0 \\ 391.63 \end{Bmatrix} + \begin{Bmatrix} +25 \\ -125 \end{Bmatrix}
$$

$$
= \begin{Bmatrix} -4.19 \\ 55.81 \end{Bmatrix} + \begin{Bmatrix} -5.87 \\ 39.16 \end{Bmatrix} + \begin{Bmatrix} 25 \\ -125 \end{Bmatrix} = \begin{Bmatrix} 14.94^{k} \\ -30.03'^{-k} \end{Bmatrix}
$$

$$
\{F\}_{21} = [k]_{22}^{1}[\beta]_{21}\{\Delta\}_2 + [k]_{21}[\beta]_{12}\{\Delta\}_1 + \{F\}_{21}^{f}
$$

$$
= \begin{Bmatrix} -5.87 \\ 78.33 \end{Bmatrix} + \begin{Bmatrix} -4.19 \\ 27.91 \end{Bmatrix} + \begin{Bmatrix} -25 \\ 125 \end{Bmatrix} = \begin{Bmatrix} -35.06^{k} \\ 231.24'^{-k} \end{Bmatrix}
$$

Member 2-3: $\{\Delta\}_2 = \dfrac{1}{EI}\begin{Bmatrix} 0 \\ 391.63 \end{Bmatrix}$ $\{\Delta\}_3 = \dfrac{1}{EI}\begin{Bmatrix} 0 \\ -1,549.88 \end{Bmatrix}$

$$
\{F\}_{23} = [k]_{22}^{3}[\beta]_{23}\{\Delta\}_2 + [k]_{23}[\beta]_{32}\{\Delta\}_3 + \{F\}_{23}^{f}
$$

$$
= \begin{Bmatrix} -2.61 \\ 52.09 \end{Bmatrix} + \begin{Bmatrix} 10.34 \\ -103.38 \end{Bmatrix} + \begin{Bmatrix} 36 \\ -180 \end{Bmatrix} = \begin{Bmatrix} 43.73^{k} \\ -231.29'^{-k} \end{Bmatrix}
$$

$$
\{F\}_{32} = [k]_{33}^{2}[\beta]_{32}\{\Delta\}_3 + [k]_{32}[\beta]_{23}\{\Delta\}_2 + \{F\}_{32}^{f}
$$

$$
= \begin{Bmatrix} 10.34 \\ -206.13 \end{Bmatrix} + \begin{Bmatrix} -2.61 \\ 26.12 \end{Bmatrix} + \begin{Bmatrix} -36 \\ 180 \end{Bmatrix} = \begin{Bmatrix} -28.27^{k} \\ 0'^{-k} \end{Bmatrix}
$$

The shear and moment diagrams are now easily constructed, as shown in Example 14.2.

14.7 SELF-STRAINING PROBLEMS

Self straining occurs when a structure is subjected to internal strains and a resulting state of stress in the absence of externally applied loads. An example of self straining is the support settlement that was described in Section 14.2. Other self-straining situations result from temperature variations or member lack of fit, in which a member of erroneous length or alignment is forced to fit during the fabrication process.

For all of these situations, there is a distinct difference in the nature of the response for statically determinate as opposed to statically indeterminate structures. In the statically determinate case, the induced movements are uninhibited, and the structure merely assumes a distorted configuration without the inducement of stresses. For example, consider the simply supported beam shown in Fig. 14.5a. Figures 14.5b, c,

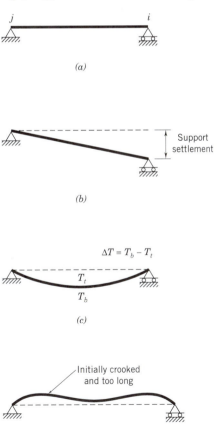

Figure 14.5 Induced movements for statically determinate beam. (a) Simply supported beam. (b) Support settlement. (c) Temperature gradient. (d) Lack of fit.

and *d* show, respectively, the responses of the structure to support settlement, temperature gradient, and fabrication error. In each case, the structure is deformed, but there are no internal stresses and no reactions are induced. For a statically indeterminate case, however, the response is quite different. The propped cantilever beam of Fig. 14.6*a* serves as an example; Figs. 14.6*b, c,* and *d* depict the responses to support settlement, temperature gradient, and fabrication error, respectively. Here, internal stresses result, and in each case there is an associated set of self-equilibrating external reactions. The structure itself is serving to inhibit the deformation; the structure is "straining against itself," or it is a s*elf-straining problem.*

A comparison of Figs. 14.5 and 14.6 shows why the reactions and the associated internal stresses are induced for the statically indeterminate system. For each of the determinate cases given in Fig. 14.5, there are no forces induced through the induced movement. However, for each of the indeterminate cases, an end moment, along with a set of equilibrating vertical reactions, has to be introduced in order to impose the zero-rotation boundary condition at point *j* for the structure of Fig. 14.6.

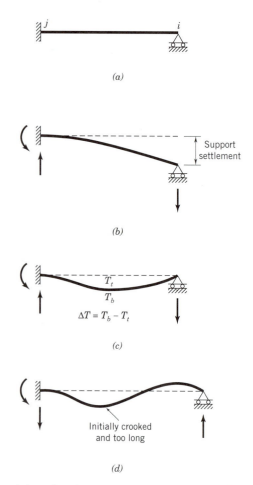

Figure 14.6 Self-straining of statically indeterminate beam. (*a*) Propped cantilever beam. (*b*) Support settlement. (*c*) Temperature gradient. (*d*) Lack of fit.

The formal analysis approach for the self-straining problem resembles that which is used when loads are applied between node points. We will limit our consideration here to planar structures without torsion, and Fig. 14.7a shows a typical member ij. It is initially assumed that the span is fixed at its ends, as shown in Fig. 14.7b. Of course, this condition does not satisfy the prescribed boundary conditions. The member is now severed at end i and the member-end displacements caused by the self-straining action are allowed to accrue at end i. These are shown in Fig. 14.7c as the vector

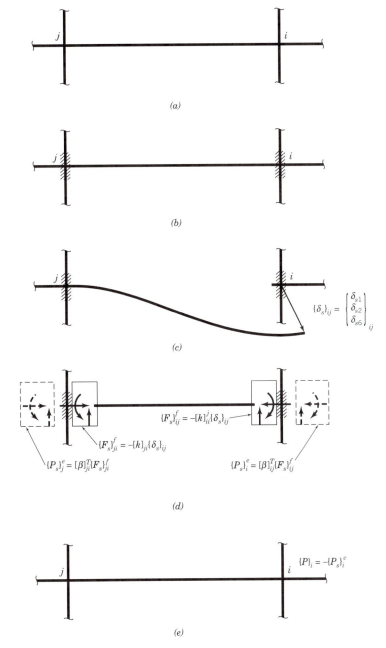

$$\{\delta_s\}_{ij} = \begin{Bmatrix} \delta_{s1} \\ \delta_{s2} \\ \delta_{s6} \end{Bmatrix}_{ij}$$

(a)

(b)

$$\{F_s\}_{ij}^f = -[k]_{ii}^j\{\delta_s\}_{ij}$$

$$\{F_s\}_{ji}^f = -[k]_{ji}\{\delta_s\}_{ij}$$

$$\{P_s\}_j^e = [\beta]_{ji}^T\{F_s\}_{ji}^f$$

$$\{P_s\}_i^e = [\beta]_{ij}^T\{F_s\}_{ij}^f$$

(c)

(d)

$$\{P\}_i = -\{P_s\}_i^e$$

(e)

Figure 14.7 Loading for self-straining problem. (*a*) Planar member ij. (*b*) Fixity for member ij. (*c*) Accrued displacements at severed i end. (*d*) Member fixed-end forces and equivalent joint forces. (*e*) Joint loading for stiffness analysis.

$\{\delta_s\}_{ij}$. If these displacements are restrained, that is, if the displacements $-\{\delta_s\}_{ij}$ are applied, then a set of fixed-end forces is imposed at joints i and j as shown in Fig. 14.7d. These fixed-end forces are given by

$$\{F_s\}^f_{ij} = -[k]^j_{ii}\{\delta_s\}_{ij} \tag{14.45}$$

and

$$\{F_s\}^f_{ji} = -[k]_{ji}\{\delta_s\}_{ij} \tag{14.46}$$

where $\{F_s\}^f_{ij}$ and $\{F_s\}^f_{ji}$ are the fixed-end forces induced at ends i and j, respectively, through the removal of $\{\delta_s\}_{ij}$, and the $[k]$ matrices are as defined by Eqs. 13.26 and 13.27.

In order to sustain these fixed-end forces, there must be an equivalent set of structure forces as shown in Fig. 14.7d. According to the presentation of Section 14.6, these equivalent forces are

$$\{P_s\}^e_i = [\beta]^T_{ij}\{F_s\}^f_{ij} \tag{14.47}$$

where $\{P_s\}^e_i$ includes the equivalent sustaining forces at joint i, and $[\beta]_{ij}$ is the compatibility matrix. Of course, in the overall structure, more than one member may join at node i and, therefore,

$$\{P_s\}^e_i = \sum_i [\beta]^T_{ij}\{F_s\}^f_{ij} \tag{14.48}$$

where j ranges over all members framing into joint i.

Because the given structure of Fig. 14.7a cannot provide these sustaining forces, they must be removed through a standard joint-loaded stiffness analysis, as noted in Fig. 14.7e, in which the loads at joint i are

$$\{P\}_i = -\{P_s\}^e_i \tag{14.49}$$

If some components of $\{P_s\}^e_i$ can be sustained by the support at point i, then a reduced vector of structure forces corresponding to the kinematic degrees of freedom, $\{\tilde{P}_s\}^e_i$, is applied as described in Section 14.6.

The general analysis procedure follows that given in Section 14.5, and the member-end forces are determined by Eq. 14.44 in which $\{F\}^f_{ij}$ is replaced by $\{F_s\}^f_{ij}$.

14.7.1 Fabrication Error

For the fabrication error, or lack of fit, problem, the vector $\{\delta_s\}_{ij}$ includes the displacements at the i end relative to the j end that indicate the components of improper length or alignment. This vector of displacement introduces fixed-end forces at both ends of the faulty member in accordance with Eqs. 14.45 and 14.46.

14.7.2 Temperature Variations

For temperature effects, $\{\delta_s\}_{ij}$ includes the displacements that accrue at the i end relative to the j end in Fig. 14.7c when the temperature variations occur. For the planar structure under consideration, these displacements are shown in Fig. 14.8c for the temperature gradient specified in Fig. 14.8a. These displacements are readily determined by applying the moment–area methods to the curvature diagram (equivalent to an M/EI diagram) given in Fig. 14.8b. Substitution of these displacements in Eqs. 14.45 and 14.46 along with appropriate $[k]$ matrices from Eq. 13.26 gives

$$\{F_s\}^f_{ij} = -\begin{bmatrix} \dfrac{EA}{l} & 0 & 0 \\[2ex] 0 & \dfrac{12EI_z}{l^3} & -\dfrac{6EI_z}{l^2} \\[2ex] 0 & -\dfrac{6EI_z}{l^2} & \dfrac{4EI_z}{l} \end{bmatrix} \begin{Bmatrix} T_{\text{avg}}\,\alpha l \\[2ex] -\dfrac{\alpha \Delta T l^2}{2h} \\[2ex] -\dfrac{\alpha \Delta T l}{h} \end{Bmatrix} = \begin{Bmatrix} -EA\,\alpha T_{\text{avg}} \\[2ex] 0 \\[2ex] \dfrac{EI_z\,\alpha \Delta T}{h} \end{Bmatrix} \quad (14.50)$$

and

$$\{F_s\}^f_{ji} = -\begin{bmatrix} \dfrac{EA}{l} & 0 & 0 \\[2ex] 0 & \dfrac{12EI_z}{l^3} & -\dfrac{6EI_z}{l^2} \\[2ex] 0 & -\dfrac{6EI_z}{l^2} & \dfrac{2EI_z}{l} \end{bmatrix} \begin{Bmatrix} T_{\text{avg}}\,\alpha l \\[2ex] -\dfrac{\alpha \Delta T l^2}{2h} \\[2ex] -\dfrac{\alpha \Delta T l}{h} \end{Bmatrix} = \begin{Bmatrix} -EA\,\alpha T_{\text{avg}} \\[2ex] 0 \\[2ex] -\dfrac{EI_z\,\alpha \Delta T}{h} \end{Bmatrix} \quad (14.51)$$

where α is the coefficient of thermal expansion, $\Delta T = T_b - T_t$ and $T_{\text{avg}} = (T_b + T_t)/2$, in which T_b and T_t are the temperatures at the bottom and top flanges of the beam, respectively. In applying these equations to any member, T_b and T_t must be carefully interpreted with respect to the assigned i and j of the member ends in order to avoid a mistake in signs.

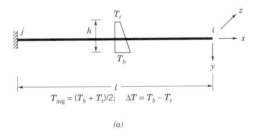

(a)

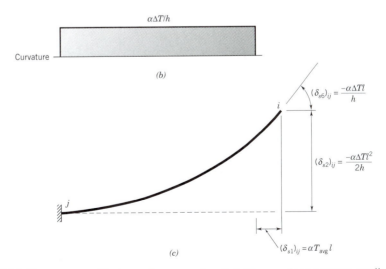

(c)

Figure 14.8 Temperature effects on planar member ij. (*a*) Imposed temperature gradient. (*b*) Induced member curvatures. (*c*) Accrued displacements at end j.

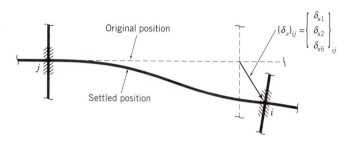

Figure 14.9 Settlement problem.

14.7.3 Settlement

For the settlement problem, the $(\delta_\sigma)_{ij}$ vector includes the settlement displacement components that occur at the i end of all members framing into joint i as shown in Fig. 14.9. These displacements will cause fixed-end moments at both ends of all members that frame into joint i. However, since these displacements are *imposed* rather than restrained, the fixed-end forces for a typical member ij are given by

$$\{F_s\}_{ij}^f = [k]_{ii}^j\{\delta_s\}_{ij} \tag{14.52}$$

and

$$\{F_s\}_{ji}^f = [k]_{ji}\{\delta\}_{ij} \tag{14.53}$$

The structure joint forces needed to sustain these member fixed-end forces are then given by Eq. 14.48, and the forces for a standard joint-loaded stiffness analysis are given by Eq. 14.49.

EXAMPLE 14.5

Determine the member-end forces for the structure of Example 14.3 if the column is subjected to the temperature gradient shown.

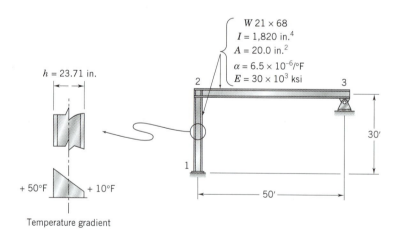

The solution proceeds as in Example 14.3. The equations of equilibrium are as follows:

$$\{P\}_I = [K]_{I,I}\{\Delta\}_I \tag{14.9}$$

where

$$\{P\}_{\mathrm{I}} = \left\{ \begin{Bmatrix} P_1 \\ P_2 \\ P_6 \end{Bmatrix}_2 \\ \{P_6\}_3 \end{Bmatrix} ; \qquad \{\Delta\}_{\mathrm{I}} = \left\{ \begin{Bmatrix} \Delta_1 \\ \Delta_2 \\ \Delta_6 \end{Bmatrix}_2 \\ \{\Delta_6\}_3 \end{Bmatrix}$$

$$[K]_{\mathrm{I,I}} = \begin{bmatrix} 0.032{,}14 & 0 & -0.006{,}67 & 0 \\ 0 & 0.053{,}0 & 0.002{,}40 & 0.002{,}40 \\ -0.006{,}67 & 0.002{,}40 & 0.213 & 0.040{,}0 \\ 0 & 0.002{,}40 & 0.040{,}0 & 0.080{,}0 \end{bmatrix}$$

Self-Straining Fixed-End Forces and Equivalent Structure Forces

Fixed-end forces

$$\{F_s\}_{ij}^f = \begin{Bmatrix} -EA\alpha T_{\mathrm{avg}} \\ 0 \\ \dfrac{EI\alpha\Delta T}{h} \end{Bmatrix} ; \qquad \{F_s\}_{ji}^f = \begin{Bmatrix} -EA\alpha T_{\mathrm{avg}} \\ 0 \\ \dfrac{-EI\alpha\Delta T}{h} \end{Bmatrix} \qquad (14.50;\ 14.51)$$

Member 1-2: For member orientation with respect to the selection of i and j (see Fig. 14.8),

$$\Delta T = T_b - T_t = 50 - 10 = 40°\,\mathrm{F}$$

$$T_{\mathrm{avg}} = (50 + 10)/2 = 30°\,\mathrm{F}$$

$$\{F_s\}_{12}^f = \begin{Bmatrix} -30 \times 10^3 \times 20 \times 6.5 \times 10^{-6} \times 30 \\ 0 \\ \dfrac{30 \times 10^3 \times 1{,}820 \times 6.5 \times 10^{-6} \times 40}{23.71} \end{Bmatrix} = \begin{Bmatrix} -117.0^k \\ 0 \\ 598.73''^{-k} \end{Bmatrix}$$

$$= \begin{Bmatrix} -117.0^k \\ 0 \\ 49.89'^{-k} \end{Bmatrix} ; \qquad \{F_s\}_{21}^f = \begin{Bmatrix} -117.0^k \\ 0 \\ -49.89'^{-k} \end{Bmatrix}$$

Member 2-3:

$$\{F_s\}_{23}^f = \{0\}; \qquad \{F_s\}_{32}^f = \{0\}$$

Equivalent Structure Forces

$$\{P_s\}_i^e = \sum_i [\beta]_{ij}^T \{F_s\}_{ij}^f \tag{14.48}$$

$$\{P_s\}_1^e = [\beta]_{12}^T \{F_s\}_{12}^f = \begin{bmatrix} 0 & -1 & 0 \\ 1 & 0 & 0 \\ 0 & 0 & 1 \end{bmatrix} \begin{Bmatrix} -117.0 \\ 0 \\ 49.89 \end{Bmatrix} = \begin{Bmatrix} 0 \\ -117.0^k \\ 49.89'^{-k} \end{Bmatrix}$$

$$\{P_s\}_2^e = [\beta]_{21}^T \{F_s\}_{21}^f = \begin{bmatrix} 0 & 1 & 0 \\ -1 & 0 & 0 \\ 0 & 0 & 1 \end{bmatrix} \begin{Bmatrix} -117.0 \\ 0 \\ -49.89 \end{Bmatrix} = \begin{Bmatrix} 0 \\ 117.0^k \\ -49.89'^{-k} \end{Bmatrix}$$

$$\{P_s\}_3^e = \{0\}$$

For the stiffness analysis,

$$\{P\}_i = -\{P_s\}_i^e \tag{14.49}$$

where only the components corresponding to the kinematic degrees of freedom enter $\{P\}_I$ of Eq. 14.9.

Determination of Displacements

$$\{\Delta\}_I = [K]_{I,I}^{-1}\{P\}_I \tag{14.10}$$

$$\begin{Bmatrix} \begin{Bmatrix} \Delta_1 \\ \Delta_2 \\ \Delta_6 \end{Bmatrix}_2 \\ \{\Delta_6\}_3 \end{Bmatrix} = [K]_{I,I}^{-1} \begin{Bmatrix} 0 \\ -117.0 \\ 49.89 \\ 0 \end{Bmatrix} = \frac{1}{EI} \begin{Bmatrix} 56.91 \\ -2,216.77 \\ 274.25 \\ -70.62 \end{Bmatrix}$$

Member Forces

$$\{F\}_{ij} = [k]_{ii}^j [\beta]_{ij}\{\Delta\}_i + [k]_{ij}[\beta]_{ji}\{\Delta\}_j + \{F_s\}_{ij}^f \tag{14.44}$$

Member 1-2: $\{\Delta\}_1 = \{0\}$; $\{\Delta\}_2 = \dfrac{1}{EI} \begin{Bmatrix} 56.91 \\ -2,216.77 \\ 274.25 \end{Bmatrix}$

$$\{F\}_{12} = EI \begin{bmatrix} 0.052,9 & 0 & 0 \\ 0 & 0.000,444 & -0.006,67 \\ 0 & -0.006,67 & 0.066,7 \end{bmatrix} \begin{bmatrix} 0 & -1 & 0 \\ 1 & 0 & 0 \\ 0 & 0 & 1 \end{bmatrix} \cdot \frac{1}{EI} \begin{Bmatrix} 56.91 \\ -2,216.77 \\ 274.25 \end{Bmatrix} + \begin{Bmatrix} -117.0 \\ 0 \\ 49.89 \end{Bmatrix}$$

$$= \begin{Bmatrix} 117.27 - 117.00 \\ -1.80 + 0 \\ 17.91 + 49.89 \end{Bmatrix} = \begin{Bmatrix} 0.27^k \\ -1.80^k \\ 67.80'^{-k} \end{Bmatrix}$$

$$\{F\}_{21} = \left\{ \begin{array}{c} 117.27 - 117.00 \\ -1.80 + 0 \\ 36.90 - 49.89 \end{array} \right\} = \left\{ \begin{array}{c} 0.27^k \\ -1.80^k \\ -13.80'^{-k} \end{array} \right\}$$

$$\text{Member 2-3: } \{\Delta\}_2 = \frac{1}{EI} \left\{ \begin{array}{c} 56.91 \\ 2{,}216.77 \\ 274.25 \end{array} \right\}; \qquad \{\Delta\}_3 = \frac{1}{EI} \left\{ \begin{array}{c} 0 \\ 0 \\ -70.62 \end{array} \right\}$$

$$\{F\}_{23} = \left\{ \begin{array}{c} -1.80^k \\ -0.27^k \\ 13.80'^{-k} \end{array} \right\}; \qquad \{F\}_{32} = \left\{ \begin{array}{c} -1.80^k \\ -0.27^k \\ 0'^{-k} \end{array} \right\}$$

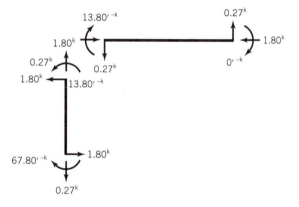

Deflected Structure

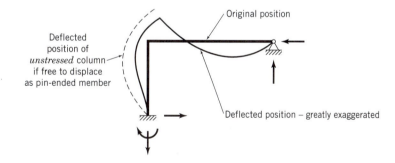

Note: The member forces and reactions result from forcing column from unstressed position to fit compatibly into frame.

14.8   DIRECT STIFFNESS METHOD

Equations 14.26 and 14.27 give expressions for the structure stiffness submatrices that collectively make up the total structure stiffness matrix; however, these submatrices remain clearly identified with member ij. That is, they represent the contribution of member ij to the total structure stiffness matrix.

On the right-hand side of each equation, the $[k]$ matrix is sandwiched between a post-multiplication by a $[\beta]$ matrix and a premultiplication by a $[\beta]^T$ matrix. By tracing through the derivation, we find that the postmultiplier ensures compatibility at the end of the member corresponding to the second subscript on the stiffness matrix, whereas the premultiplier relates to equilibrium between the structure and member forces at the end of the member corresponding to the first subscript on the stiffness matrix. These operations on $[k]$ provide for the transformation of the member stiffnesses from the member coordinate system to the structure coordinate system. Therefore, the left-hand side of Eqs. 14.26 and 14.27 are member stiffness matrices that are expressed in the structure coordinate system.

In the so-called *direct stiffness method,* emphasis is placed on the member stiffness matrices in the structure coordinate system. In fact, the right-hand sides of Eqs. 14.26 and 14.27 are frequently expanded by substituting specific forms for $[k]$ and $[\beta]$ for different structure classifications. For example, for a planar truss member, $[k]_{ii}^j$ and $[k]_{ij}$ are extracted from Eq. 13.26, and $[\beta]_{ij}$ and $[\beta]_{ji}$ are taken from Section 14.4 to produce the following:

$$[K]_{ii}^j = [\beta]_{ij}^T[k]_{ii}^j[\beta]_{ij}$$

$$= \begin{bmatrix} \cos\theta \\ \sin\theta \end{bmatrix} \begin{bmatrix} \dfrac{EA}{l} \end{bmatrix} [\cos\theta \quad \sin\theta] = \dfrac{EA}{l} \begin{bmatrix} \cos^2\theta & \sin\theta\cos\theta \\ \sin\theta\cos\theta & \sin^2\theta \end{bmatrix} \tag{14.54}$$

$$[K]_{ij} = [\beta]_{ij}^T[k]_{ij}[\beta]_{ji}$$

$$= \begin{bmatrix} \cos\theta \\ \sin\theta \end{bmatrix} \begin{bmatrix} \dfrac{EA}{l} \end{bmatrix} [-\cos\theta \quad -\sin\theta] = \dfrac{EA}{l} \begin{bmatrix} -\cos^2\theta & -\sin\theta\cos\theta \\ -\sin\theta\cos\theta & -\sin^2\theta \end{bmatrix} \tag{14.55}$$

Similarly, for planar frame members, the appropriate $[k]$ matrices are taken from Eq. 13.26, and the $[\beta]$ matrices are taken from Section 14.4, from which the following equations result:

$$[K]_{ii}^j = [\beta]_{ij}^T[k]_{ii}^j[\beta]_{ij}$$

$$= \begin{bmatrix} \cos\theta & -\sin\theta & 0 \\ \sin\theta & \cos\theta & 0 \\ 0 & 0 & 1 \end{bmatrix} \begin{bmatrix} \dfrac{EA}{l} & 0 & 0 \\ 0 & \dfrac{12EI_z}{l^3} & \dfrac{-6EI_z}{l^2} \\ 0 & \dfrac{-6EI_z}{l^2} & \dfrac{4EI_z}{l} \end{bmatrix} \begin{bmatrix} \cos\theta & \sin\theta & 0 \\ -\sin\theta & \cos\theta & 0 \\ 0 & 0 & 1 \end{bmatrix}$$

$$= \begin{bmatrix} \dfrac{EA}{l}\cos^2\theta + \dfrac{12EI_z}{l^3}\sin^2\theta & \dfrac{EA}{l}\sin\theta\cos\theta - \dfrac{12EI_z}{l^3}\sin\theta\cos\theta & \dfrac{6EI_z}{l^2}\sin\theta \\ \dfrac{EA}{l}\sin\theta\cos\theta - \dfrac{12EI_z}{l^3}\sin\theta\cos\theta & \dfrac{EA}{l}\sin^2\theta + \dfrac{12EI_z}{l^3}\cos^2\theta & -\dfrac{6EI_z}{l^2}\cos\theta \\ \dfrac{6EI_z}{l^2}\sin\theta & -\dfrac{6EI_z}{l^2}\cos\theta & \dfrac{4EI_z}{l} \end{bmatrix}$$

$$\tag{14.56}$$

$$[K]_{ij} = [\beta]_{ij}^{T}[k]_{ij}[\beta]_{ji}$$

$$= \begin{bmatrix} \cos\theta & -\sin\theta & 0 \\ \sin\theta & \cos\theta & 0 \\ 0 & 0 & 1 \end{bmatrix} \begin{bmatrix} \dfrac{EA}{l} & 0 & 0 \\ 0 & \dfrac{12EI_z}{l^3} & \dfrac{-6EI_z}{l^2} \\ 0 & \dfrac{-6EI_z}{l^2} & \dfrac{2EI_z}{l} \end{bmatrix} \begin{bmatrix} -\cos\theta & -\sin\theta & 0 \\ \sin\theta & -\cos\theta & 0 \\ 0 & 0 & 1 \end{bmatrix}$$

$$= \begin{bmatrix} -\dfrac{EA}{l}\cos^2\theta - \dfrac{12EI_z}{l^3}\sin^2\theta & -\dfrac{EA}{l}\sin\theta\cos\theta + \dfrac{12EI_z}{l^3}\sin\theta\cos\theta & \dfrac{6EI_z}{l^2}\sin\theta \\ -\dfrac{EA}{l}\sin\theta\cos\theta + \dfrac{12EI_z}{l^3}\sin\theta\cos\theta & -\dfrac{EA}{l}\sin^2\theta - \dfrac{12EI_z}{l^3}\cos^2\theta & -\dfrac{6EI_z}{l^2}\cos\theta \\ -\dfrac{6EI_z}{l^2}\sin\theta & \dfrac{6EI_z}{l^2}\cos\theta & \dfrac{2EI_z}{l} \end{bmatrix}$$

$$(14.57)$$

In the preceding equations, θ is the angle measured clockwise from the structure X axis to the member x axis at the i end of the member. At the j end, θ is replaced by $(180° + \theta)$ and, therefore, the signs of all the trigonometric functions change. Similar equations can be developed for other structure types, such as space trusses, grids, and space frames.

Once the structure stiffness submatrices have been established, the total structure stiffness matrix is assembled according to Eqs. 14.15 and 14.16. In the direct stiffness method, these equations are seen as the mechanism for superimposing the contribution from each member to the whole of the structure stiffness matrix. Each contribution, as well as the total structure stiffness matrix, is expressed in the structure coordinate system.

For hand calculations, Eqs. 14.54 through 14.57 offer no real advantages over Eqs. 14.26 and 14.27. However, for computer applications, the expanded forms are best because the repetitive matrix multiplications are avoided.

14.9 FURTHER OBSERVATIONS ABOUT STIFFNESS EQUATIONS

The arguments used in Section 13.5 to prove that a member stiffness matrix is symmetrical are applicable for a structure as a whole. Therefore, the total structure stiffness matrix given in Eq. 14.5 is a symmetrical matrix. The significance of this fact is mainly practical as it is related to computational considerations.

It is also clear from examining Eq. 14.15 that the stiffness matrix is not fully populated but is rather somewhat sparse or weakly populated. Specifically, consider the equations corresponding to $\{P\}_i$. These equations will possess only terms associated with the displacements $\{\Delta\}_i$ and the displacements at the distant ends of members framing into point i, which in this case include $\{\Delta\}_m \cdots \{\Delta\}_r$. Figure 14.10 demonstrates this arrangement based on a planar truss structure in which there are two kinematic degrees of freedom at each point. The range over which these terms are present in the stiffness matrix is referred to as the *bandwidth* as noted in Fig. 14.10. Because of the limited size of

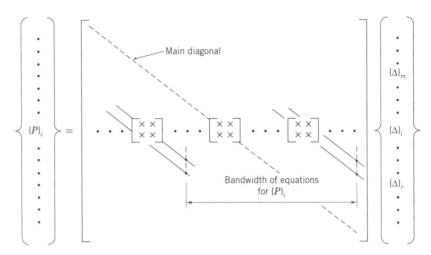

Figure 14.10 Composition of equilibrium equations.

the matrix portrayed in this figure, the relationship of the bandwidth to the total array of kinematic degrees of freedom is not evident; however, for large structures the sparseness of the stiffness matrix is more evident.

In some of the solution techniques used in computers, it is advantageous to minimize the bandwidth of the equations. In these cases, it is best to concentrate the nonzero terms as close to the main diagonal as possible, and this can be accomplished by a careful numbering of the kinematic degrees of freedom.

14.10 SUGGESTED PROBLEMS

14.1 through 14.17 Use the stiffness method to analyze each of the structures shown below for the loading conditions indicated. In each case, determine the structure displacements and the corresponding member forces for the given nodal numbering system.

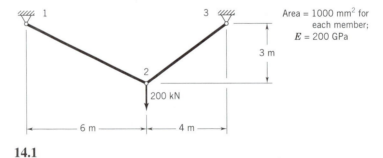

14.1

14.2

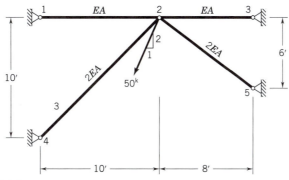

14.3

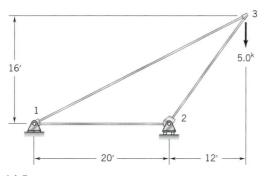

14.4

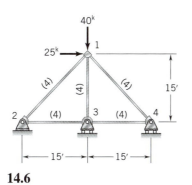

14.5

14.6

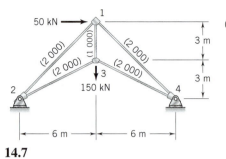

Area in parentheses
(mm²) on each member;
$E = 200$ GPa

14.7

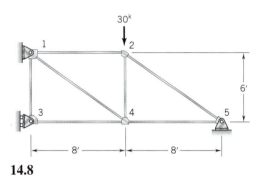

Area = 2.5 in.² for
each member;
$E = 29 \times 10^3$ ksi

14.8

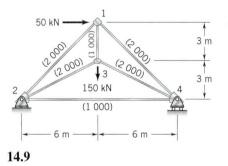

Area in parentheses
(mm²) on each member;
$E = 200$ GPa

14.9

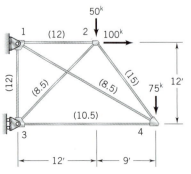

Area in parentheses (in.²)
on each member;
$E = 29 \times 10^3$ ksi

14.10

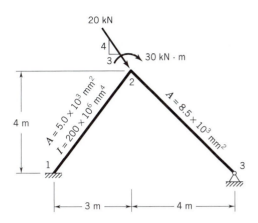

14.11

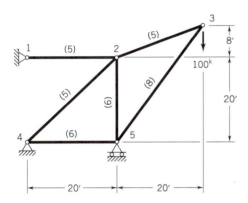

14.12

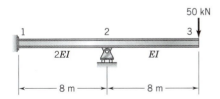

14.13

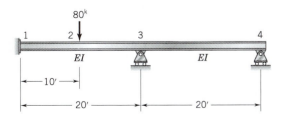

14.14

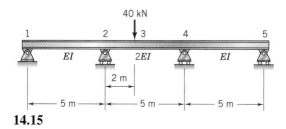

14.15

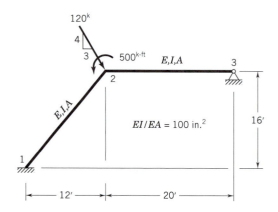

14.16

14.17 Ignore axial flexibilities; *EI* is the same for each member.

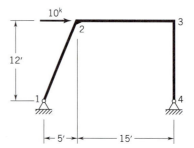

14.18 Consider the truss structure shown in which *A* and *E* are the same for all members.

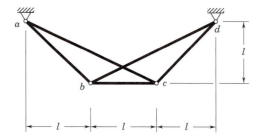

(**a**) Assemble the total and reduced stiffness matrices.

(**b**) For the application of equal vertical (downward) loads of magnitude *P* at points *b* and *c*, use symmetry considerations to reduce the order of the reduced stiffness matrix, and calculate the resulting displacements at points *b* and *c*.

(c) For the application of equal horizontal (rightward) loads of magnitude H at points b and c, use conditions of antisymmetry to reduce the order of the reduced stiffness matrix, and calculate the resulting displacements at points b and c.

(d) If members ac and bd are removed, demonstrate through a singularity test that the structure is unstable for general loading.

(e) If the modified structure of part (d) is subjected to equal vertical (downward) loads of magnitude P at points b and c, show that a solution is possible and determine the resulting displacements at points b and c.

14.19 Determine the structure displacements and the corresponding member forces for the grid structure shown.

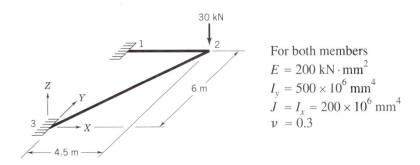

For both members
$E = 200 \text{ kN} \cdot \text{mm}^2$
$I_y = 500 \times 10^6 \text{ mm}^4$
$J = I_x = 200 \times 10^6 \text{ mm}^4$
$\nu = 0.3$

Note: *XY* plane is horizontal and the local *xy* plane for each member lies within the global *XY* plane.

14.20 Determine the displacements of joint b if members ab and bc each have a torsional stiffness of GJ/L and flexural stiffness of EI/L for bending out of the plane of the structure. Assume $GJ = 0.25 \, EI$ and $L = 6$ m.

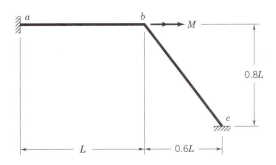

14.21 Consider the beam structure shown below.

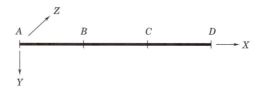

For each set of boundary conditions and prescribed load components, indicate the matrix format of the total and reduced equilibrium equations.

Use an "x" to identify each nonzero element of the stiffness matrix, express the individual displacement terms in symbolic form, and make the appropriate substitutions for the load terms.

Consider response in the *XY* plane only, including torsion about the *X* axis. Comment on any decoupling of individual displacements which allows for the solution of a subset of equations.

Note: Loads in kips, moments in ft-kips.

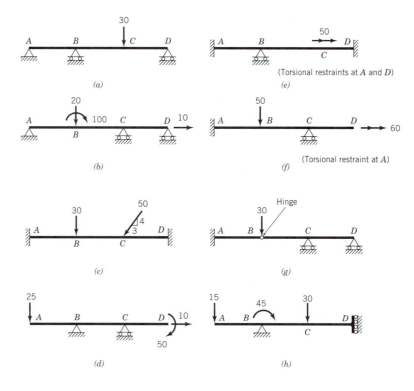

14.22 Consider the frame structure shown below.

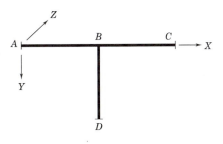

Apply the instructions that are given in Problem 21 to the structures and loading conditions that follow.

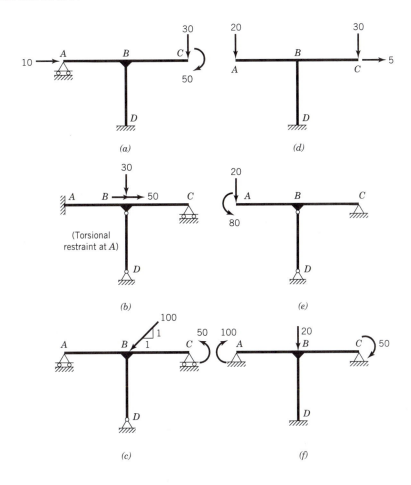

14.23 through 14.27 Use the stiffness method to analyze each of the structures shown below for the loading conditions indicated. In each case, determine the structure displacements and the corresponding member forces for the given nodal numbering system.

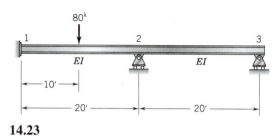

14.23

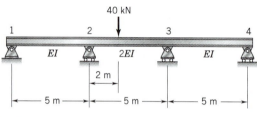

14.24

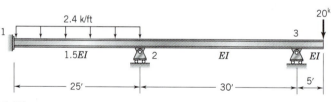

14.25

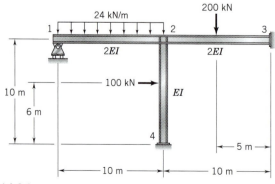

14.26

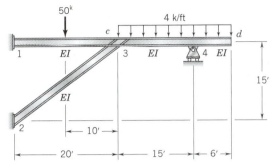

14.27

14.28 through 14.31 Determine the structure displacements, member forces, and reactions for each of the structures and conditions of support movements described below.

14.28 The structure of Problem 2 with the following displacements: joint 3, 0.5 in. downward and 0.2 in. right; joint 4, 0.3 in. downward. Disregard the loading on the structure.

14.29 The structure of Problem 6 with the following vertical support settlements: point 2, 0.50 in. upward; point 3, 0.75 in. downward; point 4, 0.50 in. downward. Disregard the loading on the structure.

14.30 The structure shown below with the loads given in addition to a settlement at point 3 of 0.012 m downward. Take $EI = 700\,000$ kN $\cdot$ m^2.

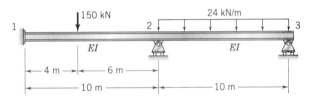

14.31 The structure shown below with the following pattern of support settlements: Point 1, 15 mm downward; point 2, 25 mm downward; point 3, 0.000 5 radian counterclockwise. Express your results in terms of EI.

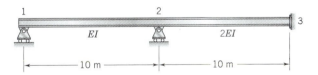

14.32 through 14.35 Determine the structure displacements and member forces for each of the structures and temperature variations given below.

14.32 The structure of Problem 2 if members 1-2 and 2-4 experience a 40° F increase in temperature. Disregard the loads, and take $\alpha_t = 0.000,006,5/°$ F.

14.33 The structure of Problem 9 if members 1-2 and 1-4 experience a 30° C increase in temperature. Disregard the loads, and take $\alpha_t = 0.000\,012/°$ C.

14.34 The structure of Problem 30 if both members experience a temperature gradient of $\Delta T = T_b - T_t = 30°$ C. Take $\alpha_t = 0.000\,012/°$ C, $h = 40$ mm, and $EI = 500\,000$ kN $\cdot$ m^2.

14.35 The structure shown below with the following temperature gradients:

(a) $T_t = -40°$ F, $T_b = -10°$ F for span 1–2 only

(b) $T_t = -40°$ F, $T_b = -10°$ F for both spans

For both cases, $E = 30 \times 10^6$ psi, $I = 1,500$ in.^{4}, $\alpha_t = 0.000,006,5/°$ F, and $h = 15$ in.

14.36 through 14.38 Determine the structure displacements and member forces for each of the structures and lack-of-fit situations described below.

14.36 The structure shown below if, in addition to the applied load, members 2-7 and 5-7 are each 0.5 in. too short but are forced to fit.

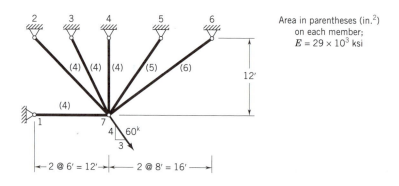

14.37 The structure of Problem 25 in which member 2-3 has the initial crookedness shown but is forced to fit. Do not include the applied loads. Take $EI = 30 \times 10^6$ k-in^2. *Hint:* Convert initial crookedness to end slopes only.

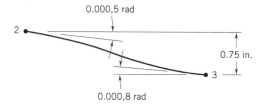

14.38 The structure of Problem 31 in which member 1-2 has the initial crookedness shown but is forced to fit. Take $EI = 400\ 000$ kN · m^2.

Chapter 15

Flexibility Method

Lake Point Tower, Chicago, Illinois (courtesy Portland Cement Association).

15.1 FUNDAMENTAL CONCEPTS OF FLEXIBILITY METHOD

It has already been seen that the concept of compatibility plays a vital role in the analysis of statically indeterminate structures. In fact, the satisfaction of the conditions of compatibility has been a requirement of each of the methods of indeterminate structural analysis that has been presented. In some methods, these conditions are satisfied in a subtle fashion; however, the so-called compatibility methods are clearly based on the simultaneous satisfaction of a number of equations that express the compatibility conditions throughout the structure. The solution of these compatibility equations results in the determination of the redundant forces. For this reason, these methods are sometimes referred to as force methods.

When compatibility methods are expressed in a general matrix formulation, the governing equations involve the structure flexibility matrix, which, in turn, results from a synthesis of the individual member flexibility matrices. Thus, the matrix method, which evolves from a generalization of the compatibility formulation, is called the *flexibility method*.

15.2 OVERVIEW OF FLEXIBILITY METHOD

Consider the externally statically indeterminate beam structure shown in Fig. 15.1*a*. In this case, loads are applied only at points that are identified as nodes on the structure. The redundant support forces are represented by $\{P\}_{\text{II}}$ so that

$$\{P\}_{\text{II}}^{T} = \{(P_2)_4(P_1)_5(P_2)_5(P_6)_5\} \tag{15.1}$$

When the constraints associated with these redundant forces are removed from the structure, the statically determinate primary structure shown in Fig. 15.1*b* remains, and the forces associated with the free displacements of the original structure are given by $\{P\}_{\text{I}}$ in the form

$$\{P\}_{\text{I}}^{T} = \{(P_1)_2(P_2)_2(P_6)_2(P_1)_3(P_2)_3(P_6)_3(P_1)_4(P_6)_4\} \tag{15.2}$$

In Eqs. 15.1 and 15.2, the first subscript on each P term identifies the force component in accordance with Fig. 13.1, and the second subscript gives the node on the structure.

For the primary structure, the relationships between the displacements and the forces at the displaceable points (nodes 2 through 5) are given by

$$\{\Delta\} = [D]\{P\} \tag{15.3}$$

where $[D]$ is the structure flexibility matrix. This equation can be partitioned in the form

$$\begin{Bmatrix} \{\Delta\}_{\text{I}} \\ \{\Delta\}_{\text{II}} \end{Bmatrix} = \begin{bmatrix} [D]_{\text{I,I}} & [D]_{\text{I,II}} \\ [D]_{\text{II,I}} & [D]_{\text{II,II}} \end{bmatrix} \begin{Bmatrix} \{P\}_{\text{I}} \\ \{P\}_{\text{II}} \end{Bmatrix} \tag{15.4}$$

where $\{\Delta\}_{\text{I}}$ and $\{\Delta\}_{\text{II}}$ are defined so that

$$\{\Delta\}_{\text{I}}^{T} = \{(\Delta_1)_2(\Delta_2)_2(\Delta_6)_2(\Delta_1)_3(\Delta_2)_3(\Delta_6)_3(\Delta_1)_4(\Delta_6)_4\} \tag{15.5}$$

$$\{\Delta\}_{\text{II}}^{T} = \{(\Delta_2)_4(\Delta_1)_5(\Delta_2)_5(\Delta_6)_5\} \tag{15.6}$$

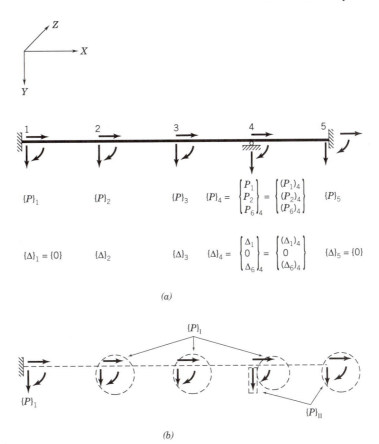

Figure 15.1 Typical statistically indeterminate structure. *(a)* Structure loads and displacements. *(b)* Primary structure with applied forces and redundant forces.

Once the redundant forces $\{P\}_{\mathrm{II}}$ are determined, they are used in conjunction with the applied loads $\{P\}_{\mathrm{I}}$ and the laws of statics to determine the reactions for the primary structure at node 1.

Equation 15.4 is not limited to the problem of Fig. 15.1, but is general, where $\{P\}_{\mathrm{I}}$ corresponds to the applied force components, $\{P\}_{\mathrm{II}}$ represents the redundant reaction components, and $\{\Delta\}_{\mathrm{I}}$ and $\{\Delta\}_{\mathrm{II}}$ give the corresponding displacement components, respectively.

Equation 15.4 can be expanded to give

$$\{\Delta\}_{\mathrm{I}} = [D]_{\mathrm{I,I}}\{P\}_{\mathrm{I}} + [D]_{\mathrm{I,II}}\{P\}_{\mathrm{II}} \tag{15.7}$$

$$\{\Delta\}_{\mathrm{II}} = [D]_{\mathrm{II,I}}\{P\}_{\mathrm{I}} + [D]_{\mathrm{II,II}}\{P\}_{\mathrm{II}} \tag{15.8}$$

Because the displacements at the redundant reactions are zero for the given structure, as shown in Fig. 15.1*a*, Eq. 15.8 becomes

$$\{0\} = [D]_{\mathrm{II,I}}\{P\}_{\mathrm{I}} + [D]_{\mathrm{II,II}}\{P\}_{\mathrm{II}} \tag{15.9}$$

This equation can now be solved for the redundant forces to give

$$\{P\}_{\mathrm{II}} = -[D]_{\mathrm{II,II}}^{-1}[D]_{\mathrm{II,I}}\{P\}_{\mathrm{I}} \tag{15.10}$$

If the displacements at the load points are desired, Eq. 15.10 is substituted into Eq. 15.7 to obtain

$$\{\Delta\}_{\mathrm{I}} = ([D]_{\mathrm{I,I}} - [D]_{\mathrm{I,II}}[D]_{\mathrm{II,II}}^{-1}[D]_{\mathrm{II,I}})\{P\}_{\mathrm{I}} \tag{15.11}$$

With the redundant forces, $\{P\}_{II}$, as determined from Eq. 15.10, and the given loads, $\{P\}_I$, the reaction forces $\{P\}_1$ can be determined from statics.

If the support displacements are not zero, then $\{\Delta\}_{II}$ must reflect the presence of these displacements, and Eq. 15.10 will include a term on the right-hand side that is a direct function of $\{\Delta\}_{II}$.

Before the analysis can proceed according to the sequence outlined in this section, it is necessary to generate the structure flexibility matrix $[D]$. The next sections of this chapter address this topic.

15.3   EQUILIBRIUM MATRICES

In the previous section, attention was focused on an externally statically indeterminate structure. This orientation will continue for now with the treatment of internal redundancies being deferred until Section 15.6.

Consider again the statically determinate primary structure that results when the redundant reaction forces are removed from the original structure. The structure of Fig. 15.1b serves as an example. Such a structure can be analyzed through the systematic application of the equations of equilibrium. These equations, which evolve from joint equilibrium considerations, take the form

$$\{P\} = [C]\{F\} \tag{15.12}$$

In this equation, $\{P\}$ embraces all of the applied forces that act on the primary structure, which are divided into two subvectors: $\{P\}_I$ includes the applied forces on the original structure, and $\{P\}_{II}$ includes the redundant reaction components. The $\{F\}$ vector includes the submatrices $\{F\}_{ij}$ for the i end of each member; the forces at the j end need not be included since they can be determined by applying statics to member ij. The $[C]$ matrix is, of course, the *global statics matrix*.

Since the primary structure is statically determinate and stable, $[C]$ is a nonsingular square matrix. Solving for the member forces, we obtain

$$\{F\} = [C]^{-1}\{P\} = [b]\{P\} \tag{15.13}$$

where $[b]$ is the equilibrium matrix that relates the member forces to the applied forces. In an expanded form, Eq. 15.13 can be written as

$$\{F\} = [[b]_I [b]_{II}]\begin{Bmatrix} \{P\}_I \\ \{P\}_{II} \end{Bmatrix} = [b]_I\{P\}_I + [b]_{II}\{P\}_{II} \tag{15.14}$$

In this form, it is clear that the member forces result from two separate contributions: $[b]_I\{P\}_I$ gives the member forces that result from the applied forces, whereas $[b]_{II}\{P\}_{II}$ gives the member forces that are induced from the redundant reaction forces.

15.4   GENERATION OF STRUCTURE FLEXIBILITY MATRIX

For each individual member in the structure, the strain energy can be expressed as

$$U_{ij} = \tfrac{1}{2}\{F\}_{ij}^T\{e\}_{ij} \tag{15.15}$$

where $\{e\}_{ij}$ contains the displacements at the i end relative to the j end of the member ij. Since the member in Fig. 13.2 is fixed at the j end, $\{e\}_{ij} = \{\delta\}_{ij}$, and thus Eq. 13.13 can be written as

$$\{e\}_{ij} = [d]_{ii}^{j}\{F\}_{ij} \tag{15.16}$$

Substituting Eq. 15.16 into Eq. 15.15, we find

$$U_{ij} = \tfrac{1}{2}\{F\}_{ij}^T [d]_{ii}^j \{F\}_{ij} \tag{15.17}$$

For the entire structure

$$U = \tfrac{1}{2}\{F\}^T [d]_u \{F\} \tag{15.18}$$

where

$$\{F\}^T = \{\{F\}_{12}\{F\}_{23} \cdots \{F\}_{ij} \cdots\} \tag{15.19}$$

and $\{d\}_u$ is a matrix containing the individual member flexibility matrices along the diagonal. This is called the *unassembled flexibility matrix* and has the form

$$[d]_u = \begin{bmatrix} [d]_{11}^2 & & & & & \\ & [d]_{22}^3 & & & & \\ & & \cdot & & & \\ & & & \cdot & & \\ & & & & [d]_{ii}^j & \\ & & & & & \cdot \\ & & & & & & \cdot \end{bmatrix} \tag{15.20}$$

Substitution of Eq. 15.13 into Eq. 15.18 gives

$$U = \tfrac{1}{2}\{P\}^T [b]^T [d]_u [b]\{P\} \tag{15.21}$$

Equation 15.21 gives the total strain energy as derived from member considerations. In terms of structure forces and displacements, the strain energy can also be expressed as

$$U = \tfrac{1}{2}\{P\}^T \{\Delta\} \tag{15.22}$$

Substituting Eq. 15.3 into Eq. 15.22 we obtain

$$U = \tfrac{1}{2}\{P\}^T [D]\{P\} \tag{15.23}$$

Comparison of Eqs. 15.21 and 15.23 reveals that

$$[D] = [b]^T [d]_u [b] \tag{15.24}$$

In a partitioned form, Eq. 15.24 becomes

$$[D] = \begin{bmatrix} [b]_{\mathrm{I}}^T \\ [b]_{\mathrm{II}}^T \end{bmatrix} [d]_u [[b]_{\mathrm{I}}[b]_{\mathrm{II}}]$$

$$= \begin{bmatrix} [b]_{\mathrm{I}}^T [d]_u [b]_{\mathrm{I}} & [b]_{\mathrm{I}}^T [d]_u [b]_{\mathrm{II}} \\ \hline [b]_{\mathrm{II}}^T [d]_u [b]_{\mathrm{I}} & [b]_{\mathrm{II}}^T [d]_u [b]_{\mathrm{II}} \end{bmatrix} \tag{15.25}$$

where $[b]_{\mathrm{I}}$ and $[b]_{\mathrm{II}}$ are defined by Eq. 15.14.

Comparing Eqs. 15.4 and 15.25, we have

$$[D]_{\mathrm{I,I}} = [b]_{\mathrm{I}}^{T}[d]_{u}[b]_{\mathrm{I}}$$

$$[D]_{\mathrm{I,II}} = [b]_{\mathrm{I}}^{T}[d]_{u}[b]_{\mathrm{II}}$$

$$[D]_{\mathrm{II,I}} = [b]_{\mathrm{II}}^{T}[d]_{u}[b]_{\mathrm{I}} \qquad (15.26)$$

$$[D]_{\mathrm{II,II}} = [b]_{\mathrm{II}}^{T}[d]_{u}[b]_{\mathrm{II}}$$

Application of Eqs. 15.26 yields the submatrices that collectively make up the structure flexibility matrix of Eq. 15.3.

15.5 APPLICATION OF THE FLEXIBILITY METHOD

In applying the flexibility method, an orderly procedure is employed, which is outlined by the following steps:

1. Number the nodal points of the structure. This must include all load points, support points, and points where two or more members join together. This procedure implicitly introduces structure force and displacement vectors in the form of Eq. 13.1 and member force and displacement vectors as described by Eq. 13.2.

2. Select the redundant forces, $\{P\}_{\mathrm{II}}$. This defines the primary structure and automatically establishes the arrangements of $\{P\}_{\mathrm{I}}$, $\{\Delta\}_{\mathrm{I}}$, and $\{\Delta\}_{\mathrm{II}}$ as indicated by Eq. 15.4.

3. Generate the matrices $[b]_{\mathrm{I}}$ and $[b]_{\mathrm{II}}$ of Eq. 15.14 from equilibrium considerations for the primary structure. Here, it is emphasized that the $\{F\}$ vector includes the member-end force vectors at only one end of each member.

4. Establish the individual member flexibility matrices based on Eq. 13.12 and form the unassembled flexibility matrix $[d]_{u}$ in accordance with Eq. 15.20.

5. Develop the submatrices $[D]_{\mathrm{II,I}}$ and $[D]_{\mathrm{II,II}}$ of the structure flexibility matrix from Eqs. 15.26.

6. Determine the redundant forces $\{P\}_{\mathrm{II}}$ from Eq. 15.10. The other reactions on the primary structure can be determined from statics.

7. Calculate the final member-end forces from Eq. 15.14. The forces at the opposite end of each member can be determined by statics.

8. If the nodal displacements are desired, develop the submatrices $[D]_{\mathrm{I,I}}$ and $[D]_{\mathrm{I,II}}$ from Eq. 15.26 and obtain the desired displacements from Eq. 15.11.

For certain classes of structures, the force and displacement vectors for both the members and the structure are simplified. For instance, planar truss members have a single member-end force and displacement component, and the structure force and displacement vectors at each joint contain only two components. Similarly, for planar beam-type structures that are subjected to transverse load only, the member and structure force and displacement vectors will exclude axial components. In these simplified cases, the complete member flexibility matrix given in Eq. 13.12 must be modified so as to include only the member actions being considered. This is illustrated in the example problems that follow.

Also, care must be taken to ensure that the rows of the individual $[d]_{ii}^{j}$ matrices are arranged so that when $[d]_{u}$ is formed, there is a correspondence between the rows of $[d]_{u}$ and the rows of the member force matrix $\{F\}$.

EXAMPLE 15.1

Determine the bar forces for the truss structure shown. Assume *EA* is the same for each member.

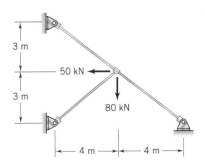

Selection of Node Points

Structure Forces and Displacements

$$\{P\}_1 = \begin{Bmatrix} P_1 \\ P_2 \end{Bmatrix}_1 = \begin{Bmatrix} (P_1)_1 \\ (P_2)_1 \end{Bmatrix} = \begin{Bmatrix} (P_1)_1 \\ 0.75(P_1)_1 \end{Bmatrix}; \qquad (\Delta)_1 = \begin{Bmatrix} \Delta_1 \\ \Delta_2 \end{Bmatrix}_1 = \{0\}$$

$$\{P\}_2 = \begin{Bmatrix} P_1 \\ P_2 \end{Bmatrix}_2 = \begin{Bmatrix} (P_1)_2 \\ (P_2)_2 \end{Bmatrix} = \begin{Bmatrix} -50 \\ 80 \end{Bmatrix}; \qquad \{\Delta\}_2 = \begin{Bmatrix} \Delta_1 \\ \Delta_2 \end{Bmatrix}_2 = \begin{Bmatrix} (\Delta_1)_2 \\ (\Delta_2)_2 \end{Bmatrix}$$

$$\{P\}_4 = \begin{Bmatrix} P_1 \\ P_2 \end{Bmatrix}_4 = \begin{Bmatrix} (P_1)_4 \\ (P_2)_4 \end{Bmatrix} = \begin{Bmatrix} (P_1)_4 \\ 0.75(P_1)_4 \end{Bmatrix}; \qquad (\Delta)_4 = \begin{Bmatrix} \Delta_1 \\ \Delta_2 \end{Bmatrix}_4 = \{0\}$$

$$\{P\}_3 = \begin{Bmatrix} P_1 \\ P_2 \end{Bmatrix}_3 = \begin{Bmatrix} (P_1)_3 \\ (P_2)_3 \end{Bmatrix} = \begin{Bmatrix} (P_1)_3 \\ -0.75(P_1)_3 \end{Bmatrix}; \qquad \{\Delta\}_3 = \begin{Bmatrix} \Delta_1 \\ \Delta_2 \end{Bmatrix}_3 = \{0\}$$

Member Forces and Displacements

$$(F_1)_{ij};\ (\delta_1)_{ij} \quad \longleftarrow \quad \underset{i}{\rule{0pt}{0pt}} \rule{7cm}{1pt} \underset{j}{\rule{0pt}{0pt}} \quad \longrightarrow \quad (F_1)_{ji} = (F_1)_{ij};\ (\delta_1)_{ji}$$

$$\text{Member 1-2:} \quad \{F\}_{12} = (F_1)_{12}; \quad \{\delta\}_{12} = (\delta_1)_{12}$$

$$\text{Member 3-2:} \quad \{F\}_{32} = (F_1)_{32}; \quad \{\delta\}_{32} = (\delta_1)_{32}$$

$$\text{Member 4-2:} \quad \{F\}_{42} = (F_1)_{42}; \quad \{\delta\}_{42} = (\delta_1)_{42}$$

Selection of Redundants

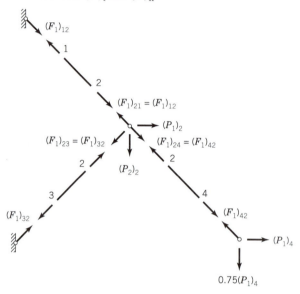

Select $\{P\}_{II} = \{P\}_4 = \begin{Bmatrix} (P_1)_4 \\ 0.75(P_1)_4 \end{Bmatrix}$; $\quad \{\Delta\}_{II} = \{\Delta\}_4 = \{0\}$

Primary structure

$\{P\}_I = \{P\}_2 = \begin{Bmatrix} (P_1)_2 \\ (P_2)_2 \end{Bmatrix}$; $\quad \{\Delta\}_I = \{\Delta\}_2 = \begin{Bmatrix} (\Delta_1)_2 \\ (\Delta_2)_2 \end{Bmatrix}$

Generation of [b] Matrices

$$\{F\} = [b]_I\{P\}_I + [b]_{II}\{P\}_{II} \tag{15.14}$$

Primary structure—loaded with $\{P\}_I$ and $\{P\}_{II}$

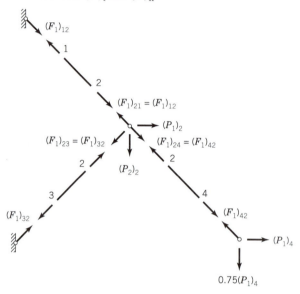

Note: Member force at j end is expressed in terms of member force at i end.

Equilibrium applied to joints 2 and 4 leads to

$$(P_1)_2 - 0.8(F_1)_{12} - 0.8(F_1)_{32} + 0.8(F_1)_{42} = 0$$

$$(P_2)_2 - 0.6(F_1)_{12} + 0.6(F_1)_{32} + 0.6(F_1)_{42} = 0$$

$$(P_1)_4 - 0.8(F_1)_{42} = 0$$

In the form of Eq. 15.12, we have

$$\{P\} = [C]\{F\}; \quad \begin{Bmatrix} (P_1)_2 \\ (P_2)_2 \\ (P_1)_4 \end{Bmatrix} = \begin{bmatrix} 0.8 & 0.8 & -0.8 \\ 0.6 & -0.6 & -0.6 \\ 0 & 0 & 0.8 \end{bmatrix} \begin{Bmatrix} (F_1)_{12} \\ (F_1)_{32} \\ (F_1)_{42} \end{Bmatrix} \tag{15.12}$$

Solving for the member forces, we obtain

$$\{F\} = [b]\{P\} = [[b]_I [b]_{II}] \begin{Bmatrix} \{P\}_I \\ \{P\}_{II} \end{Bmatrix} \tag{15.14}$$

$$\begin{Bmatrix} (F_1)_{12} \\ (F_1)_{32} \\ (F_1)_{42} \end{Bmatrix} = \begin{bmatrix} 0.625 & 0.833 & \vdots & 1.25 \\ 0.625 & -0.833 & \vdots & 0 \\ 0 & 0 & \vdots & 1.25 \end{bmatrix} \begin{Bmatrix} (P_1)_2 \\ (P_2)_2 \\ (P_1)_4 \end{Bmatrix}$$

$$\underbrace{\qquad\qquad}_{[b]_I} \quad \underbrace{\qquad}_{[b]_{II}}$$

Member Flexibilities

$$\{\delta\}_{ij} = [d]_{ii}^{j}\{F\}_{ij} \tag{13.13}$$

In this case

$$\{\delta_1\}_{ij} = \left[\frac{l}{EA}\right]_{ij}(F_1)_{ij} \tag{13.12}$$

Member 1-2: $[d]_{11}^{2} = \left[\dfrac{l}{EA}\right] = \left(\dfrac{5}{EA}\right)$

Member 3-2: $[d]_{33}^{2} = \left(\dfrac{5}{EA}\right)$

Member 4-2: $[d]_{44}^{2} = \left(\dfrac{5}{EA}\right)$

$$[d]_u = \frac{1}{EA}\begin{bmatrix} 5 & 0 & 0 \\ 0 & 5 & 0 \\ 0 & 0 & 5 \end{bmatrix} \tag{15.20}$$

Structure Flexibility Submatrices

$$[D]_{\text{II},\text{I}} = [b]_{\text{II}}^{T}[d]_u[b]_{\text{I}} \tag{15.26}$$

$$[D]_{\text{II},\text{I}} = [1.25 \ \ 0 \ \ 1.25][d]_u\begin{bmatrix} 0.625 & 0.833 \\ 0.625 & -0.833 \\ 0 & 0 \end{bmatrix} = \frac{1}{EA}[3.91 \ \ 5.21]$$

$$[D]_{\text{II},\text{II}} = [b]_{\text{II}}^{T}[d]_u[b]_{\text{II}} \tag{15.26}$$

$$[D]_{\text{II},\text{II}} = [1.25 \ \ 0 \ \ 1.25][d]_u\begin{bmatrix} 1.25 \\ 0 \\ 1.25 \end{bmatrix} = \frac{1}{EA}[15.63]$$

Calculation of Redundants

$$\{P\}_{\text{II}} = -[D]_{\text{II},\text{II}}^{-1}[D]_{\text{II},\text{I}}\{P\}_{\text{I}} \tag{15.10}$$

$$\{P\}_{\text{II}} = -\frac{EA}{15.63}\cdot\frac{1}{EA}\begin{bmatrix} 3.91 & 5.21 \end{bmatrix}\begin{Bmatrix} -50 \\ 80 \end{Bmatrix} = -14.16 \text{ kN}$$

Final Bar Forces

$$\{F\} = [b]_{\text{I}}\{P\}_{\text{I}} + [b]_{\text{II}}\{P\}_{\text{II}} \tag{15.14}$$

$$\begin{Bmatrix} (F_1)_{12} \\ (F_1)_{32} \\ (F_1)_{42} \end{Bmatrix} = \begin{bmatrix} 0.625 & 0.833 \\ 0.625 & -0.833 \\ 0 & 0 \end{bmatrix}\begin{Bmatrix} -50 \\ 80 \end{Bmatrix} + \begin{bmatrix} 1.25 \\ 0 \\ 1.25 \end{bmatrix}\{-14.16\} = \begin{Bmatrix} 17.69 \\ -97.89 \\ -17.70 \end{Bmatrix} \text{ kN}$$

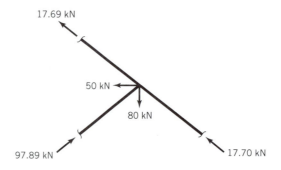

17.69 kN

50 kN

80 kN

97.89 kN

17.70 kN

EXAMPLE 15.2

Determine the member-end forces for the structure shown, and construct the shear and moment diagrams. EI = constant.

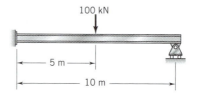

100 kN

5 m

10 m

Selection of Node Points

Structure Forces and Displacements

$$\{P\}_1 = \begin{Bmatrix} P_2 \\ P_6 \end{Bmatrix}_1 = \begin{Bmatrix} (P_2)_1 \\ (P_6)_1 \end{Bmatrix}; \quad \{P\}_2 = \begin{Bmatrix} P_2 \\ P_6 \end{Bmatrix}_2 = \begin{Bmatrix} (P_2)_2 \\ (P_6)_2 \end{Bmatrix}; \quad \{P\}_3 = \begin{Bmatrix} P_2 \\ P_6 \end{Bmatrix}_3 = \begin{Bmatrix} (P_2)_3 \\ (P_6)_3 \end{Bmatrix}$$

$$\{\Delta\}_1 = \begin{Bmatrix} \Delta_2 \\ \Delta_6 \end{Bmatrix}_1 = \begin{Bmatrix} 0 \\ 0 \end{Bmatrix}; \quad \{\Delta\}_2 = \begin{Bmatrix} \Delta_2 \\ \Delta_6 \end{Bmatrix}_2 = \begin{Bmatrix} (\Delta_2)_2 \\ (\Delta_6)_2 \end{Bmatrix}; \quad \{\Delta\}_3 = \begin{Bmatrix} \Delta_2 \\ \Delta_6 \end{Bmatrix}_3 = \begin{Bmatrix} 0 \\ (\Delta_6)_3 \end{Bmatrix}$$

Member Forces and Displacements

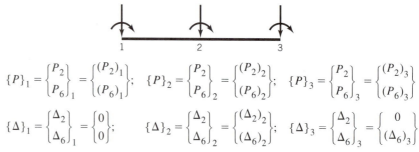

$(F_2)_{ji}$

$(F_6)_{ji}$

$(F_6)_{ij}$

$(F_2)_{ij}$

Member 1-2: $\{F\}_{12} = \begin{Bmatrix} F_2 \\ F_6 \end{Bmatrix}_{12} = \begin{Bmatrix} (F_2)_{12} \\ (F_6)_{12} \end{Bmatrix}$

$$\{\delta\}_{12} = \begin{Bmatrix} \delta_2 \\ \delta_6 \end{Bmatrix}_{12}$$

Member 2-3: $\{F\}_{23} = \begin{Bmatrix} F_2 \\ F_6 \end{Bmatrix}_{23} = \begin{Bmatrix} (F_2)_{23} \\ (F_6)_{23} \end{Bmatrix}$

$$\{\delta\}_{23} = \begin{Bmatrix} \delta_2 \\ \delta_6 \end{Bmatrix}_{23}$$

Selection of Redundants

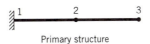

Primary structure

Select $\{P\}_{\text{II}} = \{(P_2)_3\};$ $\{\Delta\}_{\text{II}} = \{(\Delta_2)_3\} = \{0\}$

Then

$$\{P\}_{\text{I}} = \left\{ \begin{array}{c} (P_2)_2 \\ (P_6)_2 \\ (P_6)_3 \end{array} \right\} = \left\{ \begin{array}{c} 100 \\ 0 \\ 0 \end{array} \right\}; \quad \{\Delta\}_{\text{I}} = \left\{ \begin{array}{c} (\Delta_2)_2 \\ (\Delta_6)_2 \\ (\Delta_6)_3 \end{array} \right\}$$

Generation of [b] Matrices

$$\{F\} = [b]_{\text{I}}\{P\}_{\text{I}} + [b]_{\text{II}}\{P\}_{\text{II}} \tag{15.14}$$

Primary structure loaded with $\{P\}_{\text{I}}$ and $\{P\}_{\text{II}}$:

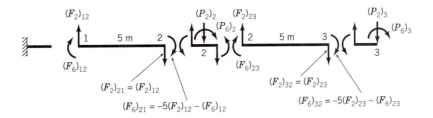

Note: Member forces at j end are expressed in terms of member forces at i end.

Equilibrium applied at joints 2 and 3 leads to

$$(P_2)_2 - (F_2)_{12} + (F_2)_{23} = 0$$
$$(P_6)_2 + 5(F_2)_{12} + (F_6)_{12} - (F_6)_{23} = 0$$
$$(P_2)_3 - (F_2)_{23} = 0$$
$$(P_6)_3 + 5(F_2)_{23} + (F_6)_{23} = 0$$

In the form of Eq. 15.12, with $\{P\}$ partitioned according to $\{P\}_{\text{I}}$ and $\{P\}_{\text{II}}$, we have

$$\left\{ \begin{array}{c} \{P\}_{\text{I}} \\ \{P\}_{\text{II}} \end{array} \right\} = [C]\{F\} = \left\{ \begin{array}{c} (P_2)_2 \\ (P_6)_2 \\ (P_6)_3 \\ \overline{(P_2)_3} \end{array} \right\} = \left[\begin{array}{cccc} 1 & 0 & -1 & 0 \\ -5 & -1 & 0 & 1 \\ 0 & 0 & -5 & -1 \\ 0 & 0 & 1 & 0 \end{array} \right] \left\{ \begin{array}{c} \left\{ \begin{array}{c} F_2 \\ F_6 \end{array} \right\}_{12} \\ \left\{ \begin{array}{c} F_2 \\ F_6 \end{array} \right\}_{23} \end{array} \right\} \tag{15.12}$$

Solving for the member forces, we have

$$\{F\} = [b]\{P\} = [[b]_I [b]_{II}] \begin{Bmatrix} \{P\}_I \\ \{P\}_{II} \end{Bmatrix}$$

$$\begin{Bmatrix} \begin{Bmatrix} F_2 \\ F_6 \end{Bmatrix}_{12} \\ \begin{Bmatrix} F_2 \\ F_6 \end{Bmatrix}_{23} \end{Bmatrix} = \begin{bmatrix} 1 & 0 & 0 & \vdots & 1 \\ -5 & -1 & -1 & \vdots & -10 \\ 0 & 0 & 0 & \vdots & 1 \\ 0 & 0 & -1 & \vdots & -5 \end{bmatrix} \begin{Bmatrix} (P_2)_2 \\ (P_6)_2 \\ (P_6)_3 \\ \hdashline (P_2)_3 \end{Bmatrix} \qquad (15.14)$$

$$\underbrace{\phantom{\begin{matrix}1 & 0 & 0\end{matrix}}}_{[b]_I} \qquad \underbrace{\phantom{\begin{matrix}1\end{matrix}}}_{[b]_{II}}$$

Member Flexibilities

$$\{\delta\}_{ij} = [d]_{ii}^j \{F\}_{ij} \qquad (13.13)$$

In this case

$$\begin{Bmatrix} \delta_2 \\ \delta_6 \end{Bmatrix}_{ij} = \begin{bmatrix} \dfrac{l^3}{3EI} & \dfrac{l^2}{2EI} \\ \dfrac{l^2}{2EI} & \dfrac{l}{EI} \end{bmatrix}_{ij} \begin{Bmatrix} F_2 \\ F_6 \end{Bmatrix}_{ij} \qquad (13.12)$$

Member 1-2: $\quad [d]_{11}^2 = \dfrac{l}{EI} \begin{bmatrix} 14.7 & 12.5 \\ 12.5 & 5.0 \end{bmatrix}$

Member 2-3: $\quad [d]_{22}^3 = \dfrac{l}{EI} \begin{bmatrix} 41.7 & 12.5 \\ 12.5 & 5.0 \end{bmatrix}$

$$[d]_u = \dfrac{1}{EI} \begin{bmatrix} 41.7 & 12.5 & 0 & 0 \\ 12.5 & 5.0 & 0 & 0 \\ 0 & 0 & 41.7 & 12.5 \\ 0 & 0 & 12.5 & 5.0 \end{bmatrix} \qquad (15.20)$$

Structure Flexibilities

$$[D]_{II,I} = [b]_{II}^T [d]_u [b]_I \qquad (15.26)$$

$$[D]_{II,I} = [1 \;\; -10 \;\; 1 \;\; -5][d]_u \begin{bmatrix} 1 & 0 & 0 \\ -5 & -1 & -1 \\ 0 & 0 & 0 \\ 0 & 0 & -1 \end{bmatrix} = \dfrac{1}{EI}[104.2 \;\; -37.5 \;\; -50]$$

$$[D]_{II,II} = [b]_{II}^T [d]_u [b]_{II} \qquad (15.26)$$

$$[D]_{II,II} = [1 \;\; -10 \;\; 1 \;\; -5][d]_u \begin{bmatrix} 1 \\ -10 \\ 0 \\ -5 \end{bmatrix} = \dfrac{1}{EI}[333.4]$$

Determination of Redundants

$$\{P\}_{\text{II}} = -[D]_{\text{II,II}}^{-1}[D]_{\text{II,I}}\{P\}_{\text{I}} \tag{15.10}$$

$$\{P\}_{\text{II}} = -\frac{EI}{333.4} \cdot \frac{1}{EI}\begin{bmatrix} 104.2 & -37.5 & -50 \end{bmatrix}\begin{Bmatrix} 100 \\ 0 \\ 0 \end{Bmatrix} = -31.3 \text{ kN}$$

Member-End Forces

$$\{F\} = [b]_{\text{I}}\{P\}_{\text{I}} + [b]_{\text{II}}\{P\}_{\text{II}}$$

$$\begin{Bmatrix} \begin{Bmatrix} F_2 \\ F_6 \end{Bmatrix}_{12} \\ \begin{Bmatrix} F_2 \\ F_6 \end{Bmatrix}_{23} \end{Bmatrix} = \begin{bmatrix} 1 & 0 & 0 \\ -5 & -1 & -1 \\ 0 & 0 & 0 \\ 0 & 0 & -1 \end{bmatrix}\begin{Bmatrix} 100 \\ 0 \\ 0 \end{Bmatrix} + \begin{bmatrix} 1 \\ -10 \\ 1 \\ -5 \end{bmatrix}\{-31.3\} = \begin{Bmatrix} +68.7 \text{ kN} \\ -187.0 \text{ kN} \cdot \text{m} \\ -31.3 \text{ kN} \\ +156.5 \text{ kN} \cdot \text{m} \end{Bmatrix} \tag{15.14}$$

Shear and Moment Diagrams

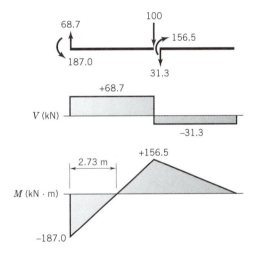

Note: Start with the forces at the left end of any member and use the static relations for shear and moment to obtain the forces at the right end of the member. These should be consistent with the forces at the left end of the next member.

EXAMPLE 15.3

Determine the member-end forces for the frame structure given.

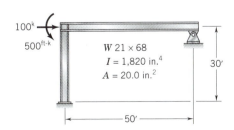

Selection of Node Points

Structure Forces and Displacements

$$\{P\}_2 = \begin{Bmatrix} P_1 \\ P_2 \\ P_6 \end{Bmatrix}_2 = \begin{Bmatrix} (P_1)_2 \\ (P_2)_2 \\ (P_6)_2 \end{Bmatrix} = \begin{Bmatrix} 100 \\ 0 \\ -500 \end{Bmatrix}; \qquad \{\Delta\}_2 = \begin{Bmatrix} \Delta_1 \\ \Delta_2 \\ \Delta_6 \end{Bmatrix}_2 = \begin{Bmatrix} (\Delta_1)_2 \\ (\Delta_2)_2 \\ (\Delta_6)_2 \end{Bmatrix}$$

$$\{P\}_3 = \begin{Bmatrix} P_1 \\ P_2 \\ P_6 \end{Bmatrix}_3 = \begin{Bmatrix} (P_1)_3 \\ (P_2)_3 \\ (P_6)_3 \end{Bmatrix} = \begin{Bmatrix} (P_1)_3 \\ (P_2)_3 \\ 0 \end{Bmatrix}; \qquad \{\Delta\}_3 = \begin{Bmatrix} \Delta_1 \\ \Delta_2 \\ \Delta_6 \end{Bmatrix}_3 = \begin{Bmatrix} (\Delta_1)_3 \\ (\Delta_2)_3 \\ (\Delta_6)_3 \end{Bmatrix} = \begin{Bmatrix} 0 \\ 0 \\ (\Delta_6)_3 \end{Bmatrix}$$

$$\{P\}_1 = \begin{Bmatrix} P_1 \\ P_2 \\ P_6 \end{Bmatrix}_1 = \begin{Bmatrix} (P_1)_1 \\ (P_2)_1 \\ (P_6)_1 \end{Bmatrix}; \qquad (\Delta)_1 = \begin{Bmatrix} \Delta_1 \\ \Delta_2 \\ \Delta_6 \end{Bmatrix}_1 = \begin{Bmatrix} (\Delta_1)_1 \\ (\Delta_2)_1 \\ (\Delta_6)_1 \end{Bmatrix} = \begin{Bmatrix} 0 \\ 0 \\ 0 \end{Bmatrix}$$

Member Forces and Displacements

$$\text{Member 1-2:} \quad \{F\}_{12} = \begin{Bmatrix} F_1 \\ F_2 \\ F_6 \end{Bmatrix}_{12} = \begin{Bmatrix} (F_1)_{12} \\ (F_2)_{12} \\ (F_6)_{12} \end{Bmatrix}; \quad \{\delta\}_{12} = \begin{Bmatrix} \delta_1 \\ \delta_2 \\ \delta_6 \end{Bmatrix}_{12} = \begin{Bmatrix} (\delta_1)_{12} \\ (\delta_2)_{12} \\ (\delta_6)_{12} \end{Bmatrix}$$

$$\text{Member 2-3:} \quad \{F\}_{23} = \begin{Bmatrix} F_1 \\ F_2 \\ F_6 \end{Bmatrix}_{23} = \begin{Bmatrix} (F_1)_{23} \\ (F_2)_{23} \\ (F_6)_{23} \end{Bmatrix}; \quad \{\delta\}_{23} = \begin{Bmatrix} \delta_1 \\ \delta_2 \\ \delta_6 \end{Bmatrix}_{23} = \begin{Bmatrix} (\delta_1)_{23} \\ (\delta_2)_{23} \\ (\delta_6)_{23} \end{Bmatrix}$$

Selection of Redundants

$$\text{Select} \quad \{P\}_{\text{II}} = \begin{Bmatrix} (P_1)_3 \\ (P_2)_3 \end{Bmatrix}; \quad \{\Delta\}_{\text{II}} = \begin{Bmatrix} (\Delta_1)_3 \\ (\Delta_2)_3 \end{Bmatrix} = \begin{Bmatrix} 0 \\ 0 \end{Bmatrix}$$

Primary structure

$$\{P\}_{\text{I}} = \begin{Bmatrix} (P_1)_2 \\ (P_2)_2 \\ (P_6)_2 \\ (P_6)_3 \end{Bmatrix} = \begin{Bmatrix} 100 \\ 0 \\ -500 \\ 0 \end{Bmatrix}; \quad \{\Delta\}_{\text{I}} = \begin{Bmatrix} (\Delta_1)_2 \\ (\Delta_2)_2 \\ (\Delta_6)_2 \\ (\Delta_6)_3 \end{Bmatrix}$$

Generation of [b] Matrices

$$\{F\} = [b]_{\text{I}}\{P\}_{\text{I}} + [b]_{\text{II}}\{P\}_{\text{II}} \tag{15.14}$$

Primary structure loaded with $\{P\}_{\text{I}}$ and $\{P\}_{\text{II}}$:

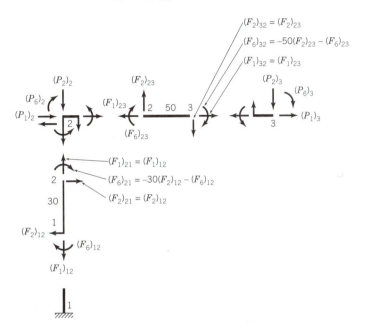

Note: Member forces at j end are expressed in terms of member forces at i end.

Equilibrium applied at joints 2 and 3 leads to

$$(P_1)_2 - (F_2)_{12} + (F_1)_{23} = 0$$
$$(P_2)_2 + (F_1)_{12} + (F_2)_{23} = 0$$
$$(P_6)_2 + 30(F_2)_{12} + (F_6)_{12} - (F_6)_{23} = 0$$
$$(P_1)_3 - (F_1)_{23} = 0$$
$$(P_2)_3 - (F_2)_{23} = 0$$
$$(P_6)_3 + 50(F_2)_{23} + (F_6)_{23} = 0$$

In the form of Eq. 15.12, with $\{P\}$ partitioned according to $\{P\}_{\text{I}}$ and $\{P\}_{\text{II}}$, we have

$$\left\{ \begin{matrix} \{P\}_{\text{I}} \\ \{P\}_{\text{II}} \end{matrix} \right\} = [C]\{F\} = \left\{ \begin{matrix} (P_1)_2 \\ (P_2)_2 \\ (P_6)_2 \\ (P_6)_3 \\ \overline{(P_1)_3} \\ (P_2)_3 \end{matrix} \right\} = \begin{bmatrix} 0 & 1 & 0 & -1 & 0 & 0 \\ -1 & 0 & 0 & 0 & -1 & 0 \\ 0 & -30 & -1 & 0 & 0 & 1 \\ 0 & 0 & 0 & 0 & -50 & -1 \\ 0 & 0 & 0 & 1 & 0 & 0 \\ 0 & 0 & 0 & 0 & 1 & 0 \end{bmatrix} \left\{ \begin{matrix} F_1 \\ F_2 \\ F_6 \\ F_1 \\ F_2 \\ F_6 \end{matrix} \right\}_{\substack{12 \\ 23}} \tag{15.12}$$

Solving for the member forces, we have

$$\{F\} = [b]\{P\} = [[b]_I [b]_{II}] \begin{Bmatrix} \{P\}_I \\ \{P\}_{II} \end{Bmatrix}$$

$$
\begin{Bmatrix}
\begin{Bmatrix} F_1 \\ F_2 \\ F_6 \end{Bmatrix}_{12} \\[2ex]
\begin{Bmatrix} F_1 \\ F_2 \\ F_6 \end{Bmatrix}_{23}
\end{Bmatrix}
=
\left[
\begin{array}{cccc:cc}
0 & -1 & 0 & 0 & 0 & -1 \\
1 & 0 & 0 & 0 & 1 & 0 \\
-30 & 0 & -1 & -1 & -30 & -50 \\
\hdashline
0 & 0 & 0 & 0 & 1 & 0 \\
0 & 0 & 0 & 0 & 0 & 1 \\
0 & 0 & 0 & -1 & 0 & -50
\end{array}
\right]
\begin{Bmatrix}
(P_1)_2 \\ (P_2)_2 \\ (P_6)_2 \\ (P_6)_3 \\ (P_1)_3 \\ (P_2)_3
\end{Bmatrix}
\qquad (15.14)
$$

$$\underbrace{\qquad\qquad\quad}_{[b]_I} \quad \underbrace{\qquad}_{[b]_{II}}$$

Member Flexibilities

$$\{\delta\}_{ij} = [d]_{ii}^j \{F\}_{ij} \qquad (13.13)$$

In this case

$$
\begin{Bmatrix} \delta_1 \\[1.5ex] \delta_2 \\[1.5ex] \delta_6 \end{Bmatrix}_{ij}
=
\begin{bmatrix}
\dfrac{l}{EA} & 0 & 0 \\[1.5ex]
0 & \dfrac{l^3}{3EI} & \dfrac{l^2}{2EI} \\[1.5ex]
0 & \dfrac{l^2}{2EI} & \dfrac{l}{EI}
\end{bmatrix}_{ij}
\begin{Bmatrix} F_1 \\[1.5ex] F_2 \\[1.5ex] F_6 \end{Bmatrix}_{ij}
\qquad (13.12)
$$

$$
[d]_u = \begin{bmatrix} [d]_{11}^2 & \\ & [d]_{22}^3 \end{bmatrix} = \frac{1}{EI}
\begin{bmatrix}
18.9 & 0 & 0 & 0 & 0 & 0 \\
0 & 9{,}000 & 450 & 0 & 0 & 0 \\
0 & 450 & 30 & 0 & 0 & 0 \\
0 & 0 & 0 & 31.5 & 0 & 0 \\
0 & 0 & 0 & 0 & 41{,}667 & 1{,}250 \\
0 & 0 & 0 & 0 & 1{,}250 & 50
\end{bmatrix}
\qquad (15.20)
$$

Structure Flexibilities

$$[D]_{II,I} = [b]_{II}^T [d]_u [b]_I \qquad (15.26)$$

$$[D]_{II,I} = \frac{1}{EI}
\begin{bmatrix}
9{,}000 & 0 & 450 & 450 \\
22{,}500 & 18.9 & 1{,}500 & 2{,}750
\end{bmatrix}$$

$$[D]_{II,II} = [b]_{II}^T [d]_u [b]_{II} \qquad (15.26)$$

$$[D]_{II,II} = \frac{1}{EI}
\begin{bmatrix}
9{,}032 & 22{,}500 \\
22{,}500 & 116{,}686
\end{bmatrix}$$

Determination of Redundants

$$\{P\}_{\text{II}} = -[D]^{-1}_{\text{II,II}}[D]_{\text{II,I}}\{P\}_{\text{I}} \tag{15.10}$$

$$\{P\}_{\text{II}} = -\frac{EI}{5.476 \times 10^8}\begin{bmatrix} 116{,}686 & -22{,}500 \\ -22{,}500 & 9{,}032 \end{bmatrix}$$

$$\times \frac{1}{EI}\begin{bmatrix} 9{,}000 & 0 & 450 & 450 \\ 22{,}500 & 18.9 & 1{,}500 & 2{,}750 \end{bmatrix}\begin{Bmatrix} 100 \\ 0 \\ 0 \\ -500 \\ 0 \end{Bmatrix} = \begin{Bmatrix} -82.2 \\ +3.0 \end{Bmatrix} \text{kips}$$

Member-End Forces

$$\{F\} = [b]_{\text{I}}\{P\}_{\text{I}} + [b]_{\text{II}}\{P\}_{\text{II}}$$

$$\begin{Bmatrix} \begin{Bmatrix} F_1 \\ F_2 \\ F_6 \end{Bmatrix}_{12} \\ \begin{Bmatrix} F_1 \\ F_2 \\ F_6 \end{Bmatrix}_{23} \end{Bmatrix} = \begin{bmatrix} 0 & -1 & 0 & 0 \\ 1 & 0 & 0 & 0 \\ -30 & 0 & -1 & -1 \\ 0 & 0 & 0 & 0 \\ 0 & 0 & 0 & 0 \\ 0 & 0 & 0 & -1 \end{bmatrix}\begin{Bmatrix} 100 \\ 0 \\ -500 \\ 0 \end{Bmatrix} + \begin{bmatrix} 0 & -1 \\ 1 & 0 \\ -30 & -50 \\ 1 & 0 \\ 0 & 1 \\ 0 & -50 \end{bmatrix}\begin{Bmatrix} -82.2 \\ +3.0 \end{Bmatrix} \tag{15.14}$$

$$= \begin{Bmatrix} -3.0^k \\ +17.8^k \\ -184.0'^{-k} \\ -82.2^k \\ +3.0^k \\ -150.0'^{-k} \end{Bmatrix}$$

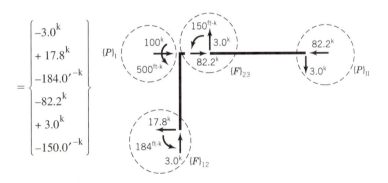

15.6 INTERNALLY REDUNDANT STRUCTURES

The procedure developed in Sections 15.2 through 15.4 and illustrated through the examples following Section 15.5 is predicated on the structure's being statically indeterminate externally. For that case, the vector $\{P\}_{\text{II}}$ included the redundant structure forces, and the vector $\{P\}_{\text{I}}$ contained the structure forces corresponding to the free displacements on the original structure.

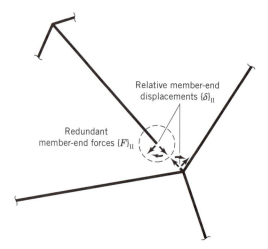

Figure 15.2 Severed redundant member.

If the structure is internally redundant, the vector $\{F\}$, which includes the member-end forces at one end of each member, must be partitioned into two parts—vector $\{F\}_{II}$ contains the redundant member-end forces, and $\{F\}_I$ includes the remaining member-end forces for the primary structure.

When the redundant members are cut at one of their ends, the member-end forces associated with these members become external to the primary structure. However, compatibility requires that the displacements of each severed member end, with respect to the node from which it is separated, must be zero. These relative displacements are collected in the vector $\{\delta\}_{II}$, as shown for a planar frame member in Fig. 15.2.

In this case, the flexibility equations for the primary structure are expressed in the form

$$\left\{ \frac{\{\Delta\}_I}{\{\delta\}_{II}} \right\} = \left[\begin{array}{c|c} [D]_{I,I} & [D]_{I,II} \\ \hline [D]_{II,I} & [D]_{II,II} \end{array} \right] \left\{ \frac{\{P\}_I}{\{F\}_{II}} \right\} \tag{15.27}$$

where $\{\Delta\}_I$ contains the free structure displacements on the primary structure and $\{P\}_I$ includes the corresponding structure forces. Expanding these equations and recalling that $\{\delta\}_{II} = \{0\}$, we obtain

$$\{F\}_{II} = - [D]_{II,II}^{-1}[D]_{II,I}\{P\}_I \tag{15.28}$$

and

$$\{\Delta\}_I = ([D]_{I,I} - [D]_{I,II}[D]_{II,II}^{-1}[D]_{II,I})\{P\}_I \tag{15.29}$$

Equilibrium conditions applied to the primary structure give

$$\{F\}_I = [b]_I\{P\}_I + [b]_{II}\{F\}_{II} \tag{15.30}$$

Following the procedures of Section 15.4, we find

$$[D]_{I,I} = [b]_I^T[d]_{uI}[b]_I$$
$$[D]_{I,II} = [b]_I^T[d]_{uI}[b]_{II}$$
$$[D]_{II,I} = [b]_{II}^T[d]_{uI}[b]_I \tag{15.31}$$
$$[D]_{II,II} = [b]_{II}^T[d]_{uI}[b]_{II} + [d]_{uII}$$

where $[d]_{uI}$ and $[d]_{uII}$ are unassembled stiffness matrices for the members included in vectors $\{F\}_I$ and $\{F\}_{II}$, respectively.

The application of the flexibility method to internally redundant forces follows the general procedure outlined in Section 15.5, except that the redundants are now the member forces $\{F\}_{II}$ instead of the structure forces $\{P\}_{II}$, and $[d]_u$ is separated into $[d]_{uI}$ and $[d]_{uII}$.

For a structure that is indeterminate both externally and internally, the procedures developed separately are combined. Here $[b]_{II}$ must be partitioned to include the effects of the redundant reactions $\{P\}_{II}$ and the redundant member forces $\{F\}_{II}$.

15.7 LOADS APPLIED BETWEEN NODE POINTS

Up until this point in our discussion of the flexibility method, structure forces have been admissible only at the node points. This restriction will now be relaxed.

Two approaches can be employed. In the first approach, the same basic procedure is used that was applied in the stiffness formulation. This method was described in Section 14.6, and it consists of the superposition of two separate solutions: a fixed-jointed analysis, which includes the effects of the member loads acting between the nodes, provides the first solution; the second solution is a standard joint-loaded analysis in which the loads include the applied structure forces and the negative of certain equivalent structure forces. In applying this approach in a flexibility analysis, the fixed-jointed analysis is completed as before, but the joint-loaded analysis is carried out by the flexibility method. For the latter part of the solution, the general approach is the same as that outlined in Section 15.5, with the exceptions that nodal numbering need not include the load points, and the final member-end forces are given by

$$\{F\} = [b]_I\{P\}_I + [b]_{II}\{P\}_{II} + \{F\}^f \tag{15.32}$$

where $\{F\}^f$ includes the fixed-end forces corresponding to the member ends included in $\{F\}$.

This approach is not a pure flexibility analysis since stiffness considerations are necessary in determining the fixed-end forces of the fixed-jointed analysis.

The second approach for treating loads between the nodes has the elegance of being a pure flexibility formulation. Our discussion will center on a planar structure in flexure, but it could be extended readily to include more complicated structures. We initially consider the fixed-ended member shown in Fig. 15.3a, in which the member is subjected to member-end forces at end i along with the transverse loading along the span. The free-end displacements result from two contributions: there is the contribution from the member-end forces, $\{\delta\}_{ij}$, which is shown in Fig. 15.3b and is given by Eq. 13.13, and there is the contribution from the transverse loading, $\{\delta_0\}_{ij}$, which is shown in Fig. 15.3c; the components of this contribution are given in the table inside the back cover for several loading conditions. Therefore, the total member-end displacements at end i, $\{\delta^T\}_{ij}$, are given by

$$\{\delta^T\}_{ij} = [d]_{ii}^j\{F\}_{ij} + \{\delta_0\}_{ij} \tag{15.33}$$

Associated with the transverse load, there is a set of reaction forces at end j, $\{F\}_{ji}^s$, that are determined from statics. These forces are shown in Fig. 15.3c, and the individual components are given in the table for an assortment of loading conditions.

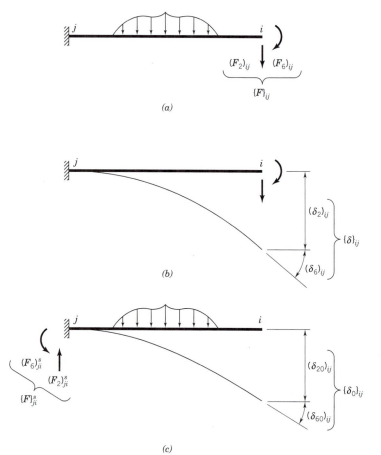

Figure 15.3 Forces and displacements on planar member. *(a)* End forces and transverse forces between nodes. *(b)* Displacements from end forces only. *(c)* Displacements and reactions for transverse load only.

We now return to the arguments that were used in Section 15.4. The strain energy for member ij is again given by Eq. 15.15; however, in this case, the relative displacements, $\{e\}_{ij}$, are given by $\{\delta^T\}_{ij}$ of Eq. 15.33. Therefore, the strain energy for member ij becomes

$$U_{ij} = \tfrac{1}{2}\{F\}_{ij}^{T}[\,d\,]_{ii}^{j}\{F\}_{ij} + \tfrac{1}{2}\{F\}_{ij}^{T}\{\delta_0\}_{ij} \tag{15.34}$$

and for the entire structure, the total strain energy is

$$U = \tfrac{1}{2}\{F\}^{T}[d_u]\{F\} + \tfrac{1}{2}\{F\}^{T}\{\delta_0\} \tag{15.35}$$

where $\{F\}$ contains the ordered array of the member-end forces at the i end of all members, $\{\delta_0\}$ includes the displacement components at the i ends that are caused by transverse loading in the order in which the respective force components are given in $\{F\}$, and $[d_u]$ is as defined in Eq. 15.20.

For an externally redundant structure, we know from Eq. 15.13 that

$$\{\,F\,\} = [b]\{P\} = [b]_{\mathrm{I}}\{P\}_{\mathrm{I}} + [b]_{\mathrm{II}}\{P\}_{\mathrm{II}} \tag{15.36}$$

where $[b]$ is the equilibrium matrix, $\{P\}_{\mathrm{I}}$ includes the applied forces on the primary structure, and $\{P\}_{\mathrm{II}}$ contains the redundant reaction forces.

Substitution of Eq. 15.36 into Eq. 15.35 gives

$$U = \tfrac{1}{2}\{P\}^T[b]^T[d_u][b]\{P\} + \tfrac{1}{2}\{P\}^T[b]^T\{\delta_0\} \tag{15.37}$$

or

$$U = \tfrac{1}{2}\{P\}^T([b]^T[d_u][b]\{P\} + [b]^T\{\delta_0\}) \tag{15.38}$$

However, as was noted in Eq. 15.22, the strain energy can be expressed in terms of structural quantities as

$$U = \tfrac{1}{2}\{P\}^T\{\Delta\} \tag{15.39}$$

Comparison of Eqs. 15.38 and 15.39 reveals that

$$\{\Delta\} = [b]^T[d_u][b]\{P\} + [b]^T\{\delta_0\} \tag{15.40}$$

Partitioning Eq. 15.40 according to the redundant and nonredundant components of $\{P\}$, we obtain

$$\begin{Bmatrix} \{\Delta\}_{\mathrm{I}} \\ \{\Delta\}_{\mathrm{II}} \end{Bmatrix} = \begin{bmatrix} [b]_{\mathrm{I}}^T \\ [b]_{\mathrm{II}}^T \end{bmatrix} [d_u][[b]_{\mathrm{I}}[b]_{\mathrm{II}}] \begin{Bmatrix} P_{\mathrm{I}} \\ P_{\mathrm{II}} \end{Bmatrix} + \begin{bmatrix} [b]_{\mathrm{I}}^T \\ [b]_{\mathrm{II}}^T \end{bmatrix} \{\delta_0\} \tag{15.41}$$

Expanding Eq. 15.41 and invoking Eqs. 15.26, we find

$$\{\Delta\}_{\mathrm{I}} = [D]_{\mathrm{I,I}}\{P\}_{\mathrm{I}} + [D]_{\mathrm{I,II}}\{P\}_{\mathrm{II}} + [b]_{\mathrm{I}}^T\{\delta_0\} \tag{15.42}$$

$$\{\Delta\}_{\mathrm{II}} = [D]_{\mathrm{II,I}}\{P\}_{\mathrm{I}} + [D]_{\mathrm{II,II}}\{P\}_{\mathrm{II}} + [b]_{\mathrm{II}}^T\{\delta_0\} \tag{15.43}$$

The individual $[D]$ matrices are defined in Eqs. 15.26.

For nonyielding supports, $\{\Delta\}_{\mathrm{II}} = \{0\}$, and we can then solve for $\{P\}_{\mathrm{II}}$ from Eq. 15.43. This leads to

$$\{P\}_{\mathrm{II}} = -[D]_{\mathrm{II,II}}^{-1}[D]_{\mathrm{II,I}}\{P\}_{\mathrm{I}} - [D]_{\mathrm{II,II}}^{-1}[b]_{\mathrm{II}}^T\{\delta_0\} \tag{15.44}$$

The member forces are then determined from Eq. 15.36, and the structure displacements are given by Eq. 15.42.

In the application of Eq. 15.44, $\{P\}_{\mathrm{I}}$ is given by

$$\{P\}_{\mathrm{I}} = \{P\}_{\mathrm{I}}^a - \{P\}_{\mathrm{I}}^e \tag{15.45}$$

where $\{P\}_{\mathrm{I}}^a$ are the applied force components associated with the kinematic degrees of freedom of the primary structure, and $\{P\}_{\mathrm{I}}^e$ are the corresponding components of the equivalent structure forces that are needed to sustain the static reactions, $\{F\}_{ji}^s$, of the transversely loaded member. In accordance with the procedure of Section 14.6, the equivalent structure forces at joint j are

$$\{P\}_j^e = \sum_i [\beta]_{ji}^T\{F\}_{ji}^s \tag{15.46}$$

where $[\beta]_{ji}$ is the compatibility matrix for member ij at joint j, and i ranges over all of the members framing into joint j. Only those components associated with the kinematic degrees of freedom of the primary structure are included in $\{P\}_{\mathrm{I}}^e$.

The table inside the back cover gives the components of $\{\delta_0\}_{ij}$ and $\{F\}_{ji}^s$ for several transverse load cases. Signs are not assigned to these cases; the signs for forces and displacements must be assigned according to the member coordinate system and the orientation of the member within the structure. This table includes only the information needed for planar flexural members. Axial and torsional effects, as well as flexure in the xz plane, can be added to $\{\delta_0\}_{ij}$ and $\{F\}_{ji}^s$ as needed.

15.7.1 Internal Redundants

For internally redundant structures, the same basic procedure is used; however, in this instance, Eq. 15.36 is replaced by Eq. 15.30, and the $[D]$ matrices are defined by Eq. 15.31.

For structures with both internal and external redundancies, the same general procedures are employed. Here, $[b]_{II}$ is partitioned to include the effects of both the redundant reactions $\{P\}_{II}$ and the redundant member forces $\{F\}_{II}$.

15.8 SELF-STRAINING PROBLEMS

The general nature of self-straining problems is discussed in detail in Section 14.7, and the reader is encouraged to review the introductory portion of that section.

Recall that member ij was severed at one end, and the self-straining action was allowed to accumulate displacements at the i end with respect to the j end. These displacements are shown as $\{\delta_s\}_{ij}$ in Fig. 15.4 and play the same role as $\{\delta_0\}_{ij}$ in Section 15.7 for the member subjected to transverse loads.

Therefore, the procedure of Section 15.7 is applicable with the $\{\delta_0\}$ being replaced by $\{\delta_s\}$. There is one exception to the procedure. Since in the self-straining problem there are no reactions developed at the j end of the severed member, there are no equivalent structure forces. Thus, the load vector $\{P\}_I$ is composed of applied load components only.

The precise makeup of the $\{\delta_s\}$ vector for each of the self-straining problems is treated in the sections that follow.

15.8.1 Fabrication Errors

For fabrication errors, or the lack-of-fit problem, the vector $\{\delta_s\}_{ij}$ includes the displacements at the i end relative to the j end that reflect the components of improper length or alignment.

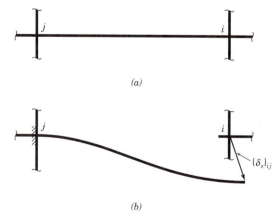

(a)

(b)

Figure 15.4 Self-straining displacements. (a) Span ij. (b) Accumulated displacements for severed member.

15.8.2 Temperature Variation

For temperature effects, the vector $\{\delta_s\}_{ij}$ includes the displacements that accrue at the i end of the severed member relative to the j end when temperature variations occur. For a flexural element (including axial effects), these were determined in Section 14.7.2. With the aid of Fig. 14.8, it is clear that

$$\{\delta_s\}_{ij} = \begin{Bmatrix} \delta_{s1} \\ \delta_{s2} \\ \delta_{s6} \end{Bmatrix} = \begin{Bmatrix} T_{\text{avg}}\alpha l \\ -\dfrac{\alpha\Delta T l^2}{2h} \\ -\dfrac{\alpha\Delta T l}{h} \end{Bmatrix} \tag{15.47}$$

This equation must be used with care. The quantities T_b and T_t, which are needed to establish ΔT, must be interpreted with respect to the assigned locations of i and j.

15.8.3 Settlement

In the settlement problem, support displacements are imposed. Accordingly, the imposed support displacements are included in $\{\Delta\}_{\text{II}}$ of Eq. 15.8. As a result, Eq. 15.10 for $\{P\}_{\text{II}}$ is augmented by the term $[D]_{\text{II,II}}^{-1}\{\Delta\}_{\text{II}}$.

15.9 CHOICE OF REDUNDANTS

Application of the flexibility method, as was also the case with the compatibility methods of Chapter 10, requires that the analyst select the reactions and member forces that are to be considered as the redundants. There are two considerations: first, the redundants must be selected so that the primary structure is stable; and second, their selection must be such that the accuracy of the solution is ensured.

The question of stability can be resolved in a straightforward fashion. One must merely apply the tests of stability to the primary structure to make certain that it is stable. Mathematically, stability can be tested by checking to see that $[C]$ of Eq. 15.12 possesses an inverse. Only if the inverse exists is it possible to determine the member forces from the structure forces in accordance with Eq. 15.13.

The problem of solution accuracy is more subtle. Regardless of the choice of redundants, the order of $[D]_{\text{II,II}}$ in Eqs. 15.10 or 15.28 is the same. Also, the final results, theoretically, should be the same regardless of the selection of redundants, as long as the primary structure is stable. However, it happens that the arrangement of the elements, as well as their relative magnitudes, will have a bearing on the accuracy of $[D]_{\text{II,II}}^{-1}$ because of limitations in the numerical procedures that are employed. To obtain an accurate inverse, it is best that $[D]_{\text{II,II}}$ be *sparsely banded* and *well-conditioned* or *diagonally strong*. That is, the matrix should possess terms that are grouped along the diagonal in as narrow a band as possible rather than being fully populated, and the terms along the diagonal should be larger numerically than those that are off the diagonal.

These factors can be illustrated by the example continuous beam given in Fig. 15.5a. If the internal reactions are taken as the redundants, then

$$[D]_{\text{II,II}} = \begin{bmatrix} D_{11} & D_{12} & D_{13} & D_{14} \\ D_{21} & D_{22} & D_{23} & D_{24} \\ D_{31} & D_{32} & D_{33} & D_{34} \\ D_{41} & D_{42} & D_{43} & D_{44} \end{bmatrix} \tag{15.48}$$

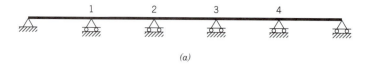

(a)

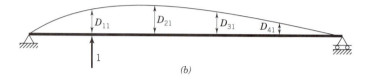

(b)

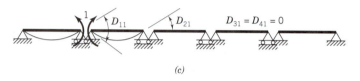

(c)

Figure 15.5 Choice of redundants. *(a)* Continuous beam. *(b)* Selected flexibility coefficients with internal reactions as redundants. *(c)* Selected flexibility coefficients with support moments as redundants.

This array is fully populated, and it is clear from Fig. 15.5*b* that $D_{21} > D_{11}$ and $D_{31} \simeq D_{11}$. Thus, the matrix is neither banded nor well conditioned. However, if the support moments are taken as the redundants, we have

$$[D]_{\text{II,II}} = \begin{bmatrix} D_{11} & D_{12} & 0 & 0 \\ D_{21} & D_{22} & D_{23} & 0 \\ 0 & D_{32} & D_{33} & D_{34} \\ 0 & 0 & D_{43} & D_{44} \end{bmatrix} \tag{15.49}$$

Here, the matrix is banded and, as shown in Fig. 15.5*c*, $D_{11} > D_{21}$, which indicates a well-conditioned matrix.

The contrast between the two cases would be more sharply defined if there were more spans. The case where the internal reactions are taken as the redundants would lead to a fully populated and poorly conditioned matrix, regardless of the order of the array. However, the case with the support moments as redundants would have a well-conditioned matrix with a bandwidth of three, regardless of the number of redundants.

Similar considerations would have to be employed for other structures to determine which forces should be taken as the redundants, and herein lies one of the major disadvantages of compatibility or flexibility methods. For complex structures, the question of stability can be complicated, and the optimum selection of the redundants for solution accuracy can be elusive.

15.10 COMPARISONS BETWEEN FLEXIBILITY AND STIFFNESS METHODS

The flexibility and stiffness methods have been presented here as formalizations of the compatibility and equilibrium methods, respectively, that formed the basis of Part Four of the book. They are equivalent in that each provides a complete analysis of a structure in which equilibrium and compatibility are fully satisfied. However, the sequence in which these requirements are met is different; the flexibility method considers compatibility explicitly, with equilibrium secondarily woven into the solution, whereas the stiffness method gives primary attention to equilibrium, with compatibility being implicitly satisfied. Of course, both methods reflect the inclusion of the appropriate material characteristics and the governing boundary conditions.

15.11 SUGGESTED PROBLEMS

15.1 through 15.7 Use the flexibility method to analyze each of the structures described below for the prescribed loading conditions. In each case, determine the member forces and the nodal displacements for the given nodal numbering system.

15.1 The structure and loading of Problem 14.3.

15.2 The structure and loading of Problem 14.4.

15.3 The structure and loading of Problem 14.13.

15.4 The structure and loading of Problem 14.14.

15.5 The structure and loading of Problem 14.15.

15.6 The structure and loading of Problem 14.16.

15.7 The structure and loading of Problem 14.17.

Appendix

ANSWERS TO SELECTED PROBLEMS

Students should not cultivate the habit of working for predetermined answers, because such an arrangement is highly artificial. Instead, they should seek independent checks to verify their results and should try to develop an ability to sense whether answers appear to be reasonable. However, in the learning process, the immediate reinforcement of a correct answer has some value. For this reason, selected answers to *most* even-numbered problems are given in the following tabulation. Answers are also included for unique problem types, which appear only as odd-numbered problems.

CHAPTER 1

2. **(a)** Beam 1   0.44 k/ft
    Beam 3   0.75 k/ft
    Girder A   0.25 k/ft and two 6.2 k concentrated loads
    Girder B   0.25 k/ft and two 12.4 k concentrated loads
    Column A1   13.6 k
    Column B1   23.5 k
    Column B4   44.7 k

    (b) Beam 1   0.25 k/ft
    Beam 3   0.50 k/ft
    Girder A   4.1 k at 1/3 points
    Girder B   8.2 k at 1/3 points
    Column A1   6.2 k
    Column B1   12.4 k
    Column B4   24.8 k

4. Triangular load with intensity varying from 0 at point b to 22.0 kN/m at point a.

CHAPTER 3

1. **(b)** Determinate, stable
    (d) Indeterminate, stable
    (f) Determinate, stable
    (h) Determinate, stable
    (j) Determinate, stable
    (l) Unstable
    (n) Unstable
    (p) Determinate, stable
    (r) Indeterminate
    (t) Indeterminate
    (v) Determinate, stable
    (x) Unstable

2. $R_{ax} = 0$; $R_{ay} = 8$ k $\rightarrow$; $R_{by} = 8$ k $\uparrow$

4. $R_{bx} = 0$; $R_{ay} = 150$ kN $\uparrow$; $R_{by} = 150$ kN $\uparrow$

6. $R_{ax} = 0$; $R_{ay} = 163.3$ kN $\uparrow$; $R_{by} = 186.7$ kN $\uparrow$

8. $R_{ax} = 0$; $R_{ay} = 67.9$ kN $\uparrow$; $R_{by} = 167.1$ kN $\uparrow$

10. $R_{bx} = 5$ k $\leftarrow$; $R_{ay} = 23.3$ k $\uparrow$; $R_{by} = 16.7$ k $\uparrow$

12. $R_{ax} = 40$ kN $\rightarrow$; $R_{ay} = 46.7$ kN $\uparrow$; $R_{bx} = 40$ kN $\leftarrow$; $R_{by} = 53.3$ kN $\uparrow$
14. $R_{bx} = 16$ k $\leftarrow$; $R_{ay} = 52.8$ k $\uparrow$; $R_{by} = 5.2$ k $\uparrow$
16. $R_{ax} = 0$; $R_{ay} = 80$ kN $\uparrow$; $M_A = 480$ kN $\cdot$ m; $R_{by} = 0$
18. $R_{ax} = 0$; $R_{ay} = 33$ kN $\uparrow$; $R_{by} = 80$ kN $\uparrow$; $R_{cy} = 115$ kN $\uparrow$
20. $R_{ax} = 32.5$ kN $\leftarrow$; $R_{ay} = 223.33$ kN $\uparrow$; $M_a = 800$ kN $\cdot$ m; $R_{bx} = 32.5$ kN $\rightarrow$; $R_{by} = 26.67$ kN $\uparrow$
22. $R_{ax} = 0$; $R_{ay} = 160$ k $\uparrow$; $M_a = 1,900$ k-ft; $R_{cx} = 45$ k $\leftarrow$; $R_{cy} = 30$ k $\uparrow$
24. $R_{ax} = 113.33$ kN $\rightarrow$; $R_{ay} = 82.5$ kN $\uparrow$; $R_{bx} = 93.3$ kN $\leftarrow$; $R_{by} = 97.5$ kN $\uparrow$
26. $R_{ay} = 0$; $R_{bx} = 8$ k $\leftarrow$; $R_{by} = 25$ k $\uparrow$
28. $R_{ax} = 30$ k $\leftarrow$; $R_{ay} = 10$ k $\downarrow$; $R_{cy} = 70$ k $\uparrow$; $M_c = 700$ k-ft
30. $R_{ax} = 240$ kN $\leftarrow$; $R_{ay} = 180$ kN $\downarrow$; $R_{by} = 180$ kN $\uparrow$
32. $R_{ax} = 0$; $R_{ay} = 30$ k $\uparrow$; $R_{az} = 0$; $M_{ax} = 62.5$ k-ft; $M_{ay} = 0$; $M_{az} = 150$ k-ft
34. $R_{ax} = 0$; $R_{ay} = 16$ kN $\uparrow$; $R_{az} = 4$ kN; $M_{ax} = 6$ kN $\cdot$ m; $M_{ay} = 24$ kN $\cdot$ m; $M_{az} = 48$ kN $\cdot$ m
36. $R_{az} = 5$ k; $R_{fx} = 8$ k $\leftarrow$; $M_{ay} = 80$ k-ft; $R_{ay} = 55$ k $\uparrow$; $R_{fy} = 30$ k $\downarrow$; $M_{az} = 720$ k-ft
38. $R_{ax} = 1.5$ kN $\leftarrow$; $R_{ay} = 6.75$ kN $\uparrow$; $R_{bx} = 6.5$ kN $\leftarrow$; $R_{cy} = 9.5$ kN $\uparrow$; $R_{dy} = 0.75$ kN $\uparrow$; $R_{dz} = 5$ kN
40. $M_{ax} = 50$ k-ft; $M_{ay} = 350$ k-ft; $R_{ay} = 17.92$ k $\uparrow$; $R_{fx} = 15$ k $\leftarrow$; $R_{fy} = 17.08$ k $\uparrow$; $R_{fz} = 10$ k
42. $R_{ax} = -116.66$ k; $R_{ay} = -100$ k; $R_{bx} = -33.34$ k; $R_{by} = 0$; $R_{cy} = 100$ k; $R_{cz} = 0$ (signs based on assumed directions in problem statement)

CHAPTER 4

1. **(b)** External: determinate, stable; Internal: indeterminate, stable
   (d) External: unstable; Internal: determinate, stable
   (f) External: determinate, stable; Internal: determinate, stable; Compound
   (h) External: determinate, stable; Internal: unstable
   (j) External: determinate, stable; Internal: determinate, stable; Compound
   (l) External: determinate, stable; Internal: determinate, stable; Compound
   (n) External: determinate, stable; Internal: geometrically unstable
   (p) External: determinate, stable; Internal: determinate, stable; Complex
   (r) External: determinate, stable; Internal: determinate, stable; Complex
   (t) External: determinate, stable; Internal: determinate, stable; Simple
2. $F_{ab} = -153.6$ kN; $F_{bc} = -96$ kN; $F_{af} = 96$ kN; $F_{fg} = 96$ kN; $F_{bf} = 120$ kN; $F_{bg} = 0$
4. $F_{ab} = 0.96$ k; $F_{bc} = -1.2$k; $F_{cd} = -0.96$ k; $F_{ac} = 1.8$ k; $F_{ed} = 0.6$ k; $F_{be} = -0.96$ k; $F_{ec} = 0.96$ k
6. $F_{ab} = -18$ k; $F_{bc} = -18$ k; $F_{cf} = -62.2$ k; $F_{ad} = 36.8$ k; $F_{bd} = 0$; $F_{dc} = 5.6$ k; $F_{de} = 22.0$ k; $F_{ef} = 22.0$ k; $F_{ec} = 40.0$ k
8. $F_{ae} = +131.99$ kN; $F_{ef} = +120.94$ kN; $F_{bc} = -99.99$ kN; $F_{cg} = -65.05$ kN

10. $F_{ab} = -83.09$ k; $F_{cd} = 77.50$ k; $F_{io} = 0$; $F_{pe} = +44.19$ k; $F_{em} = +15.00$ k; $F_{ng} = +61.25$k

12. $F_{ab} = -6.54$ k; $F_{de} = -4.5$ k; $F_{gh} = -6.08$ k; $F_{lm} = 3.94$ k; $F_{mp} = 1.26$ k; $F_{mn} = 2.81$ k; $F_{ng} = -2.52$ k

14. $F_{ab} = 36.26$ kN; $F_{bc} = 5.53$ kN; $F_{ce} = -107.03$ kN; $F_{ad} = 131.82$ kN; $F_{de} = 33.85$ kN; $F_{bd} = -97.90$ kN; $F_{cd} = 72.23$ kN

16. $F_{ab} = 111.8$k; $F_{ae} = -66.67$k; $F_{ac} = -48.48$k; $F_{ad} = 107.32$k; $F_{bc} = -141.42$k; $F_{ec} = -105.41$ k; $F_{de} = -20.12$ k

18. $F_{ab} = -75.0$ kN; $F_{cd} = -166.65$ kN; $F_{ag} = +106.07$ kN; $F_{fg} = -313.79$ kN; $F_{eh} = +194.34$ kN; $F_{hk} = -258.76$ kN

20. $F_{ea} = 301.87$ kN; $F_{bc} = 240.0$ kN; $F_{fg} = -225.0$ kN; $F_{bh} = -67.08$ kN

22. $F_{ea} = 9.79$ k; $F_{ac} = 32.99$ k; $F_{df} = -8.25$ k; $F_{bc} = -24.0$ k

24. $F_{ea} = -18.75$ k; $F_{ag} = -43.75$ k; $F_{bc} = 15.0$ k; $F_{dh} = -25.0$ k

26. $F_{ec} = -13.63$ kN; $F_{ca} = -97.50$ kN; $F_{ab} = 78.00$ kN; $F_{bd} = -97.50$ kN; $F_{cf} = -92.31$ kN; $F_{cg} = 58.14$ kN; $F_{gd} = -34.13$ kN

28. $F_{ad} = F_{bd} = +70.71$ kN; $F_{cd} = -66.67$ kN; $F_{ed} = +86.7$ kN; $F_{ec} = -155.5$ kN

30. $F_{ab} = -47.58$ kN; $F_{bc} = +55.08$ kN; $F_{ca} = -47.58$ kN; $F_{ad} = +127.67$ kN; $F_{dc} = -72.12$ kN; $F_{db} = -72.12$kN

32. $F_{ba} = -9.60$ kN; $F_{ca} = 16.80$ kN; $F_{ae} = 12.90$ kN; $F_{ce} = -46.87$ kN; $F_{cd} = 33.60$ kN; $F_{de} = -59.39$ kN; $F_{db} = -4.80$ kN; $F_{be} = 16.10$ kN

CHAPTER 5

1. **(a)** External: determinate, stable; Overall: determinate, stable

   (b) External: indeterminate (1 degree), stable; Overall: indeterminate (1 degree), stable

   (h) External: indeterminate (1 degree), stable; Overall: indeterminate (1 degree), stable

   (j) External: determinate, stable; Overall: determinate, stable

2. *AB*: $V(x) = -0.10 x^2 + 26.67$ (k); $M(x) = -0.033x^3 + 26.67x$ (ft-k)

   BC: $V(x) = 13.33$ (k); $M(x) = -13.33x + 533.2$ (ft-k)

4. *AB*: $V(x) = 11.0 - 2.0 x$ (kN); $M(x) = -x^2 + 11.0x$ (kN · m)

   BC: $V(x) = 3.0$ (kN); $M(x) = 3.0x - 16.0$ (kN · m)

   CD: $V(x) = 17.0$ (kN); $M(x) = -17.0x + 136.0$ (kN · m)

6. *AB*: $V(x) = -0.125x^2 + 25.84$ (k); $M(x) = -0.0417x^3 + 25.84x$ (ft-k)

   BC: $V(x) = -24.16$ (k); $M(x) = -24.16x + 666.4$ (ft-k)

   CD: $V(x) = -24.16$ (k); $M(x) = -24.16x + 966.4$ (ft-k)

8. $V(x) = \dfrac{p_o l}{\pi} \cos \dfrac{\pi x}{l}$; $M(x) = \dfrac{p_o l^2}{\pi^2} \sin \dfrac{\pi x}{l}$

10. *AB*: $V(x) = -0.125x^2 + 33.33$ (k); $M(x) = -0.0417x^3 + 33.33x$ (ft-k)

   BC: $V(x) = -15x + 0.25x^2 + 183.33$ (k);
   $M(x) = -7.5x^2 + 0.0833x^3 + 183.33x - 1000$ (ft-k)

12. *V* (kN): 6.0 at left support, 0.0 between loads; − 6.0 from right load to right support

   M (kN·m): 18.0 at each load and between loads

14. *V* (kN): 6.0 between left support and load; − 6.0 between load and right support

   M (kN·m): 30.0 at load

16. *V* (k): − 1.5 at left support to right support; 6.0 to right of right support to load

   M (k-ft): − 30 at right support

18. *V* (k): 27.5 at left support; 7.5 between uniform loads; − 32.5 at right support

   M (k-ft): 175.0 at right of left uniform load; 250.0 at left of right uniform load

20. *V* (kN): 140.5 at left support; 60.5 left of 120 kN load; − 59.5 right of 120 kN load; − 134.5 at right support

   M (kN·m): 421.5 at 100 kN load; 663.5 at 120 kN load; 485 at left end of distributed load

22. *V* (k): 22.42 at left support; − 27.58 left of right support; 80.50 right of right support

   M (ft-k): 200.0 maximum between supports; − 161.0 at right support

24. *V* (kN): − 80.0 left of left support; 106.4 right of left support; − 113.6 left of right support; 60.0 right of right support

   M (kN·m): − 80 at left support; 202.9 maximum between supports; − 120.0 at right support

26. *V* (kN) − 90 right of left support; 165 right of interior support; − 15 left of right support

   M (kN·m): − 1 400 at interior support; 0 at hinge; 75 maximum between hinge and right support

28. *V* (k): 10.0 at left support; − 70.0 right of 80.0 k load; 60.0 right of second support; − 24.0 just to left of third support; 30.0 just to right of third support; − 18.0 at last support

   M (k-ft) 60.0 at 80 k load; − 360.0 at second support; − 72.0 at third support

30. On Girder

   V (kN): 12.5 at left support; − 237.5 left of right support; 100.0 right of right support

   M (kN·m): 187.5 at intermediate floor beam; − 1 000.0 at support

32. *V* (k): 75.32 at left support; 79.13 at right support
   M (k-ft): 886.4 at top; 839.3 at load on right member

34. *V* (k): left leg 6.0; top member 4.8; right leg 0.0
   M (k-ft): 96.0 at load

36. V (k): 0.25 constant on left leg; 10.75 at left end of top member; -34.25 at right end of top member; 19.75 at top of right leg; 12.25 at bottom of right leg

   M (ft-k): 0 at left pin; 3.75 at upper left corner; 23.00 maximum on top member; 0 at hinge; 172.5 at upper right corner; 0 at right pin

38. V (k): 0 at bottom of left leg; -20 constant from midpoint of left leg to upper end; 30 at right end of top member; -10 at left end of top member; 10 constant on right leg

   M (ft-k): 0 on left leg to midpoint; -200 at upper left corner; 25 maximum on top member; 0 at hinge; 200 at fixed-end support

40. Moments (ft-k) for each span: Left span: -51.4, 55.7, -77.2; Center span: -77.2, 58.69, -46.4; Right span: -46.4, 0

42. ab: $M_{ax} = M_{bx} = -62.5$ k-ft; $M_{az} = -150.0$ k-ft; $M_{bz} = 0.0$ k-ft; $V_y = 30.0$ k

   bc: $M_{bx} = M_{cx} = -62.5$ k-ft; $M_{bz} = M_{cz} = 0.0$ k-ft; $V = 0.0$ k

   cd: $M_{cz} = -62.5$ k-ft; $M_{dx} = 0.0$; $V_{cy} = 25.0$ k

44. ab: $M_{ay} = 0$; $M_{by} = -40$ k-ft

   bc: $M_{bz} = -115$ k-ft; $M_{cz} = 165$ k-ft; $V_z = -40$ k

   cd: $F_x = -40$ k; $M_{cx} = M_{dx} = -165$ k-ft

   de: $M_{dy} = 240$ k-ft; $M_{ey} = 0$ k-ft; $V_y = 27.5$ k

   ef: $M_{ez} = 240$ k-ft; $M_{fz} = 0$ k-ft; $V_z = 0$ k

46. ab: $F_x = 10.0$ k; $M_x = -50.88$ k-ft; $V_y = 225.0$ k;
   $M_{az} = -147.5$ k-ft; $M_{bz} = -35.0$ k-ft
   $M_{ay} = 0$ k-ft; $M_{by} = -100.0$ k-ft

   cd: $F_x = 20.0$ k; $M_x = 5.0$ k-ft; $V_{cy} = 17.5$ k; $V_{dy} = 10.0$ k;
   $M_{cz} = 9.17$ k-ft; $M_{dz} = 55.0$ k-ft; $V_z = 10.0$ k;
   $M_{cy} = -100.0$ k-ft; $M_{dy} = -50.0$ k-ft

   ef: $F_x = 10.0$ k; $M_x = 15.0$ k-ft; $V_y = 5.0$ k;
   $M_{ez} = -25.0$ k-ft; $M_{fz} = 0.0$

CHAPTER 6

Note: Influence lines for statically determinate structures are composed of straight-line segments.

2. I.L. R_B (kN/kN): $+1.0$ across entire structure
   I.L. M_B (kN·m/kN): -5.0 at A to 0.0 at B
   I.L. V_C (kN/kN): -1.0 from A to C
   I.L. M_C (kN·m/kN): -2.0 at A to 0.0 at C

4. I.L. R_{Ay} (kN/kN): $+1.0$ from A to C; then to 0.0 at B
   I.L. $R_{Ax} =$ I.L. R_{Bx} (kN/kN): zero from A to C; then to 0.5 at B
   I.L. M_A (kN·m/kN): 0.0 at A to -3.0 at C; then to 0.0 at B

6. I.L. R_{Ay} (kN/kN): $+1.0$ at A to -1.0 at C
I.L. R_{Dy} (kN/kN): 0.0 at A to 1.0 at C
I.L. M_F (kN $\cdot$ m/kN): 0.0 at A to $+3.0$ at C

8. I.L. R_{Cy} (k/k): 0.0 at A to $+0.643$ at B; then to 0.0 at D
I.L. R_{Cx} (k/k): 0.0 at A to 0.179 at B; then to $+0.500$ at D
I.L. V_E (k/k): 0.0 at A to $+0.357$ at B; then to -0.714 left of E; $+0.286$ right of E to 0.0 at D

10. I.L. R_{Ay} (k/k): $+1.0$ at A to -0.250 at right end
I.L. V_C (k/k): 0.0 at A to $+0.625$ left of stringer discontinuity; $+0.375$ right of stringer discontinuity to -0.250 at right end

12. I.L. R_{Ay} (kN/kN): $+1.0$ at B to 0.0 at E
I.L. R_{Ax} (kN/kN): 0.0 at B to -0.556 at C; then to 0.0 at E
I.L. M (top of BA) (kN $\cdot$ m/kN): 0.0 at B to -3.33 at C; then to 0.0 at E

14. I.L. R_{ay} (k/k): $+1.0$ at a to 0.0 at b
I.L. R_{dy} (k/k): -0.333 at a to $+1.50$ at b; then to 0.0 at c
I.L. M_f (ft-k/k): $+1.0$ at a to -1.50 at b; then to $+1.50$ at f; then to 0.0 at c

16. I.L. R_{Dy} (k/k): zero from A to C; then to 1.5 at E
I.L. F_{AB} (k/k): zero from A to B; then to 2.0 at C; then to -1.75 at E
I.L. F_{FG} (k/k): 0.0 at A to -2.24 at C; then to $+1.12$ at E

18. I.L. R_{L0} (k/k): $+1.0$ at L_0 to -0.667 at L_5
I.L. F_{U1L1} (k/k): 0.0 at L_0 to 0.667 at L_1; then to -0.667 at L_5
I.L. F_{U2L3} (k/k): 0.0 at L_0 to -0.943 at L_2; then to 0.0 at L_3; then to -0.943 at L_5

20. I.L. R_{fy} (kN/kN): $+1.0$ at a to $+0.625$ at c; then to 0.0 at e
I.L. R_{dy} (kN/kN): 0.0 at a to $+0.375$ at c; then to $+1.0$ at e
I.L. R_{fx} = I.L. R_{kx} (kN/kN): 0.0 at a to $+0.625$ at c; then to 0.0 at e

22. $R_B = 483.75$ k; negative $V_E = -116.25$ k; negative $M_E = -1,425.00$ k-ft

24. $R_{Ax} = 176.46$ kN; positive $M_E = 226.74$ kN $\cdot$ m; positive $V_E = 144.75$ kN

26. Envelope Ranges: Shears (k), -10.0 to -27.5 left of point 3, $+2.6$ to $+30.8$ right of point 3, -6.7 to -24.9 left of point 7, $+12.0$ to $+33.0$ right of point 7; Moments (ft-k), -25.0 to -68.8 at point 4, -36.0 to -99.0 at point 7

28. Envelope Ranges: Shears (k), -7.5 to -20 left of point 3, $+18.2$ to $+48.6$ right of point 3, -6.7 to -24.5 at point 9; Moments (ft-k), -18.8 to -50.0 left of point 3, -40.2 to -107.2 right of point 3, 0 at point 9

CHAPTER 7

2. $u_c = 0.96$ mm $\rightarrow$; $v_c = 4.44$ mm $\downarrow$

4. $u_E = 0.016,1$ in. $\rightarrow$; $v_E = 0.092,2$ in. $\downarrow$

6. $u_D = 1.17$ mm $\rightarrow$; $v_D = 8.37$ mm $\downarrow$

8. $u_E = 0.053,0$ in. $\leftarrow$; $v_E = 0.309$ in. $\downarrow$

10. $u_C = 0.089,8$ in. $\rightarrow$; $v_C = 0.096,8$ in. $\downarrow$

12. $u_C = 15$ mm $\rightarrow$; $v_C = 10$ mm $\downarrow$

14. $u_B = 0.458$ mm $\downarrow$; $v_B = 0.670$ mm $\rightarrow$

16. $u_d = 0.376$ in. $\leftarrow$; $v_d = 0.297$ in. $\downarrow$

18. $v_E = 0.106,6$ in. $\downarrow$

20. $\Delta_f = 0.262$ in. $\swarrow$ (along sloping surface)

22. $\Delta T_{AC} = 53.4$ °F

24. $u_D = 30$ mm $\leftarrow$

26. Shorten BC by 0.667 in.

28. $\{\Delta\}^T = \{-0.096,8 \quad 0.089,8 \quad -0.010,4\}$ in.

30. $\{\Delta\}^T = \{-7.5 \quad 15.0 \quad -10.0 \quad 15.0 \quad 17.5\}$ mm

32. $v_C = 20.84$ mm $\downarrow$

CHAPTER 8

2. $$\theta(x) = \frac{1}{EI}\left[\frac{p_o x^5}{60 l^2} - \frac{p_o l x^2}{6} - \frac{p_o l^2 x}{4}\right]$$

$$\Delta(x) = \frac{1}{EI}\left[\frac{p_o x^6}{360 l^2} - \frac{p_o l x^3}{18} - \frac{p_o l^2 x^2}{8}\right]$$

4. Answers for segment AB: $\theta = \dfrac{1}{EI}\left(12.92 x^2 - \dfrac{5}{480}x^4 - 2927\right)$

$$y = \frac{1}{EI}\left(4.31 x^3 - \frac{5}{2400}x^5 - 2927 x\right)$$

6. $\Delta_B = 26.1$ mm $\downarrow$; $\theta_B = 0.006\,30$ rad $\searateched$

8. $\Delta_C = 0.701$ in. $\downarrow$; $\Delta_E = 0.855$ in. $\uparrow$; $\theta_B = 0.005,84$ rad

10. $\Delta_{max} = \Delta_e = 164,014\ P/EI$ (k-in.^{3}) $\downarrow$ at 9.48 ft left of point d; $P_{max} = 212.18$ kips

12. $\Delta_B = 24.16$ mm $\downarrow$; $\theta_B = 0$ rad

14. $\Delta_D = 26.0$ mm $\downarrow$

16. $\Delta = 0.036,6$ in. $\downarrow$; $\theta = 0.000,234$ rad

18. Span AB, midspan $\Delta = 60.76$ mm $\downarrow$; Span CD, midspan $\Delta = 34.72$ mm $\downarrow$

20. $\Delta_f = 0.218$ in. $\downarrow$; $\theta_f = 0.005,57$ rad

22. $\theta_A = 0.012\,36$ rad ; $(\Delta_D)_H = 108.1$ mm $\rightarrow$

24. $(\Delta_b)_H = 0.552$ in. $\rightarrow$; $\theta_b = 0.001,72$ rad ; $(\Delta_c)_V = 0.034,5$ in. $\downarrow$

26. $(\Delta_B)_H = (\Delta_C)_H = 0.049,7$ in. $\rightarrow$; $\theta_D = 0.000,276$ rad

28. Same answers as for Problem 6

30. Δ_{max} = 3.727 mm $\downarrow$ at 4.472 m left of B

32. Same answer as for Problem 14.

34. Same answers as for Problem 16.

36. Check points on diagrams: θ_E = 0.002,15 rad $\diagdown\!\!\!\!\!\times$; Δ_E = 0.231 in. $\downarrow$

38. Check point on deflection diagram: Δ_{max} same as for Problem 30.

40. Check points on diagrams: Δ_C = 0.795 in. $\uparrow$; θ_E = 0.002,48 rad $\diagup\!\!\!\!\!\diagdown$

42. Δ_C = 3.0 mm $\downarrow$

44. Δ_B = 76.8 mm $\rightarrow$

46. Δ_D = 0.174 in. $\downarrow$

48. Δ_C = 0.004,15P in. /k $\downarrow$

50. Δ_B = 1.99 in. $\downarrow$; Angle change at B = 0.017,4 rad $\diagup\!\!\!\!\!\diagdown$

52. Δ_{CL} = 0.033,5 in. $\downarrow$; θ_{CL} = 0.000,234 rad $\diagdown\!\!\!\!\!\times$

54. Δ_e = 0.13 in. $\leftarrow$

56. Δ_c = 81.8 mm $\rightarrow$

58. Δ_{bH} = 0.55 in. $\rightarrow$; θ_b = 0.001,72 rad $\diagdown\!\!\!\!\!\times$

60. Δ_{BH} = Δ_{CH} = 0.049,7 in. $\rightarrow$

62. Δ_C = 1.24 in. $\downarrow$

64. Δ_{CL} = 0.153 in. $\downarrow$

66. Δ = 6,718.9/AE k-ft $\rightarrow$

68. Δ_{ex} = 9.97 in. $\rightarrow$; Δ_{ey} = 18.38 in. $\downarrow$; Δ_{ez} = 0.006,90 in. $\swarrow$

70. Δ_{BH} = 81.80 mm $\rightarrow$; Δ_{EV} = 12.08 mm $\downarrow$

72. Δ_{BH} = 81.8 mm $\rightarrow$; Δ_{EV} = 12.08 mm $\downarrow$

CHAPTER 10

2. V_a = 36.14 k; V_c = $-$ 31.36 k; M_a = $-$ 296.73 k-ft; M_b = 132.89 k-ft

4. V_a = 145.6 kN; V_c = $-$ 80.9 kN; M_b = $-$ 293.6 kN·m; M_c = $-$ 103.1 kN·m

6. $V_{b,beam}$ = 2.0 k; $V_{c,beam}$ = $-$ 8.0 k; M_{bc} = 18.75 k-ft; $M_{under\ 10k\ load}$ = 33.75 k-ft; M_{cb} = $-$ 26.25 k-ft

8. F_{AC} = $-$ 21.56 kN; F_{CB} = $-$ 92.27 kN; F_{BD} = $-$ 77.75 kN; F_{DA} = $-$ 77.75 kN; F_{DC} = 80.45 kN

10. F_{AB} = 71.39 k; F_{BC} = 107.70 k; F_{BD} = 40.47 k; F_{BE} = 11.39k; F_{CE} = $-$ 172.05 k; F_{DE} = $-$ 100.00 k

12. F_{AC} = 17.92 k; F_{AD} = F_{DF} = $-$ 16.00 k; F_{CD} = $-$ 20.00 k; F_{CE} = F_{EG} = $-$ 4.44 k; F_{CF} = 22.36 k; F_{EF} = F_{GH} = 0 k; F_{FG} = $-$ 72.11 k; F_{FH} = 44.00 k; F_{AB} = 8.00 k

14. V_a = 50.4 k; V_c = $-$ 80.4 k; $M_{under\ 150\ k\ load}$ = 201.4 k-ft; M_b = $-$ 396.5 k-ft

16. V_a = 49.8 k; V_c = $-$ 79.8 k; $M_{under\ 150\ k\ load}$ = 199.2 k-ft; M_b = $-$ 402.0 k-ft

18. $V_{ab} = 0.000\ 098\ 2\ EI;\ V_{bc} = -0.000\ 267\ EI;\ M_b = 0.000\ 982\ EI;$
    $M_c = -0.001\ 691\ EI$

20. $F_{AB} = 53.12\ \text{k};\ F_{CD} = -46.57\ \text{k};\ F_{DE} = 46.59\ \text{k};\ F_{AC} = 0\ \text{k};\ F_{BD} = -69.85\ \text{k};$
    $F_{AD} = 116.43\ \text{k};\ F_{BE} = 66.43\ \text{k}$

22. $F_{AC} = 160.23\ \text{k};\ F_{AD} = F_{DF} = -143.06\ \text{k};\ F_{CD} = -20.00\ \text{k};$
    $F_{CE} = F_{EG} = 137.87\ \text{k};\ F_{CF} = 22.36\ \text{k};\ F_{EF} = F_{GH} = 0\ \text{k};\ F_{FG} = -72.11\ \text{k};$
    $F_{FH} = -83.06\ \text{k};\ F_{AB} = 71.53\ \text{k}$

24. $F_{AC} = -22.17\ \text{kN};\ F_{CB} = -92.88\ \text{kN};\ F_{BD} = -76.78\ \text{kN};$
    $F_{DA} = -76.78\ \text{kN};\ F_{DC} = 81.31\ \text{kN}$

26. $F_{AB} = -141.35\ \text{k};\ F_{CD} = 50.66\ \text{k};\ F_{DE} = -50.69\ \text{k};\ F_{AC} = 0;\ F_{BD} = 76.01\ \text{k};$
    $F_{AD} = -126.69\ \text{k};\ F_{BE} = 176.69\ \text{k}$

28. $M_a = -186.76\ \text{k-ft};\ M_b = -1.47\ \text{k-ft}$

30. $M_b = -402\ \text{kN} \cdot \text{m}$

32. $M_b = 0.000\ 982\ EI\ \text{kN} \cdot \text{m};\ M_c = -0.001\ 691\ EI\ \text{kN} \cdot \text{m}$

34. $M_a = 729.30\ \text{k-ft};\ M_b = 197.67\ \text{k-ft}$

36. $M_a = -11.76\ \text{k-ft};\ M_b = -23.53\ \text{k-ft}$

38. $F_{AB} = 196.62\ \text{k};\ F_{DC} = -104.61\ \text{k};\ F_{AD} = 76.64\ \text{k};\ F_{BC} = 33.30\ \text{k};$
    $F_{AC} = 97.50\ \text{k};\ F_{BD} = -108.43\text{k}$

40. $F_{AB} = 16.87\ \text{kN};\ F_{EF} = -3.48\ \text{kN};\ F_{FD} = -3.55\ \text{kN};\ F_{AD} = 18.28\ \text{kN};$
    $F_{CB} = -16.31\ \text{kN}$

42. $F_{AB} = 311.18\ \text{k};\ F_{DC} = -39.15\ \text{k};\ F_{AD} = -34.17\ \text{k};\ F_{BC} = -200.81\ \text{k};$
    $F_{AC} = 142.10\ \text{k};\ F_{BD} = 115.13\text{k}$

44. $\{\Delta\}_2^T = \dfrac{1}{EI}\ \{11{,}165 \quad 38{,}663\}(l\ \text{in ft})$

CHAPTER 11

2. $M_{ab} = 0\ \text{kN} \cdot \text{m};\ M_{ba} = 293.6\ \text{kN} \cdot \text{m};\ M_{bc} = -293.6\ \text{kN} \cdot \text{m};$
   $M_{cb} = 103.2\ \text{kN} \cdot \text{m}$

4. $M_{ab} = -186.8\ \text{k-ft};\ M_{ba} = 1.4\ \text{k-ft};\ M_{bc} = -1.5\ \text{k-ft};\ M_{cb} = 100.0\ \text{k-ft};$
   $M_{cd} = -100.0\ \text{k-ft}$

6. $M_{ab} = -8.66\ \text{kN} \cdot \text{m};\ M_{ba} = 108.27\ \text{kN} \cdot \text{m};\ M_{bc} = -108.27\ \text{kN} \cdot \text{m};$
   $M_{cb} = 240\ \text{kN} \cdot \text{m}$

8. $M_{ab} = 0\ \text{k-ft};\ M_{ba} = 144.6\ \text{k-ft};\ M_{bc} = 144.6\ \text{k-ft};\ M_{cb} = 0\ \text{k-ft}$

10. $M_{ab} = 0\ \text{kN} \cdot \text{m};\ M_{ba} = 201.73\ \text{kN} \cdot \text{m};\ M_{bc} = -201.73\ \text{kN} \cdot \text{m};$
    $M_{cb} = 259.15\ \text{kN} \cdot \text{m}$

12. $M_{ab} = -176.0\ \text{k-ft};\ M_{ba} = +109.4\ \text{k-ft};\ M_{bc} = -104.4\ \text{k-ft};\ M_{cb} = -52.2\ \text{k-ft};$
    $M_{bd} = -4.3\ \text{k-ft};\ M_{db} = 0\ \text{k-ft}$

14. $M_{ab} = -133.7\ \text{k-ft};\ M_{ba} = 107.7\ \text{k-ft};\ M_{bc} = -93.8\ \text{k-ft};\ M_{bc} = -13.9\ \text{k-ft};$
    $M_{cb} = 72.0\ \text{k-ft};\ M_{cb} = -6.9\ \text{k-ft}$

16. $M_{ab} = -273.3\ \text{k-ft};\ M_{ba} = -171.6 - \text{k-ft};\ M_{bc} = 171.6\ \text{k-ft};\ M_{cb} = 0\ \text{k-ft}$

18. $M_{ab} = 0$ kN·m; $M_{ba} = 119.22$ kN·m; $M_{bc} = -119.24$ kN·m;
$M_{cb} = -284.62$ kN·m

20. $M_{ab} = -297.1$ k-ft; $M_{ba} = -182.84$ k-ft; $M_{bc} = 182.84$ k-ft; $M_{cb} = 91.42$ k-ft

22. $M_{ab} = 0$ kN·m; $M_{ba} = 95.9$ kN·m; $M_{bc} = -95.9$ kN·m; $M_{cb} = 77.5$ kN·m;
$M_{cd} = -77.5$ kN·m; $M_{dc} = -18.4$ kN·m

24. $M_{ab} = -438.1$ kN·m; $M_{ba} = -366.3$ kN·m; $M_{bc} = 366.3$ kN·m;
$M_{cb} = 301.9$ kN·m; $M_{ce} = -106.2$ kN·m; $M_{ec} = 0$ kN·m;
$M_{cd} = -195.6$ kN·m; $M_{dc} = 0$ kN·m

26. $M_{ab} = 0.4$ k-ft; $M_{ba} = 5.8$ k-ft; $M_{bc} = -5.8$ k-ft; $M_{cb} = 100.6$ k-ft;
$M_{cd} = -100.6$ k-ft; $M_{dc} = 81.5$ k-ft

28. $M_{ab} = -74.0$ kN·m; $M_{ba} = -55.0$ kN·m; $M_{bc} = 55.0$ kN·m;
$M_{cb} = 204.9$ kN·m; $M_{cd} = -204.9$ kN·m; $M_{dc} = -149.0$ kN·m

30. $M_{ab} = 0$ k-ft; $M_{ba} = -100.4$ k-ft; $M_{bc} = 100.4$ k-ft; $M_{cb} = 244.1$ k-ft;
$M_{cd} = -244.1$ k-ft; $M_{dc} = -106.1$ k-ft

32. $M_{ab} = 0$ kN·m; $M_{ba} = -346.2$ kN·m; $M_{bc} = +346.2$ kN·m;
$M_{cb} = -1\,176.9$ kN·m

34. $M_{ab} = -60.8$ k-ft; $M_{ba} = 36.5$ k-ft; $M_{bc} = -16.2$ k-ft; $M_{cb} = -8.1$ k-ft;
$M_{bd} = -20.3$ k-ft; $M_{db} = 0$ k-ft

36. $[0.9EI]\{\Delta_1\} = \{P_1\}$

38.
$$\begin{Bmatrix} P_1 \\ P_2 \\ P_3 \end{Bmatrix} = \begin{Bmatrix} -100 \\ 0 \\ 0 \end{Bmatrix} = \frac{EI}{1000} \begin{bmatrix} 6.604 & 9.14 & 25.14 \\ 9.14 & 348.56 & 94.28 \\ 25.14 & 94.28 & 188.56 \end{bmatrix} \begin{Bmatrix} \Delta_1 \\ \Delta_2 \\ \Delta_3 \end{Bmatrix} \; (l \text{ in ft})$$

CHAPTER 12

2. Same answers as for Problem 11.2.

4. Same answers as for Problem 11.4.

6. Same answers as for Problem 11.6.

8. Same answers as for Problem 11.8.

10. Same answers as for Problem 11.10.

12. Same answers as for Problem 11.12.

14. Same answers as for Problem 11.14.

16. Same answers as for Problem 11.16.

18. Same answers as for Problem 11.18.

20. $M_{ab} = M_{dc} = 0$ kN·m; $M_{ba} = -30.0$ kN·m; $M_{bc} = 30.0$ kN·m;
$M_{cb} = 30.0$ kN·m; $M_{cd} = -30.0$ kN·m

22. $M_{ab} = M_{dc} = 0$ kN·m; $M_{ba} = 11.3$ kN·m; $M_{bc} = -11.3$ kN·m;
$M_{cb} = 41.3$ kN·m; $M_{cd} = -41.3$ kN·m

24. $M_{ab} = 0$ k-ft; $M_{ba} = -65.34$ k-ft; $M_{bc} = 65.34$ k-ft; $M_{cb} = 253.2$ k-ft;
$M_{cd} = -253.2$ k-ft; $M_{dc} = -326.6$ k-ft

26. $M_{ab} = -45.5$ k-ft; $M_{ba} = 7.1$ k-ft; $M_{bc} = -7.1$ k-ft; $M_{cb} = +95.0$ k-ft;
$M_{cd} = -15.0$ k-ft; $M_{dc} = -26.6$ k-ft

28. $M_{ab} = -298.3$ kN·m; $M_{ba} = -245.6$ kN·m; $M_{bc} = 245.6$ kN·m;
$M_{cb} = 333.2$ kN·m; $M_{ce} = -333.2$ kN·m; $M_{ec} = -394.8$ kN·m;
$M_{ed} = 145.8$ kN·m; $M_{de} = -72.9$ kN·m; $M_{ef} = 249.1$ kN·m;
$M_{fe} = 170.1$ kN·m; $M_{fg} = -170.1$ kN·m; $M_{gf} = -230.8$ kN·m

30. $M_{ab} = 0$ k-ft; $M_{ba} = -441.4$ k-ft; $M_{bc} = 441.8$ k-ft; $M_{cb} = 470.0$ k-ft

32. $M_{ab} = 0$ kN·m; $M_{ba} = -532.4$ kN·m; $M_{bc} = 532.4$ kN·m;
$M_{cb} = -326.7$ kN·m; $M_{cd} = 326.7$ kN·m; $M_{dc} = 205.7$ kN·m

34. $M_{ab} = 0$ k-ft; $M_{ba} = 437.5$ k-ft; $M_{bc} = -437.5$ k-ft; $M_{cb} = 437.5$ k-ft;
$M_{cd} = -437.5$ k-ft; $M_{dc} = 0$ k-ft

36. $M_{ab} = 0$ kN·m; $M_{ba} = 25.0$ kN·m; $M_{bc} = -25.0$ kN·m; $M_{cb} = 25.0$ kN·m;
$M_{cd} = -25.0$ kN·m; $M_{dc} = 0$ kN·m

CHAPTER 13

2.
$$\begin{Bmatrix} F_1 \\ F_2 \\ F_3 \\ F_4 \end{Bmatrix} = \begin{bmatrix} \dfrac{4EI}{l} & \dfrac{2EI}{l} & 0 & 0 \\ \dfrac{2EI}{l} & \dfrac{4EI}{l} & 0 & 0 \\ 0 & 0 & \dfrac{EA}{l} & \dfrac{EA}{l} \\ 0 & 0 & \dfrac{EA}{l} & \dfrac{EA}{l} \end{bmatrix} \begin{Bmatrix} \delta_1 \\ \delta_2 \\ \delta_3 \\ \delta_4 \end{Bmatrix}$$

4. $[k]_r[d] = \begin{bmatrix} \dfrac{EA}{l} & 0 \\ 0 & \dfrac{4EI}{l} \end{bmatrix} \begin{bmatrix} \dfrac{l}{EA} & 0 \\ 0 & \dfrac{l}{4EI} \end{bmatrix} = \begin{bmatrix} 1 & 0 \\ 0 & 1 \end{bmatrix}$

6. $[k] =$
$$\begin{bmatrix} \dfrac{EA}{l} & 0 & 0 & \dfrac{EA}{l} & 0 & 0 \\[2ex] 0 & \dfrac{4EI}{l} & -\dfrac{6EI}{l^2} & 0 & \dfrac{2EI}{l} & -\dfrac{6EI}{l^2} \\[2ex] 0 & -\dfrac{6EI}{l^2} & \dfrac{12EI}{l^3} & 0 & -\dfrac{6EI}{l^2} & \dfrac{12EI}{l^3} \\[2ex] \dfrac{EA}{l} & 0 & 0 & \dfrac{EA}{l} & 0 & 0 \\[2ex] 0 & \dfrac{2EI}{l} & -\dfrac{6EI}{l^2} & 0 & \dfrac{4EI}{l} & -\dfrac{6EI}{l^2} \\[2ex] 0 & -\dfrac{6EI}{l^2} & \dfrac{12EI}{l^3} & 0 & -\dfrac{6EI}{l^2} & \dfrac{12EI}{l^3} \end{bmatrix}$$

8. $[d] =$
$$\begin{bmatrix} \dfrac{l}{3EI} & -\dfrac{l}{6EI} & 0 \\[2ex] -\dfrac{l}{6EI} & \dfrac{l}{3EI} & 0 \\[2ex] 0 & 0 & \dfrac{l}{EA} \end{bmatrix}$$

$[b] = [0 \ \ 0 \ \ 1]$

$[k] =$ given by Eq. 13.47, results agree with Problem 2.

CHAPTER 14

2. $\{\Delta\}_I^T = \{(\Delta_1)_2 (\Delta_2)_2\} = \{-0.025,0 \ \ -0.200\}$ ft; $\{F\}_{12} = 24.10$ k; $\{F\}_{23} = -29.17$ k; $\{F\}_{24} = -30.06$ k

4. $\{\Delta\}_I^T = \{(\Delta_1)_2 (\Delta_2)_2\} = \dfrac{1}{EA}\{-72.25 \ \ 326.2\}$ (l in ft); $\{F\}_{12} = -7.22$ k; $\{F\}_{42} = -39.8$ k; $\{F\}_{23} = 9.03$ k; $\{F\}_{52} = -27.6$ k

6. $\{\Delta\}_I^T = \{(\Delta_1)_1 (\Delta_2)_1 (\Delta_1)_3 (\Delta_1)_4\} = \{0.006,8 \ \ 0.004,0 \ \ 0.002,2 \ \ 0.004,4\}$ ft; $\{F\}_{12} = 11.04$ k; $\{F\}_{13} = -30.60$ k; $\{F\}_{14} = -2.40$ k; $\{F\}_{23} = 17.21$ k; $\{F\}_{34} = 17.21$ k

8. $\{\Delta\}_I^T = \{(\Delta_1)_2(\Delta_2)_2(\Delta_2)_3(\Delta_1)_4(\Delta_2)_4\} =$
$\{-0.002,17 \quad 0.008,57 \quad 0 \quad -0.001,12 \quad 0.007,32\}$ ft; $\{F\}_{12} = -19.73$ k;
$\{F\}_{13} = 0$ k; $\{F\}_{14} = 25.34$ k; $\{F\}_{24} = -15.20$ k; $\{F\}_{25} = -24.66$ k;
$\{F\}_{34} = -10.14$ k; $\{F\}_{45} = 10.14$ k

10. $\{\Delta\}_I^T = \{(\Delta_1)_2(\Delta_2)_2(\Delta_2)_3(\Delta_1)_4(\Delta_2)_4\} =$
$\{0.082 \quad 0.239 \quad 0.032 \quad -0.831 \quad 0.381\}$ in.; $\{F\}_{12} = 197.4$ k;
$\{F\}_{13} = 76.5$ k; $\{F\}_{14} = 98.5$ k; $\{F\}_{23} = -108.1$ k; $\{F\}_{24} = 34.6$ k;
$\{F\}_{34} = -100.7$ k

12. $\{\Delta\}_I^T = \{(\Delta_1)_2(\Delta_2)_2(\Delta_1)_3(\Delta_2)_3(\Delta_1)_5\} =$
$\dfrac{1}{E}\{285.6 \quad -38.0 \quad 1763.3 \quad 2407.0 \quad -333.3\}$ (l in ft, A in in.^{2});
$\{F\}_{12} = 71.3$ k; $\{F\}_{23} = 107.7$ k; $\{F\}_{45} = -100.0$ k; $\{F\}_{53} = -172.0$ k;
$\{F\}_{42} = 40.45$ k; $\{F\}_{25} = 11.4$ k

14. $\{\Delta\}_I^T = \{(\Delta_2)_2(\Delta_6)_2(\Delta_6)_3(\Delta_6)_4\} =$
$\dfrac{1}{EI}\{4,762.32 \quad 142.88 \quad -571.44 \quad 285.68\}$ (l in ft);
$\{F\}_{12}^T = \{48.58$ k $- 257.14$ k-ft$\}$; $\{F\}_{23}^T = \{-31.43$ k $\quad 228.60$ k-ft$\}$;
$\{F\}_{34} = \{4.28$ k $\quad -85.72$ k-ft$\}$

16. $\{\Delta\}_I^T = \{(\Delta_1)_2(\Delta_2)_2(\Delta_6)_2(\Delta_6)_3\} =$
$\dfrac{1}{EA}\{204.91 \quad 692.42 \quad -4.29 \quad -49.79\}$ (l in ft);
$\{F\}_{12}^T = \{-21.55$ k $\quad 93.34$ k $\quad -911.97$ k-ft$\}$;
$\{F\}_{23}^T = \{-10.25$ k $\quad -22.75$ k $\quad 454.93$ k-ft$\}$

18. **(b)** $\{\Delta\}_b^T = \dfrac{PL}{AE}\{-0.149 \quad 2.321\}$

      (c) $\{\Delta\}_b^T = \dfrac{HL}{AE}\{1.549 \quad -0.610\}$

20. $\{\Delta\}_I^T = \{(\Delta_3)_2(\Delta_4)_2(\Delta_5)_2\} = \dfrac{M}{EI}\{-4.229 \quad 3.103 \quad -0.213\}$ (l in m)

24. $\{\Delta\}_I^T = \{(\Delta_6)_1(\Delta_6)_2(\Delta_6)_3(\Delta_6)_4\} = \dfrac{1}{EI}\{-9.371 \quad 18.743 \quad 15.543 \quad 7.771\}$;
$\{F\}_{12}^T = \{-2.249 \quad 0\}$; $\{F\}_{23}^T = \{24.38 \quad -11.245\}$;
$\{F\}_{34}^T = \{1.865 \quad -9.325\}$

26. $\{\Delta\}_I^T = \{(\Delta_6)_1(\Delta_6)_2\} = \dfrac{1}{EI}\{303.93 \quad -107.86\};$

$\{F\}_{12}^T = \{96.47 \text{ kN} \quad 0 \text{ kN} \cdot \text{m}\}; \{F\}_{23}^T = \{112.94 \text{ kN} \quad -336.29 \text{ kN} \cdot \text{m}\};$

$\{F\}_{24}^T = \{-58.33 \text{ kN} \quad 100.86 \text{ kN} \cdot \text{m}\}$

28. $\{\Delta_1\}_2 = 0.025 \text{ in.}; \{\Delta_2\}_2 = -0.200 \text{ in.};$

$\{F_1\}_{12} = -64.6\text{k}; \{F_1\}_{32} = 0; \{F_1\}_{42} = -64.4\text{k}$

30. $\{F\}_{12}^T = \{49.8 \quad 0\}; \{F\}_{21}^T = \{-100.2 \quad 402\}; \{F\}_{23}^T = \{160.2 \quad -402\};$

$\{F\}_{32}^T = \{-79.8 \quad 0\}$ in kN and kN $\cdot$ m

32. $\{F\}_{12} = -33.5 \text{ kN}; \{F\}_{32} = 46.1 \text{ kN}; \{F\}_{42} = -33.5 \text{ kN}$

34. $\{F\}_{12}^T = \{-675 \quad 0\}; \{F\}_{21}^T = \{-675 \quad 6750\}; \{F\}_{23}^T = \{675 \quad -6750\};$

$\{F\}_{32}^T = \{675 \quad 0\}$ in kN and kN $\cdot$ m

36. $\{F\}_{71} = 14.6 \text{ k}; \{F\}_{72} = 211.5 \text{ k}; \{F\}_{73} = -122.1 \text{ k}; \{F\}_{74} = -160.4 \text{ k};$

$\{F\}_{75} = 271.8 \text{ k}; \{F\}_{76} = -97.1 \text{ k}$

38. $\{F\}_{12}^T = \{-87.6 \quad 0\}; \{F\}_{21}^T = \{-8.76 \quad 87.36\}; \{F\}_{23}^T = \{13.08 \quad -87.36\};$

$\{F\}_{32}^T = \{13.08 \quad -43.60\}$ in kN and kN $\cdot$ m

CHAPTER 15

2. Same answers for Problem 14.4.

4. Same answers for Problem 14.14.

6. Same answers for Problem 14.16.

Index

Frequently Used Area Properties

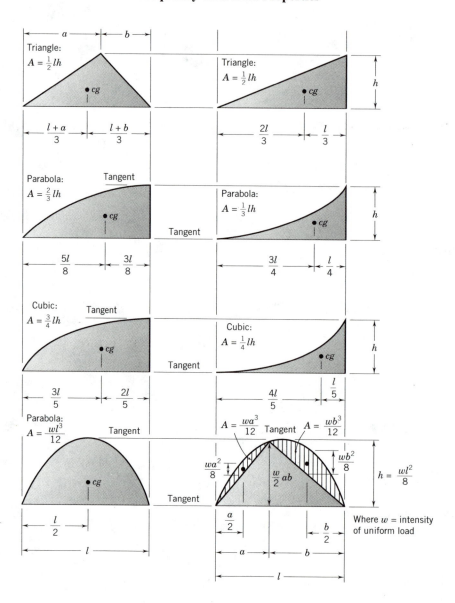

Triangle:
$A = \frac{1}{2}lh$

Triangle:
$A = \frac{1}{2}lh$

$\frac{l+a}{3}$ $\frac{l+b}{3}$ $\frac{2l}{3}$ $\frac{l}{3}$

Parabola:
$A = \frac{2}{3}lh$ Tangent

Parabola:
$A = \frac{1}{3}lh$

$\frac{5l}{8}$ $\frac{3l}{8}$ $\frac{3l}{4}$ $\frac{l}{4}$

Cubic:
$A = \frac{3}{4}lh$ Tangent

Cubic:
$A = \frac{1}{4}lh$

$\frac{3l}{5}$ $\frac{2l}{5}$ $\frac{4l}{5}$ $\frac{l}{5}$

Parabola:
$A = \frac{wl^3}{12}$ Tangent

$A = \frac{wa^3}{12}$ Tangent $A = \frac{wb^3}{12}$

$\frac{wa^2}{8}$ $\frac{w}{2}ab$ $\frac{wb^2}{8}$ $h = \frac{wl^2}{8}$

$\frac{l}{2}$ $\frac{a}{2}$ $\frac{b}{2}$

Where w = intensity of uniform load

l a b l